Silicon-Based Optoelectronic Materials

MATERIALS RESEARCH SOCIETY SYMPOSIUM PROCEEDINGS VOLUME 298

Silicon-Based Optoelectronic Materials

Symposium held April 12–14, 1993, San Francisco, California, U.S.A.

EDITORS:

M.A. Tischler
ATM
Danbury, Connecticut, U.S.A.

R.T. Collins
IBM T.J. Watson Research Center
Yorktown Heights, New York, U.S.A.

M.L.W. Thewalt
Simon Fraser University
Burnaby, British Columbia

G. Abstreiter
Walter Schottky Institut
Technische Universität München
Garching, Germany

MATERIALS RESEARCH SOCIETY
Pittsburgh, Pennsylvania

CAMBRIDGE UNIVERSITY PRESS
Cambridge, New York, Melbourne, Madrid, Cape Town,
Singapore, São Paulo, Delhi, Mexico City

Cambridge University Press
32 Avenue of the Americas, New York NY 10013-2473, USA

Published in the United States of America by Cambridge University Press, New York

www.cambridge.org
Information on this title: www.cambridge.org/9781107409538

Materials Research Society
506 Keystone Drive, Warrendale, PA 15086
http://www.mrs.org

First published 1993
First paperback edition 2012

Single article reprints from this publication are available through
University Microfilms Inc., 300 North Zeeb Road, Ann Arbor, MI 48106

CODEN: MRSPDH

ISBN 978-1-107-40953-8 Paperback

This work relates to Department of Navy Grant N00014-93-1-0444 issued by the Office of Naval
Research. The United States Government has a royalty-free license throughout the world in all
copyrightable material contained herein.

The views, opinions, and/or findings contained in this report are those of the author(s) and should
not be construed as an official Department of the Army position, policy, or decision, unless so
designated by other documentation.

This work was supported by the Air Force Office of Scientific Research under Grant Number
AFOSR-F 49620-93-1-0383.

Contents

*Invited Paper

*Invited Paper

*Invited Paper

*Invited Paper

Preface

Although silicon is at the heart of the microelectronics revolution, its low optical efficiency has limited its use in optoelectronic applications. The potential significance of combining communications and display technology with microelectronics technology has generated considerable activity directed at developing a silicon-compatible optoelectronic material. The last few years have seen some interesting and potentially important advances in this area. Symposium B was organized as a forum for the various groups studying the physics, materials science, processing and applications of silicon-based optoelectronic materials to present their most recent results in this rapidly growing field.

Talks were organized into five basic areas: $Si_{1-x}Ge_x$, rare earth-doped silicon (this session was organized jointly with symposium E, Rare Earth Doped Semiconductors), silicon nanoparticles, porous silicon and applications. Many of the key research groups in each of these areas were represented at the meeting. This proceedings volume provides an excellent overview of the work reported during the symposium and of the field in general. Much of the work in this field has been directed at either understanding the physics of light emission in these systems, or at making efficient optical devices from Si-based materials. It is also clear from the meeting that much work remains before the mechanisms of luminescence in many of these materials are completely understood or an efficient, stable room-temperature LED which is compatible with Si VLSI technology is developed. At the same time, in nearly all of the materials systems considered, significant progress has been made on both fronts. For example, room-temperature electroluminescent devices made from many of the materials are discussed in this proceedings. With the rapid pace at which this field is proceeding, one can speculate that a day in which silicon will have more of a presence in the arena of optical applications may not be that far away.

This symposium benefited from the support of the Army Research Office, Office of Naval Research, and the Air Force Office of Scientific Research. Their support is greatly appreciated.

M.A. Tischler
R.T. Collins
M.L.W. Thewalt
G. Abstreiter

MATERIALS RESEARCH SOCIETY SYMPOSIUM PROCEEDINGS

Prior Materials Research Society Symposium Proceedings
available by contacting Materials Research Society

Silicon-Based Optoelectronic Materials

SiGe/Si QUANTUM WELLS BY MBE : A PHOTOLUMINESENCE STUDY

D.C. HOUGHTON[†], N.L. ROWELL, J.-P. NOEL, G. AERS, M. DAVIES,
A. WANG and D.D. PEROVIC[*]
Institute for Microstructural Sciences, National Research Council, Ottawa, Ontario, Canada,
K1A OR6. [†]Visiting Professor, RCAST, Univ of Tokyo, Komaba, Tokyo 171,
Japan.[*]Department of Metallurgy and Materials Science, University of Toronto, Toronto,
Ontario, Canada M5S 1A4

ABSTRACT

Strained $Si_{1-x}Ge_x$ quantum wells and multi-quantum wells, synthesized by solid source e-beam evaporated MBE on Si(100) substrates have been studied by low temperature photoluminescence (PL) spectroscopy. Phonon resolved transitions originating from excitons bound to shallow impurities in $Si_{1-x}Ge_x$ layers were observed over the temperature range 2K to 100K and used to characterize $Si_{1-x}Ge_x/Si$ heterostructures. Thin $Si_{1-x}Ge_x$ quantum wells exhibited phonon-resolved PL spectra, similar to bulk material, but shifted in energy due to strain, quantum well width and Ge fraction. In single quantum wells confinement shifts up to ~200 meV were observed (1.2 nm wells with $x = 0.38$) and NP linewidths down to 1.37 meV were obtained. The confinement shifts were modeled by hole confinement in $Si_{1-x}Ge_x$ wells. An annealing study was performed to investigate the role of Si-Ge interdiffusion on luminescence. Both the geometrical shape and optical emission of the quantum well were found to significantly change through intermixing. In addition to near edge luminescence a broad band of intense luminescence was obtained from several $Si_{1-x}Ge_x/Si$ heterostructures. Some layers exhibited both types of PL spectra. However, the broad PL band (peak energy ~120meV below the strained bandgap) was predominant when the alloy layer thickness was greater than 2 - 10nm, depending on x, growth temperature, and substrate surface preparation. The strength of the broad PL band was correlated with the areal density of strain perturbations (~10^9cm^{-2} per quantum well corresponding to a spacing of 300nm in the plane of the well; local lattice dilation ~ 1.5 nm in diameter) observed in plan-view TEM. The first few wells of MQW exhibited only band edge luminescence as was revealed by etching off the upper MQW periods. In addition, post growth anneals at temperatures in the range 700°C to 1100°C were found to enhance band edge luminescence, while the broad luminescence band decayed to zero intensity. Interdiffusion at these these temperatures has been shown to dramatically change the QW shape and consequently interfacial asperities would be expected to disappear, consequently only shallow phonon resolved luminescence is observed in PL after annealing. The influence of PL measurement parameters such as excitation power density and PL sample temperature on the relative strengths of band edge versus broad band luminescence were also consistent with the presence of exciton traps at sites of reduced bandgap.

INTRODUCTION

Rapid progress toward efficient electroluminescence and photodetection in Si-based heterostructures has stimulated considerable interest in recent years [1-6]. Si-based optoelectronic devices would allow monolithic integration of mature Si technology with optical signal processing. However, Si and epitaxial $Si_{1-x}Ge_x/Si(100)$ alloys are indirect semiconductors and band structure perturbations are necessary to enhance luminescence efficiencies to practical levels. Room temperature quantum efficiencies in Si remain low (~10^{-4} photons emitted per electron hole pair created). The fundamental limitation is the low probability for radiative recombination of electrons and holes. Many approaches have been used in the quest for efficient luminescence from Si based heterostructures, which in essence involve localisation of an electron and hole, and have met with various levels of success. Isoelectronic centres such as Be[7] and S[8] and others including C, In, Cu, Li, Tl and Se provide binding sites for excitons which then undergo a radiative transition characteristic of the centre. Defect complexes such as those in irradiated C doped[9] Si exhibit sharp features in photoluminescence and electroluminescence. Zone folding in atomic layer superlattices, e.g. Si_mGe_n[10,11], and rare earth doping, e.g. Er in Si and $Si_{1-x}Ge_x$

alloys have also been explored[12-14]. The simultaneous requirements of band structure engineering and superlative crystal quality extend a considerable challenge to the materials scientist / crystal grower. It is a formidable task involving the controlled removal of non-radiative point defects and complexes, controlled introduction of radiative centers, and simultaneous maintenance of coherently strained interfaces without extended defect injection. Recent progress in the study of luminescence (wavelength range ~1.0 -1.7 µm) from $Si_{1-x}Ge_x$ alloys and multiquantum wells grown by MBE and CVD will be discussed below.

In the last few years intense activity has been focused on the optical and electronic properties of epitaxial Si and $Si_{1-x}Ge_x$ alloys. The components necessary for a practical Si based optoelectronic chip have also been demonstrated with varying quantum efficiencies; the $Si_{1-x}Ge_x$/Si digital optoelectronic switch [15], the p-i-n diode ($\lambda <\approx 1.3\mu m$)[16] and IR detector [17]($\lambda \approx 3-14\mu m$) illustrate possibilities for photodetectors, while the $Si_{1-x}Ge_x$ / Si ($\lambda \approx 1.5\mu m$) LED[5,18,19] porous silicon [6] and Si (Er) LED's [12,13] have shown some promise for electroluminescence at temperatures approaching 300K. $Si_{1-x}C_x$ and $Si_{1-x-y}Ge_xC_y$ alloys have recently been synthesized[20] and may permit further band structure tailoring with the additional benefit of strain compensation in Si-Ge-C multilayer devices. $Si_{1-x}Ge_x$ /Si and Si based dielectric multilayer waveguides (optimized for 1.3 or 1.5µm) may be readily fabricated to act as conduits for optical signals in any future Si-based opto-electronic integrated circuit[21].

We first reported PL and electroluminescence from heteroepitaxial, strained layers prepared by solid source MBE in ultrahigh vacuum[1,5]. The luminescence spectra differed significantly from phonon-resolved, near-bandgap PL observed in unstrained, bulk material[22,23] and coherently strained $Si_{1-x}Ge_x$ synthesized by CVD techniques[3,4]. MBE-grown $Si_{1-x}Ge_x$ typically exhibits a broad, often intense PL band which was consistently shifted below the strained alloy bandgap energy by ~ 100 meV for a wide range of Ge concentrations. The first observation of phonon-resolved, near-bandgap PL from strained $Si_{1-x}Ge_x$ layers was reported by Terashima et al.[2] for MBE-grown material and by us[3] for material grown by rapid thermal chemical vapor deposition (RTCVD). The present paper is a study of the PL mechanisms in fully-strained $Si_{1-x}Ge_x$/Si heterostructures grown by solid source molecular beam epitaxy (MBE). The extent of confinement shift for various QW widths and Ge fractions and the influence of QW shape (through RTA anneal cycles) on PL spectra is described in detail. We also present a model that defines the role of growth induced compositional or structural inhomogeneities which give rise to the broad, often intense PL band[1,24,25] frequently reported for MBE $Si_{1-x}Ge_x$ in contrast to phonon-resolved, near-bandgap luminescence. This unified treatment draws together many of the previous experimental observations and compares the relative strengths of broad PL bands and near-bandgap transitions as a function of growth conditions, $Si_{1-x}Ge_x$/Si heterostructure dimensions and post growth thermal treatment.

EXPERIMENT

$Si_{1-x}Ge_x$/Si heterostructures were epitaxially deposited from solid source (electron beam evaporated) MBE, without intentional doping in an ultrahigh vacuum deposition chamber with a background pressure during growth ~ 1 x 10^{-9} mbar. Just before introduction into the vacuum loadlock, Si(100) substrate wafers were either oxide-passivated by a 50 minute UV-ozone treatment or H-passivated by dipping in 5% HF. Surface oxide was then desorbed by heating the wafer to ~ 900 °C for 600s under a Si flux equivalent to 0.01 nm/s growth rate, whereas H was desorbed simply by heating to ~600 °C. The Si growth rate was 0.1 nm/s and the Ge growth rate was set to achieve a given value of x. Both rates were controlled with a florescence rate monitor and calibrated with a quartz crystal thickness monitor during growth of the Si buffer layer. The multiquantum well (MQW) structures were designed to exhibit a range of metastability (effective "critical thickness") with respect to misfit dislocation injection[26]. A diluted Schimmel etch (4 parts 48% HF : 5 parts 0.3M CrO_3 at 300°K) was used to controllably remove Si and $Si_{1-x}Ge_x$ layers; typical etch rate 100nm in ~15s. Post growth anneals were carried out using a rapid thermal annealer in a dynamic nitrogen ambient at temperatures up to 1100°C.

The $Si_{1-x}Ge_x$ well thickness and Ge fraction were determined using x-ray double-crystal diffraction and dynamical rocking curve analysis which yielded an absolute uncertainty in x of ± 0.02. QW thicknesses were also obtained by rocking curve analysis of MQW with an accuracy of ± 2Å, and these dimensions were confirmed by XTEM measurements. Once well thickness and Ge fraction had been accurately determined for a MQW structure using x-ray and TEM analysis etching could then be employed to produce a $Si_{1-x}Ge_x$ SQW, of known dimensions and x, for PL study.

TEM specimens were prepared for both plan-view and cross-sectional imaging by conventional ion milling techniques. Samples were examined in either a Philips EM 430 operated at 250 kV or in a Hitachi 800 at 200 kV.

The PL experiments were carried out with a modified Fourier transform spectrophotometer, which comprised an argon ion laser source to excite the samples, a cryostat for sample cooling, and a spectrometer to spectrally analyze the emitted light. Photodetection was obtained by a cooled germanium detector. The cryostat used was a variable temperature model with the samples mounted strain-free in liquid helium for temperatures between 1.8 and 4.5 K and in helium gas from 4.2 to 300 K. With an unfocused laser beam, the excitation power density was varied from 0 to 10 W/cm^2 over an area of 0.1 cm^2. By focusing the laser, power densities approximately three orders of magnitude higher were obtained.

Fig 1

PL spectrum from a single $Si_{0.83}Ge_{0.17}$, 77Å quantum well, taken at 2K.

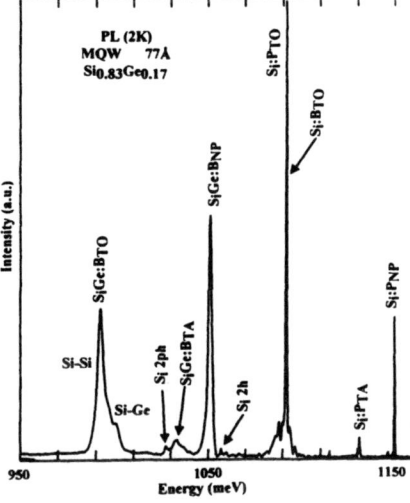

SiGe QUANTUM WELL LUMINESCENCE

Fig 1 is a 2 K PL spectrum obtained from a $Si_{1-x}Ge_x$ MQW sample after chemical etching to remove all but one quantum well (77 Å thick with x = 0.17). This procedure removes interferences from other wells, allows comparison with Si substrate peaks and establishes PL nomenclature. The dominant substrate peak is $Si:P^1_{TO}$, which refers to the annihilation of <u>one</u> exciton bound to an isolated <u>P</u> impurity atom in <u>Si</u> and the concomitant creation of a transverse optical (TO) phonon to conserve momentum. As grown, this sample also exhibits a $Si:B^1_{TO}$ peak from the 200 Å thick buffer layers which disappears when the buffer layers are etched off. Other weak substrate peaks include a two-hole process and a two-phonon process involving TO plus a zone-center phonon , as have been discussed previously[27].

The luminescence arising from the $Si_{0.83}Ge_{0.17}$ quantum well is due to B-bound exciton annihilation. The SiGe QW spectral shape, however, differs significantly from the substrate luminescence, as follows. First, the $Si_{0.83}Ge_{0.17}:B_{NP}$ peak (labeled simply NP on Fig 1) is the dominant quantum well peak. In an ideally pure Si or Ge lattice at 0 K, the NP transition is forbidden because of the requirement for momentum conservation. In practice, however, the presence of lattice perturbations such as shallow dopants provides localization centers for electrons or holes and a concomitant spread in the range of allowable momentum states. The alloy scattering process in solid solutions of Si and Ge leads to an even greater spread in allowable momentum states, with the result that NP transitions are not only possible but very probable, i.e. phonon-less transitions occur although the bandgap remains indirect. Thus, the same PL process

(B-bound exciton annihilation) occurring in either the SiGe alloy well or the Si substrate will give different spectral shapes. Alloy scattering leads to larger NP/phonon replica ratios and, secondly, to an increase in the PL linewidths: < 0.1 meV full-width at half-maximum for Si:B[1] NP compared to 2.1 meV for the NP $Si_{1-x}Ge_x$ counterpart in the $x = 0.17$ alloy well. At low excitation densities (~ 8 mW/cm^2), the $Si_{1-x}Ge_x$ NP peak with a FWHM of 1.37 meV is narrower than previously reported for epilayers[2-4,24,25] and bulk material[22,23]. This width is below the minimum for random bulk alloys and represents narrowest observed SiGe NP line for any epitaxial SiGe material.

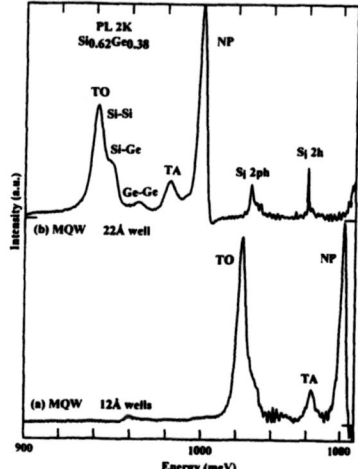

Fig 2

Comparison of PL spectra from two 10 period $Si_{0.62}Ge_{0.38}$ MQW of different well widths; SiGe QW thickness 12Å and 22Å, Si barrier width 200Å.

The sharpness of the SiGe NP line and its TO replica in Fig 1 can be used for an accurate determination of the Si-Si TO momentum-conserving phonon frequency[28]. Phonon replicas include the transverse acoustic (TA) peak, and three transverse optical (TO) peaks, one each for Ge-Ge, Si-Ge, and Si-Si bonding configurations. The measured energies of the phonon modes are 18.3, 35.3, 50.3, and 58.7 meV, respectively, in general agreement with the phonon energies determined from bulk alloys[23] although modified by the presence of strain in the wells. The relative ratios of phonon replicas depend on the alloy composition, x, and for $x = 0.17$, the TOGe-Ge peak is not large. Overall, the PL from this single well is relatively strong, with the NP peak $\sim 50\%$ as high as the substrate Si:B[1]TO line with an integrated intensity which is much larger.

Fig 2 shows PL spectra for two MQW samples of the same Ge concentration but of different thicknesses Fig 2 (a) 12Å and Fig 2 (b) 22Å. There are two main features which distinguish these spectra; the effect of quantum confinement and a change in the NP to phonon assisted transition peak ratios. The behaviour of the momentum conserving phonon modes in these SiGe quantum wells can be understood in terms of SiGe bulk properties by allowing for the extension of the exciton wavefunction into the Si regions on both sides of the quantum wells. Since the valence band offset is much larger than the near-zero conduction band offset, it is reasonable to assume that an exciton's hole is localized at a shallow impurity in a SiGe quantum well and its electron's orbit extends into the Si barriers unimpeded by the well boundary. For such a situation, it is possible to calculate an effective Ge concentration, x_{ph} ($< x$), based on the exciton's relative volumes in the SiGe and Si [23]. In doing this we assumed the exciton to be spherical and of the Bohr radius, ~ 50 Å. We also observed that the TOSi-Si phonon energy was smaller for thin well than for thicker wells of the same Ge concentration. Once again, this phenomena can be explained using the effective x for the phonons, x_{ph}.

The extent of quantum confinement shown for one composition in Fig 2, can also be seen in Fig 3 for the entire $Si_{1-x}Ge_x$-Si system. Confinement shifts are mapped out as a function of QW thickness and Ge fraction, x. An effective mass envelope calculation was used, assuming a

7

single mass (m* = 0.3) for the Si and SiGe valence bands and negligible electron confinement (ie SiGe - Si band offset [4] is taken up completely at the valence band).

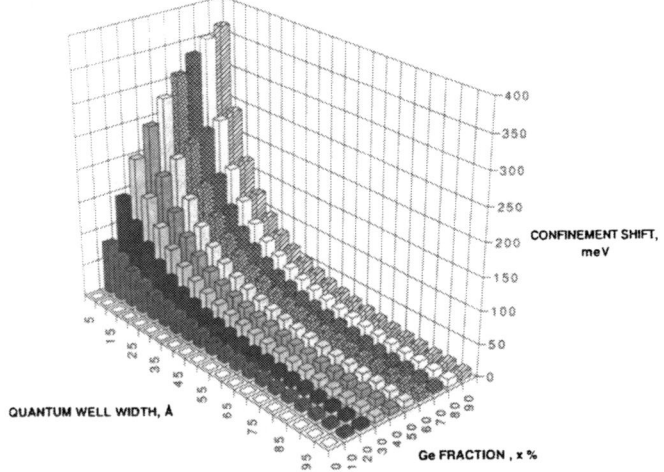

Fig 3 Confinement shift for $Si_{1-x}Ge_x$ quantum wells as a function of both well width and Ge fraction , x.

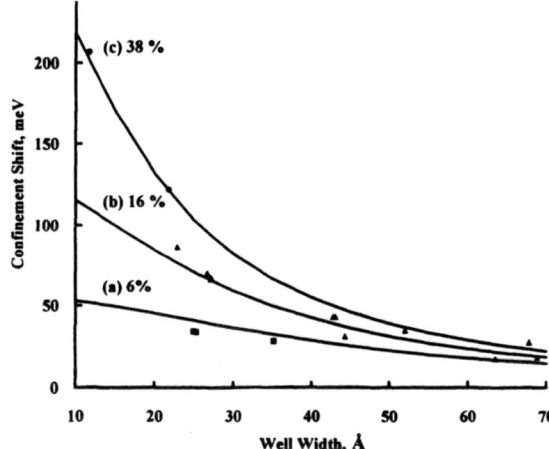

Fig 4 Experimentally observed confinement shifts compared with the predicted shifts using the model of Fig 3.

Fig. 4 shows a comparison of observed confinement shifts with the results of a simple one-band model (solid lines) for three groups of samples: (i) three with x ~ 0.06 (crosses), (ii) ten with x ~ 0.16 (triangles), and (iii) two with x ~ 0.38 (circles). The experimental confinement shifts were obtained by subtracting the unconfined, bound exciton peak energy[28,4], as calculated from the strained bandgap and exciton binding energies, from the observed NP peak position. Confinement effects on exciton binding energies, of the order of a few meV, were not included. The model used to generate the theoretical curves assumed the $Si_{1-x}Ge_x$-Si band offset to be completely taken up by the valence bands each of which was assumed to have a single effective mass[29]. This model was solved for the confinement shifts assuming finite wells for which a Schrodinger-like wave equation was used subject to the normal boundary conditions of wave function and particle current continuity. The model results of Fig. 3 used the heavy hole mass for $Si_{1-x}Ge_x$ (~0.3) and the light hole mass for the Si (0.16), values which gave the best agreement between the model and experimental results. Superlattice effects due to particle tunneling through the barriers were not significant since the Si spacers used in our MQW's were thick (~200Å) compared with the wave function $1/e$ decay length in the Si (~ 10 - 40 Å).

BROAD BAND Vs BAND EDGE LUMINESCENCE FROM MBE $Si_{1-x}Ge_x$

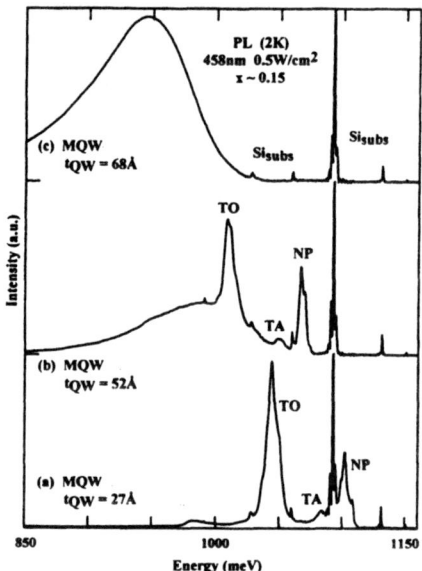

Fig 5

Transition of phonon resolved luminescence to broad band luminescence for three MQW with increase in $Si0.85Ge_{0.15}$ well thickness; (a) 27Å, (b) 52Å, (c) 68Å. PL spectra taken at 2K using λ=458nm excitation light.

Fig. 5 shows how SiGe PL evolves from phonon-resolved, near-band-edge transitions to a lower energy, broad PL band for three MQW structures grown by MBE at 600 °C. The MQW growth sequences differed only in the SiGe alloy layer growth times, giving measured quantum well thicknesses of 2.7, 5.2 and 6.8 nm for Fig. 5(a), (b), and (c), respectively. The two NP peaks of Fig 5 (a) (X_{NP}), one at 1102.6 meV and a weaker one at 1108.9 meV, indicate that compositional variations in x from well to well were ~ 0.01. The minimum NP peak width was ~ 2 meV, which is less than half the width for RTCVD material[3,4]. The NP peaks and their TA and TO phonon replicas originate from excitons bound to shallow dopant atoms (boron) in the 2.7 nm-thick $Si0.85Ge_{0.15}$ layers.

In Fig 5 (b), an increase in the $Si_{0.85}Ge_{0.15}$ layer thickness to ~5.2 nm reduces the confinement shift by ~ 25 meV and the broad PL band emerges with its high energy edge near the X_{NP} energy. Fig. 5 (c) shows that for $Si_{0.81}Ge_{0.19}$ layers with a thickness of 6.8 nm, phonon-resolved PL is no longer observed and the broad band grows in intensity and shifts further down in energy. Dilute Schimmel etching through the topmost $Si_{1-x}Ge_x$ layer and examination by Nomarski phase contrast microscopy at 1000X magnification revealed circular features in densities correlated with the intensity of the broad PL band. Plan-view TEM was then used to identify the origin of these etch features.

Figure 6(a-c) are typical TEM diffraction contrast images of the three $Si_{0.85}Ge_{0.15}$/Si multilayers whose PL spectra are shown in Fig. 5. The images are taken using plan-view projections with (001) foil normals. Under these imaging conditions, the superlattice layers cannot give rise to atomic number contrast but are delineated because of strain field contrast associated with stress relaxation effects where the strained layers emerge at a thin-foil surface[31].

Although the absolute width of a given layer depends locally on the angle between the foil normal and the electron beam direction, the relative widths of the $Si_{0.85}Ge_{0.15}$ and the Si layers for any image reflects the actual period of the superlattice. The principal contrast feature to note in Fig. 6 is fine dispersion of black-white lobes superimposed on the multilayer contrast described above. The black-white strain field contrast is characteristic of a small (< 1.5 nm) dilatational perturbation which is characteristic of a region of larger lattice parameter. The inset of Fig. 6(a) clearly shows the strain contrast symmetry perpendicular to the diffracting vector (g) near the (001) pole. Upon tilting to other zone-axes, the associated image contrast relationship with diffraction condition indicates that the perturbations are atomic-scale regions possessing plate-shaped symmetry and a larger lattice parameter than the adjacent matrix region. It should be noted that the interstitial perturbations have been observed in a wide range of both single and multiple layer heterostructures. Moreover, the strain features are never observed in the thin foil regions away from the heterostructure (i.e. in a Si capping layer) thus ruling out specimen damage-related artifacts as possible contrast sources. Most importantly, it is evident from Fig. 6(a-c) that the density of interstitial-like perturbations decreases markedly with $Si_{0.85}Ge_{0.15}$ layer thickness under otherwise identical conditions, in accordance with the marked change in PL characteristics.

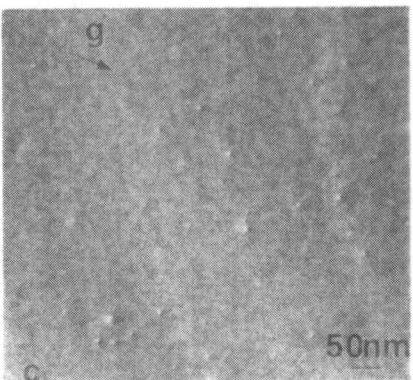 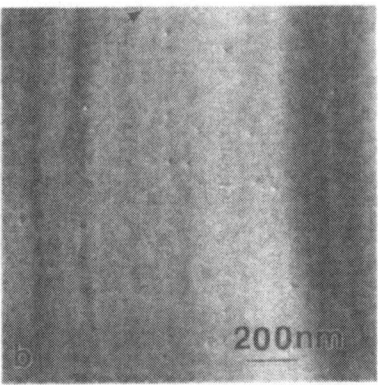

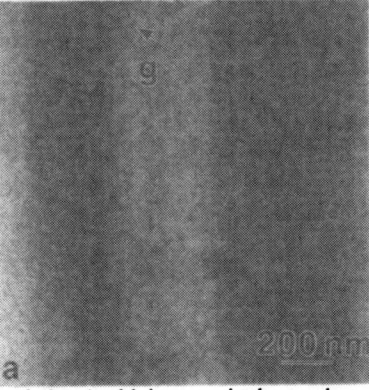

Fig 6   Plan view (**g=220**) bright field TEM diffraction contrast images of the MQW in Fig 5. Local lattice dilation gives rise to strain perturbations ~15Å in diameter and which are visible through their strain field contrast. The strain perturbation density per well (ρ, cm^{-2}) increases with increase in well width and correlates with the broad PL band intensity of Fig 5 ; (a) 27Å, $\rho = 2 \times 10^6$cm^{-2}, (b) 52Å, $\rho = 3 \times 10^7$cm^{-2}, (c) 68Å, $\rho = 7 \times 10^8$cm^{-2}.

The variation in strain perturbation density (growth morphology) with increase in the number of grown wells is illustrated in Fig. 7 which shows two PL spectra for a MQW which consisted of 6.8nm thick $Si_{83}Ge_{17}$ wells. These spectra were both taken at 2 K with 458 nm laser excitation. The lower spectrum (a) was taken from an as-grown sample and displays both the PR and broad PL peaks. Spectrum (b) was obtained from a sample which had been etched for 70 s in a dilute Schimmel etch leaving one quantum well intact. As can be seen by comparing Figs 7(a) and (b), the overall intensity of the PR peaks has been reduced but, more importantly, etching has completely removed the broad PL peak, consistent with the formation of platelets which occurred during epitaxy when the effective stress was relatively high; i.e. in those layers farthest from the substrate. The effect of etching is to remove these outer (last deposited) layers leaving only layers grown under conditions of low effective stress, hence low strain perturbation density and no broad PL signal.

Fig 7(b)   PL of the MQW in (a) after etching back to leave a single (i.e. the first deposited) QW.

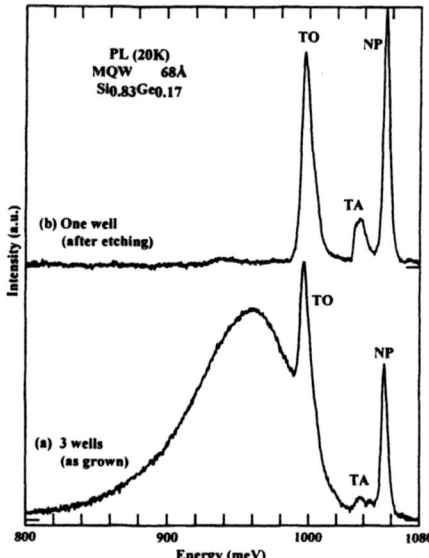

Fig 7(a)   PL of as deposited 10 period $Si_{0.83}Ge_{0.17}$ 68Å /Si 200Å MQW .

The morphological transition which occurs when a certain strain/thickness combination (corresponding to a critical strain energy density) is exceeded is a two stage process[28]. It is probable that a strain mediated breakdown in 2D stepflow occurs during epitaxial growth of a strained SiGe alloy. The second stage of this process is the decoration of these growth perturbations by segregation of Ge (or Si). The strain field around these features can be detected directly by TEM and indirectly by PL spectroscopy. This inhomogeneity in strain field around atomic scale perturbations at SiGe/Si interfaces could also directly (without the need for Ge segregation), give rise to the strain contrast seen in plan view TEM micrographs, e.g. Fig. 6. The localized bandgap reduction that would occur at a Ge-rich platelet is a result of two effects which can be calculated; interfacial band discontinuity and strain induced bandgap narrowing[28, 30].

Further evidence to support this morphological model for SiGe broad band luminescence was obtained from annealing MQW samples which exhibited broad band and band edge luminescence in the as deposited condition. Fig. 8 depicts two PL spectra from the $Si_{83}Ge_{17}$ /Si MQW of Fig 7. As deposited, this MQW exhibits both a broad band of luminescence ~100meV below the expected bandgap and sharp but weak phonon resolved features close to the band edge, Fig. 8 (a). On annealing at 900°C for 2500s the broad band of PL disappears and the shallow luminescence dominates the PL spectrum. No defect related luminescence was observed, indicating that coherent SiGe/Si interfaces are preserved with negligible misfit dislocation injection. These observations can be readily explained in the context of strain mediated morphological changes taking place during stained layer epitaxy, which can be subsequently removed by interdiffusion. To ensure that significant interdiffusion was taking place at these annealing conditions a series of single QW structures were also annealed and the position of their X_{NP} lines monitored as a function of anneal time and temperature. The luminescence energy corresponding to a transition to the confined state in the QW shifted predictably to higher energy as interdiffusion changed the Ge profile in the well. The extent of this shift confirmed that considerable interdiffusion had taken place while TEM and defect -reveal etching detected no evidence for misfit dislocation activity.

Fig 8 (b) PL spectrum of (a) after RTA anneal at 900°C for 2500s.

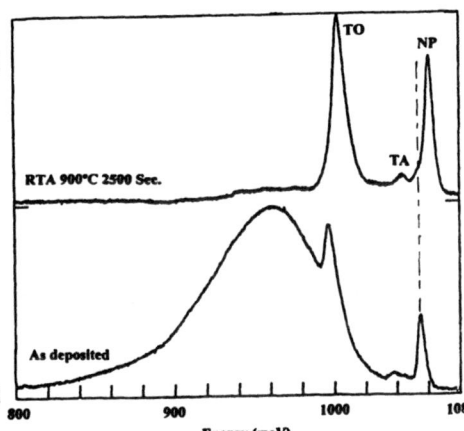

Fig 8 (a) PL spectrum of as deposited 3 period MQW (wells 68Å thick of $Si_{0.83}Ge_{0.17}$).

Fig. 9 shows the temperature dependence of the PL spectrum for a 3 period MQW taken with 100 mW of 458 nm exciting laser light. This "transition sample" (as e.g. Fig 5(b)) exhibits both near band edge luminescence and broad band PL. It consisted of three wells, 68Å thick, of $Si_{0.83}Ge_{0.17}$ and in trace (a) we see evidence for at least two no-phonon peaks. As the temperature was increased from 2 to 20 K there were two very noticeable changes in the spectra: (a) the bound exciton no phonon peaks became a single free exciton no phonon peak as observed in Fig. 9 (b) the broad PL peak became significantly strong. As the temperature was increased further the broad PL peak was observed to predominate, persisting to temperatures above 75 K, a trend which is consistent with the assignment of this line to a (Ge-rich) strain perturbation.

Although the growth asperities themselves are small, their strain fields are relatively large enhancing the capture of excitons [32]. Exciton diffusion in the volume between platelets is important because platelet elastic displacement fields are much large (~1 μm) than the strain fields imaged in TEM and phonon resolved PL will not be observed if the regions between the elastic displacement fields are not large enough to include any shallow dopant atoms. Exciton diffusion effects are apparent in the variation with temperature of PL spectra like Fig. 9, where the phonon-resolved PL disappears by 20K while the broad PL intensity persists to ~70K. This behavior at higher temperatures occurs as shallow bound excitons become free excitons that diffuse within the SiGe quantum wells and fall into the localized regions of strain reduced bandgap, where the decay time and radiative efficiency [33] are observed to be quite large in the absence of the usual competing non-radiative channels. Such persistence to high temperature of exciton related PL is expected in quantum well systems where exciton binding is increased, confirmed by saturation behavior with excitation intensity.

The variation of PL character in "transition samples", such as in Fig 10 (cf Fig 5 (b)), with incident power also provides indirect evidence for the presence of saturable exciton sinks where the density per unit volume of these centres is comparable with the exciton density. An increase in exciton density through increasing the power density produces phonon resolved luminescence from excess excitons free to migrate and self annihilate at dopant atom sites in regions of the SiGe QW free of strain perturbations.

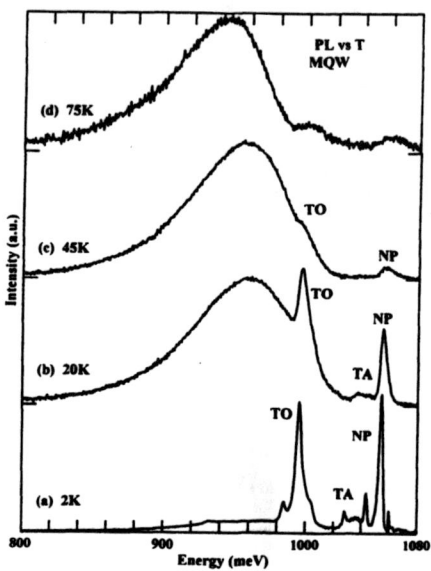

Fig 9 Temperature dependence of PL for a three period MQW (68Å, $Si_{0.83}Ge_{0.17}$: 200Å, Si), showing the predominance of the broad peak with increase in PL measurement temperature.

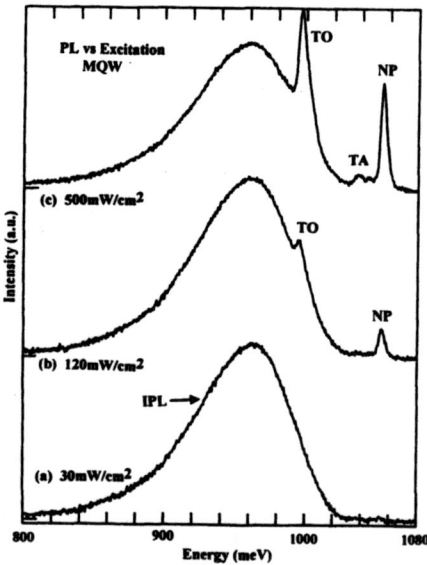

Fig. 10 shows the tendency of the broad band, (strain perturbation-related) emission to saturate with pump power, hence exciton density. In the three traces the excitation density was varied from approximately 20 to 300 mW/cm^2 for 458 nm light at 20 K for the MQW of Fig 9. The incremental behaviour of PR peak intensity with laser intensity is fairly linear but the straight line does not pass through the origin, instead there are some pump intensities for which the PR peak height is zero, c.f. trace (a). On the other hand, the broad PL intensity saturates quite readily with pump intensity, barely increasing from 70 to 300 mW/cm^2.

SUMMARY

We have deposited coherent $Si_{1-x}Ge_x$ epilayer and multiple quantum well samples by MBE on Si(100) substrates and we have characterized these samples by low temperature PL spectroscopy and TEM. Phonon-resolved transitions originating from excitons bound to shallow impurities (boron) were observed by PL in addition to a broad band of intense luminescence. The shallow luminescence shifted reproducibly and systematically to higher energies with increased quantum confinement as observed for a wide range of well widths and Ge fractions. The blueshift was quantitatively consistent with hole confinement in SiGe wells whose effective thickness is reduced by interfacial roughening from the nominal dimensions.

The broad PL band, typical of MBE SiGe, was predominant when the alloy layer thickness was greater than 40 - 100 Å, depending on x (or the cumulative strain energy density in a multi layer structure). The strength of the broad peak was correlated with the areal density (up to $\sim 10^9$ cm^{-2}) of strain perturbations (local lattice dilation ~ 15 Å in diameter) observed in plan view TEM. The role of MBE growth parameters in determining optical properties was investigated by changing the quantum well thickness, composition and surface preparation. The transition from phonon-resolved, near-bandgap luminescence in thin layers to the broad PL band typical of thick layers was explained by a change in growth morphology. Band edge luminescence typical of coherent SiGe strained layers can be restored by annealing in the range 700 to 900°C or by etching away the uppermost MQW layers which contain the highest density of growth asperities. The dominance of the broad luminescence band at relatively high PL measurement temperatures >~75K is consistent with shallow bound excitons diffusing to regions of lower bandgap(the strain perturbation) before self annihilation. The saturation effect, observed for the broad luminescence when exciton density and strain perturbation density are comparable, is further indirect evidence for saturable exciton sinks, characteristic of the MBE growth process.

REFERENCES

1. J.-P. Noël, N.L. Rowell, D.C. Houghton, and D.D. Perovic, Appl. Phys. Lett. **57**, 1037 (1990).
2. K. Terashima, M. Tajima, and T. Tatsumi, Appl. Phys. Lett. **57**, 1925 (1990).
3. J.C. Sturm, H. Manoharan, L.C. Lenchyshyn, M.L.W. Thewalt, N.L. Rowell, J-P. Noël and D.C. Houghton, Phys. Rev. Lett. **66**, 1362 (1991).
4. D.J. Robbins, L.T. Canham, S.J. Barnett, A.D. Pitt, and P. Calcott, J. Appl. Phys. **71**, 1407 (1992).
5. N.L. Rowell, J-P. Noël, D.C. Houghton and M. Buchanan, Appl. Phys. Lett. **58**, 957 (1991).
6. L.T. Canham, Appl. Phys. Lett. **57**, 1046 (1990).
7   R.A. Modavis, D.G. Hall, J. Bevk, B.S. Freer, L.C. Feldman and B.E. Weir, Appl. Phys. Lett. **57**, 954, (1990)
8. P.L.Bradfield, T.G. Brown and D.G. Hall, Appl. Phys. Lett. **55**, 100 (1989).
9. L.T. Canham, K.G. Barraclough and D.H. Robbins, Appl. Phys. Lett. **51**, 1509 (1987).
10. R. Zachai, K. Eberl, G. Abstreiter, E. Kasper and H. Kibbel, Phys. Rev. Lett. **64**, 1055 (1990).
11. M.A. Gell, Phys. Rev. **B38**, 7535 (1988).
12. Y H. Xie, E.A Fitzgerald and YJ Mii, J. Appl Phys. **70**, 3223 (1991)
13. H.Ennen , G. Pomrenke, A. Axmann, K. Eisele, W.Haydle and J. Schneider Appl. Phys.Lett. **46** 381 (1985).
14. E. Efeoglu, J.H Evans, T.E. Jackman, B Hamilton, D.C. Houghton, J.M. Langer, A.R. Peaker, DD Perovic, I. Poole, N Ravel, P Hemment and C.W. Chan, Semiconductor Science and Technology **7**, 00, (1993).
15. S.J. Kovacic, J.G. Simmons, K Song, J.-P. Noël and D.C. Houghton, IEEE Electron. Device Lett. **12**, 439 (1991).
16. D.V. Lang, R. People, J.C. Bean, and A.M. Sergent, Appl. Phys. Lett.**47**, 1333 (1985).
17. HC Liu, Lujian Li, J-M Baribeau, M Buchanan, JG Simmons, J. Appl.Phys. **71**, 2039 (1992).
18. DJ Robbins, P Calcott and WY Leong, Appl Phys Lett **59**, 1350 (1991).

19. S Fukatsu, N Usami, T Chinzei, Y Shiraki, A Nishida and K Nakagawa, Jap J Appl. Phys. **31**, L1015 (1992).
20. K Eberl and SS Iyer, Appl. Phys. Lett. **60**, 3033 (1992).
21. DC Houghton, J-P Noel and NL Rowell , MRS Symp Proc Vol **220**, p 299.1991
22. G.S. Mitchard and T.C. McGill, Phys. Rev. B **25**, 5351 (1981).
23. J. Weber and M.I. Alonso, Phys. Rev. B **40**, 5683 (1989).
24. J. Spitzer, K. Thonke, R. Sauer, H. Kibbel, H.-J. Herzog and E. Kasper, Appl. Phys. Le 1729 (1991).
25. T.D. Steiner, R.L. Hengehold, Y.K. Yeo, D.J. Godbey, P.E. Thompson, and G.S. Pom J. Vac. Sci. and Technol. **B10**, 924 (1992).
26. D.C. Houghton, J. Appl. Phys. **70**, 2136 (1991).
27. R. B. Young and N. L. Rowell, Proc. SPIE **1145**, 80 (1989).
28. NL Rowell, J-P Noel, DC Houghton, A Wang, LC Lenchyshyn, MLW Thewalt and DD Perovic, to be published J. Appl. Phys. 1993.
29. C. Weisbuch and B. Vinter, *Quanutm Semiconductor Structures,* (Academic, San Diego, USA 1991).
30. J.-P. Noël, N. L. Rowell, D.C. Houghton, A. Wang, and D.D. Perovic, Appl. Phys. Lett. **61**, 690 (1992).
31. D.D. Perovic and G.C. Weatherly, Ultramicroscopy **35**, 271 (1991).
32. PL Gourley and JP Wolfe, Phys Rev. **B20**, 3319 (1979)
33. L.C. Lenchyshyn, M.L.W. Thewalt, J.C. Sturm, PV Schwartz, EJ Prinz, N.L. Rowell, J-P. Noël and D.C. Houghton, Appl Phys Lett. **60**, 3174 (1992).

VISIBLE PHOTOLUMINESCENCE FROM SI$_{1-x}$GE$_x$ QUANTUM WELLS

T.W. STEINER[1], L.C. LENCHYSHYN[1], M.L.W. THEWALT[1],
D.C. HOUGHTON[2], J.-P. NOËL[2], N.L. ROWELL[2],
J.C. STURM[3], AND X. XIAO[3]
1 Simon Fraser University, Physics Department, Burnaby, British Columbia, Canada.
2 National Research Council Canada, Ottawa, Ontario, Canada.
3 Princeton University, Electrical Engineering Department, Princeton, NJ.

ABSTRACT

We have observed photoluminescence from strained SiGe quantum well layers at energies approximately equal to twice the SiGe band-gap energy. This luminescence is caused by the simultaneous recombination of two electron hole pairs yielding a single photon. Detection of luminescence at twice the band-gap has been previously used in Si to observe luminescence originating from electron-hole droplets, biexcitons, bound multiexciton complexes and polyexcitons. Time resolved spectra at twice the band-gap have been obtained from our SiGe samples prepared by molecular beam epitaxy (MBE) as well as rapid thermal chemical vapor deposition (RTCVD). This new luminescence clearly distinguishes multiexciton or dense e-h plasma processes from single exciton processes such as bound excitons, free excitons or localized excitons, which are difficult to separate in the usual near-infrared luminescence.

INTRODUCTION

The observation of infrared (IR) luminescence from quantum confined excitons in SiGe quantum wells has been achieved by several groups in the past two years[1-9]. Detailed investigation of the luminescence as a function of excitation density and temperature as well as time-resolved measurements indicate that several processes are responsible for the observed luminescence. Impurity bound excitons, free excitons, localized excitons and biexcitons can all play a role creating considerable complexity in the IR spectrum. However, as has been shown for bulk Si the visible luminescence technique allows the unambiguous identification of luminescence due to multiexcitonic species or dense plasmas[10,11]. This technique requires the simultaneous recombination of two electron hole pairs resulting in a photon with an energy close to twice the band gap. Such a recombination process is extremely unlikely and can occur only for a complex containing two or more electron hole pairs such as a biexciton. Ordinary bound excitons for example cannot give rise to any visible photoluminescence. It is important to appreciate that this is not a non-linear optical effect which, if it were allowed, would generate green photoluminescence from all excitonic species. We have observed visible luminescence in a variety of SiGe samples proving that biexcitons contribute significantly to the near edge luminescence from SiGe quantum wells.

EXPERIMENTAL

The SiGe samples were grown either by Rapid Thermal Chemical Vapour Deposition (RTCVD) or by Molecular Beam Epitaxy (MBE), as described previously [12,6]. The nominal growth parameters are used here to describe the samples. The actual parameters are expected to be within ± 10% and ± 1 nm for the Ge fraction and SiGe thickness, respectively. The IR luminescence was excited using an Ar ion laser and the photoluminescence (PL) spectra were measured using a Bomem DA8 Fourier transform interferometer with an InGaAs detector.

The visible luminescence was excited at 740 nm using a Ti:sapphire laser. The visible luminescence was dispersed by a single 3/4m spectrometer with a 300 groove/mm grating blazed at 500 nm and then imaged onto an imaging photomultiplier (ITT / Surface Science Laboratories Mepsicron) mounted on the spectrometer camera port [13]. The parallel collection capability of this photomultiplier, as well as its extremely low dark count, are essential in order to observe the extremely weak visible PL [11]. Furthermore, by picking off a timing signal from the photomultiplier [13], and cavity dumping the Ti:sapphire laser, time resolved measurements of the luminescence can be made with nanosecond resolution . By combining the output from the timing electronics with that from the photomultiplier position computer a complete three dimensional manifold of the luminescence can be collected with wavelength along the x-axis, time along the y-axis and intensity along z.

The samples were immersed in superfluid liquid helium. The dewar, spectrometer and optics were all enclosed in a light tight enclosure. The laser was filtered at the entrance of the enclosure using a long pass glass filter to remove any visible components from the beam. In order to reject the laser excitation a holographic rejection filter centered at 756 nm was used in conjunction with a stack of short pass glass filters.

RESULTS

The visible PL from an MBE SiGe sample with three quantum wells of width 2.8, 4.2 and 8.4 nm is shown in Fig. 1. The spectrum clearly exhibits contributions from all three wells as well as a small contribution from the electron-hole droplet (EHD) in the substrate at higher energy. For comparison purposes the IR spectrum at half the energy scale is plotted in the upper part of the figure. Note that the ordinary bound exciton (BE) luminescence from the substrate, which dominates the IR spectrum, is absent from the visible PL spectrum. The phonon replicas of the quantum well luminescence are also absent in the visible spectrum since the visible no-phonon (NP) transitions do not require the participation of a phonon [11]. Phonon less recombination can occur when the two constituent excitons have their electrons in opposite conduction band valleys resulting in zero net crystal momentum. Consequently, the intensity of phonon replicas should be much weaker than principal lines in the visible. The features in the visible spectrum are found to line up with the NP quantum well features in the IR after taking into account the factor of two difference in energy.

In Fig. 2 the visible and IR spectra of 6 different SiGe quantum well samples is shown. Sample a) contains three quantum wells but only two are seen in luminescence. The visible luminescence intensity as a function of excitation density in this sample was found to be close to linear over two orders of magnitude indicating that biexciton decay is the dominant quantum well recombination mechanism under high excitation conditions.

Infrared Photoluminescence Energy (meV)

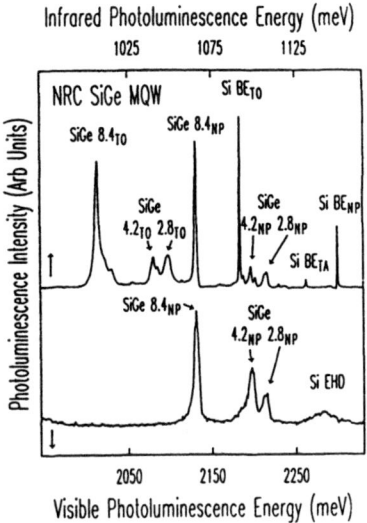

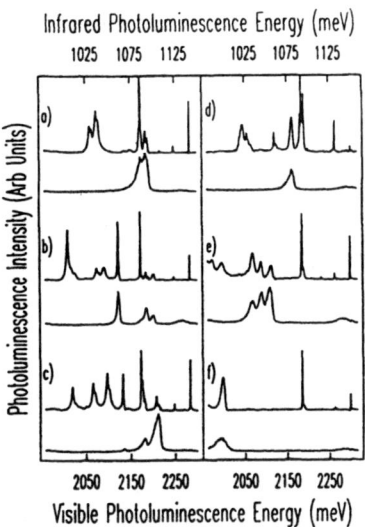

Figure 1. Comparison of the IR (top) and visible (bottom) spectra of a 3 well (2.8, 4.2, 8.4 nm, 14% Ge) MBE sample. The arrows point to the corresponding energy scale. The sample temperature was ≈ 1.7K. For the IR and visible spectra the excitation was ≈ 50 mW / cm^2 at 514.5 nm and ≈ 50 W / cm^2 at 740 nm respectively. Features labeled BE are bound exciton recombination from the Si substrate. Features labeled TO are TO phonon replicas of the corresponding no-phonon (NP) lines.

Figure 2. Comparison of the IR (upper) and visible (lower) spectra for 6 SiGe quantum well samples.
a) MBE, 2.8, 4.2, 8.4 nm, 14% Ge
b) MBE, 2.8, 4.2, 8.4 nm, 14% Ge
c) MBE, 8.4, 4.2, 2.8 nm, 14% Ge
d) MBE, 1.2 nm, 38% Ge
e) MBE 12 periods nominally identical
f) RTCVD 1.5 nm, 35% Ge

In other words the shallow alloy fluctuations are saturated with localized excitons at very low excitation densities in such samples (see Lenchyshyn et. al., these proceedings), so that at higher excitation the biexciton density increases linearly with excitation power. Sample b) is the same as the one presented in Fig. 1. Sample c) is similar to sample b); it also contains three wells of width 2.8, 4.2 and 8.4 nm. However, c) has the narrowest well closest to the substrate while b) has the widest well closest to the substrate. Since most of the excitons formed by the excitation originate in the substrate the well closest to the substrate is expected to have the greatest luminescence intensity. These expectations are borne out by the relative intensities of luminescence peaks shown in Fig. 2 b) and c). Sample d) has only a single very thin well which is 1.2 nm wide. Sample e) is a multiple quantum well sample in which all the well widths were intended to be the same. The spectra however shows three distinct peaks indicating that either the well widths or the Ge fraction is not uniform. The last sample f) is a single quantum well RTCVD sample whereas the others were all grown by MBE. This indicates that biexciton processes are important in samples grown by both techniques.

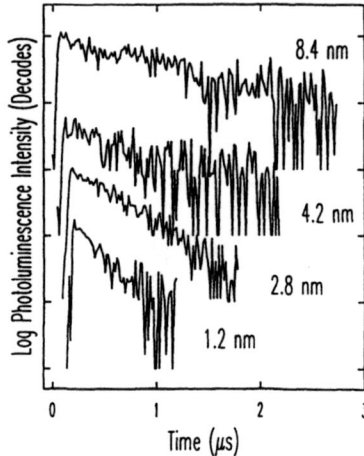

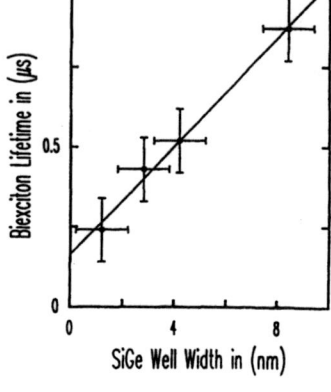

Figure 3. Biexciton lifetimes as determined from samples b) c) and d). The excitation used was cavity dumped 740 nm radiation with an average power of $\approx$ 4W / cm^2 at a repetition rate of 250 Khz and a pulse width of $\approx$ 10 ns. The sample temperature was $\approx$ 1.7 K.

Figure 4. Biexciton lifetimes as a function of well width. The points are from the best fits to the data in Figure 3. The straight line is the best fit to the points.

By averaging together all counts over the spectral width of the individual peaks in the 3D luminescence manifold visible luminescence decay curves can be extracted for each well. The results are displayed in Fig. 3 and indicate that the biexciton lifetime decreases with decreasing well width. This is qualitatively as expected given the increased exciton-exciton overlap as the wavefunctions are increasingly confined and the consequent greater recombination probability. The observed lifetime as a function of well width is shown in Fig. 4 where the lifetime has been determined by the best straight line fit to the data in Fig. 3. The values obtained for the biexciton lifetimes $\approx 0.5\mu$s are close to the fast lifetime components observed in the IR quantum well luminescence under high excitation conditions in the same samples (Lenchyshyn et. al., these proceedings).

CONCLUSIONS

We have clearly observed visible luminescence from a variety of SiGe quantum well samples. These results prove that a major component of the quantum well luminescence under high excitation conditions and low sample temperatures is due to biexcitons. Biexciton lifetime measurements indicate that the biexciton lifetime decreases with decreasing quantum well width.

REFERENCES

1. K. Terashima, M. Tajima, and T. Tatsumi, Appl. Phys. Lett. **57**, 1925 (1990)
2. J.C Sturm, H. Manoharan, L.C. Lenchyshyn, M.L.W. Thewalt, N.L. Rowell, J.-P. Noël, and D.C. Houghton, Phys. Rev. Lett. **66**, 1362 (1991).
3. D.J. Robbins, L.T. Canham, S.J. Barnett, A.D. Pitt, and P. Calcott, J. Appl. Phys. **71**, 1407, (1992).
4. V. Arbet-Engels, J.M.G. Tijero, A. Manissadjian, K.L. Wang, and V. Higgs, Appl. Phys. Lett., **61**, 2586 (1992).
5. N. Usami, S. Fukatsu, and Y. Shiraki, Appl. Phys. Lett. **61**, 1706 (1992).
6. J.-P. Noël, N.L. Rowell, D.C. Houghton, A. Wang, and D.D. Perovic, Appl. Phys. Lett. **61**, 690 (1992).
7. T.D. Steiner, R.L. Hengehold, Y.K. Yeo, D.J. Godbey, P.E. Thompson, and G.S. Pomrenke, J. Vac. Sci. Technol. **B10**, 924 (1992).
8. E.R. Glaser, T.A. Kennedy, D.J. Godbey, P.E. Thompson, K.L. Wang, and C.H. Chern, Phys. Rev. **B47**, 1305 (1993)..
9. J. Spitzer, K. Thonke, R. Sauer, H. Kibbel, H.-J. Herzog, and E. Kasper, Appl. Phys. Lett. **60**, 1729 (1992).
10. W. Schmid, Phys. Rev. Lett. **45**, 1726, (1980)
11. M.L.W. Thewalt and W.G. McMullan, Phys. Rev. **B30**, 6232, (1984)
12. J.C. Sturm, P.V. Schwartz, E.J. Prinz, and H. Manoharan, J. Vac. Sci. Technol. **B9**, 2011 (1991).
13. W.G. McMullan, S. Charbonneau and M.L.W. Thewalt, Rev. Sci. Instrum. **58**, 1626 (1987)

LUMINESCENCE STUDIES OF MBE GROWN SI/SIGE QUANTUM WELLS

M. Gail , J. Brunner , U. Menczigar, A. Zrenner and G. Abstreiter
Walter Schottky Institut, Am Coulombwall, D-8046 Garching, Germany

ABSTRACT

We report on detailed luminescence studies of MBE grown $Si/Si_{1-x}Ge_x$ quantum well structures. Both well width and composition is varied over a wide range. Bandgap photoluminescence is observed for all samples grown at elevated temperatures. The measured bandgap energies are in good agreement with subband calculations based on effective mass approximation and taking into account the segregation of Ge atoms during growth. Diffusion is found to limit quantum well (QW) growth with Ge-contents above 35% at high temperatures. The photoluminescence signals are detected up to about 100K and can be attributed to interband transitions of free excitons. We also present investigations of the exciton binding energy as a function of well width and composition. The observed shift of the exciton binding energy is compared with results of a variational calculation. A distinct onset in photocurrent and electroluminescence up to 200 K are observed in quantum well diodes.

INTRODUCTION

Extensive luminescence studies of SiGe-quantum wells have been performed recently. There is a lot of interest for strained $Si/Si_{1-x}Ge_x$ heterostructures, because the energy gap can be adjusted to the 1.3 - 1.6 μm range by varying the Ge content and the well width. Intense bandgap related photoluminescence was reported by several groups using rapid thermal chemical vapour deposition (RTCVD) [1] or molecular beam epitaxie (MBE) [2][3]. Recently also electroluminescence near 1.3 μm was demonstrated by Robbins et al. [4] using RTCVD-grown samples and by Fukatsu et al. [2]. For most of the MBE grown samples it was essential to use high growth temperatures (about 700 °C) in order to achieve efficient radiative recombination related to the bandgap of the SiGe-QWs [5][6]. In this letter we present PL of QWs varying over a wide range of well width and composition and discuss the difficulties arising from growth at high temperatures.

We have grown single and multi quantum wells (SQW, MQW) by conventional MBE on intrinsic Si(100)-substrate. In the MQW-samples the QWs were separated by 30 nm Si and repeated 5 times. The well width L varied from nominally 2 monolayers Ge up to 333 Å $Si_{1-x}Ge_x$ and the Ge-content in the alloy QWs was altered from x=0.15 to 0.53. The temperature of growth was 700 °C. Some of the samples were produced as p-i-n structures on p-substrate ($5 \cdot 10^{18} cm^{-3}$) with an intrinsic MQW-region and a Sb doped top-layer ($4 \cdot 10^{19} cm^{-3}$). PL measurements were performed using the 457 nm line of an argon ion laser. The power density was typically 0.8 W/cm². Signals were recorded by standard lock-in technique using a 640 mm grating monochromator and a liquid-nitrogen cooled Ge detector. The samples were mounted on a copper block cooled by liquid helium.

PL-SPECTRA OF QWs

Fig. 1 shows the PL-spectra at 5 K of MQWs with various Ge-contents. The emission attributed to the SiGe-QWs shifts to lower energy with increasing Ge-content in the alloy. This shift saturates at higher x. For the PL spectrum of the sample with x=0.19 the lines correspond to the no phonon transition (NP) and the phonon replicas involving TA, TO_{Ge-Ge}, TO_{SiGe} and TO_{SiSi} phonons [1]. The line width broadens with increasing Ge-content. For x=0.43 and x=0.53 additional a broad luminescence about 100 meV below the SiGe-bandgap superimposes the bandgap luminescence. For the sample with x=0.53 luminescence at positions 0.81 eV (D_1) and 0.87 eV (D_2) appears, which is well known as dislocation related luminescence [9].

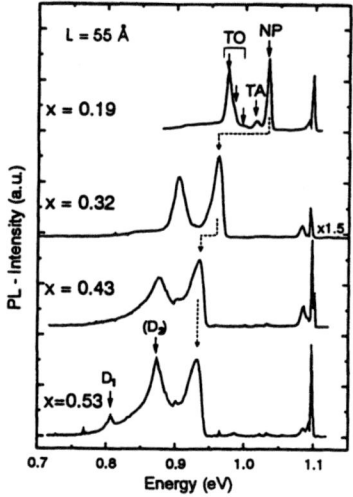 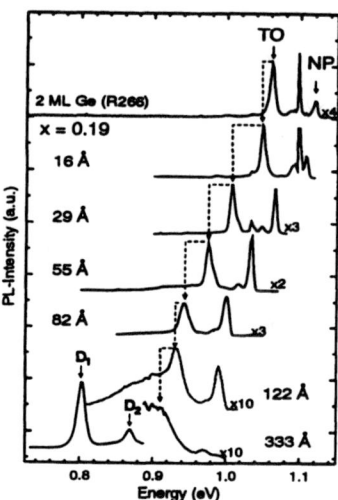

Fig. 1: PL spectra of MQWs (well width 55 Å) grown at 700 °C

Fig. 2: PL spectra of $Si_{0.81}Ge_{0.19}$-QWs for various well widths. The topmost sample was grown as MQW with QWs of nominally 2 ML Ge

In Fig. 2 PL-spectra of $Si_{0.81}Ge_{0.19}$ QWs with various well widths are depicted. The topmost spectrum (R266) shows the emission of a MQW, where one QW consists of 2 monolayers Ge. Due to less confinement of the carriers the emission of the QWs shifts to lower energy with increasing well width. For the largest well width again the broad luminescence band and dislocation related peaks appear.

The broad feature about 100 meV beneath the SiGe-bandgap is similar as often observed luminescence in MBE-samples grown at lower temperatures [10][11]. Its origin is not clear up to now. It may be attributed to defects in the alloy composition or clustering. In our case this alloy band appears with increasing Ge-content and well width, before the defect related lines D_1 and D_2 indicate partial relaxation of the strained SiGe layers. The positions of NP-lines for x=0.19 agree with effective mass calculations for the SiGe-bandgap, if one takes into account segregation effects during growth at such high temperature [7].

The dependence of the SiGe-bandgap on the composition is compared in Fig. 3 to calculated energy gaps based on [12][13]. The filled symbols indicate the energetic position of the QW-NP lines, which were corrected for confinement energy in a potential well slightly deformed by segregation as illustrated by the dotted line in the inset. The composition is taken from the growth conditions. Up to 30% Ge the experimental data agree well with calculation. At higher Ge-content the bandgap levels off. We attribute this deviation from the expected bandgap to diffusion of the Ge in the SiGe-alloy layers during growth of the subsequent layers. The filled triangular symbols in Fig. 3 are the result of SQWs, which are not as long at the elevated temperature. The bandgap is consequently lower due to shorter annealing time.

The activation energy for Ge diffusion in SiGe-alloys was found to be strongly dependent on the Ge-content in the alloy [14]. A drastic change in the diffusion constant was observed beyond a Ge concentration of 35%. We described diffusion in our QWs by a numerical solution of Fick's second law with a diffusion constant $D(x)$ dependent on the Ge-content x

$$D(x) = D_0 e^{-\frac{E_a(x)}{kT}} \tag{1}$$

using the activation energies $E_a(x)$ from [14] with $D_0 = 3.3 \cdot 10^{-4} m^2/s$. The effect of diffusion

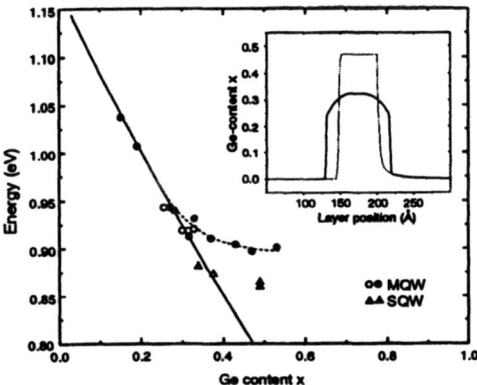

Fig. 3: Positions of NP lines corrected for confinement; filled symbols: x as nominally grown, hollow symbols: x as calculated taking diffusion into account; solid line represents calculations based on [12][13]. The inset shows the effect of diffusion of a Si$_{0.53}$Ge$_{0.47}$-alloy at a growth temperature of 700 °C and an annealing time of 6500 s.

on the concentration profile of a QW is shown in the inset of Fig. 3 (solid line), starting from a profile given by the dotted line and assuming an annealing time of 6500 s. The change of activation energy with x has two consequences. The diffusion profile is not a gaussian, because of the much smaller diffusion in the adjacent Si layers. Secondly the concentration profile changes quite fast until the topmost Ge content in the QW reaches about 30% and then nearly stops. We calculated this diffusion for all samples. The experimental values were then corrected by the Ge content and the confinement obtained in the QWs after diffusion (hollow symbols in Fig. 3). Then the resulting bandgaps fit quite well to the theoretical expected values. We observed no remarkable change of the bandgap of alloys below 30% Ge content. Rutherford backscattering measurements confirm that the total amount of Ge in the QWs agrees within 3% with the nominal Ge content given by the Ge flux. TEM pictures on the other hand show a broadening of SiGe wells in agreement with our diffusion model. As a consequence of this diffusion, also the 2 ML Ge well results in a 10 Å QW with x=0.25 with a bandgap luminescence at 1.11 eV.

EXCITON BINDING ENERGY

Temperature dependent PL experiments were performed in order to get information on the exciton binding energy in the Si$_{1-x}$Ge$_x$ QWs. With increasing temperature the PL lines of the SiGe-QWs exhibit a characteristic high energy Boltzmann tail (Fig. 4). We attribute the PL lines to recombination of free excitons (FE), whose kinetic energy is increased with temperature. The luminescence from the SiGe-QWs persists up to nearly the same temperature as the signal of the Si substrate, which indicates a comparable exciton binding energy. Even in high quality SiGe-MQWs we expect that the luminescence is dominated by shallow bound excitons (BE) at liquid-He temperature [1] due to interface roughness, fluctuations in alloy composition [15] and impurities. Despite of the fact we can resolve this BE only in one of our samples (Fig. 5) we believe that they exist in all our QWs, but with a dissociation energy less than half of the NP line width.

The temperature dependence of the integrated luminescence intensity allows to determine the dissociation energy for BE by fitting the data to the expression [16]:

$$\frac{I}{I_0} = \frac{1}{1 + g_0 e^{-\frac{E_0}{kT}} + g_1 e^{-\frac{E_1}{kT}}} \tag{2}$$

The first term in the denominator (g_0, E_0) corresponds to the dissociation of the BE into FE, which determines the temperature dependence at low temperatures, and g_1, E_1 relates to the dissociation to free carriers at higher temperatures. Fig. 6 shows least square fits to the

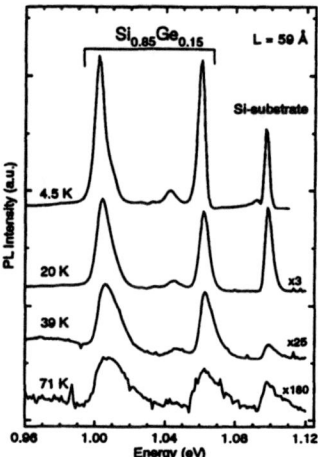

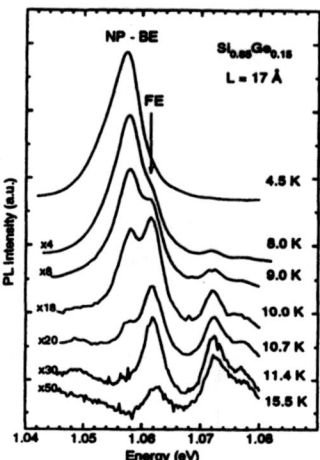

Fig. 4: PL spectra of a MQW with L=59 Å at different temperatures; the boltzmann shape is characteristic for FE recombination

Fig. 5: PL spectra of SQW width L=17 Å; a luminescence take-over from BE to FE recombination is observed;

temperature dependence of the integrated luminescence using Eq. 2. The data can be fitted quite well over the whole measured temperature range.

A least square fit with only one dissociation term is presented in the inset of Fig. 6 by the dotted line for comparison. It describes well the low temperature part of the intensity dependence and leads to a binding energy of about 4 meV. We believe that this energy corresponds to the dissociation energy of the BE to FE. As emphasised by Blakemore [17] it is necessary to determine first the slope of the temperature dependence at high temperatures and fit the data with this slope. The result of such a procedure with also only one dissociation term is shown by the dashed line in the inset of Fig. 6. This fit deviates from experiment in the temperature regime where the slope changes. As demonstrated in [17] for defect state statistics the consideration of additional excited states between the level of BE and free carriers can explain the difference. Using both terms in the denominator of Eq. 2 for the fitting procedure results in the solid line in the inset, which is an excellent fit to the data.

The binding energy of the FE in the SiGe-QWs is given by the difference $E_1 - E_0$. The values obtained by this procedure are shown in Fig. 7 for various QW widths and Ge-contents. The dissociation energies for BE to FE were found to be 2-4 meV for x=0.19 and 4-6 meV for x=0.33 meV. A maximum binding energy of 12.9 ± 1 meV at a well width of 82 Å is observed in the $Si_{0.81}Ge_{0.19}$ alloys . For $Si_{0.67}Ge_{0.33}$ we find the highest binding energy of 29.5 ± 2 meV at 57 Å. The FE binding energy decreases quite strongly to smaller layer thickness and more moderately to wider layer widths. This behaviour is in principle predicted for a type I quantum well. A type II band structure would result in a monotonically increasing binding energy with smaller well width due to the spatial separation of electrons and holes [20]. Furthermore the binding energy of the QW excitons is enlarged beyond this in bulk SiGe, which is expected to be 11.0 meV and 9.4 meV for x=0.19 and x=0.33 respectively [18] (indicated by arrows in Fig. 7). In type II QWs the binding energy is expected to be smaller than in bulk.

The lines in Fig. 7 represent the results given by a variational calculation according to the approach of Greene et al. [19] and Bastard et al. [20]. We use the trial wave function

$$\psi = N f_e(z_e) f_h(z_h) e^{-\frac{r}{\lambda}} \qquad (3)$$

The functions f_e and f_h are numerical solutions for finite square well potentials, λ is the trial parameter and N a normalisation constant. The variational binding energy is given by subtracting

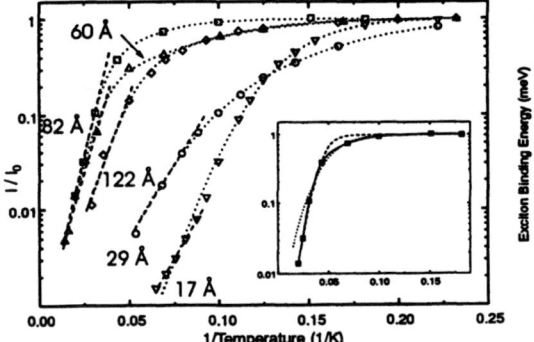
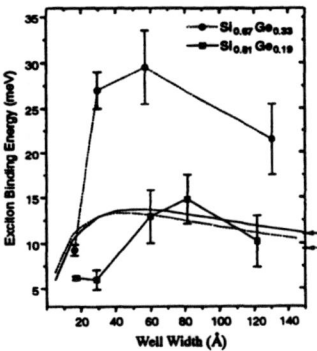

Fig. 6: Temperature dependence of the integrated luminescence for different well widths (x=0.19); exciton binding energy is determined by the slope at high temperature and a
fit by Eq. 2; the inset shows different fitting procedures

Fig. 7: Binding energies of free excitons versus well width for x=0.19 and
x=0.33; lines show result of variational
calculation;

from the variational solution of $H\psi = E\psi$ the subband energies of electron and heavy hole in their
finite square wells. For 1s ground states this variational approach agrees very well with the results
of [19] in AlGaAs/GaAs QWs despite of the much more simple variational parameterization.

Our variational calculation assumes for x=0.19 and x=0.33 a type I QW with a small offset
in the conduction band (26 meV and 37 meV respectively). Consequently the electron wave
function is widely spread over the QW. We compute for both Ge-contents a maximal exciton
binding energy of about 13 meV. The maximum of binding energy is found at 50 Å and 40
Å respectively. For large well thickness our calculation approaches the values given for bulk
alloys. The basic dependence of the binding energy versus QW-width agrees with the calculation,
particularly in the case of the $Si_{0.81}Ge_{0.19}$. The larger binding energies determined experimentally
for the samples with higher Ge content is not explained by our calculations, though the overall
dependence is similar.

ELECTROLUMINESCENCE

We have also studied p-i-n structures with built-in intrinsic $Si_{1-x}Ge_x$-MQWs. The Ge-content
of the 5 SiGe-layers varied between x=0.27 and x=0.44. Ti/Au contacts were prepared to the
top layer with a contact area of $300 \times 200\mu m$ leaving a window of $180 \times 160\mu m$ in the center. No
mesa etching was done, but the contact was surrounded by a non-transparent guard ring.

Photocurrent, photoluminescence and electroluminescence measurements were performed.
Fig. 8 shows the spectra obtained for the samples with x=0.27 and x=0.37. The onset of photocurrent agrees quite well with the bandgap luminescence from PL and EL. The EL is shifted
about 16 meV to higher energy compared to the PL. This is probably due to the high current
injection conditions. We observe both NP and TO line in PL an EL. In EL we do not find any
defect or substrate related luminescence. Therefore the injected carriers recombine radiatively
only in the SiGe-QWs. This holds up to 200 K (see Fig. 9). In Fig. 10 the measured EL intensity
as emitted through the window and as collected externally in an angle of 46° is plotted versus
injected electrical power.

This work was supported in part by ESPRIT under contract P7128 and by the Siemens AG.

26

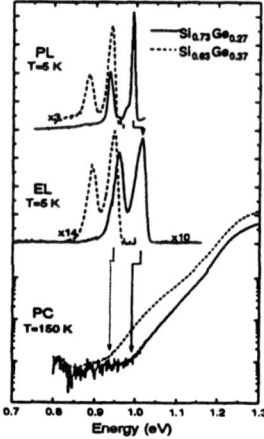 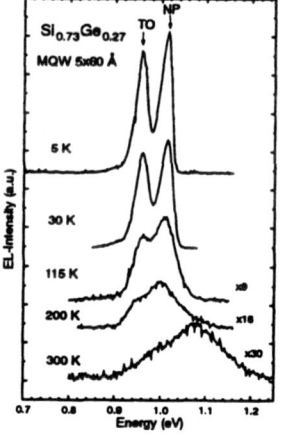 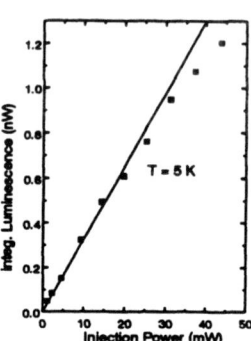

Fig. 8: Comparison of photo-luminescence (PL), electrolumi-nescence (EL) and photocur-rent (PC) of p-i-n MQWs for x=0.27 and x=0.37.

Fig. 9: EL spectra of a MQW (x=0.27); injection current was 20 mA for 4 K and raised up to 75 mA at 300 K (this was considered in magnification).

Fig. 10: Integrated luminescence as measured externally with an angle of 46° without spectral de-composition versus electrical in-jection power;

REFERENCES

1. J.C. Sturm, H. Manoharan, L.C. Lenchyshyn, M.L.W. Thewalt, N. L. Rowell, J.-P. Noël, and D.C. Houghton, Phys. Rev. Lett. **66**, 1362 (1991)
2. S. Fukatsu, H. Yoshida, N. Usami, A. Fujiwara, Y. Takahashi and Y. Shiraki, Thin Solid Films **222** 1 (1992)
3. M. Wachter, K. Thonke, R. Sauer, F. Schäffler, H.-J. Herzog, and E. Kasper, Thin Solid Films **222** 10 (1992)
4. D. J. Robbins, P. Calcott, and W. Y. Leong, Appl. Phys. Lett. **59** 1350 (1991)
5. J. Brunner, U. Menczigar, M. Gail, E. Friess, G. Abstreiter, Thin Solid Films **222** 27 (1992)
6. N. Usami, S. Fukatsu, and Y. Shiraki (submitted to Appl. Phys. Lett.)
7. J. Brunner, J. Nützel, M. Gail, U. Menczigar and G. Abstreiter, presented at AVS Chicago 1992 (unpublished)
8. J. Weber and M. I. Alonso, Phys. Rev. B **40**, 5683 (1989)
9. R. Sauer, J. Weber, and J. Stolz, E. R. Weber, K.-H. Ksters, and H. Alexander, Appl. Phys. A **36**, 1 (1985)
10. J.-P. Noël, N. L. Rowell, D. C. Houghton and D. D. Perovic, APL **57**, 1037 (1990)
11. J. Spitzer, K. Thonke, R. Sauer, H. Kibbel, H.-J. Herzog, E. Kasper, APL **60**, 1729 (1992)
12. C. G. Van de Walle and R. M. Martin, Phys. Rev. B **34**, 5621 (1986)
13. L. Colombo, R. Resta, and S. Baroni, Phys. Rev. B **44**, 5572 (1991)
14. G. L. McVay and A. R. DuCharme, Phys. Rev. B **9**, 627 (1974)
15. L.C. Lenchyshyn, M.L. Thewalt, J.C. Sturm, P.V. Schwartz, E.J. Prinz, N.L. Rowell, J.-P. Noël, D.C. Houghton
16. E. W. Williams and H. B. Bebb, Semiconductors and Semimetals, edited by R. K. Willardson and A. C. Beer, Vol. 8 (Academic Press, New York, 1972), p. 321
17. J. S. Blakemore, Semiconductor Statistics (Pergamon Press, London, 1962), p. 140
18. D. J. Robbins, L. T. Canham, S. J. Barnett, A. D. Pitt, and P. Calcott, J. Appl. Phys. **71**, 1407 (1992)
19. R. L. Greene, K. K. Bajaj and D. E. Phelps, Phys. Rev. B **29**, 1807 (1984)
20. G. Bastard, E. E. Mendez, L. L. Chang and L. Esaki, Phys. Rev. B **26**, 1974 (1982)

X-RAY INVESTIGATION OF STRAIN RELAXATION IN SHORT-PERIOD Si_mGe_n SUPERLATTICES USING RECIPROCAL SPACE MAPPING

P.HAMBERGER*, E.KOPPENSTEINER*, G.BAUER*, H.KIBBEL**, H.PRESTING**, E.KASPER** AND A.PESEK#
*Johannes Kepler University, Institut für Halbleiterphysik, A-4040 Linz, Austria.
**Daimler Benz AG, Forschungszentrum, D-7900 Ulm, Germany.
#Johannes Kepler University, Forschungsinstitut für Optoelektronik, A-4040 Linz, Austria.

ABSTRACT

The optoelectronic properties of Si_mGe_n strained layer superlattices (SLS´s) depend strongly on the structural perfection. We used double crystal and triple axis x-ray diffractometry to characterize the structural properties of short period Si_9Ge_6 SLS´s grown on about 1μm thick step-graded SiGe alloy buffers. As grown SLS´s and samples annealed subsequently at 550°C, 650°C and 780°C for 60 min were investigated. Precise strain data were extracted from two-dimensional reciprocal space maps around (004) and (224) reciprocal lattice points. These data were used as refined input parameters for the dynamical simulation of the integrated intensity along the q‖[004] direction. Annealing causes interdiffusion as indicated by the decreasing superlattice (SL)-satellite peak intensities and by the change of the Si/Ge thickness ratio . However, the full width at half maximum of the SL satellite peaks does not change significantly with annealing up to 650°C. The in-plane SL lattice constant in both samples is increased only slighty by annealing (< $9x10^{-3}$ Å). Consequently the interface intermixing due to interdiffusion is the main cause for the shift of the luminescence energy to higher values in these annealed samples.

INTRODUCTION

Within the last two years substantial progress has been reported in the field of SiGe molecular beam epitaxy (MBE).[1] High quality Si/SiGe heterostructures and Si_mGe_n short period SLS´s (m,n number of monolayers) have been grown on rather thick linearly graded[2] or step-graded[3] SiGe alloy buffers instead of thin (200Å) single step buffers with constant Ge content. If graded buffers are used, a reduction of the number of threading dislocations by several orders of magnitude has been shown to occur.[2] Short-period strain symmetrized Si_mGe_n superlattices have recently found considerable attention too, because of their sharp and intense luminescence in the near infrared.[3] For further processing it is of interest to study the stability of the electronic structure with respect to thermal treatment. Since strained layer superlattices are metastable a number of investigations was devoted to a study of the structural stability of Si/Ge structures using x-ray diffraction techniques.[4,5] So far the lateral strain, the decrease of peak intensities and the full width at half maximum (FWHM) of x-ray diffractograms of SiGe buffers and SL´s were investigated using double crystal x-ray diffraction (DCD).[6-9] Recently, the use of triple axis diffractometry (TAD) has been introduced for the structual analysis of complex epitaxial multilayer stuctures, which allows the determination of the intensity distribution around reciprocal lattice points (RELP´s).[10,11] The relative position of the RELP maxima reflects the strain status in the corresponding layer, while the shape of RELP´s gives detailed information on structural imperfections like strain gradients[12], mosaicity and dislocation density[13,16] and interface roughness.[10,14] Baribeau et al[5] have shown that Si_mGe_n short period SL´s exhibit an appreciable strain relaxation due to thermal treatment at 600 - 800°C. However the SL´s investigated were grown on rather thin $Si_{1-x}Ge_x$ alloy buffers (200 nm). For Si_mGe_n superlattices grown on thick graded SiGe buffers on which a nearly freestanding superlattice is deposited (in-plane lattice constant of the top most buffer layer equal to the mean in-plane lattice constant of the SL), it can be expected that after annealing the strain status does not change significantly.

ROCKING CURVES AND RECIPROCAL SPACE MAPS

DCD and TAD rocking curves and reciprocal space maps around both symmetric (004) and asymmetric (224) RELP's have been employed to investigate the strain status and structural perfection of two short period Si_mGe_n structures with nominally m=6, n=4 (sample B2416) and m=9, n=6 (sample B2512), respectively, which have been grown both by MBE on thick step-graded buffers. In the following data for the sample B2512 will be presented. For this sample a 100nm thick Si layer was grown on top of the substrate followed by the step-graded buffer (B1), in which the Ge content was increased in 12 steps by 3% per 50nm up to a total thickness of 650nm (i.e., to a nominal Ge content of 39%). During the buffer growth the temperature was decreased continuously from 600 to 520°C. Subsequently, a 500nm thick $Si_{0.6}Ge_{0.4}$ alloy buffer layer (B2) was grown at 500°C. Prior to the growth of the SL a monolayer of antimony was deposited as a surfactant. The superlattice consists of 145 periods. On top of the SL a 1 nm thick Si cap layer has been deposited.

The diffractometer uses $CuK_{\alpha 1}$ radiation , a Bartels-type 4-crystal monochromator[15] in the Ge(220) setting in the primary beam, and in the diffracted beam either slits with detector opening angles of 180 and 360 arcsec (DCD optics) or a two-reflection Ge(220) channel-cut analyser crystal (TAD optics, detector opening angle: 12 arcsec). A reciprocal space map is obtained by scanning a number of $\omega/2\Theta$ scans (radial from the origin (000) in reciprocal space, parallel to q∥[hkl], h,k,l being Miller's indices) at different ω-positions of the sample (ω-scan: transverse along a circle with center (000) in reciprocal space, near the RELP perpendicular to q∥[hkl]). In real space ω is the angle between the incident beam and the sample surface, and 2Θ the angle between incident and diffracted beam.

The (004) Bragg reflection was measured along q∥[004] (q: wave vector in reciprocal space) using DCD optics before annealing and the results are shown for the Si_9Ge_6 SL in Fig.1. The insert in Fig. 1 shows the nominal structural parameters . The Bragg-reflection peaks from the graded buffer B1 are lying in between the Si(004) substrate and the zero order SL peak (SL0). The Bragg-peak of buffer B2 with constant Ge content coincides with SL0. The SL0 and SL-1 peaks have FWHM values of 325 and 377 arcsec, respectively. In addition a dynamical simulation is shown with parameters given in Table I.

Figure 2 shows (004) and (224) reciprocal space maps in the vicinity of the substrate, buffer and SL0 RELP's. The deformation of the substrate RELP's (Figs.2a,b), which should have the shape of circles for undistorted substrate crystals, along the direction of the Ewald sphere is an artefact known as "analyser streak" for TAD. This effect appears at high intensity RELP's and does not influence the buffer and SL RELP's of interest. In the symmetric (004) maps of the unannealed sample the centers of buffer and SL RELP's nearly coincide with the direction q∥[110]=0 (i.e. q⊥[004]=0) through the center of the substrate RELP. The FWHM

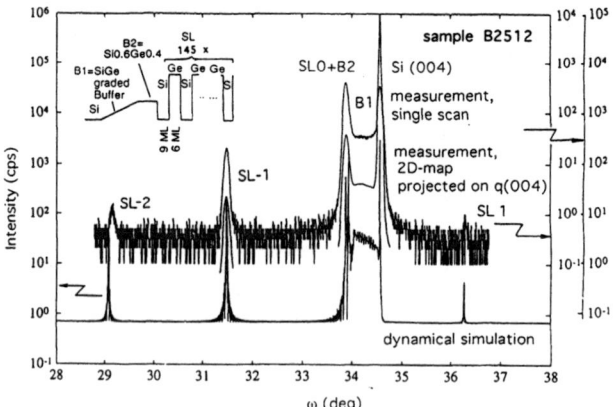

Fig.1: Rocking curve of Si_9Ge_6 , SL (004) Bragg reflection. center curve: integrated intensity and lower curve: dynamical simulation.Position of SL-2 is not well reproduced in dynamical simulation due to the use of Takagi -Taupin approximation. Insert shows sample with stepgraded buffer (B1) and buffer B2 with constant Ge -content.

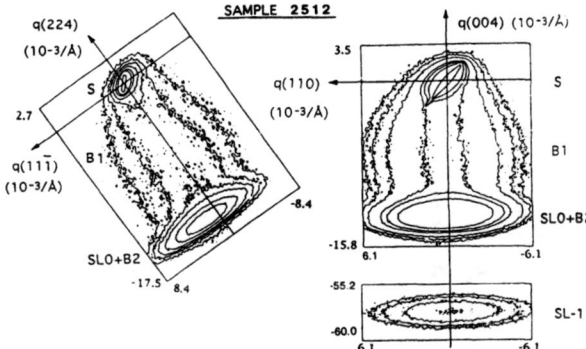

Fig.2: Reciprocal space maps around (004) and (224) reciprocal lattice points of Si_9Ge_6 SL . (004) plot: shows Si substrate, buffer, SL0 and SL-1 peaks. (224) plot: shows relaxed buffer and SL0 -intensity contours the center of which is shifted slightly with respect to the [224] direction through Si substrate peak maximum.

values of SL0 and SL-1 in the direction perpendicular to $q\|[004]$ are considerably higher than in the direction parallel to the growth direction (see Fig.1), namely 1270 and 1380 arcsec. In the asymmetric (224) map the RELP maxima of the buffer B1 lie along the direction $q\|[224]$ through the center of the (224) substrate RELP, while the centers of the (224)SL0 RELP is lying closely below this line indicating that the SL is not completely strain symmetrized. The FWHM´s of SL0 perpendicular to $q\|[224]$ is 1330 arcsec. The lattice constants a_p (p: parallel) and a_n (n: normal to the growth plane) of the Si and Ge layers in the SL were calculated from the positions of the intensity maxima of the (004) and (224) SL0 reflections (Fig.2) with an accuracy of ±0.001Å using the procedure outlined in Refs.12 and 16 and are listed in Table I.

After growth, the sample was annealed for 1h at 550°C, subsequently for 1h at 650°C, and finally for 1h at 780°C. For a precise determination of the change in the strain status due to annealing TAD optics have been used to record symmetrical (004) and asymmetrical (224) reciprocal space maps after each annealing step, similar to the measurements of the as-grown sample described above (Fig.2). In order to demonstrate the sensitivity of this method, a detail of the (004) reciprocal space maps is shown in Fig.3. The left-hand side of this figure shows RELP´s of the sample after annealing at 650°C and the right-hand side similar data after a subsequent annealing at 780°C. Due to the latter annealing step the relative distance between the substrate Si(004)- and the SL0(004) RELP-maxima along $q\|[004]$ has been reduced by 0.002Å, whereas both the peak intensity and the shape of the SL0 RELP, which are mainly determined by the FWHM perpendicular and parallel to the direction $q\|[004]$, remain unchanged within the accuracy reached here. The clear change in the relative positions of substrate- and SL0-RELP shows that TAD reciprocal space mapping offers the possibility to determine *minute changes in the strain status* in the SL stack with respect to the Si substrate with high accuracy.

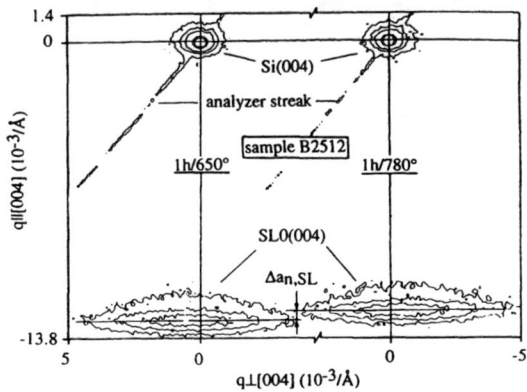

Fig.3: Reciprocal space maps around (004) reciprocal lattice points of Si_9Ge_6 SL, anneald at T= 650°C (left side) and subsequently at T=780°C (right side). Shift of (004) SL0 RELP with respect to that of Si substrate indicates minute strain relaxation of a_n of SL.

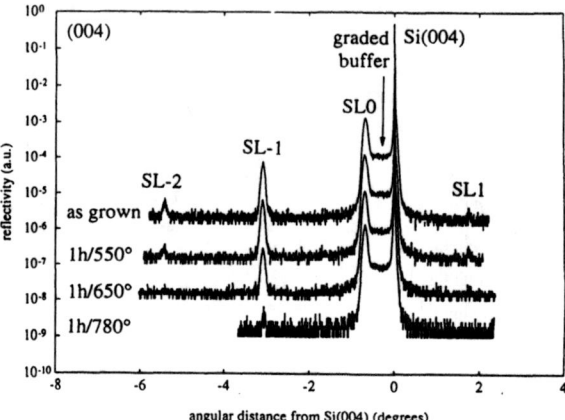

Fig.4: X-ray reflectivity vs ω/2Θ for the Si_9Ge_6 SL, (004) Bragg reflection: upper trace: as grown sample, than three traces for three subsequent annealing steps as indicated.

The ω/2Θ-scans measured after each annealing step are compared to the rocking curve of the as-grown sample in Fig.4. It is evident that the SL satellites SL-2 and SL+1 completely disappear and that the intensity of SL-1 is decreasing continuously. Especially the last annealing step at 780°C causes the intensity of SL-1 to decrease close to the background level, while the peak intensity of SL0 remains unchanged. Figure 5 shows the average Ge profiles within one period, which have been incorparated in the dynamical simulation model to reproduce the measured SL satellite peak intensities.

INTERDIFFUSION AND STRAIN

From the reciprocal space maps in Figure 2, it turnes out that the SL in sample B2512 is nearly freestanding, because the distance of the SL0 maximum from the q∥[224] direction is small. The residual lateral strain in the SL is -0.190% for the as grown sample and -0.046% after the last annealing step at 780°C. Furthermore all portions of the graded SiGe buffer are fully relaxed, because the diffuse scattering around their (224)RELP's is symmetrical to the q∥[224] direction. We want to point out that the buffer RELP's in the samples have higher FWHM's in the direction perpendicular to q∥[224] and q∥[004] and the FWHM's of the SL RELP's are significantly lower compared to similar samples grown on single step SiGe alloy buffers,[16] which means that in samples grown on thick graded buffers, the dislocation densities are much smaller in the SLS stack.

Table I: effect of annealing on structural parameters of sample B2512 (Si_9Ge_6)

annealing parameters	layer	a_p(Å)	a_n(Å)	t_{Si} / t_{Ge}	FWHM(") SL0	FWHM(") SL-1
as grown	Si	5.509	5.371	1.57	325	377
	Ge	5.509	5.770			
1h/550°C	Si	5.515	5.366	1.50	323	379
	Ge	5.515	5.765			
1h/650°C	Si	5.515	5.366	1.50	330	357
	Ge	5.515	5.765			
1h/780°C	Si	5.517	5.365	1.49	356	-
	Ge	5.517	5.764			

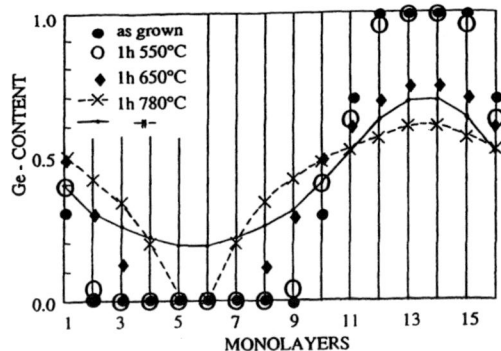

Fig.5 : Ge profiles as obtained from dynamical simulation of data shown in Fig.4 for the as grown Si_9Ge_6 sample and after three annealing steps as marked by symbols. Dashed line and full line show two profiles which both reproduce the SL satellite intensity ratio in the measurement after the last annealing step at 780°C.

With the strain status determined by reciprocal space mapping (Tab.I), we employ the dynamical diffraction theory in order to determine the periods and the thickness ratio of the Si and Ge layers in the SL, using just the positions of the SL satellite extrema but not their intensities. Both the strain status and the relative thickness ratio influence the position of the SL0 peak. The thickness ratios are listed in Table I for the as grown sample as well as after the subsequent annealing steps, using the values for a_p and a_n determined by the reciprocal space maps. In fact, not the rocking curves measured by single $\omega/2\Theta$ scans along $q\|[004]$ (topmost curve in Fig.1) have been fitted, but the integrated intensities perpendicular to this direction obtained from the reciprocal space maps (second curve from top in Fig.1).[12] However, we would like to point out that in the Si_6Ge_4 SLS (B2416) the values for a_p and a_n after annealing at T = 550°C for 1 hour are identical to those of the as grown SLS.

The fact that all the SL satellite intensities except that of SL0 decrease with increasing annealing temperature (Fig.4), indicates interdiffusion.[17-21] In order to model the effect of interdiffusion in such a short period SL, we treat this effect as an increase of interface roughness in growth direction by steps of one monolayer thickness. The lateral interface roughness is expected to be dominated by the mosaic block size and the density of still existing threading dislocations in the SL stack. It turns out that the SL satellite intensities of the as grown sample can be described assuming that the interface roughness is about two monolayers (Fig.4).The Ge profiles used in the dynamical simulation, which reproduce the measured SL-2, SL-1, SL0, and SL+1 satellite intensities are shown in Fig. 5. The Si core layers are present up to a annealing step at T=780°C if profiles similar to those found previously in short period Si_mGe_n SLS[19] were used. However with just two SL peaks present (SL0, SL-1) the fit to their intensities is not unique and in Fig.5 another profile which fits the x-ray data for the annealing step at T=780°C is shown as well. Experimentally it is found that only for the last annealing step at 780°C the linewidths increase significantly, which indicates an increase of the dislocation density in the SL stack. Near infrared photoluminescence measurements on the same sample[3,18], showed additional luminescence peaks after the last annealing step, attributed to defect lines probably caused by dislocations.

For a precise determination of the Ge-profile in the interface other theories than the dynamical diffraction theory of undistorted crystals have to be applied.[13] The problem for these particular samples is that the diffuse scattered intensity distribution near the SL RELP's is strongly dominated by dislocations. Moreover, the interface roughness can be much better evaluated from higher order SL satellites. Nevertheless, whenever the FWHM values of the SL-1 RELP exceed the linewidth of the SL0 RELP, this difference has to be attributed to interface roughness. We would like to mention that the structural stability of Si/SiGe structures and of Si_mGe_n SL's with respect to thermal annealing has been investigated by a variety of methods including x-ray diffraction[6], transmission electron microscopy[17] and Raman scattering from optical and folded acoustical phonon branches,[19] apart from luminescence. Apart from recent investigations,[3] usually samples grown on thin buffers and thus with completely different strains were investigated, so that a direct comparison of our results with those published previously is difficult. A microscopic interdiffusion model for the evaluation of the composition profile of as grown and annealed Si_8Ge_8 SLS´s was used by Schorer et al.[19], who started with a similar initial interface roughness of about 2 monolayers, comparably to our case.

CONCLUSION

The structural parameters of short period strained layer Si_9Ge_6 superlattices grown on about 1 μm thick step graded SiGe alloy buffers on top of (001) Si substrates have been analysed using DCD and TAD single scans and reciprocal space mapping around symmetrical (004) and asymmetrical (224) reciprocal lattice points of the superlattice and the buffer. The x-ray data confirm the results obtained from photoluminescence[3,18] on annealed short period SL´s grown on a thick step-graded buffer. Annealing the Si_9Ge_6 SL at temperatures from 550°C to 780°C causes a slight decrease of the small residual in plane strain. The graded bufferlayers are completely unaffected by the annealing process. The main effect of annealing is the interdiffusion as evidenced by the decrease of the SL intensities. From these intensities the composition profiles were obtained through a simple fitting procedure. The application of the graded bufferlayer growth concept allows to tailor the desired strain status in the SiGe short period superlattices, which is essential for thermal stability and optoelectronic applications.

Work supported by FWF Project No.9119 and ESPRIT Basic Research Action 7128.

REFERENCES

1. For a recent review see e.g.:W. Presting, A. Kibbel, M. Jaros, R.M.Turton,U.Menczigar, G.Abstreiter, and H.G.Grimmeiss, Semicond.Sci.Technol. **7**, 1127 (1992).
2. E.A.Fitzgerald, Y.-H.Xie, M.L.Green, D.Brasen, A.R.Kortan, J.Michel, Y.-J. Mii, and B.E.Weir, Appl.Phys.Lett. **59**, 811 (1991).
3. U.Menczigar, G.Abstreiter, J.Olajos, J.Engvall, H.Gimmeiss, H.Kibbel, E.Kasper, and H.Presting, Phys.Rev.B (1993), in print
4. S.J.Chang and K.L.Wang, Appl.Phys.Lett. **54**(13) 1253 (1989).
5. J.M.Baribeau, R.Pascual, and S.Saimoto, Appl.Phys.Lett. **57**(15) 1502 (1990).
6. D.C.Houghton, D.D.Perovic, J.M.Baribeau, and G.Weatherly, J.Appl.Phys. **67**,1850 (1990).
7. F.K.LeGoues, J.A.Ott, K.Eberl, and S.S.Iyer, Appl.Phys.Lett. **61**, 174 (1992).
8. W.Koschinski, K.Dettmer, and F.R.Kessler, J.Appl.Phys. **72**, 471 (1992).
9. P.J.Wang, M.S.Goorsky, B.S.Meyerson, and F.LeGoues, Appl.Phys.Lett. **59**, 814(1991).
10. P.F.Fewster, J.Appl.Cryst. **22**, 64 (1989); **24**, 178 (1991).
11. F.Schäffler, D.Többen, H.J.Herzog, G.Abstreiter, and B.Holländer, Semicond.Sci. Technol. **7**, 260 (1992).
12. E.Koppensteiner, G.Springholz, P.Hamberger, and G.Bauer, J.Appl.Phys. submitted
13. V.Holy, J.Kubena, E.Abramof, K.Lischka, A.Pesek and E. Koppensteiner, J.Phys.D (1993), in print.
14. V.Holy, J.Kubena and K.Ploog, phys.stat.sol.(b) **162**, 347 (1990).
15. W.J.Bartels, J.Vac.Sci.Technol. **B1** (2), 338 (1983).
16. E.Koppensteiner, P.Hamberger, G.Bauer, A.Pesek, H.Kibbel, H.Presting, and E.Kasper, Appl.Phys.Lett., **62** (15) 1 (1993).
17. S.S.Iyer and F.K.LeGoues, J.Appl.Phys. **65** (12) 4693 (1989).
18. U.Menczigar, G.Abstreiter, private communication.
19. R.Schorer, E.Friess, K.Eberl and G.Abstreiter, Phys.Rev.B **44** 1772 (1991).
20. G.L.McVay and A.R.DuCharme Phys.Rev.B **9**, 627 (1974).
21. R.L.Headrick, J.-M.Baribeau, D.J.Lockwood, T.E.Jackman, M.J.Bedzyk, Appl.Phys.Lett. **62**, 687 (1993).

Enhanced Band-gap Luminescence in strain-symmetrized $(Si)_m(Ge)_n$ Superlattices

U. Menczigar, G. Abstreiter, H. Kibbel [*], H. Presting [*], and E. Kasper [*].
Walter Schottky Institut, TU München, Am Coulombwall, D-8046 Garching, Germany.
*Daimler Benz AG, Forschungszentrum, Wilhelm Runge Str. 11, D-7800 Ulm, Germany.

Abstract

We report on band-gap luminescence in short period, strain symmetrized $(Si)_m(Ge)_n$ superlattices grown on relaxed, step-graded $Si_{1-x}Ge_x$ alloy buffer layers. The dislocation density in the superlattices, which were grown at 500°C using Sb as a surfactant, is reduced by 2-3 orders of magnitude compared with superlattices grown on thin, partly relaxed $Si_{1-x}Ge_x$ buffer layers. Due to the improved quality of the superlattices, well defined band-gap luminescence could be observed which is for a $(Si)_6(Ge)_4$ superlattice strongly enhanced compared with a $Si_{0.6}Ge_{0.4}$ alloy reference sample. The measured band-gap energies compare well with theoretical predictions. To study the influence of interdiffusion of the Si- and Ge-layers on the band-gap of the superlattices, the samples were annealed and studied with photoluminescence and Raman spectroscopy. An increasing band-gap and a decreasing luminescence efficiency was found with increasing intermixing of the layers. These experimental results are well described with an interdiffusion model of the layers in conjunction with an effective mass calculation.

A. Introduction

$(Si)_m(Ge)_n$ strained layer superlattices (SLSs) attracted much attention because of their capacity to function as an emitter/detector in the infrared region from 1.55μm to 1.3μm. This field of research was stimulated by an early theoretical investigations of Gnutzmann and Clausecker[1] who predicted an enhanced oscillator strength for the fundamental band-gap transition in a short period $(Si)_m(Ge)_n$ SLS compared with the indirect semiconductors Si and Ge. The basic idea is the folding of the bandstructure into the reduced superlattice Brillouin zone[1]. For a period length of about 10 monolayers (ML) the 2-fold degenerate Δ minima in growth direction [$\Delta(2)$], which are located in bulk Si at 0.85 in Γ–X direction of the Brillouin-zone, are folded back to the Γ point. An enhanced oscillator strength compared with the indirect fundamental band-gap transitions in the Si and Ge is expected for this so-called quasi-direct interband transition between the valence-band (VB) and the folded $\Delta(2)$ states. First observations of superlattice-induced interband transitions in a superlattice quantum-well[2] grown pseudomorphically on Si(100) resulted in a large number of bandstructure calculations[3,4,5]. These calculations indeed predict an enhanced oscillator strength between the VB and the folded $\Delta(2)$ states. The fundamental band-gap for superlattices grown directly on Si(100) substrate, however, was found to be due to the four-fold degenerate non-folded Δ minima perpendicular to the growth direction [Δ(4)]. The Δ(2) minima are only lower in energy than the Δ(4) minima if the Si layers are under a lateral tensile strain. This can be achieved if the superlattice is grown on a substrate or buffer layer with a lateral lattice constant larger than that of pure Si. A $Si_{1-x}Ge_x$ alloy layer which is grown on Si beyond its critical thickness and partly relaxed to an intermediate lattice constant between those of Si and Ge provides such a buffer[6]. The lateral lattice constant can be chosen over a wide range between the lattice constants of Si and Ge by varying the Ge content and (or) the thickness of the partly relaxed $Si_{1-x}Ge_x$ layer. Results of an effective mass (EM) calculation[7] for the energies of the interband transitions in a $(Si)_6(Ge)_4$ SLS grown on Si, $Si_{0.6}Ge_{0.4}$ and Ge substrate are shown in fig. 1.

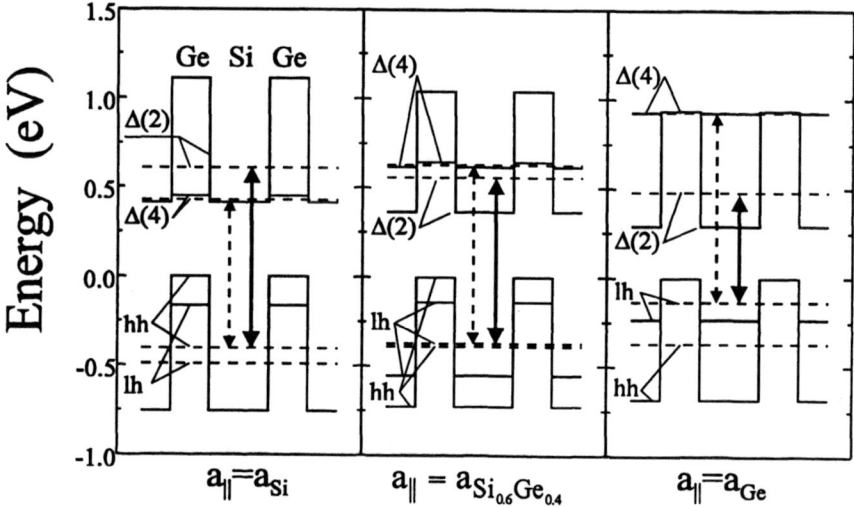

Fig.1   Band-edges in the Si and Ge layers (solid lines) and mini-band edges (dashed lines) for lh, hh, Δ(2), and Δ(4) states in a $(Si)_6(Ge)_4$ SLS grown on different substrates. The arrows mark the transitions between the VB and CB states.

The solid lines are the calculated band-edges in the Si- and Ge layers using the theoretical VB offsets of Ref. [8] and the deformation potentials of Ref. [9]. The calculated band-edges of the superlattice minibands, using an EM calculation[7], are shown as dashed lines in fig. 1. The lowest interband transition in the SLS is an indirect transition between the non-folded Δ(4) states and the hh states if the SLS is grown directly on Si. In addition, the overall CB minimum is located in the Si substrate which causes an out-diffusion of injected electrons into the substrate[3]. A $(Si)_6(Ge)_4$ superlattice grown on Si substrate is therefore a semiconductor with an indirect fundamental band-gap both in k-space *and* in real space. In order to achieve a quasi-direct fundamental band-gap, the superlattice has to be grown on a substrate with a larger lateral lattice constant than bulk Si. Only in this case, the zone-folded $\Delta(2)$ states are lower in energy than the non-folded $\Delta(4)$ states. This is shown in Fig. 1 for a $(Si)_6(Ge)_4$ SLS grown on a $Si_{0.6}Ge_{0.4}$ alloy buffer layer. According to our calculation, the band-gap of the SLS is reduced compared with the corresponding $Si_{0.6}Ge_{0.4}$ alloy due to ordering of the alloy into a layered structure. In a strain symmetrized $(Si)_6(Ge)_4$ SLS both the CB minimum and the VB maximum are expected to be located in the SLS region[3]. If the $(Si)_6(Ge)_4$ SLS is grown pseudomorphic on Ge the band-gap is further reduced. The overall VB minimum, however, is localized in the Ge substrate[3]. In this case, the fundamental transition is indirect in real space. The most favourite choice of substrate is therefore a $Si_{0.6}Ge_{0.4}$ alloy buffer layer. In this case, the strain symmetrized superlattice is expected to have a quasi-direct band-gap. Due to the strain symmetrization in the Si and Ge layers the SLS can be grown pseudomorphic to the alloy buffer without limitations of a critical thickness[6].

A serious problem which comes up with the growth of thin partly relaxed buffer layers is the high density of threading dislocations which are created by the relaxation of the $Si_{1-x}Ge_x$ buffer layer. Typical dislocation densities for partly relaxed $Si_{1-x}Ge_x$ buffer layers are in the order of 10^{10} to 10^{11} cm^{-2} and limit the use of such buffer layers for electrical and optical devices. Photoluminescence studies by R. Zachai et al. on $(Si)_m(Ge)_n$ superlattices which were grown on partly relaxed $Si_{1-x}Ge_x$ buffers layers[10] were discussed controversially mainly because of the high dislocation density which was present in the samples at that time[11]. During the last two years, however, there was an enormous progress in the synthesis of relaxed $Si_{1-x}Ge_x$ layers on Si. The dislocation density in thick $Si_{1-x}Ge_x$ buffer layers grown with a step graded[12] or linearly graded[13] Ge concentration profile was found to be reduced by about 4 orders of magnitude compared with thin, partly relaxed alloy layers. Well defined band-gap luminescence in MBE-grown pseudomorphic $Si_{1-x}Ge_x$ alloys on Si(100) was recently observed by different groups[14,15,16] at growth temperatures above 500°C. Until recently, $(Si)_m(Ge)_n$ superlattices could not be synthesized at such high temperatures because of the intermixing of the Si and Ge layers due to the segregation of Ge atoms during growth. However, as shown by Fujita et al.[17], the segregation of Ge atoms can be efficiently suppressed by depositing only one monolayer (ML) of Sb prior to the growth of the superlattice. Based on these developments an enormous improvement in the quality of strain-symmetrized Si_mGe_n SLS's was achieved[18]. Well defined band-gap PL was measured in these samples[19] and electroluminescence was observed in similar structures up to room temperature[20]. Here we report on strongly enhanced band-gap PL in a $(Si)_6(Ge)_4$ and $(Si)_9(Ge)_6$ SLS grown on a step-graded $Si_{1-x}Ge_x$ alloy buffer using Sb as surfactant. The results clearly support the earlier results using samples of lower quality[10,16,21] and are in contradiction to the theoretical work of ref. [4,11].

B. Experimental

The $(Si)_9(Ge)_6$, $(Si)_6(Ge)_4$, $(Si)_3(Ge)_2$ SLS's, and the $Si_{0.6}Ge_{0.4}$ alloy buffer reference sample were grown with MBE at Daimler research center[18]. The buffer layer consists of a step-graded alloy buffer with a thickness of 6500Å, followed by a 5000Å-thick $Si_{0.6}Ge_{0.4}$ alloy. In the graded region of the buffer the Ge content was increased step-wise by 3%/500Å, while the temperature was continuously lowered from 600°C down to 520°C. The residual lateral strain in the topmost $Si_{0.6}Ge_{0.4}$ alloy was determined by x-ray diffraction (XRD) to be -0.03%. The SLS's with an entire thickness of 2000Å [$(Si)_3(Ge)_2$], 1500Å [$(Si)_6(Ge)_4$], and 3000Å [$(Si)_9(Ge)_6$] were deposited on top of the alloy buffer at T=500°C. Prior to the growth of the SLS's a monolayer of Sb was deposited to act as a surfactant. The $(Si)_6(Ge)_4$ and $(Si)_9(Ge)_6$ SLS's were characterized in detail with XRD and Raman spectroscopy. The $(Si)_6(Ge)_4$ was also investigated with transmission electron microscopy (TEM)[22]. The dislocation density, as determined with cross-sectional TEM[22], is 10^7 cm^{-2} which is about 2-3 orders of magnitude lower compared with previously studied Si_mGe_n SLS's[10,16,21] grown on partly relaxed alloy buffer layers. The thickness of the Si and Ge layers (d_{Si}, d_{Ge}), the period length (d_{SLS}), the lateral strain in the Si layers (ε_{Si}) and the average Ge content (x_{SLS}) for the nominal $(Si)_6(Ge)_4$ and $(Si)_9(Ge)_6$ SLS's, as determined by XRD, are given in table 1.

Sample	nominal	d_{Si} [Å]	d_{Ge} [Å]	d_{SLS} [Å]	ε_{Si} [%]	x_{SLS} [%]
B2416	$(Si)_6(Ge)_4$	9.2±0.5	6.9±0.5	16.1±0.5	1.5±0.15	41±2
B2512	$(Si)_9(Ge)_6$	11.8±0.5	10.4±0.5	22.2±0.5	1.7±0.17	45±2

Tab. 1 Structural data for sample B2416 and B2512

The PL was excited by the 457nm line of an Ar$^+$ laser. The excitation power density varied between ~ 0.05 and 50mW/mm^2. The samples were mounted in a variable temperature, He cryostat. PL signals were analysed with a 64cm grating monochromator and detected by either a Ge or an InAs detector, in both cases cooled by liquid nitrogen. Annealing was performed in a quartz furnace under vacuum conditions (p<10^{-6} mbar).

C. Results and discussion

The PL spectra for the $(Si)_3(Ge)_2$, $(Si)_6(Ge)_4$, $(Si)_9(Ge)_6$ SLS's, and the $Si_{0.6}Ge_{0.4}$ alloy reference sample are plotted in fig. 2 on the same scale. The luminescence intensities are normalized to the thickness of the superlattice and the thickness of the alloy layer with constant Ge content for the alloy reference sample, respectively. The PL spectra for the $(Si)_3(Ge)_2$ SLS and the $Si_{0.6}Ge_{0.4}$ alloy are almost identical which gives clear evidence that the $(Si)_3(Ge)_2$ superlattice is more like an alloy due to the intermixing of the Si and Ge layers at growth temperatures of 500°C. The PL-Intensity for the $(Si)_3(Ge)_2$ SLS, however, is already enhanced by about a factor of 10 compared with the alloy reference sample. The PL-signals for the $(Si)_3(Ge)_2$ SLS and the $Si_{0.6}Ge_{0.4}$ alloy can be attributed to a no-phonon (NP) transition and an associated phonon replica involving a transverse optical Si-Si mode (TO^{Si-Si}), respectively[23].

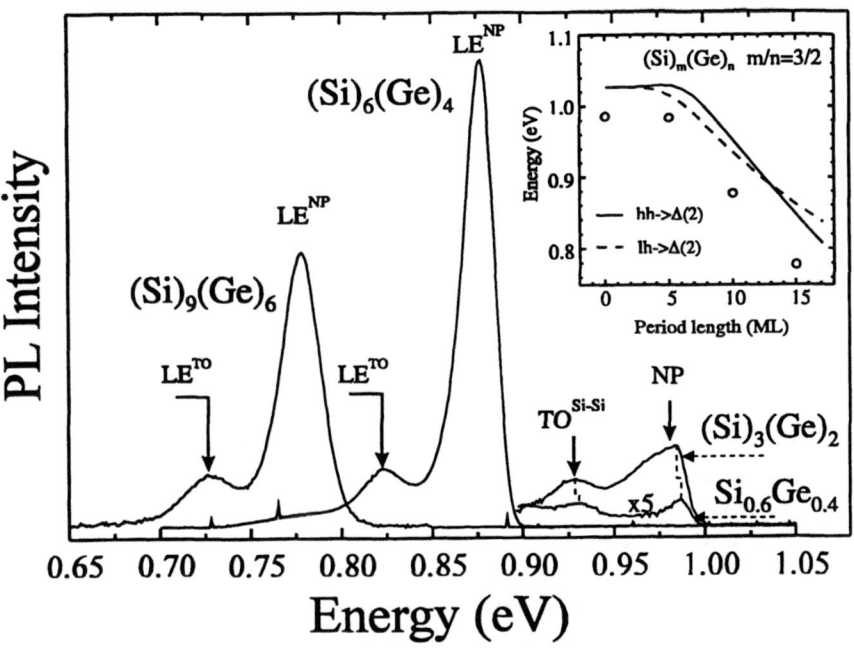

FIG. 2. PL spectra for $(Si)_9(Ge)_6$, $(Si)_6(Ge)_4$, $(Si)_3(Ge)_2$ SLS's, and a $Si_{0.6}Ge_{0.4}$ alloy. The inset shows the calculated energies of the interband transitions as a function of the period length.

For the random alloy sample the energy position for the NP line is in excellent agreement with the calculated band-gap energy of the alloy taking into account the built-in lateral strain. For the $(Si)_6(Ge)_4$ and the $(Si)_9(Ge)_6$ SLS's much stronger PL signals are observed at 0.877eV and 0.778eV, respectively. As shown below, these PL signals can be attributed to a NP transition of localized excitons (LE^{NP}). Associated phonon replica (LE^{TO}) are also observed at 0.825eV and 0.726eV, respectively. The normalized PL intensity is strongest for the $(Si)_6(Ge)_4$ SLS. For both $(Si)_6(Ge)_4$ and $(Si)_9(Ge)_6$ the $\Delta(2)$ minimum is folded close to the Γ point which should result in an enhanced oscillator strength. The reduced PL intensity and band-gap energy for $(Si)_9(Ge)_6$ compared with $(Si)_6(Ge)_4$ can be explained by the staggered band-line-up of the CB and VB edges in the Si and Ge layers (see fig.1). The holes (electrons) are more confined in the Ge (Si) layers with increasing period length. The separation of electrons and holes in real space with increasing period length results in an decreasing oscillator strength and a reduced band-gap. The calculated energy of the interband transitions between the lh and hh states of the VB and the $\Delta(2)$ states of the CB for strain symmetrized $(Si)_m(Ge)_m$ SLS's with $m/n=3/2$ are shown in the inset of fig. 2 and compared with the energy positions of the LE^{NP} and LE^{TO} lines. In this calculation, the Si concentration (x_{Si}) in the Si layers and the Ge concentration in the Ge layers (x_{Ge}) of the SLS's are assumed to be reduced with decreasing period length due to the intermixing of the Si and Ge layers during growth at T=500°C. The actual concentration profile in the SLS is not known. x_{Si} is assumed to depend on the Si layer thickness (d_{Si}) as

$$x_{Si} = 1 - \frac{0.4}{1 + e^{(d_u - 4.0\text{Å})/1.0\text{Å}}} \tag{1}$$

The x_{Ge} can be deduced from Eq. 1. For the intermixed Ge and Si layers, the effective masses, deformation potentials (DP), and band offsets used for the calculation were linearly interpolated between those of Si and Ge. The energy positions of the NP lines for the alloy and the $(Si)_3(Ge)_2$ SLS are almost identical, which indicates that the $(Si)_3(Ge)_2$ SLS is more like an alloy due to the intermixing. The systematic shift of the NP lines for $(Si)_6(Ge)_4$ and $(Si)_9(Ge)_6$ to lower energies with increasing period length is well described by the calculation. Deviations for the SLS's from the nominal structural data result in slightly different calculated band-gap energies. These deviations, however, are of the same order of magnitude as the theoretical errors arising from the uncertainties in deformation potentials given in the literature.

In order to attribute the PL signals to the band-gap of the superlattices, the samples were annealed for 1h at 550°C, 600°C, 650°C, 700°C and 750°C. One specimen was completely alloyed by subsequent annealing for 1 h at 600°C, 650°C and 780°C. The PL spectra for the $(Si)_6(Ge)_4$ SLS after each annealing step are shown in fig. 3. After the first annealing step, the PL efficiency is enhanced by a factor of 5. The increasing PL intensity observed after the first step may be explained by a reduction of the density of non-radiative recombination centers due to thermal annealing. Further annealing results in an increasing energy position of the LE^{NP} and the LE^{TO} lines. At higher annealing temperatures, broad and weak PL signals are observed below the LE^{NP} lines of the SLS. The origin of these PL bands is not clear. For the completely alloyed sample intense defect lines (D_1, D_2) appear in the spectrum which are due to dislocations generated by the relief of the residual strain of about -0.3%[24]. A further broad peak appears at 0.977eV (L) which looks similar to PL signals observed by Noël et al. in Si/SiGe quantum well after thermal annealing[25]. The LE^{NP} and LE^{TO} lines can be attributed to the band-gap related transition of the superlattice which is shown in the inset of fig. 3 were the energies of the PL signals are compared with the calculated band-gap as a function of the Ge concentration in the Ge layers. The assumed concentration profile based on a interdiffusion model of R. Schorer et al.[26]. In the initial stage of the interdiffusion the Si atoms are more easily diffusing into the Ge layers than Ge atoms into the Si layers. With increasing degree of interdiffusion, the intermixing of the Si layers with

Ge atoms is more pronounced. Starting with a $(Si)_m(Ge)_n$ SLS with a Si layer thickness d_{Si}^0, the Ge concentration in the "Si layers" (x_{Si}) is assumed to depend on the thickness d_{Si} as

$$x_{Si} = \frac{n}{m+n} \cdot \cos^2\left(\frac{\pi}{2} \cdot \frac{d_{Si}}{d_{Si}^0}\right) \qquad (2)$$

We still assume a step-like concentration profile with increasing interdiffusion. The Ge and Si content in the layers, however, is reduced and given by Eq. (2). The Ge concentration in the Ge layers in the course of interdiffusion of the superlattice was determined by a detailed analysis of the Raman spectra taken after each annealing step[27].

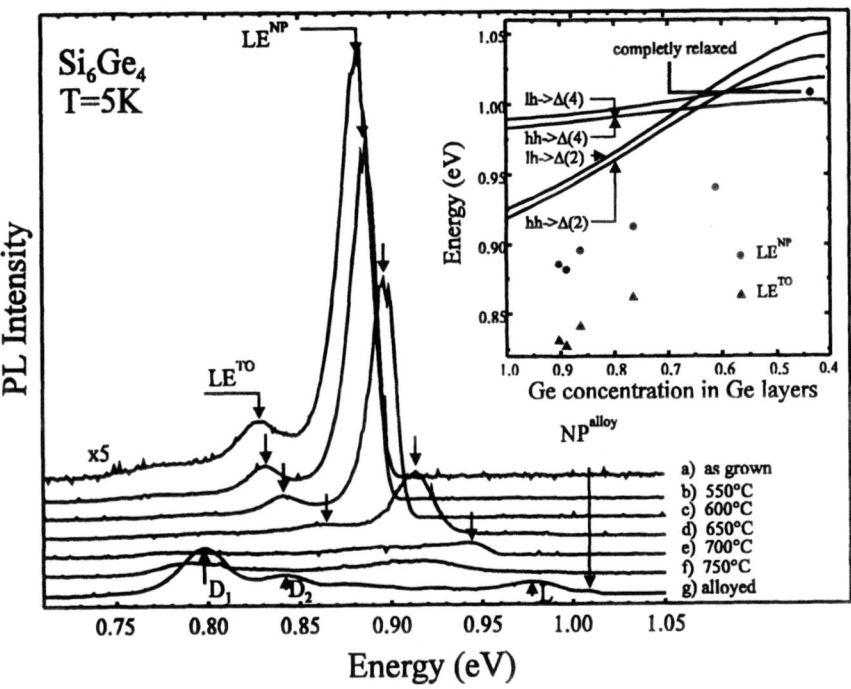

FIG. 3 PL spectra for the $(Si)_6(Ge)_4$ SLS. Spectrum a) for the as-grown sample, spectrum b) ,c), d), e) and f) after annealing for 1h at 550°C, 600°C, 650°, 700°C, and 750°C, respectively. Spectrum g) for the completely annealed sample. The inset shows results of an EM calculation. The circles mark the energy position of the NP line, the triangle the energy position of the TO replica.

The energy of the superlattices phonons are strongly affected by the Ge content in the individual layers[28], the layer thicknesses and the lateral strain[29]. Based on Eq. (2) the phonon energies in the superlattice can be calculated as a function of the Ge concentration in the Si and Ge layers. The Ge concentration in the Si and Ge layers can now be determined by the comparison of the measured Raman spectra with the calculated phonon energies. The shift of the PL signals is well described with the interdiffusion model in conjunction with the EM calculation (inset of fig. 3). The energy difference between LE^{NP} and LE^{TO}, which is 52meV for both Si_6Ge_4 and Si_9Ge_6, does not change with interdiffusion. This gives clear evidence that also the LE^{TO} line can be assigned to transitions between the VB and $\Delta(2)$ states across the SLS bandgap. The energy difference between the LE^{NP} and LE^{TO} line of 52meV compares well with the energy of Si-Ge optical phonons measured with Raman spectroscopy. The LE^{TO} signal could therefore be attributed to a phonon replica due to local Si-Ge modes at the Si/Ge interface. The weak PL line observed at 1.008eV (NP^{alloy}) is due to a NP transition in the $Si_{0.59}Ge_{0.41}$ alloy[23] as the SLS, with an average Ge concentration of 41%, is finally completely interdiffused. The abrupt change in the energy position of the NP line can be explained by the relief of the residual strain in the interdiffused layer. The excitonic band-gap energy of 1.013eV for an unstrained $Si_{0.59}Ge_{0.41}$ alloy[23] compares well with the energy position of the excitonic NP^{alloy} transition which is observed at 1.008eV.

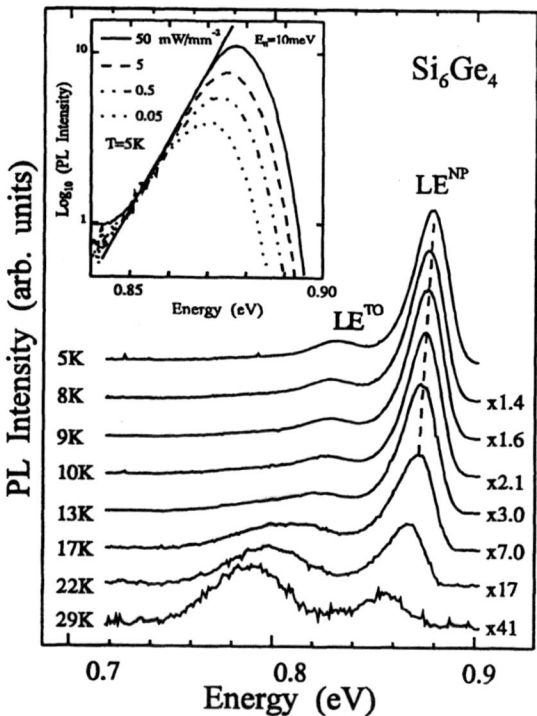

FIG. 4 PL spectra for a Si_6Ge_4 SLS at different temperatures. The inset shows PL spectra at T=5K taken with different excitation power densities.

The decreasing PL-intensities for the NP lines with increasing degree of interdiffusion of the layers are in agreement with the observed intensity ratio for the NP line in a random alloy and a $(Si)_6(Ge)_4$ SLS (see fig. 1). In order to identify the origin of the observed PL signals we performed PL measurements at different temperatures. PL spectra for the $(Si)_6(Ge)_4$ superlattice as function of temperature are shown in fig. 4. The dependence of the PL spectra on the excitation power density is shown in the inset of fig. 4. The peak positions of the LE[NP] and LE[TO] signals are shifting to lower energies with increasing temperature by 0.77meV/K. PL signals observed above 30K are due to defect related transitions and are similar to defect PL signals observed in samples of lower quality. At T=5K the intensity I(E) of the PL signal shows a exponential slope on the low energy side which can be fitted to $I(E) \propto e^{E/E_0}$ with $E_0 = 10$meV. Furthermore, for T=5K the PL intensity shows a pronounced saturation behaviour with increasing excitation power density P_{ex} which can be written as $I \propto (P_{ex})^{0.63}$. All these findings are consistently described in terms of the recombination of excitons localized at random potential fluctuations which are expected due to slight variations in both the strain distribution and the thicknesses of the Si and Ge layers. The PL line shape for localized excitons (LE) can be written as[30]

$$I(E,T) \propto e^{E/E_0} \cdot \frac{\tau_{tot}}{\tau_{rad}} = e^{E/E_0} \cdot \left[1 + \nu \cdot \tau_{rad} \cdot e^{-(E_c - E)/(k_B \cdot T)}\right]^{-1} \qquad (3)$$

The first term describes the density of states due to a fluctuating potential, the second term is the ratio between the total lifetime τ_{tot} and the radiative lifetime τ_{rad} of the carriers. Carriers which are thermally excited to an energy E_C, the so-called mobility edge, become mobile and can more easily recombine non-radiative. ν is the radiative lifetime times an effective frequency for thermal excitation attempts. Both the observed upward shift of the PL peak position with P_{ex} and the line shape of the LE[NP] line below the peak energy can be explained by the filling of higher states. This is also reflected in the sublinear dependence of the PL intensity on P_{ex}. Also the observed shift of the peak positions to lower energies with increasing temperature is qualitatively described by Eq. (3) [30].

The PL signals observed for the SLS's grown on the completely relaxed alloy buffer layers can be unambiguously attributed to the band-gap of the superlattice. Earlier results obtained with a sample of lower quality[10] grown on partly relaxed $Si_{1-x}Ge_x$ buffer layers gave rise to a critical comment based on a bandstructure calculation[11]. PL spectra of a $(Si)_6(Ge)_4$ superlattice grown on a thin, partly relaxed buffer layer (sample B2068) which is very similar to the samples studied in ref. [10] and a $(Si)_6(Ge)_4$ superlattice grown on a step-graded buffer layer (sample B2416) are compared in fig. 5. The samples were mounted next to each other on a sample holder and measured under identical conditions. The energy scales for the spectra spectra are shifted against each other by 33meV in order to account for the shift in the peak position of the dominant NP line due to the slightly different strain and layer thicknesses in the SLS's. This shift of 33meV compares well to the shift in the bandgap energy of the samples which is 50meV accoording to our EM calculations. The PL spectrum of the sample grown on the partly relaxed buffer layer is magnified by a factor 5. Due to the lower quality of the sample grown on the partly relaxed buffer layer, the PL signal is significantly broadened and less intense compared with the sample grown on the completely relaxed buffer layer. Due to the linewidth broadening the TO replica can not be resolved for the sample grown on the partly relaxed buffer layer. According to our EM calculation the calculated bandgap energy for a $(Si)_6(Ge)_4$ SLS with $\varepsilon_{Si}=1.4\%$ is 0.96eV which is about 120meV higher than the measured bandgap[10]. The band-gap energy for $(Si)_6(Ge)_4$ SLS with a in-plane lattice constant of 5.507Å ($\varepsilon_{Si}=1.4\%$) determined theoretically in Ref. [4,11] is 1.08eV which is 240meV higher than the measured bandgap[10].

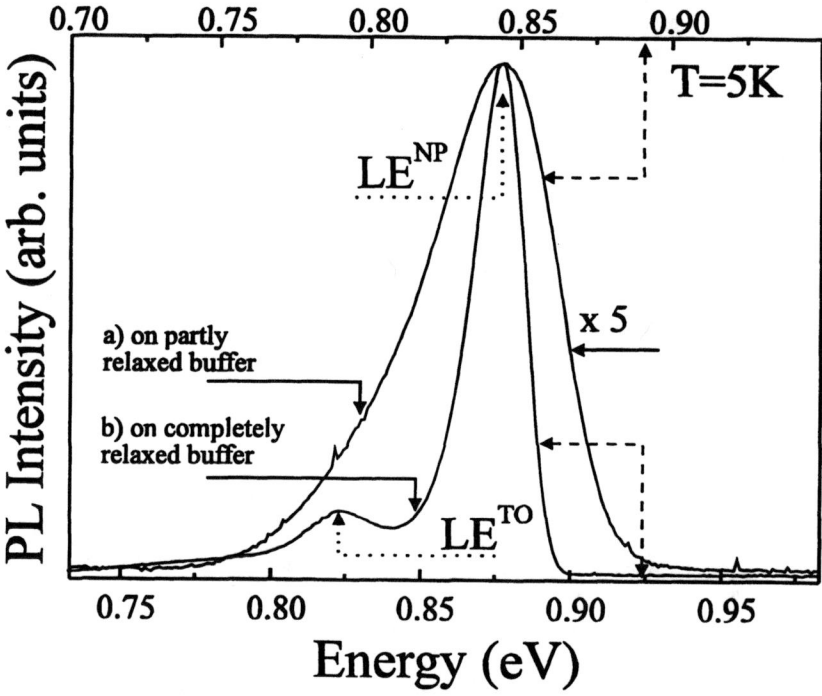

FIG. 5 PL spectra for a $(Si)_6(Ge)_4$ grown on a) a thin partly relaxed buffer layer and b) a step graded buffer layer. The energy scale of the spectra are shifted by 33meV in order to account for the slightly different strain and layer thicknesses of the SLS's.

This calculated bandgap energy[4,11] is even higher than the band-gap for a strained $Si_{0.6}Ge_{0.4}$ alloy grown on a buffer with a lateral lattice constant of 5.507Å which is calculated to be 1.00eV[16]. Theoretical band-energies calculated by different groups[3,5], however, are in good agreement with experimental data[10,16,19,21]. In a detailed absorption study[21] of $(Si)_m(Ge)_m$ SLSs with m=4,5, and 6 a similar disagreement between the measured band-gap and theoretical studies of ref. [4] was found. The PL-Signals observed in samples of lower quality by Zachai et al.[10], can therefore also be attributed the band-gap of the superlattice.

Conclusions

In conclusion we observed well defined band-gap luminescence in strain symmetrized $(Si)_m(Ge)_n$ superlattices grown on a step-graded $Si_{1-x}Ge_x$ alloy buffer layers using Sb as a surfactant. The PL intensity for a $(Si)_6(Ge)_4$ and a $(Si)_9(Ge)_6$ SLS is enhanced by about a factor 150 and 90 compared with a corresponding $Si_{0.6}Ge_{0.4}$ alloy, respectively. The comparison of the measured band-gap with an EM calculation for interdiffused superlattices confirm the interpretation of the observed PL signals. The luminescence can be attributed to the recombination of excitons localized at random potential fluctuations which are expected in a superlattice due to statistical fluctuations in strain and layer thicknesses. Finally, these results clarify the controversy in the literature about earlier results [10,11].

Acknowledgement

We acknowledge fruitful discussions with J. Olajos, L. Samuelson, R. Schorer, H. M. Polatoglou, and M. Jaros. This work was partly financed by the European Basic Research Programme ESPRIT under contract P7128.

References

1. U. Gnutzmann and K. Clausecker, Appl. Phys. 3, 9 (1974).
2. P. Pearsall, J. Bevk, L. C. Feldman, J. M. Bonar, J. P. Maennaerts, Phys. Rev. Lett. 58, 729 (1986).
3. J. Turton and M. Jaros, Mat. Sci. and Engineering B 7, 37 (1990) and references therein.
4. U. Schmid, E. Christensen, M. Alouani, and M. Cardona, Phys. Rev. B 43, 14597 (1991).
5. C. Tserbak, H. M. Polatoglou, and G. Theodorou, Phys. Rev. B, in press.
6. Kasper, H. Kibbel, H. Jorke, H. Brugger, E. Friess, and G. Abstreiter, Phys. Rev. B 38, 2599 (1988).
7. G. Bastard, Wave mechanics applied to semiconductor heterostructures, (Les Editions de Physique, Paris, 1988).
8. L. Colombo, R. Resta, and S. Baroni, Phys. Rev. B 44 (1991) 5572
9. C. G. van de Walle and R. M. Martin, Phys. Rev. B 34 (1986) 5621
10. R. Zachai, K. Eberl, G. Abstreiter, E. Kasper, and H. Kibbel, Phys. Rev. Lett. 64, 1055 (1990).
11. U Schmid, N. E. Christensen, M. Cardona, Phys. Rev. Lett. 65, 2610 (1990).
12. F. K. Le Goues, B. S. Meyerson, and F. J. Morar, Phys. Rev. Lett. 66, 2903 (1991).
13. E. A. Fitzgerald, Y. H. Xie, M. L. Green, D. Brasen, A. R. Kortan, J. Michel, Y. J. Mie, and B. E. Weir, Appl. Phys. Lett. 59, 811 (1991).

14. K. Terashima, M. Tajima, and T. Tatsumi, Appl. Phys. Lett. **57** (1990) 1925

15. J. Spitzer, K. Thonke, R. Sauer, H. Kibbel, H.-J. Herzog, and E. Kasper, Appl. Phys. Lett. **60** (1992) 1729

16. U. Menczigar, J. Brunner, E. Friess, M. Gail, G. Abstreiter, H. Kibbel, H. Presting, and E. Kasper, Thin Solid Films **222**, 227 (1992) and references therein.

17. K. Fujita, S. Fukatsu, H. Yaguchi, T. Igarashi, Y. Shiraki, and R. Ito, Silicon Molecular Beam Epitaxy, edited by J. C. Bean, E. H. C. Parker, S. S. Iyer, Y. Shiraki, E. Kasper, and K. L. Wang (Mater. Res. Soc. Proc. 220, Pittsburgh, PA, 1991).pp. 193-197.

18. H. Presting, and H. Kibbel, Thin Solid Films **222**, 215 (1992)

19. U. Menczigar, G. Abstreiter, H. Kibbel, H. Presting, E. Kasper, J. Olajos, H. G. Grimmeiss, Phys. Rev. B **47**, 4099 (1993).

20. J. Engvall, J. Olajos, H. G. Grimmeiss, H. Kibbel, E. Kasper, and H. Presting, this volume.

21. J Olajos, J. Engvall, H. G. Grimmeiss, U. Menczigar, G. Abstreiter, H. Kibbel, E. Kasper, and H. Presting, Phys. Rev. B **46**, 12857 (1992).

22. W. Jäger, D. Stenkamp, P. Ehrhart, K. Leifer, W. Sybertz, H. Kibbel, H. Presting, and E. Kasper, Thin Solid Films **222**, 221 (1992).

23. J. Weber, and M. I. Alonso, Phys. Rev. B **40**, 5683 (1989).

24. J. Weber, and M. I. Alonso, in Defect Control in Semiconductors, edited by K. Sumino, Elsevier Science Publisher B. V. (North Holland), 1453 (1990).

25. J. -P. Noël, N. L. Rowell, D. C. Houghton, and D. D. Perovic, Appl. Phys. Lett. **57**, 1037 (1990).

26. R. Schorer, E. Friess, K. Eberl, and G. Abstreiter, Phys. Rev. B **44**, 1772 (1991).

27. U. Menczigar, G. Abstreiter, H. Kibbel, H. Presting, E. Kasper, J. Olajos, H. G. Grimmeiss, D. Stenkamp, and W. Jäger, to be published.

28. Renucci, J. B. Renucci, and M. Cardona, Proc. of the 2nd Int. Conf. on Light Scattering in Solids, edited by M. Balkanski (Flammarion, Paris, 1971), pp. 326.

29. E. Friess, K. Eberl, U. Menczigar, and G. Abstreiter, Solid State Commun. **73**, 203 (1990).

30. M. Oueslati, M. Zouaghi, M. E. Pistol, L. Samuelson, H. G. Grimmeiss, and M. Balkanski, Phys. Rev. B **32**, 8220 (1985).

REDUCTION OF MISFIT DISLOCATION DENSITY IN FINITE LATERAL SIZE $Si_{1-x}Ge_x$ FILMS GROWN BY SELECTIVE EPITAXY

L. VESCAN, T. STOICA*, C. DIEKER AND H. LÜTH
*Inst. für Schicht und Ionentechnik, Forschungszentrum Jülich, 5170-Jülich,Germany
*Inst. of Physics and Technology of Materials, POB MG7, Bucharest, Romania

ABSTRACT

In $Si_{0.88}Ge_{0.12}/Si$ strained layers misfit dislocations formed during growth in small pads are generated at a significantly higher critical thickness than on extended areas, while pads of lateral size of 10 μm or smaller show no evidence of misfit dislocations at all. The SiGe layers investigated were selectively grown on patterned substrates with pad sizes from 2 μm to 1 cm. An elastic relaxation model was used to calculate the pad size dependence of the critical thickness. The main hypothesis of the model is that the density of misfit dislocations is solely affected by the elastic relaxation at the edges of small epitaxial areas. This equilibrium model is able to explain the observed absence of misfit dislocations on small pads, however it predicts a critical thickness for finite sizes much lower than the observed one.

INTRODUCTION

The role of the finite lateral size of the sample on the generation of misfit dislocations (MD) is of interest both from the perspective of material science as well as for novel devices. A reduction in MD density was observed in III-V strained systems [1], but also in SiGe/Si [2,3] when the growth area was reduced. Noble et al. [2] observed an area-dependent reduction of MD density in selectively grown $Si_{0.8}Ge_{0.2}/Si$, yet there was no obvious dominant nucleation source. In this paper, we report on similar observations on $Si_{0.88}G_{0.12}$ and present an elastic strain relaxation model for the pad size dependence of the density of misfit dislocations.

EXPERIMENTAL DETAILS

The SiGe layers investigated in this work were deposited by selective epitaxial growth using a load locked Low Pressure Chemical Vapor Deposition (LPCVD) system. The substrates used were lowly doped ($\sim 10^{13}$ cm^{-3}) p-type (001) Si. Wafers with oxide patterns were loaded into the reactor after a RCA cleaning and a HF dip. Epitaxy was performed at 700°C and 0.12 Torr using $SiCl_2H_2$, 10 % GeH_4 in He and H_2 as carrier gas. The pads were squares and rectangles with sides parallel to the <100> directions and sizes varying from 2 μm to 1 cm. The deposition sequence was: a 60 nm Si buffer, the $Si_{0.88}Ge_{0.12}$ layer and a 3 nm Si cap, with a growth rate for $Si_{0.88}Ge_{0.12}$ of 12 nm/min and for Si of 2 nm/min. The dislocations were revealed by a diluted Schimmel etch (etch rate $\sim$8 nm/s) and by Nomarski microscopy, scanning electron microscopy (SEM) and transmission electron microscopy (TEM) were used for analysing the dislocations.

Mat. Res. Soc. Symp. Proc. Vol. 298. ©1993 Materials Research Society

DENSITY OF MISFIT DISLOCATIONS: LARGE AND SMALL AREAS

Above a critical thickness MDs are formed in the SiGe/Si interface in orthogonal <110> directions. We define the *linear density of misfit dislocations* as the number of MDs per unit length on a <110> direction. In the following, we present results for large areas, for pads with area 270x175 μm^2, 115x50 μm^2, 100x30 μm^2 and 10x10 μm^2. *Fig.1* shows optical micrographs of three samples with thickness 500nm, 615nm and 730nm, respectively. The reduction of MD density for the small pads of a given thickness is clearly seen.

d_{SiGe}= 500 nm   615 nm   730 nm

large
area

pad

"a"

"b"

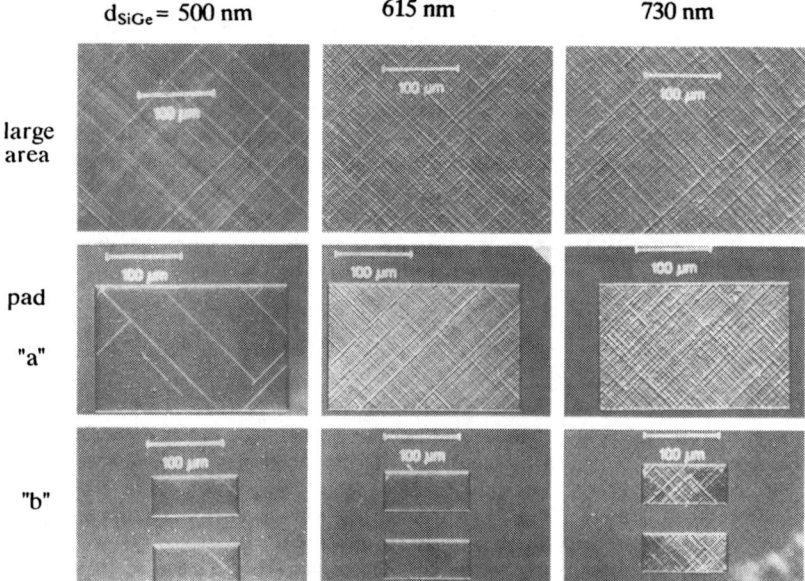

Fig.1 Nomarski micrographs of orthogonal array of interface misfit dislocations in SiGe (x=0.12) layers with different thicknesses deposited at 700°C. The upper micrographs were taken in large areas (1x1 cm²), the other two sets are for pads *a* (270x175μm^2) and *b* (115x50μm^2). The pad sides are parallel to <100> directions.

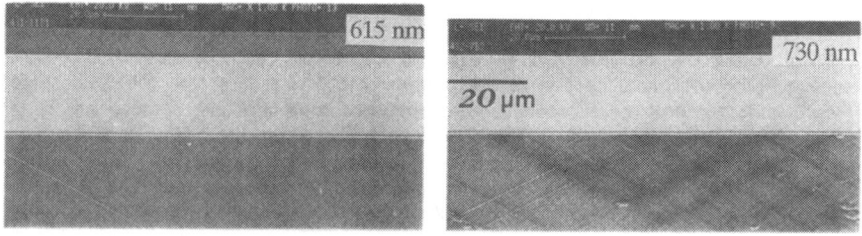

Fig.2 SEM micrographs showing 30x100 μm^2 pads with the long side horizontally. The samples were tilted ~50°. On the 615 nm sample only one MD line is seen, while the 730 nm sample shows a dense array of MD lines.

For smaller pads the resolution of the optical microscope was insufficient, so the MDs were analysed using an SEM. *Fig.2* shows $30 \times 100 \mu m^2$ pads where the MD lines can be clearly seen. The linear density of MDs as a function of layer thickness and pad size is shown in *fig.3*. The results of *figs.1-3* can be summarized as follows: 1) On *large areas* MDs are generated above a critical thickness $h_c \approx 140$ nm in agreement with previous results [4]. 2) On *pads a and b* MDs are generated at a higher critical thickness, which increases as the pad size decreases. One can see, that for pads *a* $(270 \times 175 \mu m^2)$ $h_c \sim 500$-600 nm and for pads *b* $(115 \times 50 \mu m^2)$ $h_c \sim 600$-700nm, which is much higher than the value of 140nm on the large area. 3) No MDs were observed on pads with $10 \times 10 \mu m^2$ or smaller up to the investigated layer thickness of 1260nm. 4) After onset of relaxation the density of MDs increases very fast with increasing thickness of the layer. 5) For large areas and pads *a* the density saturates at ~11 disl./μm. In *fig.4* the experimental density for large area from fig.3

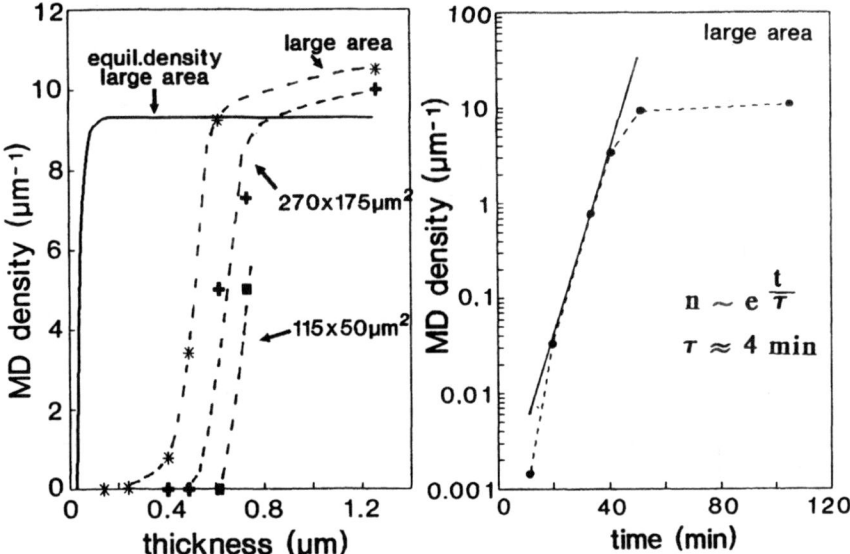

Fig.3 Linear density of MDs in $Si_{0.88}Ge_{0.12}$ as a function of layer thickness for large areas and for pads *a* $(270 \times 175 \mu m^2)$ and *b* $(115 \times 50 \mu m^2)$. The continuous curve n_e is the calculated equilibrium density for large areas.
Fig.4 Linear density of MDs on large area as a function of time.

is plotted as a function of deposition time t, assuming that the layer thickness $h = R.t$, with $R = 12$ nm/min. After onset of relaxation, an almost exponential increase of density with time is observed. This agrees with results of Dodson and Tsao [5].
Fig.5 shows a cross section TEM taken in a large area of a strongly relaxed sample (730 nm) with dislocations penetrating the substrate about $4 \mu m$, and dislocation threading to the surface. Such a microstructure was explained by LeGoues et al. [6] by a modified Frank-Read mechanism for compositionally uniform as well as for graded $Si_{0.88}Ge_{0.12}$ layers. *Fig.6* is a plan view TEM of pad *a* of sample 1260 nm thick (*fig.1*). One can see the dense array of MDs in the pad interface, but also the projection of dislocation loops which must have glided out of the pad area on {111} planes into the substrate.

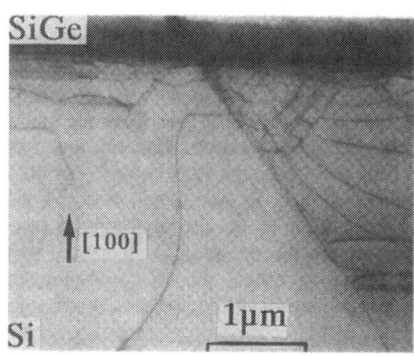 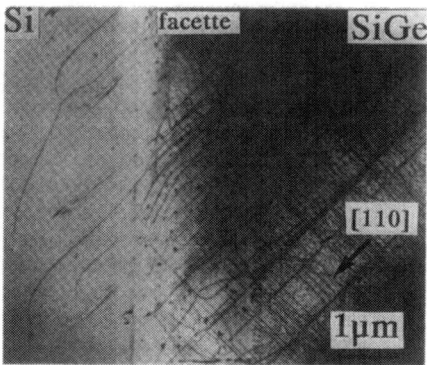

Fig.5 TEM *cross section* taken in a large area of a relaxed SiGe sample (730 nm thick, ~10 disl./μm) showing dislocation loops introduced deep in the substrate;

Fig.6 TEM *plan view* of a pad 270x175μm² (1260 nm thick, ~10 disl./μm) showing the projection in the layer plane of dislocation loops in the substrate.

ELASTIC STRAIN RELAXATION MODEL

Elastic relaxation provides a natural starting point for analysing the influence of pad size on the relaxation. The model used (*fig.7*) shows an infinite substrate in a square lattice and a finite pad elastically confined to the substrate. The lattice of the pad in the relaxed state is greater than the lattice constant of the substrate. The layer is maximally strained in the interface plane, but the pad is allowed to expand laterally as it growths in z direction. The pad has an interfacial width L in y direction. The strain ε in a finite size strained layer was calculated assuming elastic deformation and with a procedure based on an atomic force model [7]. With the formula for the strain the critical thickness h_c for finite pads was derived. In the following we describe shortly these calculations.

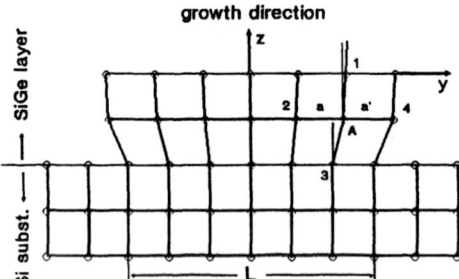

Fig.7 Schematic of an elastically relaxed SiGe layer with finite lateral size L grown on a Si substrate.

We choose the origin at the interface, at mid plane. The third dimension x is taken infinite. The film is supposed to grow localized with the strain relaxed at the edge of the pad, as in the case of free lateral growth. The strain ε(y,z) in the pad is [7]:

$$(1) \qquad \varepsilon(y,z) = -f \, e^{-\lambda z} \cos(\pi \frac{y}{L}), \qquad \lambda = \frac{\pi \gamma^{1/2}}{L}, \qquad \gamma_{SiGe} \approx \frac{C_{11}}{C_{44}} \approx 2,$$

where C_{11} and C_{44} are elastic constants of the layer and $f = 0.042x$ for $Si_{1-x}Ge_x$. Equation (1) shows, that on finite areas the strain decreases exponentially in growth direction z. Therefore, the effect of the finite size on the strain is that by decreasing the lateral dimension the strain decreases faster in z direction. A similar dependence on dimension z was obtained by Luryi and Suhir [8] for the stress, apart from the factor $\gamma^{1/2}$ (1.4 for SiGe) in the exponent. The strain energy depends on layer thickness h as $(1 - e^{-2\lambda h}) \approx (1 - e^{-gh/L})$, and thus for large L the strain energy tends to be proportional to h, but for small pads, with L comparable to h, the strain energy becomes independent of thickness. Therefore, for small pads the upper part of the layer contributes only insignificantly to the deformation energy and the thickness of the layer can be increased further without generation of misfit dislocations.

The critical thickness for a pad of interfacial width L was calculated assuming that the critical thickness is reached when the excess strain is null [5,9]. The expression which relates the critical thickness h_c to the lateral dimension L is [7]:

$$(2) \quad \frac{1 - e^{-\lambda h_c}}{\lambda} = \frac{g}{x} \ln(4h_c/b) , \qquad g \equiv \frac{b(1-\nu/4)}{4\pi(1+\nu)f_0}$$

where b is the Burgers vector, ν is the Poisson's ratio and $f_0 = 0.042$ for SiGe. The critical thickness (2) is represented in *fig.8* (with $b \approx 0.4$nm, $\nu = 0.27$). The result of these calculations is that at a given Ge concentration a critical size can be found below which misfit dislocations will never be formed. For example, for $x < 10\%$ and $L < 0.2\mu m$ the layers can be grown free of dislocations.

To calculate the density of MDs for large area we suppose that at complete relaxation the strain at dislocation lines is ≈ 0, while between two parallel dislocation lines the strain has a maximum. The strain between two MDs is supposed to vary as $\cos(\omega y)$, that is, in a similar way as the strain in a finite size layer (see(1)).

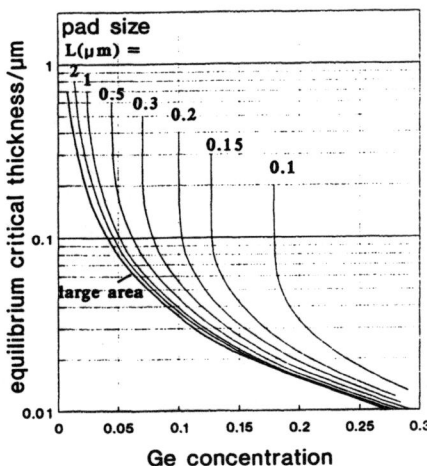

Fig.8 Calculated equilibrium critical thickness as a function of Ge concentration

From (1) results $\varepsilon = 0$ for $y = \pm L/2$. Therefore, we assume the distance between misfit dislocations to be L and neglect the width of the dislocation line. This approximation is reasonable, because we limit our analysis to $L >> a_0$ (a_0 - lattice constant). The probability for a new MD to appear is maximum at the midst between two MD lines. Therefore, we define the equilibrium linear MD density for *infinite areas* as:

$$(3) \quad n_e \equiv 1/L$$

and rewrite equation (2) as an equation for the equilibrium density for infinite area:

$$(4) \quad \frac{1 - e^{-(\pi n_e \gamma^{1/2})h}}{\pi n_e \gamma^{1/2}} = \frac{g}{x} \ln(4h/b)$$

where g is defined in (2). n_e is represented in *fig. 3* as a function of layer thickness with the same values for γ, b, and g as before. One can see the large differences for $h < 0.4\mu$m for large areas between the computed equilibrium curve and the experimental points. While for $x = 0.12$ a critical thickness of ≈ 140 nm is found experimentally, the equilibrium prediction is ≈ 25 nm.

DISCUSSION

The calculations presented above show that an equilibrium model based solely on elastic relaxation at the edges of small epitaxial areas can not explain the observed reduced MD density at sizes as high as 200μm, as it predicts a reduction of MD density only for sizes below $\sim 1\mu$m. This suggests the presence of kinetic barriers to MD nucleation not included in the equilibrium calculations. These barriers obviously prevent the minimum energy configuration to be attained. Similar conclusions have already been discussed by several authors before, see for instance [1,2,5].

CONCLUSIONS

In summary, we have investigated the dependence of the MD density on the size of the deposited pads for strained layer epitaxy of SiGe. Selectively grown LPCVD-$Si_{0.88}Ge_{0.12}$ revealed a strong reduction of misfit dislocation density when the pad size is reduced, and even complete absence of MDs for pads $\leq 2\mu$m. The dependence of the critical thickness on the pad size was calculated within a model of elastic strain relaxation. The calculations show that there is a critical size below which layers can be grown free of dislocations. Therefore, this model can explain the observed absence of misfit dislocations for pads smaller than $10 \times 10\mu m^2$. However, the reduction of misfit dislocation density on larger pads ($\sim 200\mu$m) can only be understood if the kinetics of misfit dislocation generation will be taken into account.

The authors are grateful to K. Schmidt for RBS measurements, to K. Wambach for technical assistance and H.-P. Bochem for SEM. One of us (T.S) gratefully acknowledges the financial support from the Alexander von Humboldt Foundation.

REFERENCES

[1] E.A. Fitzgerald, G.P. Pettit, R.E. Proano and D.G. Ast, J. Appl. Phys. 65, 2220 (1989).

[2] D.B. Noble, J.L. Hoyt, C.A. King and J.F. Gibbons, Appl. Phys.Lett. 56, 51 (1990).

[3] R. Hull, J.C. Bean, G.S. Higashi, M.L. Green, L. Peticolas, D. Bahnck and D. Brasen, Appl. Phys. Lett. 60, 1468 (1992).

[4] L. Vescan, W. Jäger, C. Dieker, K. Schmidt, A. Hartmann and H. Lüth, in Mechanism of Heteroepitaxial Growth, edited by M.F. Chrisholm, B.J. Garrison, R. Hull and L.J. Schowalter (Mater. Res. Soc. Proc. 263, Pittsburgh, PA, 1992), pp. 23-28.

[5] B.W. Dodson and J.Y. Tsao, Appl. Phys. Lett. 51, 1325 (1987).

[6] F.K. LeGoues, K. Eberl and S.S. Iyer, Appl. Phys. Lett. 60, 2862 (1992).

[7] T. Stoica and L. Vescan, submitted to J. Cryst. Growth.

[8] S. Luryi and E. Suhir, Appl. Phys. Lett. 49, 140 (1986).

[9] J.W. Matthews, S. Mader and T.B. Light, J. Appl. Phys. 41, 3800 (1979).

X-RAY DIFFRACTION AND REFLECTANCE, RAMAN SCATTERING AND PHOTOLUMINESCENCE CHARACTERIZATION OF THERMALLY ANNEALED EPITAXIAL SI$_{1-x}$GE$_x$ LAYERS

M. LÍBEZNÝ[*‡] I. DE WOLF[*], J. POORTMANS[*], A. VAN AMMEL[*], M. CAYMAX[*], V. HOLÝ[‡], F. VIŽĎA[‡], J. KUBĚNA[‡], K. WERNER[**] AND M. ISHIDA[‡‡]
[*]IMEC, Kapeldreef 75, B-3001 Leuven, Belgium
[‡]Dept. Solid State Physics, Masaryk Univ., Kotlářská 2, 611 37 Brno, Czech Republic
[**]DIMES, Lorentzweg 1, 2628 CJ Delft, Netherlands
[‡‡]Dept. Electrical & Electronic Eng., Toyohashi Univ. of Technology, Toyohashi 441 Japan

ABSTRACT

Si$_{1-x}$Ge$_x$ undoped strained layers with a pure Si-capping layer were grown epitaxially by UHV CVD on Si (100) substrates. The samples were subjected to thermal treatments corresponding to typical device-processing. Effects connected with thermally induced relaxations were studied by Raman scattering, XRD, photoluminescence (PL) and defect etching. The thickness of the layers was determined from X-ray reflectance measurement. Strain values for the as-grown and relaxed samples were extracted from Raman scattering, XRD measurements and defect etching and correlated. Si$_{1-x}$Ge$_x$-layer related peaks were observed in PL spectra of several samples. An explanation why those peaks are not observed in all the samples is suggested.

INTRODUCTION

Considerable attention is currently paid to the implementation of Si$_{1-x}$Ge$_x$ strained layers into standard silicon devices [1], e.g. heterojunction bipolar transistors (HBT), p-MODFET, etc. This intention requires, though, important changes in further processing steps. Any high temperature step jeopardizes the strain inherent to the layer and some relaxation can occur, producing two effects. At first, a decrease of the strain causes a change of the electronic band structure which results into a decrease of electrical performance of the device. At second, the relaxation occurs through an increased density of misfit dislocations and thus also of threading dislocations, which can have a damaging effect on the electronic devices. Therefore all the subsequent thermal steps must be limited to the very necessary amount to avoid these problems and also to minimize outdiffusion of dopants from one region to another. Some methods must be established for the evaluation of the influence of thermal processing steps. Non-destructiveness of these methods would be an additional advantage.

In this work we tried to assess possible methods for the evaluation of the strain relaxation for the case of HBT processing. The available nondestructive methods were photoluminescence (PL), X-ray diffraction (XRD) and X-ray reflectance (XRR) and Raman spectroscopy (RS). For comparison, we also used defect etching as a classical method with a clear interpretation.

EXPERIMENT

The layers were prepared on Si (100) substrates, B-doped, 2-4 Ω.cm, cleaned by RCA procedure, by the Ultra High Vacuum Chemical Vapour Deposition (UHV CVD) at 650 °C with prebake at 400 °C. The structure (Si substrate - Si buffer 150 Å - Si$_{1-x}$Ge$_x$ - Si cap) corresponds to that of the HBTs, except for the fact, that only undoped Si$_{1-x}$Ge$_x$ layers were used.

The values of x = 0.05, 0.10, 0.12, 0.16 and 0.20 were chosen. Parts of the samples were then either left as-grown (in further text assigned as A) or N$_2$-annealed in a furnace 800°C/30 min (B) or by rapid thermal processing 850°C/5 min (C) or 900°C/2 min (D). Processing times for chosen temperatures were estimated using process simulation in such a way, that for the real HBT-structure the diffusion of boron would not shift the junction to the Si-layer.

The PL setups used were a home-built system and a modified BIORAD PL6100 instrument based on a Fourier spectrometer. Excitation was provided by an Ar ion laser in the multimode regime with a band filter centered at 500 nm.

Micro-Raman experiments were performed in back-scattering configuration, using the 457.9 nm argon ion laser line.

The XRR measurements were done with the double crystal diffractometer, where the first crystal (Si, 111 symmetrical diffraction) served as a monochromator for the CuKα$_1$ line. After passing the slit

$(0.1 \times 5 \text{ mm}^2)$ the beam irradiated the sample, the incidence angle (with the sample surface) was below 2 deg. There was a narrow slit $(0.3 \times 10 \text{ mm}^2)$ in the reflected beam in front of the detector. During the measurement the sample was rotated and the detector slit was shifted so that the detector only measured the specularly reflected wave.

The XRD measurements were performed with the 224 asymmetric diffraction on the first crystal and the $CuK\alpha_1$ spectral line. We have measured the diffraction curves of the samples in the following diffractions: symmetrical 004, asymmetrical 224 (both asymmetries) and asymmetrical 115 (both asymmetries).

Defect etching was performed in 1/7-diluted Secco-etchant, typically for 30 - 90 s.

INTERPRETATION

A. Photoluminescence

$Si_{1-x}Ge_x$-near-band-gap no-phonon and TO-phonon photoluminescence peaks were observed on samples with x = 0.05 and 0.10 (Fig.1a,b).The other peaks in the PL spectra were substrate related bound excitons and also dislocation peaks. The latter were observed for some samples, mainly after annealing. In this paper we will restrict only to the first mentioned lines.

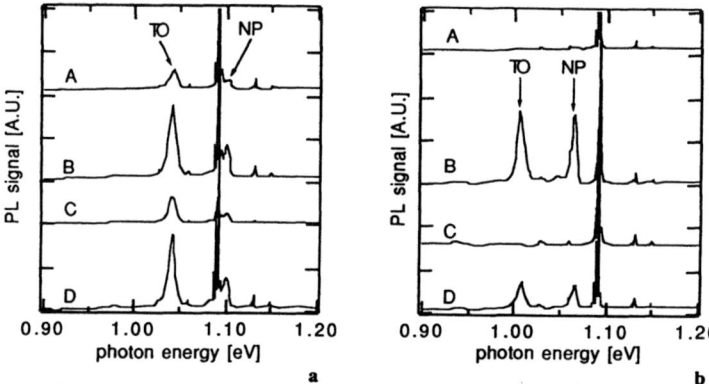

Fig. 1: PL spectra for $Si_{0.95}Ge_{0.05}$ (a) and $Si_{0.90}Ge_{0.10}$ (b) samples. NP and TO peaks originate from $Si_{1-x}Ge_x$-near-band-gap no-phonon and TO-phonon photoluminescence.

The absorption depth of exciting laser radiation for Si, resp. $Si_{1-x}Ge_x$ in this region widely exceeds the thickness of the layered structure, therefore the excitons are created largely in the substrate and then they diffuse to the $Si_{1-x}Ge_x$ layer as a consequence of the potential drop there [2]. The same potential drop also prevents a substantial influence of the surface recombination.

For the $Si_{0.90}Ge_{0.10}$ sample the peaks appeared only after the annealing (Fig.1b). It follows from the paragraph above, that the lack of the PL signal in the non-annealed sample can be caused by the presence of some kind of nonradiative recombination centres either acting in the $Si_{1-x}Ge_x$ layer itself or preventing the diffusion of the excitons, e.g. at the Si substrate - Si buffer interface. When we measured the temperature dependence of the PL signal of an annealed $Si_{0.90}Ge_{0.10}$ sample, these lines disappeared, which clearly showed a bound exciton behaviour of these lines, but no free exciton lines arose. This behaviour can be explained by the presence of some nonradiative recombination centers in $Si_{1-x}Ge_x$ layer which capture excitons freed at higher temperature. For lower temperature these excitons are bound to the boron atoms, which are unintentionally introduced there, and radiative recombination competes with the non-radiative one. Final signal is determined by capture cross-sections and concentrations of boron atoms and those non-radiative centres. During annealing more boron atoms are outdiffused from the substrate to the layer and the non-radiative recombination centres are partially annealed out. Because both these processes enhance the radiative boron bound exciton recombination, the PL lines can now appear for an appropriate annealing treatment (Fig. 1b, curves B, D).

The energies of our no-phonon peaks for both x = 0.05 and 0.10 samples reasonably agree with those given by Robbins et al. [3] for the excitons in the $Si_{1-x}Ge_x$ layers. The position of such a peak is directly related to the energy gap of the alloy [3] and any strain relief leading to a change of the energy gap could be registered with very high sensitivity by photoluminescence. But as follows from Table I, no substantial shift of the energy gap connected with the thermal treatment was observed for the $Si_{0.95}Ge_{0.05}$ sample and energies for $Si_{0.90}Ge_{0.10}$ do not differ significantly from the value of 1.070 eV given by a formula in [3].

Tab.I: PL $Si_{1-x}Ge_x$ no-phonon peak positions for $Si_{0.95}Ge_{0.05}$ and $Si_{0.90}Ge_{0.10}$ samples for different thermal treatments:

sample	NP peak energy [eV]			
	A	B	C	D
$Si_{0.95}Ge_{0.05}$	1.101	1.101	1.101	1.101
$Si_{0.90}Ge_{0.10}$	-	1.065	-	1.067

B. Raman spectroscopy

The Raman peak used for the interpretation originates from Si-Si vibrations in the $Si_{1-x}Ge_x$ film (Fig. 2). The peak 520 cm^{-1} (intensity = 100 a.u.) arises from the substrate, while the shoulders seen at the left side, correspond to Si-Si vibrations in the film. The frequency of this film peaks will be denoted ω_{Si-Si}. Besides these Raman peaks, also signals arising from vibrations between Si-Ge and Ge-Ge neighbouring atoms are detected, but because of the low Ge content in the investigated films, the intensity of these peaks is in general small. For this reason only the Si-Si vibrations are used for investigation of the different $Si_{1-x}Ge_x$ films.

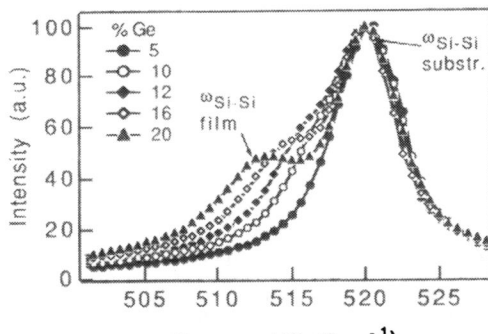

Fig. 2: Raman spectra of the Si-Si Raman peak for different x. The spectra are normalized in such a way, that the intensity of the Si-Si peak from the substrate (at 520 cm^{-1}) is 100 a.u.

The position of ω_{Si-Si} depends both on the $Si_{1-x}Ge_x$ layer composition and on the strain in this layer. ω_{Si-Si} is linearly related to x in strain-free alloys [4] and shifts to higher values in the presence of compression in the film. This is shown in Fig. 3.

If we define $a_{//}$ as the in-plane lattice constant of the strained $Si_{1-x}Ge_x$ film and a_{SiGe} as that of strain-free $Si_{1-x}Ge_x$ alloy, then the strain $\varepsilon_{//} = (a_{//} - a_{SiGe})/a_{SiGe}$.

The Raman peaks from the epitaxial $Si_{1-x}Ge_x$ films clearly have a larger frequency than the strain free frequency, indicating compressive stress in the films. In order to obtain an estimation between ω_{Si-Si} and x for strained films, a line is fitted through the data points x = 0.05, x = 0.1 and x = 0.12, with the condition $\omega_{Si-Si} = 520$ for x = 0. The result of this fit is:

$$\omega_{Si-Si} = 520 - 36\,x \tag{1}$$

and from the relation between strain, $\varepsilon_{//}$, and x: $\varepsilon_{//} = -4.18\ \%\ x$, follows that the strain induced shift:

$$\omega_{Si-Si} - \omega_{Si-Si}^{o} = -8.6\ \varepsilon_{//}\ (\%) \tag{2}$$

Only the data points for low x values are used in this fit because at larger x values the films are found to be partially relaxed due to defect formation. Relation (2) corresponds very well with the relation given by Halliwell et al. [5]: $\Delta\omega = -9.3 \pm 0.9\ \varepsilon_{//}$ (%) obtained from x-ray double-crystal diffractometry and RS experiments on $Si/Si_{1-x}Ge_x$ strained-layer superlattice structures. The data points of the samples with

$x = 0.16$ and $x = 0.20$ are shifted downward with respect to the fitted line, indicating that they are indeed partially relaxed. Using Eq. (2) it follows that this relaxation is very small: $\varepsilon_{//} = - 0.61$ % for $x = 0.16$ and $\varepsilon_{//} = - 0.77$ % for $x = 0.2$ while for fully strained films $\varepsilon_{//} = - 0.66$ % ($x = 0.16$) and $\varepsilon_{//} = - 0.84$ % ($x = 0.2$).

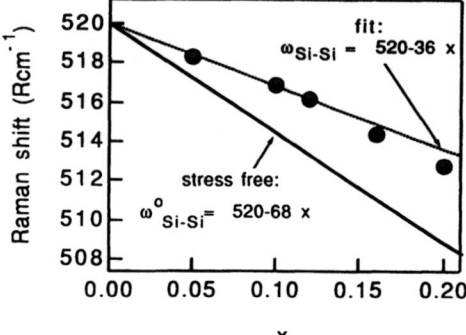

Fig. 3: Full circles: the position of Si-Si peaks, obtained by fitting, plotted as a function of the Ge content x for as-grown samples. The dotted line is a fit through this data. The full line on this figure shows the relation between $\omega_{Si\text{-}Si}$ and x as measured by [4] on strain-free $Si_{1-x}Ge_x$ alloys.

In order to study the temperature stability of the $Si_{1-x}Ge_x$ films, Raman spectra of samples before and after annealing at 800 or 850 °C are compared. The influence of annealing on the frequency of the Si-Si optical phonon peak from the film is found to be very small or zero. Only for the films with low Ge content ($x = 0.1$ and 0.12) a small decrease of $\omega_{Si\text{-}Si}$ is observed, indicating strain relaxation. The substrate peak is not affected by annealing, while the peak from the film is slightly shifted towards a lower frequency.

Using Eq. (2), $\varepsilon_{//}$ for $Si_{0.90}Ge_{0.10}$ at the different thermal treatments is calculated to be: A: $- 0.44 \pm 0.01$, B: $- 0.42 \pm 0.01$ and C: $- 0.39 \pm 0.01$.

C. X-ray reflectivity

We fitted the measured reflectivity curves with those calculated on the basis of the Fresnel formulas and Debye-Waller-like correction factors for the interfacial roughness after [6]. The refractive indices of Si and Ge have been calculated from the atomic scattering factors after [7], the index of the $Si_{1-x}Ge_x$ layer has been calculated as a mixture of both Si and Ge values. The properties of the surface of the annealed samples differ from those of the non-annealed ones probably due to a very thin adsorbed layer. In order to obtain a good fit of the theoretical and experimental reflectivity curves we assumed that the surface was covered by a monoatomic layer with variable refractive index.

Tab II: Thicknesses d_{Si} of the Si capping layer and d_{SiGe} of the $Si_{1-x}Ge_x$ layer obtained from XRR measurement and the composition x derived from XRD:

sample	x	d_{Si} [nm]	d_{SiGe} [nm]
$Si_{0.90}Ge_{0.10}$	10.2	39.6	105
$Si_{0.88}Ge_{0.12}$	12.1	37	110
$Si_{0.84}Ge_{0.16}$	16.2	39	103
$Si_{0.80}Ge_{0.20}$	20.5	36	91

The fit yielded the thicknesses $d_{1,2}$ of the Si and $Si_{1-x}Ge_x$ layers and the root-mean-square roughnesses of the free surface and the $Si\text{-}Si_{1-x}Ge_x$ interface. All the experimental curves were insensitive to the roughness of the $Si_{1-x}Ge_x\text{-}Si$ (substrate) interface. The results of the fits are in Table II.

The fits done for annealed samples showed that the thermal annealing caused no distinct changes in the thicknesses and the interface roughnesses.

D. X-RAY diffraction

The degree of strain relaxation has been studied by means of double crystal x-ray diffractometry. The strain in the layer is described by the strain tensor, whose diagonal components are ε_{xx}, ε_{xx} and ε_{zz} where x and z are the coordinates parallel and perpendicular to the surface, respectively, and e.g., $\varepsilon_{zz} = (a_\perp - a_{Si})/a_{Si}$, where $a_\perp$ is the lattice constant of the $Si_{1-x}Ge_x$ film along the z-direction, and a_{Si} is that of Si. The values of these components were determined from the distance of the substrate and layer peaks in the rocking curves after [8] (see Tab. III). The lateral misfit in all samples is comparable with the method sensitivity and the structure can be considered nearly pseudomorphic. Then, a direct connection exists of the chemical composition x of the $Si_{1-x}Ge_x$ layer with the ε_{zz} component of the strain tensor [9].

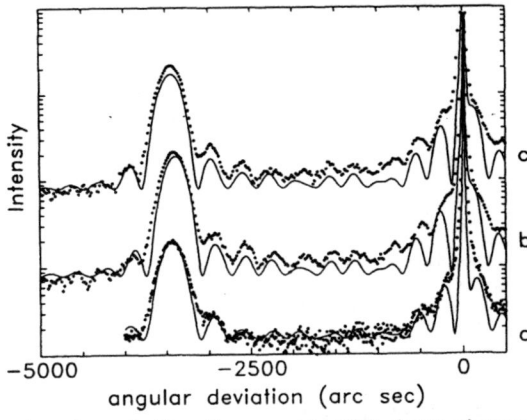

Fig. 4: Measured (points) and fitted (lines) theoretical x-ray diffraction curves at the 224 grazing incidence diffraction of the sample $Si_{0.80}Ge_{0.20}$ before annealing (curve a), after the type B (b) and D (c) annealing. The curves b and c are shifted up by a factor of 10 and 100, respectively.

According to the Vegard law, we get for (001) oriented substrates

$$x = 13.54 \, \varepsilon_{zz} \tag{3}$$

Using this formula we obtained the values of x given in Table II.

Tab III: XRD values for the ε_{zz} strain of samples subjected to a different thermal treatment. For as-grown samples XRD and Raman values are compared.

sample	ε_{zz} $[10^{-2}]$			
	A		B	D
	XRD	Raman	XRD	XRD
$Si_{0.88}Ge_{0.12}$	0.895	0.86	0.886	0.896
$Si_{0.84}Ge_{0.16}$	1.20	1.11	1.20	1.18
$Si_{0.80}Ge_{0.20}$	1.52	1.39	1.52	1.50

Fig. 4 shows the diffraction curves measured in the 224 diffraction (grazing incidence of x-rays) and those calculated on the basis of the dynamical x-ray diffraction theory (see, e.g., [10]). In the calculations we used the layer thicknesses following from the x-ray reflectivity measurements. The coincidence of the theory with the experiments is quite good, the experimental diffracted intensity is increased probably due to the diffuse x-ray scattering from the structural defects.

Using the values of $\varepsilon_{//}$ measured by RS on the as-grown samples, ε_{zz} were calculated and compared with the XRD values (Table III).

E. Defect etching

A number of selected samples was subjected to defect etching. The results were analysed by Nomarski microscopy (see Fig. 5). Typically etch pits as would be caused by etching threading dislocations are not detected, whereas line dislocations parallel to the interface are revealed as an orthogonal network of straight lines, extending to the edges.

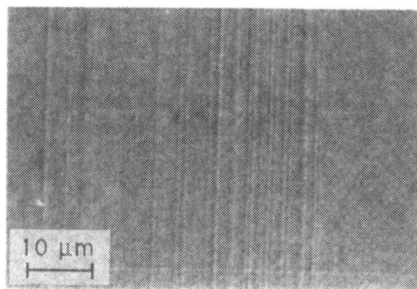

Fig. 5: Nomarski micrograph of defect - etched $Si_{0.90}Ge_{0.10}$ layer (annealed at 900°C for 2 min) showing dislocation lines.

If we assume that every line corresponds to one missing atomic plane, then we can state that the strain changes from its normal value ε to $\varepsilon_{disl} = \varepsilon - l\ D$, with l the appropriate value for the mean width of the atomic (100) planes in the fully relaxed $Si_{1-x}Ge_x$ lattice and D the average number of dislocation lines per unit length (typically 5000/cm). The relaxation is then given by $(\varepsilon - \varepsilon_{disl})/\varepsilon$.

Based on this analysis, we can say, that the $Si_{0.95}Ge_{0.05}$ sample does not relax at all, even after thermal treatment. The $Si_{0.90}Ge_{0.10}$ sample shows dislocation lines only after annealing. The $Si_{0.88}Ge_{0.12}$ sample shows very few lines before annealing (typically 4 lines/mm), but relaxes clearly during annealing. The samples with x = 0.16 and 0.20 have not been examined before annealing, but do show misfit dislocations after annealing.

We have found a pronounced influence of the way of the sample cleaving on the resulting relaxation during annealing, which can be explained by dislocations nucleated at the edges. E.g., in some samples, there was an order of magnitude difference in D for the two <110> directions in (100) plane. In any case, the relaxation as measured by this method never exceeded 2% and has an average value of about 1%.

CONCLUSION

The relaxation behaviour of Si-$Si_{1-x}Ge_x$ multilayer structures as a result of thermal annealing typical for the HBT processing was studied. The $Si_{0.95}Ge_{0.05}$ layer was found to remain stable for all thermal annealing applied. The $Si_{0.90}Ge_{0.10}$ as-grown sample was completely strained, while some relaxation occured during annealing. The $Si_{0.88}Ge_{0.12}$, $Si_{0.84}Ge_{0.16}$ and $Si_{0.80}Ge_{0.20}$ layers were partially relaxed already after growth.

From all the used nondestructive methods, PL offers the advantage to measure a quantity which is directly related to the bandgap, but it can be problematic to obtain a PL signal. The change in the bandgap for $Si_{0.95}Ge_{0.05}$ sample was found to be smaller than 1 meV. XRD is a very accurate tool for strain measurements, but the extraction of strain values for a layered structure becomes complicated. Probably for this reason no substantial relaxation was observed. Raman spectroscopy is sufficiently sensitive for the assessment of a significant strain relaxation for samples with x>0.10 and micro-Raman arrangement makes this technique interesting for an analysis on a micro-scale. Defect etching has proved to be a very sensitive technique, but it is destructive. The relaxation as evaluated by the defect etching did not exceed 2%.

REFERENCES

1. S.C. Jain and W. Hayes, Semicond. Sci. Technol. 6, 547 (1991)
2. D. Dutartre, G. Brémond, A. Souifi and T. Benyattou, Phys. Rev. B 44, 11 525 (1991)
3. D.J. Robbins, L.T. Canham, S.J. Barnett, A.D. Pitt and P. Calcott, Appl. Phys. Lett. 71, 1407 (1992)
4. M.I. Alonso Carmona, PhD Thesis, Max-Planck Institut für Festkörperforschung, Stuttgart, 1989
5. M.A.G. Halliwel, M.H. Lyons, S.T. Davey, M. Hockly, C.G. Tuppen and C.J. Gibbings, Semicond. Sci. Technol. 4, 10 (1989)
6. D.K.G. de Boer, Phys. Rev. B 44, 489 (1991) and citations therein
7. Int. Tables for X-Ray Crystallography, Vol. IV, Kynoch Press, Birmingham, 1974
8. J.C.P. Chang, J. Chen, J.M. Fernandez, H.H. Wieder, and K.L. Kavanagh, Appl. Phys. Lett. 60, 1129 (1992)
9. V. Holý, J. Kubèna and K. Ploog, Phys. Status Solidi (b) 162, 347 (1990)
10. W.J. Bartels, J. Hornstra and D.J.W. Lobeek, Acta Cryst. A42, 539 (1986)

INTRINSIC OPTICAL AND ELECTRICAL PROPERTIES OF STRAIN-ADJUSTED SHORT-PERIOD Si_mGe_n SUPERLATTICES

JANOS OLAJOS*, JESPER ENGVALL*, HERMANN G. GRIMMEISS,* ERICH KASPER**, HORST KIBBEL**, AND HARTMUT PRESTING**

*Dept of Solid State Physics, Lund University, Box 118, S-221 00 LUND, SWEDEN

**Daimler-Benz AG, Research Center, Wilhelm Runge Strasse 11 D-7800 ULM, GERMANY

ABSTRACT

Interband optical transitions are observed in a series of strain-adjusted, short-period Si/Ge superlattices by means of photocurrent spectroscopy, infrared absorption, photo (PL)- and electroluminescence (EL). The onsets of the interband absorption in the energy range of 0.7 - 0.9 eV are in good agreement with the observed PL and EL. Bandgap-related EL is observed in mesa diodes at room temperature, whereas the PL disappears at about 40K. In samples, annealed at growth temperatures (550^oC) and higher, a systematic shift of the bandgap is observed which is discussed in terms of a process involving interdiffusion of the Si and Ge atoms. Photocurrent measurements at low temperatures support the model from PL studies suggesting that the photogenerated electrons are immobile in the SLS at low temperatures and have to be thermally ionized from shallow levels.

INTRODUCTION

The luminescence from Si/Ge short period, strained layer superlattices (SLS's) has attracted considerable interest over the last years[1-2], but not until recently [3] have there been samples available with good enough quality concerning dislocation densities, interface sharpness etc. such that reliable studies of the intrinsic optical properties could be performed in these structures. The breakthrough during the last years in the improvement of the PL properties of the Si_mGe_n SLS has been a result of the ability in reducing the dislocation density during growth. This is acomplished in several ways e.g. by growing the buffer with continously increasing Ge concentration [4], so called step-graded or linearly graded bufferlayers. A further improvement is achieved by using surfactants during growth. The main goal of this paper is to examine in more detail the physical processes involved in the optical transitions around the fundamental bandgap of the strain-adjusted Si/Ge SLS's. We present results from experiments comprising photoconductivity (PC) and electroluminescence (EL) that strongly support previous PL results.

Experimental:

The samples used in this work consists of a series of strain-adjusted Si_mGe_n SLS with m/n ratios of nominally 9/6, 6/4, and 3/2 respectively. as well as a Si_6Ge_4 SLS p-n junction diode grown in a Si-MBE chamber. For the undoped samples the following layer structure was used: The buffer layer consisted of a step graded alloy buffer with a thickness of 6500Å, followed by a 5000Å thick $Si_{0.6}Ge_{0.4}$ alloy. The dislocation density, measured with cross-sectional TEM, was found to be reduced by at least 3 orders of magnitude compared to previously studied Si_mGe_n SLS´s [1] grown on partly relaxed alloy buffer layers. Typical dislocation densities are in the order of 10^6 cm^{-2}. The absorbance of these samples was measured with the parallel photoconductivity method. For this, alloyed ohmic contacts made of Au+Ti+Sb were used. The diode sample was prepared in the following configuration: First a 50nm thick Si layer was grown, Sb-doped by Secondary implantation (DSI) to more than 10^{17}cm-3. Therafter a 100 nm thick $Si_{0.50}Ge_{0.50}$ buffer layer was grown at 575°C, and was followed by a 145 period Si_6Ge_4 superlattice (200nm), grown at 500°C. Both the buffer and superlattice growth were preceded by deposition of a monolayer (ML) of Sb. The resulting n-doping of the buffer and the SLS were slightly different, $0.6 \cdot 10^{17}$cm-3 and $2 \cdot 10^{17}$cm-3, respectively, as deduced from SIMS and C-V analysis. The p-layer was grown as a $Si_{0.60}Ge_{0.40}$ alloy, doped by boron coevaporation in excess of $1 \cdot 10^{20}$cm-3. The strain in the Si-layers of the SLS as determined with X-ray diffraction was found to be 1.64%, which is the strain symmetrized value for a nominal SLS composition. The samples with p$^+$n junctions were structured into mesa diodes and Au/Cr ohmic contacts were fabricated with a window to allow the light to pass. The spectroscopic measurements were performed using either a Bomem DA3.02 Fourier Transform (FTIR) spectrometer or a double grating monochromator. The EL and PL spectra were detected with a liquid-nitrogen-cooled North Coast Ge-detector. For the EL measurements, the resolution was low, typically 750 Å.

Results and discussion

1) Undoped superlattices

In a recent paper by Menzcigar et.al. [3] the PL properties of a series of strain-adjusted Si_mGe_n SLS were examined. An enhanced photo-luminescence was observed in the superlattices as compared to a Si/Ge alloy and arguments were given for a excitonic radiant recombination via localized excitons, bound to potential fluctuations in the superlattice. In fig 1. the PL from a Si_9Ge_6 SLS is

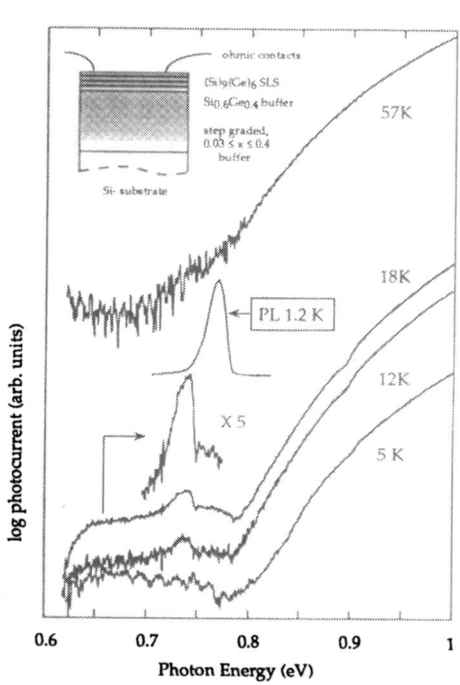

Fig. 1 PL and photocurrent spectra of a Si_9Ge_6 superlattice. The exciton is observed in the photocurrent spectra only in a narrow temperature range. The photocurrent spectra are displaced to facilitate comparison.

compared with spectra obtained with photocurrent spectroscopy (PC) at various temperatures. At higher energies in the PC- spectra the onset of interband transitions is observed for all temperatures. In this region, the photocurrent is proportional to the absorption coefficient and thus reflects the joint density of states in the superlattice layers[2]. This onset shifts to higher energies in samples with decreasing superlattice period-length but identical m/n ratio, i.e. Ge-concentration, indicating that indeed band-to-band, and not defect-related transitions are observed. The PL shows the same shift [3] and it is accordingly possible to attribute also this signal to (near-)bandgap recombination. The luminescence peak exhibits

an assymetric lineshape and further characteristics such as anomal dependence on temperature and excitation power that are consistent with a localized exciton [5]. In the photocurrent spectra, at lower energies, a weak and broad peak is seen at 12-18K. This feature is *positive* indicating that carriers excited at this energy give rise to an increase in the current. It is interesting to note that peak is absent at very low temperatures, 5K, but instead a weak, negative feature is discerned. Such behaviour is sometimes observed in defect spectroscopy when a transition to a excited bound state is seen as positive peak in photocurrent measurements. This is a two-step process and involves a phonon that thermally excites the carrier into the continuum (photothermal-ionization spectroscopy, PTIS) and gives rise to a positive photocurrent for some temperatures. The peak in the PC spectra is seen at the same energy as the PL and has the same lineshape, indicating that the same transition is involved. These findings strongly supports that there exists a bound state, below the bandgap, in the superlattices that favours the luminescence as a recombination channel *at low temperatures.*

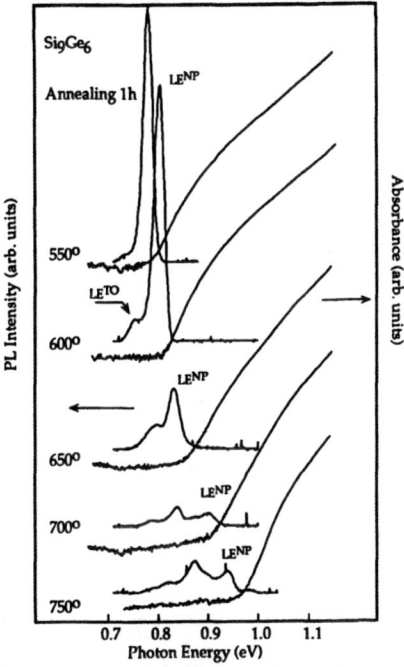

Fig 2. PC and PL spectra of a Si_9Ge_6 superlattice as a function of annealing temperature after heat-treatment for 1 h. The spectra are displaced in vertical direction. (same scale)

Additional support for the assignment of the optical transitions as being interband related was found in an annealing study of the superlattices. In Fig.2 the absorbance and photoluminescence of a nominal Si_9Ge_6 superlattice is shown after heat-treatment for 1 h at various temperatures. It is clearly seen that the onset as well as the PL shifts to higher energies with increasing annealing temperature. The structural behaviour of Si/Ge superlattices after heat-treatment has been studied by R. Schorer et.al. [6]. It was shown that the intermixing of the atomic species in the superlattice layers after annealing started at about 550° C and it was argued that the Si atoms diffused into the Ge-layers wheadas the pure Si-layers became thinner. If one models the band-edges with the envelope-function approach [3], it is seen that the bandgap shifts to higher energies with increasing Si- concentration in the Ge layers. At about 750°C the superlattice is totally alloyed. The shift of the absorbtion onset is in good agreement with the modelled bandgap energies. From the comparision in Fig 2, it is also possible to determine the origin of the PL lines since other defect-related transitions become visible at higher annealing temperatures.

2) p-n junctions

Electroluminescence spectra at three different heat-sink temperatures: 159K, 218K and 296K, measured with the same injection current density (3 A/mm^2) are shown in Fig.3. It is seen that the signal consists of in principle two broad features with peak energies of 0.77 eV and 0.88 eV respectively, and that the relative intensity of the two peaks are temperature dependent. In order to determine the nature of the emission processes involved, the EL intensity (L) was measured as a function of the injection current(I). A typical L-I plot, obtained with a heat-sink temperature of 156K, is shown in the inset of Fig. 2 where the intensity of the 0.77 eV and the 0.88 eV- peaks are marked with full and open circles, respectively.

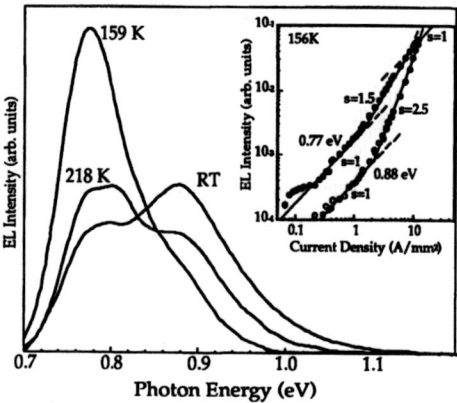

Fig.3 EL spectra measured at an injection current of 3 A/mm^2 at three heat sink temperatures. The inset shows the EL intensity measured at a heat sink temperature of 156K as afunction of injection current density for the two peaks.

Several important observations are noteworthy from this graph, namely : (1) the intensity of both peaks shows a superlinear dependence for certain current densities, (2) the onset of superlinearity occurs at similar current densities for both peaks, and (3) the slopes of the L-I lines are approximately equal to 1 before and after the superlinear region, at least for the 0.77 eV - peak. The superlinearity is characteristic of a multi-state model of recombination channels across and in the forbidden energy gap of the semiconductor[7]. In our case, this means that the two observed EL peaks originate from the same region of the sample if they compete with the same non-radiative recombination channel. Assuming that the peak at 0.78 eV originates from a transition via a defect level with an activation energy of about 100 meV, and the peak at 0.88 eV is caused by a band-to-band transition, then the defect probably acts as a recombination centre at low temperatures. At elevated temperatures the defect level is thermalized which reduces the recombination rate via this level and increases the intensity of the interband process.

Measuring the voltage over the diode allows an analysis of the ideality factor (n) of the current and the total EL intensity. The corrected L-U and I-U characteristics display an ideality factor close to one which suggests that the room temperature EL recombination as well as the total current is either a diffusion current, implying recombination in the neutral region and excluding recombination over deep levels in the space-charge region, or a (near-) bandgap recombination in the depletion region. We have difficulties in believing that the observed EL emission originates from two different defect-related transitions due to the observed temperature dependence of the relative intensities. If both EL-lines would originate from defect related transitions, the high-energy EL-peak should quench faster with elevating temperatures than the low-energy peak, which is not observed.

At thermodynamic equilibrium, assuming unit quantum efficiency and non-degenerate doping, the recombination rate R can be written as [8]:

$$R = \frac{8\pi k_B^3 T^3}{c^2 h^3} \int_0^\infty \frac{\alpha n^2 u^2}{e^u - 1} du$$

where $u = h\nu/k_B T$. With added current carriers, the total rate of recombination R_c per unit volume is given by

$$R_c = \left(\frac{np}{n_i^2}\right) \cdot R$$

This expression can be used to calculate the shape of the electroluminescence spectrum at at given temperature if α, the absorption coefficient, is known as a function of hv. It has been experimentally shown that the absorption in Si/Ge SLS structures follows the expression $\alpha \sim (h\nu - E_G)^2$, however, without the phonon-assisted branches[2] that are characteristic for indirect transitions in bulk semicon-

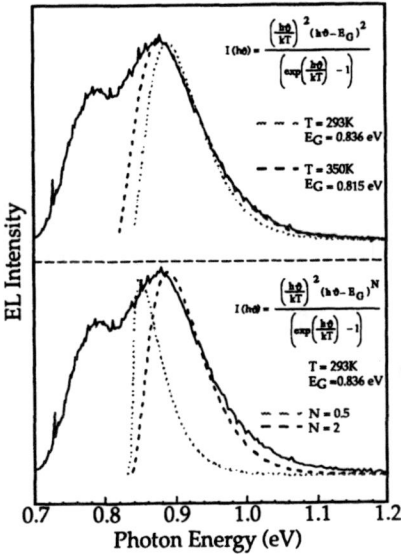

Fig.4 Room-temperature EL spectra compared with theoretically obtained curves under various conditions.

ductors. In Fig.4 the room-temperature EL spectra are replotted together with the calculated curves. In the upper part of the Fig.4 we have used the expression for the absorption as discussed above and calculated the spectrum of the peak assigned as the band-to-band recombination for the measured heat sink temperature and a temperature that gives the best fit. In the lower part of the figure we use in the calculations the band-gap energy obtained from the absorption onset together with 2 possible α (hv). The expression with N=0.5 is valid for bulk, direct transitions. It is readily seen that the EL spectrum exhibits very clearly the absorption dependence observed.

This work was partially financed by the European Basic Research programme (ESPRIT) under Contract no. P7128 and by the Bank of Sweden Tercentenary Foundation.

References

1) R. Zachai, K. Eberl, G. Abstreiter, E. Kasper, and H. Kibbel, Phys.Rev. Lett. **64**, 1055 (1990)

2) J. Olajos, J. Engvall, H. G. Grimmeiss, U. Menczigar, G. Abstreiter, H. Kibbel, E. Kasper, and H. Presting, Phys. Rev. **B 46**, 12 857 (1992)

3) U. Menczigar, G. Abstreiter, J. Olajos, H. G. Grimmeiss, H. Kibbel, H. Presting, and E. Kasper, Phys. Rev. **B 47**, 4099 (1993)

4) F. K. Le Goues, B. S. Meyerson, and F. J. Morar, Phys. Rev. Lett **66**, 2903 (1991)

5) M. Oueslati, M.Zoughai, M-E Pistol, L. Samuelsson, H.G. Grimmeiss, and M. Bal kanski, Phys. Rev. **B32**, 8220 (1985)

6) R. Schorer, E. Friess, K. Eberl, and G. Abstreiter, Phys. Rev. **B 44**, 1772 (1991)

7) H. A. Klasens, J. Phys. Chem. Solids **7**, 175 (1958)

8) W. van Roosbroeck and W. Shockley, Phys. Rev **94**, 1558 (1954)

INTRASUBBAND 1.55 μm ELECTROOPTICAL MODULATION IN SiGe ASYMMETRIC QUANTUM WELL STRUCTURES

LIONEL R. FRIEDMAN* AND RICHARD A. SOREF
USAF Rome Laboratory, RL/EROC, Hanscom Air Force Base, MA 01731
*National Science Research Council, Senior Fellow

ABSTRACT

A new, fast, intrasubband 1.55 μm electrooptic modulator in the SiGe/Si/CaF$_2$-on-Si stepped-quantum-well system is proposed and analyzed. At electric fields of ± 8 V/μm, resonant 1-3 conduction intrasubband absorption is predicted to give 18 dB of optical extinction for narrow-linewidth transitions. We expect this field effect modulator to have microwave response, plus compatibility with Si-based optoelectronic integration.

INTRODUCTION

This paper presents the design of a new Si-based optical modulator for the 1.55 μm fibre-optic communication wavelength. A novel Si-Ge quantum well structure, compatible with Si-based optoelectronic integration, is proposed. This field-effect modulator is expected to have speeds > 10 GHz. We consider near-resonant intrasubband transitions in the conduction band of narrow asymmetric quantum-well (AQW) structures at 1.55 μm., corresponding to energy differences five-to-six time larger than those in III-V QW materials [1] [2]. Valence subbands will be examined in a later publication. Here, we consider the AQW structure integrated in an nin channel waveguide modulator [3] in which the sign of the applied electric field can be reversed. Figure 1 shows the operation of the modulator [4], a redshift or blueshift of the intersubband absorption line in the $\pm$ fields (F).

The larger effective masses of silicon compared to III-V masses will yield smaller second-order optical susceptibilities than those based on III-V semiconductors; so the choice of Group IV alloys and substrate orientation must minimize the effective mass in the growth direction, m_z. The well width should be kept as large as possible consistent with the effective mass, since the susceptibility also increases with well size [2]. The AQWs are uniformly doped to populate only the ground state, and the intersubband valence-to-conduction confinement energies are widely separated, so that band-to-band absorption is avoided at 1.55 μm. To obtain resonant enhancement of the susceptibilities at 1.55 μm ($\hbar\omega$=0.8 eV), suitably high barriers must be found which can be grown on a silicon substrate. This aspect of the problem will require much additional work in materials and growth; in the present paper we will only identify potential barrier materials.

DESIGN AND FIELD RESPONSE OF AQWs

For the growth of pseudomorphic Ge$_x$Si$_{1-x}$ on a (001) silicon substrate, it is known that the conduction band offset is too small for the required internal step V$_i$ within the AQW. For growth of Si on a (001) Ge$_y$Si$_{1-y}$ substrate, $m_z = m_l = 0.916m_0$. In that case, there is sufficient conduction band offset to achieve the step, but the large mass is unfavorable as discussed earlier. the following three AQW cases have a reduced value of m_z:

Case 1, Growth of Si on (110) Ge_ySi_{1-y} : Here, the four X_4 valleys in the plane along the growth direction [(001) plane] lie lower in energy and have an effective mass [5] $m_z = 2m_tm_l/(m_t+m_l) = 0.315m_0$.

Case 2, Growth of Si on a (111) Ge_ySi_{1-y} substrate: Here, $m_z = 3m_tm_l/(m_t+2m_l) = 0.258m_0$ [5].

Case 3, Growth of germanium-rich Ge_xSi_{1-x} ($x \approx 1$) on (110) Ge substrate: For the conduction band ellipsoids at the L-point, $m_t = 0.082m_0$ and $m_l = 1.68m_0$. For the (110) surface, the two lower L_1 minima have $m_z = m_t = 0.082m_0$, while for the two higher L_2, $m_z = 3m_tm_l/(m_t+m_l) = 0.12m_0$.

With the donor concentration per quantum well of 4.85×10^{17} cm^{-3}, only the lowest subband is populated in all three cases.

For the AQWs in our MQW stack, we use stepped wells with the step at the midpoint of the well, the halfwidth of the well denoted by l. The internal potential step within the well is denoted by V_i. We solved for the electron bound state energies E_i and the normalized wavefunctions ψ_i for the i=1,2,3 subbands for both finite and infinite external barrier heights, V_0. For finite V_0, the wavefunctions are sinisoidal functions in the well region which join smoothly on to decaying exponentials in the barrier regions. From the wavefunctions we obtained the dipole matrix elements $z_{ij} = \int dz\psi_i z\psi_j$. The well and barrier dimensions, W and B, were chosen for a small detuning Δ, where $\Delta = E_3 - E_1 - \hbar\omega$. Figure 2 illustrates our Case 1 system of Si/Ge_ySi_{1-y} on a strain-relieved Ge_ySi_{1-y} buffer layer on a (110) Si substrate, for $V_0 = 1500$ meV. For the 1-3 resonance, the AQW has W/2 = 14 Å, and we take x=y=0.4. Then the tensile strain is entirely in the Si layers optimizing the step V_i : $V_i = 189$ meV [6]. In this case, the critical thickness is 40 Å for stable strain and 200 Å for metastable strain; hence the Si layer is pseudomorphic. For infinitely high barriers, we find W/2 = 17.5 Å, $E_1 = 169.5$ meV, $E_3 = 962.2$ meV, $z_{11} = 20.67$ Å, $z_{33} = 17.55$ Å, and $z_{13} = 1.01$ Å. The effect of finite V_0 will be addressed later. Note that $E_1 < V_i$, so that the i=1 wavefunction is largely confined to the right side of the well, increasing the wavefunction asymmetry. It is shown below that the shift E_3-E_1 or E_2-E_1 with field F is proportional to $z_{33}-z_{11}$ or $z_{22}-z_{11}$, respectively. We find generally that the 1-2 transition has a smaller level shift than does the 1-3 transition.

For the double resonance applicable to the linear electro-optic effect, the real part of the second order susceptibility is [2]

$$\chi^2 - (N_wfe^3/2\epsilon_0)(z_{13})^2(z_{33}-z_{11})(\Delta^2-\Gamma^2)/(\Delta^2+\Gamma^2) \qquad (1)$$

where N_w is the donor concentration in the wells, f is the fill factor f=W/W+B, e is the electronic charge, ϵ_0 is the dielectric constant of free space, and Γ is the linewidth. Noting that $\chi^{(2)}$ and r_{zz} depend on F through Δ, the integration over Δ is readily carried out to give the standard dispersive lineshapes of the index $n(\Delta)$ and absorption coefficient $\alpha(\Delta)$:

$$n(\Delta) - (N_wfe^2/2\epsilon_0n_0)(z_{13})^2\Delta/(\Delta^2+\Gamma^2)$$

$$\alpha(\Delta) - N_wfe^2/2\epsilon_0n_0)(z_{13})^2\Gamma/(\Delta^2+\Gamma^2) \qquad (2)$$

The blueshift in the energy difference $\Delta - \Delta'$, is given by $eF(z_{33}-z_{11})$.

The optimal AQW structures for Cases 2 and 3 are as follows: For Case 2, B= 24 Å, $W/2 = 12$ Å of unstrained $Ge_{0.4}Si_{0.6}$ on a strain-relieved (111) $Ge_{0.4}Si_{0.6}$ buffer layer grown on a (111) Si substrate. Here, $V_i = 143$ meV [5] and $m_z = 0.258\ m_0$. For infinite barriers, $E_1 = 158.1$ meV, $E_3 = 961.1$ meV, $z_{11} = 21.8$ Å, $z_{33} = 19.2$ Å, and $z_{13} = 0.864$ Å. Case 3 consists of B=68 Å, W/2 = 34 Å of coherently strained $Ge_{0.8}Si_{0.2}$ and W/2 = 34 Å of unstrained Ge grown upon a strain-relieved (110) Ge buffer layer on a (110) Si substrate. Here, $V_i = 90$ meV [5]. For infinite barriers, $E_1 = 138.4$ meV, $E_3 = 930.7$ meV, $z_{11} = 37.2$ Å, $z_{33} = 34.1$ Å, and $z_{13} = 1.02$ Å. Although $E^c_1 - E_{v1} > 800$ meV (lowest valence to conduction subband energy difference), the Ge substrate will absorb those photons, so a Ge-matched film of $ZnSe_{1-x}S_x$ or $Ca_{1-y}Sr_yF_2$ as a buffer layer would be preferable.

Regarding the difficult choice of high-barrier material, possibly the best lattice-matched widegap material is the cubic flourite CaF_2, an epitaxial insulator with $a_0 = 5.464$ Å (+0.6% lattice mismatch to Si), $E_g = 9.53$ eV, and $V_0 = 2$ to 4 eV [7]. This offset is the difference in the conduction band minima of CaF_2 and Si regardless of where the minima occur in the respective Brillouin zones; hence the offset at the X-points of the two materials can only be larger. By alloying with SrF_2, an insulator with an even larger E_g, this alloy can be exactly lattice matched to $Ge_{0.4}Si_{0.6}$. Another possibility is the alloy zincblende crystal $ZnSe_{1-x}S_x$ whose lattice constant can also be matched to Ge_ySi_{1-y} with the proper choice of x. The quartenary ZnMgSSe is another candidate, as are GaP and AlP.

RESULTS AND DISCUSSION

Figures 3 and 4 show the calculated values of α and n versus wavelength for three choices of electric field, F=0 and ± 8 V/μm for case 1. According to equation (2), the magnitudes of α and n are proportional to N_w and $(z_{13})^2$. The field induced changes $\Delta\alpha$ and Δn at each wavelength are obtained by subtraction of the curves of figures 3 and 4, and are shown as figures 5 and 6.

The results for cases 2 and 3 are as follows: Normalized to case 1 for $V_i = \infty$, the magnitudes of α and n for cases 2 and 3 are respectively lower by 0.73 and higher by 1.02, while the E_3-E_1 shifts with F are smaller by factors of 0.84 and 0.97.

We next consider the effect of finite rather than infinite barrier heights, V_0. Compared to case 1 and $V_0 = \infty$, for $V_0 = 3.0$, 2.0, and 1.5 eV, W/2 is smaller by factors of 0.88, 0.84, and 0.805, respectively. This is because for decreasing V_0, W/2 must be reduced to achieve the same subband energy difference, but this is compensated for in the matrix elements by the exponential penetration of the wavefunctions into the barrier regions. It appears that the upper level can almost be at the top of the barrier without any substantial reduction of the electromodulation; the use of above-the-barrier quasi-bound states will be examined at a later time. For the above three cases, the magnitudes of α and n are all within 2%, while the shifts with F are 0.98, 0.95, and 0.92, respectively.

Now we shall calculate the optical extinction ratio of the AQW modulator operating between an initial bias field of -8 V/μm and a final bias field of +8 V/μm, points A and B in figure 1. At a suitable detuning $\Delta = 3$ meV from the F=0 resonance, the initial negative bias

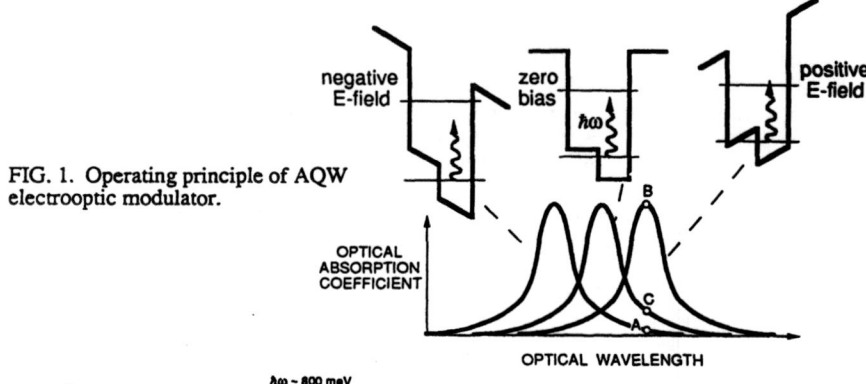

FIG. 1. Operating principle of AQW
electrooptic modulator.

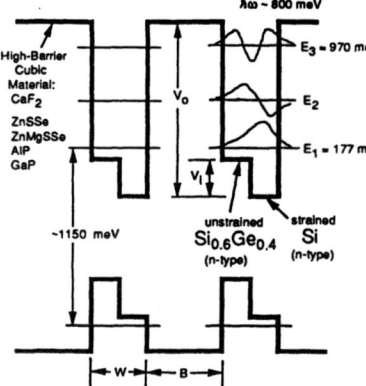

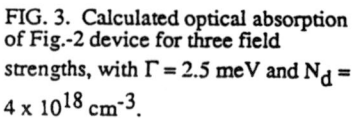

FIG. 2. Bandstructure and subband
energies of Case-1 AQW device with
$V_o = 1.5$ eV, $V_i = 189$ meV, w = b =
28.2 Å.

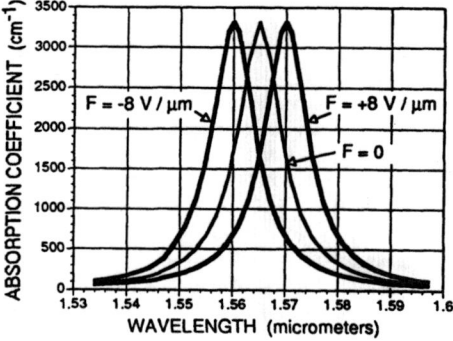

FIG. 3. Calculated optical absorption
of Fig.-2 device for three field
strengths, with $\Gamma = 2.5$ meV and $N_d =$
4×10^{18} cm^{-3}.

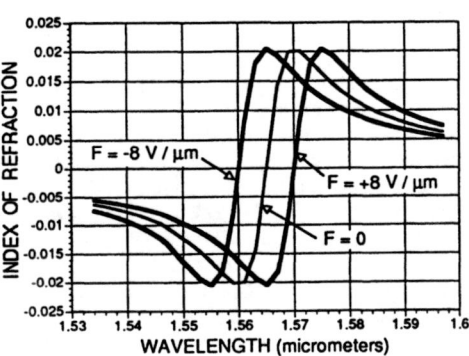

FIG. 4. Calculated index of refraction of Fig.-2 device for three field strengths, with the same Γ and N_d as in Fig. 3. The index shown is n(total) minus n_o.

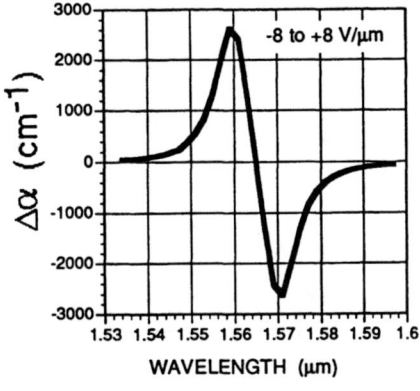

FIG. 5. Electric-field-induced change in optical absorption (F = -8 to +8 V/μm) versus wavelength, for the same parameters as in Fig. 3.

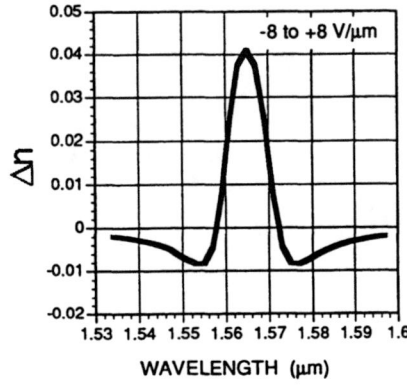

FIG. 6. Electric-field-induced change in index of refraction (F = -8 to +8 V/μm) versus wavelength, for the same parameters as in Fig. 3.

produces a lower insertion loss (A) than does the zero bias loss (C), thereby allowing a larger amplitude extinction ratio (B-A)/A as compared to (B-C)/C in figure 1. The initial absorption loss is $\mathscr{L}_i = 10 \log \exp(-\alpha_0 L)$, and we choose the initial absorption α_0 and the active modulator length L to make the device reasonably transparent at point A, namely $\mathscr{L}_i = 3$ dB. For example, with $\Gamma = 3$ meV and $N_w = 4.85 \times 10^{17}$ cm^{-3}, $\alpha_0 = 69$ cm^{-1} and L = 100 μm for case 1. Note that we can scale α_0 as desired by varying N_w. Thus it appears that the modulator can be operated at a lower doping levels where carrier screening and band renormalization effects are negligible. Next we calculate the final state attenuation $\mathscr{L}_f$(dB) and the extinction ratio R(dB)= 10 log($-\Delta\alpha L$), which depends on the linewidth Γ. Specifically, for case 1, R= 18 dB for $\Gamma = 2$ meV, R= 10.8 dB for $\Gamma = 3$ meV, and R=4.7 dB for $\Gamma =5$ meV. At present, the linewidths Γ are not known accurately, although Hertle et al [8] have cited $\Gamma = 3$ meV of silicon QWs at 8 K, increasing only to 5 to 10 meV at room temperature. Although achieving narrow linewidths is a difficult problem, these values are within the range of those required for effective electromodulation. Finally, our calculations show that electroabsorption is dominant over electrorefraction . From figures 5 and 6, the modulator length required to get π radians of optical phase shift is always larger than the modulator length needed for the extinction ratios quoted above.

Very recent work has incorporated the nonparabolicity of the conduction band. This is important because it results in an increase of the effective mass with increasing energy above the band edge. As a consequence, the well widths have to be reduced to keep the same subband energy difference, and the field shift is reduced by a factor of about 2.5. However, preliminary calculations in the valence band for heavy hole transition (for which there are no nonparabolicity corrections) gives a performance superior to that of the conduction band cited above.

ACKNOWLEDGEMENTS

The authors wish to thank Drs. Jacob Khurgin, K.L. Wang, and G. Abstreiter for very helpful discussion and correspondence.

REFERENCES

1. M.M Fejer, S.J.R. Yoo, R.L. Beyer, A. Harwit, and J.S. Harris Jr., Phys. Rev. Lett., **62**, 1041 (1989)
2. J. Khurgin, J. Opt. Soc. Am. **B6**, 1673, (1989)
3. L. Friedman and R. A. Soref, Electronics Letters, **22**, 819, (1986)
4. R.P.L. Karunisari, Y.J. Mii, and K.L. Wang, IEEE Electron Devics Letters, **11**, 227, (1990)
5. T. Ando, A.B. Fowler, and F. Stern, Rev. Mod. Phys., **54**, 460, (1982)
6. C.G. Van de Walle and R.M. Martin, Phys. Rev. **B4**, 5621, (1986)
7. L.J. Schowalter and R.W. Fathauer, CRC Critical Reviews, **15**, 367,(1989)
8. H. Hertle, F. Schaffler, A. Zrenner, G. Abstreiter, and E. Gornik, Thin Solid Films, **222**, 20, (1992)

LUMINESCENCE PROCESSES IN $Si_{1-x}Ge_x$/Si HETEROSTRUCTURES GROWN BY CHEMICAL VAPOR DEPOSITION

J.C. STURM, X. XIAO AND Q. MI
Department of Electrical Engineering, Princeton Center for Photonic & Optoelectronic Materials (POEM), Princeton University, Princeton, NJ 08544

L.C. LENCHYSHYN AND M.L.W. THEWALT
Department of Physics, Simon Fraser University, Burnaby, B.C. V5A1S6 Canada

ABSTRACT

Well-resolved band-edge exciton photoluminescence (PL) has been observed in strained $Si_{1-x}Ge_x$ heterostructures grown on Si(100) by rapid thermal chemical vapor deposition. The luminescence is due to shallow-impurity bound excitons at low temperatures (under 20K) and at higher temperatures is due to free excitons or electron-hole plasmas, depending on the pump power. The luminescence can also be electrically pumped, with both the electroluminescence and PL persisting above room temperature in samples with a sufficient bandgap offset. Loss of carrier confinement and subsequent non-radiative recombination outside the $Si_{1-x}Ge_x$ is found to be the reason for reduced PL and EL at high temperature.

I. INTRODUCTION

Strained $Si_{1-x}Ge_x$ layers commensurate on Si(100) substrates have been under intense investigation for nearly a decade for the development of silicon-based heterojunction electronic devices, and more speculatively, light emitting devices. While photoluminescence spectra from such $Si_{1-x}Ge_x$/Si structures and Si_mGe_n short period superlattices have been reported for some time [1-3], the interpretation of these initial results has been controversial [4] because of the broad features, emission energies well below expected bandgaps, and correlation of the emission peaks in some work with those of known dislocation luminescence in Si. Well resolved luminescence features of band-edge exciton recombination has been observed only in the last three years; first in thick strained layers with only 4% Ge (x = 0.04) [5] and then finally in strained layer quantum wells and superlattices with higher amounts of Ge [6].

The samples in this last work (Ref. 6) were grown by Rapid Thermal Chemical Vapor Deposition (RTCVD), not molecular beam epitaxy (MBE) as in all of the previous work. This paper first reviews the basic RTCVD technique, and then focuses on three separate issues: the basic features and mechanisms of the luminescence in such strained $Si_{1-x}Ge_x$/Si heterostructures grown by RTCVD, electroluminescence, and finally the temperature dependence of the photo- and electroluminescence.

II. RAPID THERMAL CHEMICAL VAPOR DEPOSITION

A schematic diagram of the reactor used for RTCVD is shown in Fig. 1. A single four-inch Si wafer is suspended on quartz pins without a susceptor inside a 175-mm diameter quartz tube, outside of which is a bank of tungsten halogen lamps which heat the wafer. Process gases (typically dichlorosilane, germane, diborane and phosphine in a hydrogen carrier) are introduced into one end of the reactor and removed from the other end by a simple mechanical rotary vane pump. The chamber is not ultra-high vacuum (UHV), and no pump down with a

Mat. Res. Soc. Symp. Proc. Vol. 298. ©1993 Materials Research Society

high vacuum pump is done after loading samples. However, due to the use of a load lock to prevent atmospheric contamination when loading samples, $Si_{1-x}Ge_x$ layers with low oxygen concentrations ($< 10^{18}cm^{-3}$) and high lifetime ($\geq 1\mu s$) can be routinely achieved at a growth temperature of 625 °C [7]. Although layers have been grown from 500 °C to 1200 °C, typical growth conditions (used for all work in this paper unless otherwise specified) are 625 °C for $Si_{1-x}Ge_x$ growth and 700 °C for Si. Typical growth rates under these conditions are ~ 100 Å /min. The lack of a susceptor allows fast changes (> 100 K/s) in sample temperature so that the growth temperature of each layer or interface can be optimized. The lack of a susceptor or any other hardware (except for the quartz support pins) also removes possible sources of contamination (e.g. metallic impurities, non-radiative centers, etc.) from the chamber to the maximum degree possible. This is important since the luminescence can easily be quenched by excessive non-radiative recombination. The wafer temperature is monitored *in-situ* during growth with an accuracy of a few K by the measurement of the infrared absorption in the wafer (at 1.3 μm and 1.5 μm), without any adjustable parameters such as emissivity [8]. Further growth details can be found in Ref. 9.

III. PHOTOLUMINESCENCE SPECTRA

Figure 2 shows the typical PL spectra of a single strained 33Å $Si_{0.8}Ge_{0.2}$ quantum well and of a single 500Å $Si_{0.8}Ge_{0.2}$ well (both with ~ 150Å silicon caps) at 2K and 77K. The 2K spectra are qualitatively similar to each other except for a blue shift due to quantum confinement in the narrow QW [10]. They are also similar to those observed by Weber and Alonso in their study of bulk (unstrained) $Si_{1-x}Ge_x$ alloys [11], which allows straightforward interpretation of the features. The highest energy feature results from no-phonon (NP) recombination mediated by the alloy randomness. That the feature exists similarly in both the narrow and wide wells and at 2K and 77K supports the hypothesis that this feature is not due to spatial confinement or low temperature localization effects but is indeed an intrinsic feature of the alloy. The lower energy features are phonon replicas, i.e. from transitions assisted by the emission of momentum-conserving transverse acoustic (TA) and transverse optical (TO) phonons. In the

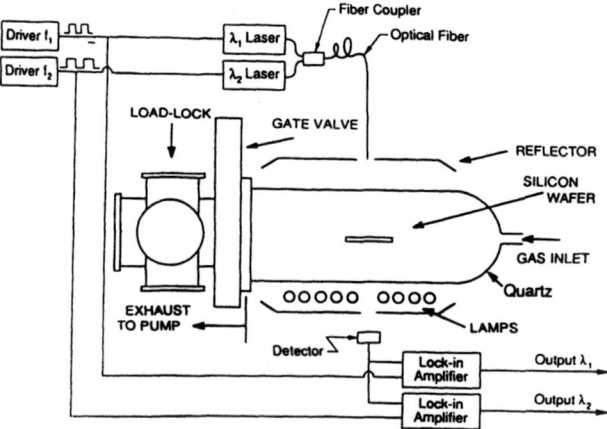

Fig. 1: Schematic cross-section of the RTCVD reactor with temperature measurement by infrared transmission.

33Å well, the narrow linewidth allows one to observe the splitting of the TO replica into various local vibrational modes (Si-Si, Si-Ge, Ge-Ge) representing the different nearest neighbor interactions. From the relative strength of the local modes one can infer the sample composition as shown in bulk material [11], although this ratio is modified in thin QW's and superlattices [6].

On the basis of its temperature dependence, excitation spectroscopy, and lifetime, luminescence at 2K is attributed to excitons bound to a shallow impurity. The background doping of these samples is typically $\sim 10^{16} \mathrm{cm}^{-3}$ and may include B or P depending on the reactor history. At higher temperatures (> 20K), the PL is due to free excitons at low pump powers (as seen in Fig. 2a) and to an electron-hole plasma at higher pump powers [12]. The characteristic feature of this electron-hole plasma is a broadening of the lineshape (especially on the low energy side) as the quasi-fermi levels move into the conduction and valence bands. Samples with similar PL at 77K have also been grown at a temperature of 550°C. This indicates that growth temperatures over 600°C are not required for observing strong band-edge luminescence features.

IV. ELECTROLUMINESCENCE

In this section electroluminescence (EL) is demonstrated by incorporating $Si_{1-x}Ge_x$ QW's in a lightly doped region between n^+ and p^+ Si layers, which inject electrons and holes respectively in forward bias. In previous EL work in $Si_{1-x}Ge_x$ structures, light emission in one case was reported at 4K in samples grown by MBE, but the emission was well below the bandgap and of uncertain origin [13]. In CVD samples with $x = 0.2$ QW, clear band-edge EL was seen, but it decreased sharply above 150K and was virtually extinct by 200K [14]. In this work we have grown a n^+-i-p^+ structure with ten $Si_{0.65}Ge_{0.35}$ QW's of width ~ 50Å in the i-region. 60 μm x 60 μm diodes were fabricated by simple mesa etching with aluminum contacts. Light was observed through a window in the top aluminum contact.

Figure 3 shows the 4K and 77K PL on this sample (from a piece not processed into diodes) as well as the EL spectrum ($I = 10$ μA) with a heat sink temperature of 80K. At 4K, the resolved NP and TO peaks show clear evidence of the band-edge exciton recombination described earlier. The peak NP energy of 890 meV is somewhat higher than that expected for a bound exciton in strained $x = 0.35$ (870 meV) [15], but this difference is within the range of expected quantum confinement effects and uncertainty in sample parameters. Although thermally broadened at 77K, the spectra are qualitatively similar, indicating a band-edge recombination mechanism (although no longer bound exciton). The magnitude of the blue shift (~ 30 meV) is not well understood: ~ 15 meV can be understood as due to the BE to FE transition and the band-filling effects described earlier; the remainder of the shift may be due to unintentional differences in the ten QW's and different wells dominating at different temperatures.

At a heat sink temperature of 80K, the 10 mA (400 Hz modulation, 50% duty cycle) EL (Fig. 3) is qualitatively similar to the 77K PL, although broader, presumably due to poor thermal contact between the sample and heat sink and consequently higher sample temperature. Therefore we infer that the EL mechanism also results from band-edge carrier recombination. At a heat sink temperature of 300K, the EL was still clearly observable with a peak at ~ 930 meV (1.3 μm), corresponding to the NP recombination in the SiGe (Fig. 4). Some emission from the TO replica of the cladding Si layers was also evident (which was much weaker at lower temperatures), but this was estimated to make up less than 10%

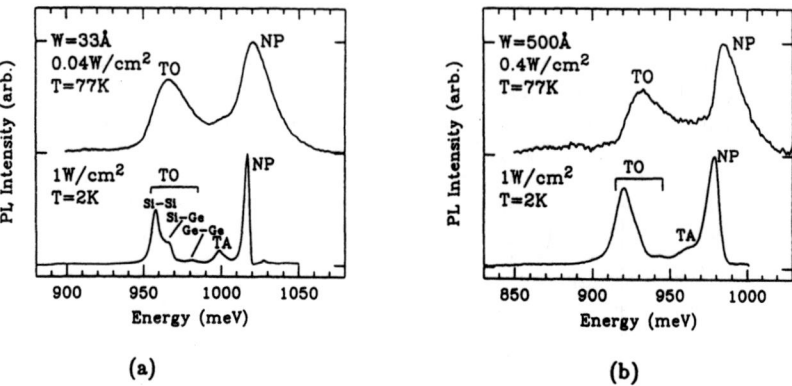

Fig. 2: PL spectra of Si/strained $Si_{1-x}Ge_x$/Si potential wells of width (a) 33Å and (b) 500Å at both 2K and 77K.

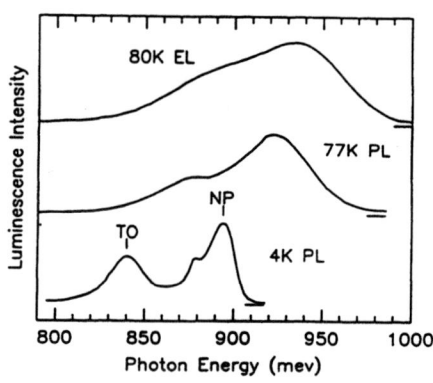

Fig. 3: PL spectra of the EL sample before processing at 4K and 77K, and the EL spectrum with 10 mA drive current and heat sink temperature of 80K.

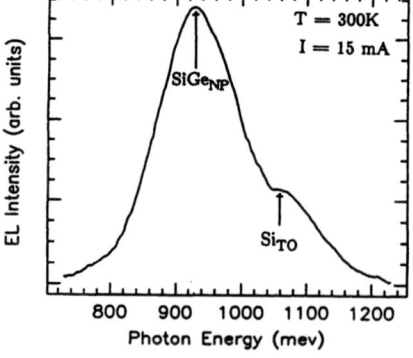

Fig. 4: EL spectra with a drive current of 15 mA at a heat sink temperature of 300K.

of the total amount of emitted light. The peak EL intensity increased linearly with drive current up to 60 mA ($\sim$ 1500 A/cm^2 assuming a uniform current distribution) above an extrapolated threshold of $\sim$ 10 mA (250 A/cm^2), and was sublinear at lower currents (Fig. 5). The weaker emission efficiency at lower drive currents is thought to be due to parasitic space-charge region recombination at defects (such as the mesa sidewalls). At 60 mA, the estimated internal quantum efficiency (after correcting the external signal for window area, solid angle, etc.) had a lower limit of 2 x 10^{-4} [16]. This number is considered a lower limit because of the considerable lateral resistance of the top p$^+$ layer, so that the current density was probably much higher under the contact area than under the window area.

PL and EL were also studied from a single 10-Å pure Ge layer (grown at 625°C) sandwiched between silicon cladding. The microstructure of the Ge layer was not explicitly observed by TEM, etc. The 10Å thickness was estimated from the measured growth rate of Ge from the growth of thick (eg. > 1000Å) Ge layers in other samples, and the Ge layer may be "islanded" and not uniform in thickness. Figure 6 shows the PL (4K and 77K) and EL (90 mA, 80K and 300K heat sink) of such structures. Whereas the room temperature EL peak of the Si$_{0.65}$Ge$_{0.35}$ QW structure was at 1.3 μm, the room temperature EL peak of the pure Ge structure was at 1.5 μm. The peak intensity at 300K increased linearly above a threshold current density of 25 A/cm^2, but the efficiency at higher drive currents was only $\sim$ 10% that of the 1.3 μm emitter. The physical origin of the EL and PL is not clear, however, due to the very broad spectrum ($\sim$ 100 meV peak) at 4K. It is possible that the origin of the luminescence in this sample is dislocations or other defects and not band-edge carriers.

V. TEMPERATURE DEPENDENCE

Except for the BE to FE transition described in Ref. 6, there is little change in the photoluminescence of most of our Si$_{1-x}$Ge$_x$ samples from 4K to 77K. Most of the decay in intensity occurs well above 77K. Fig. 7 shows the evolution of the photoluminescence Ar$^+$-ion excitation ($\sim$ 10 W/cm^2), with increasing temperature above 77K of a single quantum well for both x = 0.2 and x = 0.35. Figure 8 shows the peak intensity of the NP-line in each sample vs temperature. While the x = 0.2 PL decays sharply above 120K and is barely observable at 174K, the x = 0.35 does not decay until temperatures over 200K and is still observable at room temperature. To the best knowledge of the authors, these are the highest temperatures for which PL has been observed in such structures grown by any technique.

A simple quantitative model is now developed to explain what physical mechanism is controlling the decay of PL of higher temperatures. The PL efficiency depends on 3 factors:

$$\eta = \frac{\tau_{\text{non-rad}}}{\tau_{\text{rad}}} \cdot f_{\text{SiGe}} \tag{1}$$

where η is the internal PL efficiency, $\tau_{\text{non-rad}}$ is the non-radiative lifetime

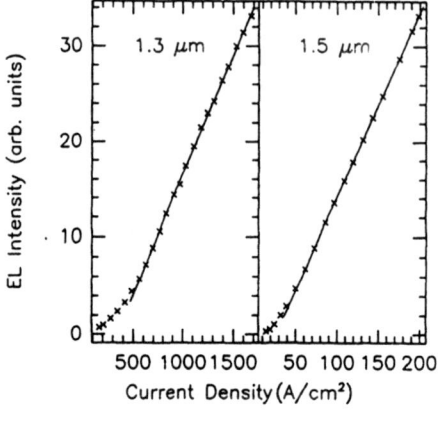

Fig. 5: Peak EL intensity vs. current density (assuming uniform current distribution) of the (a) $Si_{0.65}Ge_{0.35}$ strained QW (1.3 μm) and (b) 10Å pure Ge (1.5 μm) structures. (The vertical axis for the 1.5 μm LED is expanded by ~ 100 X relative to that for the 1.3 μm LED).

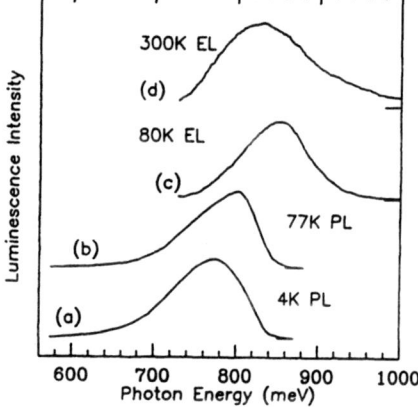

Fig 6. PL (4K and 77K) and EL (90 mA drive current, 80K and 300K heatsink) for the 10-Å Ge layer.

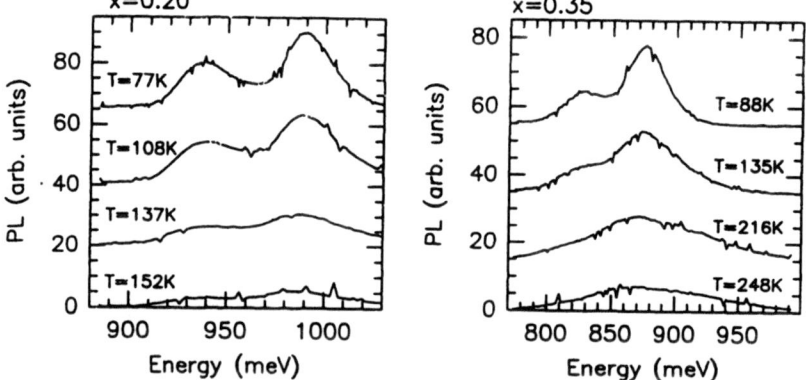

Fig. 7: PL spectra for various temperatures for a single $Si/Si_{1-x}Ge_x/Si$ quantum well with (a) x = 0.2 and (b) x = 0.35.

(assumed much lower than the radiative lifetime), τ_{rad} is the radiative lifetime, and f_{SiGe} is the fraction of carriers in the $Si_{1-x}Ge_x$ well. The radiative lifetime is due to the NP process (due to alloy scattering) and phonon-assisted transitions (predominately TO). For temperatures of 300K or less, these rates are both expected to depend little on temperature because the phonon energies are relatively large (e.g. $\sim$ 670K for the Si TO). At temperatures above 20K, at which carriers are mobile and not localized as bound excitons or otherwise, the dominant nonradiative recombination mechanism is recombination at deep levels, so that the non-radiative lifetime can be described as

$$\tau_{nonrad} = (N_T \, \sigma \, v_{th})^{-1} \qquad (2)$$

where N_T is the density of levels within the bandgap, σ is their cross-section, and v_{th} the carrier thermal velocity. This lifetime is therefore expected to have only a weak ($\sim T^{-\frac{1}{2}}$) temperature dependence. This apparently leaves f_{SiGe} as the only term which might depend exponentially on the temperature and explain our high temperature luminescence decay.

The fraction of carriers in the SiGe (f_{SiGe}) first depends on the transport of the photo-generated carriers from the Si substrate to the QW. The QW's are within the top 0.05 μm of the sample, while the absorption depth of the pump laser is $\sim 1\mu$m. We have typically observed at 2K that the TO-PL from the Si-substrate is comparable to or stronger than that of the $Si_{1-x}Ge_x$, indicating that only $\leq$ 50% of the generated carriers are collected into the QW before luminescing. By 77K however, the TO-PL from the Si is less than 5% that of the $Si_{1-x}Ge_x$, indicating that nearly all of the carriers are collected by the $Si_{1-x}Ge_x$ [17]. Therefore we can assume that at temperatures over 77K, the redistribution of carriers occurs faster than the luminescence, and that an approximate quasi-equilibrium distribution of carriers is established.

Assuming an equilibrium distribution of carriers (not limited by the transport of carriers to the well) one can describe the carrier populations over the regions of interest by flat quasi-fermi levels and a thermal distribution. Since the valence band offset is much larger than that for holes for strained $Si_{1-x}Ge_x$ on Si, the bandgap offset is most effective on holes, which in turn will attract electrons to the $Si_{1-x}Ge_x$. As a first approximation, the fraction of carriers in the $Si_{1-x}Ge_x$ can then be expressed as

$$f_{SiGe} = \frac{W_{SiGe}}{W_{SiGe} + W_{Si} \, e^{-\Delta E_v/kT}} \qquad (3)$$

where W_{SiGe} is the width of the SiGe, W_{Si} is the width of the Si region over which the carriers are distributed, and ΔE_v is the valence band offset. The above neglects any effects of band-bending and also assumes all carrier densities are non-degenerate. For the samples of Fig. 7, $W_{SiGe} = 100\mathring{A}$, and W_{Si} can be approximated by a minority carrier diffusion length (estimated at $\sqrt{D\tau} = 10cm^2/s \cdot 10^{-6}s = 3 \times 10^{-3}cm$). For x = 0.2, the valence band offset is $\sim$ 160 meV. At 100K and 200K one would then predict $f_{SiGe} \simeq$ 1.00 and 0.91, respectively. Clearly, this does not explain the large drop in the $Si_{0.8}Ge_{0.2}$ PL (down by a factor of 100 at 200K). A similar result is found for the x = 0.35 sample.

This inconsistency of the simple model with the data can be resolved by more closely examining the non-radiative lifetime, specifically if one assumes a substantially lower effective lifetime in the Si regions compared to the $Si_{1-x}Ge_x$ layers. This might not result from bulk effects, but could more likely result from a high rate of recombination at the top Si-surface or at the original substrate interface ($\sim 1\ \mu m$ beneath the QW). In this case the overall non-radiative recombination rate for the entire sample can be modelled by an average weighted lifetime, $\tau_{non-rad,avg}$:

$$\tau_{non-rad,avg}^{-1} = \frac{f_{Si}}{\tau_{non-rad,Si}} + \frac{f_{SiGe}}{\tau_{non-rad,SiGe}} \tag{4}$$

$$= \frac{W_{SiGe}}{\tau_{non-rad,SiGe}} \left(\frac{1 + \dfrac{\tau_{non-rad,SiGe} \cdot W_{Si}}{\tau_{non-rad,Si} \cdot W_{SiGe}} e^{-\Delta E_v/kT}}{W_{SiGe} + W_{Si}\, e^{-\Delta E_v/kT}} \right)$$

where the f_i and $\tau_{non-rad,i}$ are the fraction of carriers and effective lifetime in layer i. Combining this with equations (1) and (3) gives a dominant temperature dependence for the PL as

$$\eta \propto \frac{1}{1 + C \cdot e^{-\Delta E_v/kT}} \tag{5}$$

where

$$C \equiv \frac{\tau_{non-rad,SiGe} \cdot W_{Si}}{\tau_{non-rad,Si} \cdot W_{SiGe}} \tag{6}$$

This expression was fit to the data of the x = 0.2 sample in Fig. 8 using ΔE_v = 180meV and C = 5×10^6, with a reasonably good agreement as shown. That C is much larger than the expected $W_{Si}/W_{SiGe} \simeq 30\mu m/0.01\mu m \simeq 3000$ implies that the effective non-radiative lifetime in the Si is indeed much lower than that in the SiGe. Using the same C, the x = 0.35 data was fit using $\Delta E_v \simeq 310$ meV and again reasonable agreement was achieved (Fig. 7). That the fitted ΔE_v are indeed close to the known ΔE_v values (~180 meV, 270 meV respectively [15]) indicates that the valence band offset is the crucial parameter for the temperature dependence of luminescence. The conclusion of this modeling of the temperature dependence of the PL is that the luminescence decreases at high temperature because of the low effective lifetime for carriers outside the quantum well. Only a relatively few number of carriers are required to be outside of the quantum well to cause a substantial reduction in the luminescence efficiency.

The temperature dependence of the peak SiGe NP electroluminescence signal of our $Si_{0.65}Ge_{0.35}$ QW LED and that of the $Si_{0.8}Ge_{0.2}$ LED of Ref. 14 are shown in Fig. 9 along with the x = 0.2 and x = 0.35 modelling results of Fig. 8. The temperature dependence of the EL is qualitatively similar to that of the PL: sharp decay at high temperatures and higher x (more Ge) resulting in a stronger signal at high temperatures. (The significance of the pronounced feature in the x = 0.35 EL at 190K is not known.) It is clear, however, that the EL does not decay as fast at high temperature as the PL for the same x. This may be due to an extra confining effect of the p-n junction on the injected carriers. Extra confinement could suppress the size of the W_{Si} region or could prevent carriers from reaching the top Si surface. Quantitative modelling to support these effects has not been done however.

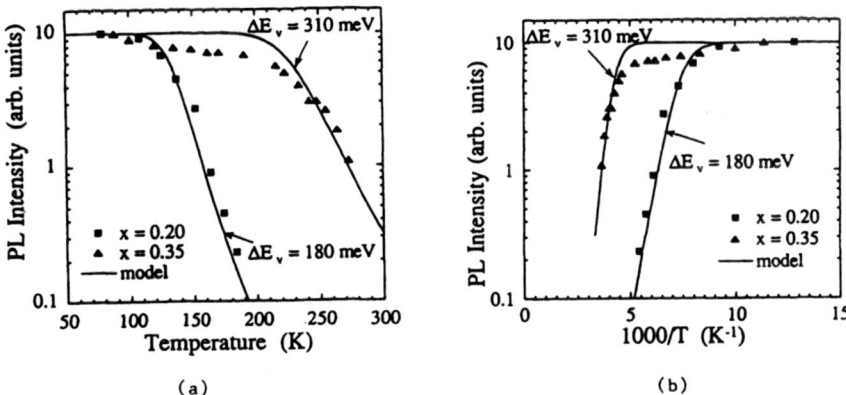

Fig. 8: Peak no phonon photoluminescence intensities vs. (a) temperature and (b) vs. inverse temperature for the x = 0.2 and x = 0.35 QW samples of Fig. 7 and fitted model results are described in the text.

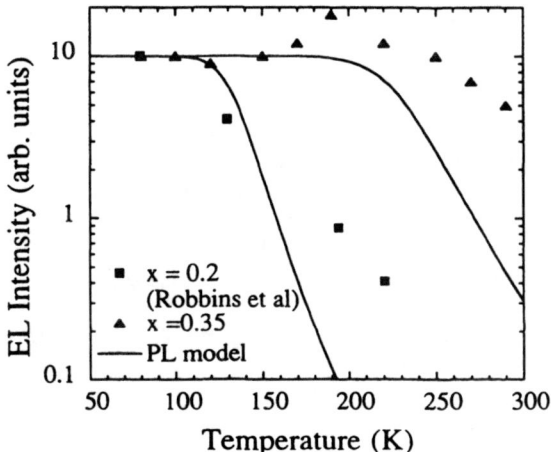

Fig. 9: Peak electroluminescence intensity vs. temperature for the $Si_{0.65}Ge_{0.35}$ QW's and the $Si_{0.8}Ge_{0.2}$ QW's (of Ref. 14), along with the model results for PL vs. temperature of Fig. 8.

VI. SUMMARY

Well resolved exciton luminescence has been observed in Si/strained $Si_{1-x}Ge_x$/Si quantum well structures grown by Rapid Thermal Chemical Vapor Deposition. Key features are a no-phonon line due to alloy randomness and a threefold splitting of the TO replica. The luminescence process can be pumped electrically as well as optically, with room temperature $1.3\mu m$ electroluminescence from the no-phonon process in $Si_{0.65}Ge_{0.35}$ quantum wells. At high temperature the luminescence decreases exponentially with an activation energy close to that of the valence band offset. This decay is thought to be due to excessive recombination in the silicon cladding layers. For $x = 0.35$, both PL and EL are visible at room temperature, but not at $x = 0.2$.

VI. ACKNOWLEDGEMENT

The authors would like to thank P.V. Schwartz, C.W. Liu, Z. Matutinovic-Krstelj, A. St. Amour, and V. Venkataraman for experimental assistance and helpful discussions. The support of NSF, ONR, and the New Jersey Commission on Science and Technology (for Princeton) and NSERC (for Simon Fraser University) is gratefully acknowledged.

VII. REFERENCES

1. J.P. Noel, N.L. Rowell, D.C. Houghton and D.D. Perovic, Appl. Phys. Lett. **57**, 1037 (1990).

2. H. Okumura, K. Miki, S. Misawa, K. Sakamoto, T. Sakamoto and S. Yoshida, Jpn. J. Appl. Phys. **28**, L1893 (1989).

3. R. Zachai, K. Eberl, G. Abstreiter, E. Kasper and H. Kibbel, Phys. Rev. Lett. **64**, 1055 (1990).

4. U. Schmid, N.E. Christensen and M. Cardona, Phys. Rev. Lett. **65**, 2610 (1990).

5. K. Terashima, M. Tajima and T. Tatsumi, Appl. Phys. Lett. **57**, 1925 (1990).

6. J.C. Sturm, H. Manoharan, L.C. Lenchyshyn, M.L.W. Thewalt, N.L. Rowell, J.P. Neol, and D.C. Houghton, Phys. Rev. Lett. **66**, 1362 (1991).

7. P.V. Schwartz and J.C. Sturm, Appl. Phys. Lett. **57**, 2004 (1990).

8. J.C. Sturm, P.M. Garone and P.V. Schwartz, J. Appl. Phys. **69**, 542 (1991).

9. J.C. Sturm, P.V. Schwartz, E.J. Prinz and H. Manoharan, J. Vac. Sci. Technol. **B9**, 2011 (1992).

10. X. Xiao, C.W. Liu, J.C. Sturm, L.C. Lenchyshyn and M.L.W. Thewalt, Appl. Phys. Lett. **60**, 2135 (1992).

11. J. Weber and M.I. Alonso, Phys. Rev. **B40**, 5683 (1989).

12. X. Xiao, C.W. Liu, J.C. Sturm, L.C. Lenchyshyn and M.L.W. Thewalt, Appl. Phys. Lett. **60**, 1720 (1992).

13. N.L. Rowell, J.P. Noel, D.C. Houghton and M. Buchana, Appl. Phys. Lett. **58**, 957 (1991).

14. D.J. Robbins, P. Calcott, W.Y. Leong, Appl. Phys. Lett. **59**, 1350 (1991).

15. C.G. van de Walle and R.M. Martin, Phys. Rev. B. **34**, 5621 (1986).

16. Q. Mi, X. Xiao, J.C. Sturm, L.C. Lenchyshyn and M.L.W. Thewalt, Appl. Phys. Lett. **60**, 3177 (1992).

17. J.C. Sturm, X. Xiao, P.V. Schwartz, C.W. Liu, L.C. Lenchyshyn and M.L.W. Thewalt, J. Vac. Sci. Technol. **B10**, 1998 (1992).

PHOTOLUMINESCENCE OF THIN SI₁₋ₓGEₓ QUANTUM WELLS

L.C. LENCHYSHYN[1], M.L.W. THEWALT[1], D.C. HOUGHTON[2], J.-P. NOËL[2],
N.L. ROWELL[2], J.C. STURM[3], AND X. XIAO[3]
1 Simon Fraser University, Physics Department, Burnaby, British Columbia, Canada.
2 National Research Council Canada, Ottawa, Ontario, Canada.
3 Princeton University, Electrical Engineering Department, Princeton, NJ.

ABSTRACT

Well-resolved band edge photoluminescence spectra were obtained from SiGe quantum wells of various widths. In addition to the usual shallow bound exciton features, we observed a highly efficient deeper luminescence process, under conditions of low excitation density, in thick SiGe quantum wells. This luminescence band can be attributed to excitons localized by fluctuations in alloy concentration. The binding energy of the localized exciton feature is found to decrease with decreasing well width. In the thinnest quantum well samples only a single luminescence feature is observed at all power levels, while in several other thin quantum well samples having very sharp lines the localized exciton feature appears at higher energy than the bound exciton. Despite these changes in the spectra, the localized exciton luminescence could be identified in all cases by its characteristic intensity saturation at low excitation power density, as well as its slow decay time (~ 1 ms). The mechanism behind the changes in the localized exciton luminescence may originate from limiting the exciton motion to two dimensions in thin wells, which at low temperatures would hinder migration to the lowest energy alloy fluctuation centers.

INTRODUCTION

Since the initial observation of well-resolved band edge photoluminescence (PL) from strained SiGe on Si just a few years ago,[1,2] many other research groups have been successful in observing similar features in SiGe grown by various techniques.[3-9] While such luminescence provides important information about the SiGe band gap and crystal quality, as well as quantum confinement effects, the exact nature of the luminescence itself has not been studied in great detail as yet. One expects that the luminescence at low temperatures should be dominated by transitions due to free excitons or excitons bound to shallow impurities, such as phosphorus or boron,since this is the case for both pure Si and pure Ge. However, SiGe is in fact quite different from Si or Ge in that it is an alloy. This is manifested in a broadening of the PL peaks (typical linewidths are a few meV) because of the random variation in band gap in different regions of the crystal. The usual momentum conservation rules, which suppress the no-phonon (NP) transitions in indirect band gap materials, are relaxed so that relatively strong NP peaks are observed for the case of the SiGe alloy.

Besides the expected bound exciton (BE) PL features, we have observed a PL band at low excitation density in thick SiGe quantum well samples that is shifted to lower energy relative to the BE by roughly 15 meV.[10] This new PL band is unique to alloys and can be attributed to excitons localized by random fluctuations in Ge concentration. This process is particularly interesting since the localization reduces the chance for decay by non-radiative channels and thereby leads to an unusually high external quantum efficiency of > 10 %.

We have recently turned our attention to PL mechanisms in thin SiGe quantum wells.[11]

Although spectrally resolved BE and localized exciton (LE) peaks were not always observed, on close examination the PL in fact appeared to be consistent with the PL processes occuring in the thicker layers. A single sharp peak (and its phonon replicas), observed in several samples with very thin SiGe quantum wells, was found to behave at low excitation power density in a manner indicative of the LE process. In addition, we have very recently obtained some new results for other thin SiGe quantum wells which in fact show two sharp NP peaks. Like the thick wells, we can interpret these two peaks as a BE and an LE transition. However, the LE peak shifts to shallower energies relative to the BE, possibly due to the more limited 2D movement of the excitons in these quantum wells. In this paper we briefly describe the LE PL for thick SiGe layers and then study how this PL appears to be modified by the restriction of the excitons to two dimensions in the case of the thin SiGe quantum wells.

RESULTS

The SiGe samples were grown either by Rapid Thermal Chemical Vapour Deposition (RTCVD) or by Molecular Beam Epitaxy (MBE), as described previously.[5,12] The nominal parameters (ie. Ge fraction and well thickness) are used here to describe the samples. The luminescence was excited using an Ar ion laser and the PL spectra were measured using a Bomem DA8 Fourier transform interferometer with an InGaAs detector. The time resolved data was obtained by pulsing the Ar laser with an acousto-optic modulator and detecting the PL with a Varian (VPM159A3) photomultiplier tube operated in photon counting mode and coupled to a 3/4m double spectrometer. The samples were typically immersed in liquid He with temperatures between 1.7 and 4.2 K. For temperatures above 4.2 K a flowing He gas Varitemp dewar was used.

The LE and BE PL features are shown in Fig. 1a) for a thick (8.3 nm) CVD $Si_{0.8}Ge_{0.2}$ quantum well by the solid and dashed curves, respectively. In each spectrum we see the usual NP transition and the TA and TO phonon replicas to lower energy. The excitation power, I_0, was chosen to minimize the broadening of the features at high excitation and was of the order of a few W cm^{-2}. The BE PL dominates at high excitation power density (I_0), however these features drop linearly with excitation density until only the LE band remains at low power ($10^{-3} I_0$). At intermediate power densities both

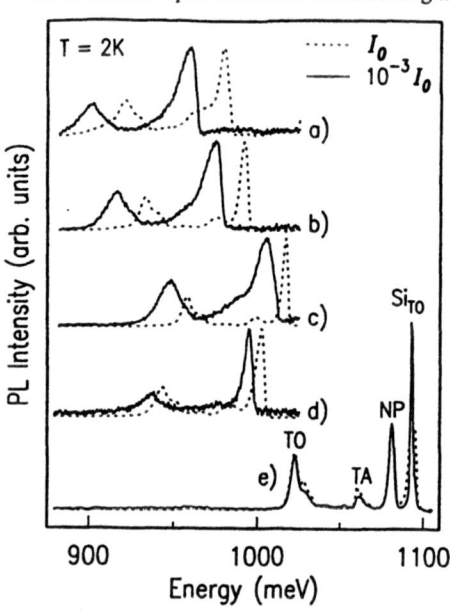

Figure 1 PL spectra for a) 8.3 nm CVD $Si_{0.8}Ge_{0.2}$, b) 5.8 nm CVD $Si_{0.8}Ge_{0.2}$, c) 3.3 nm CVD $Si_{0.8}Ge_{0.2}$, d) 1.5 nm CVD $Si_{0.65}Ge_{0.35}$, and e) 1.2 nm MBE $Si_{0.62}Ge_{0.38}$. Taken under conditions of high and low excitation power density, where I_0 is of the order of a few W cm^{-2}.

processes are resolved (see Lenchyshyn *et al*.[10]). The LE PL has been found to saturate at extremely low power densities (~ 100 μW cm⁻²), consistent with the expectation that there are a limited number of regions rich enough in Ge to act as such deep localization centers. The exponential tail to low energy reflects the exponential dependence on energy of the density of states into the forbidden gap, which follows from an Anderson model of the alloy fluctuations.[13,14] At low temperatures excitons tunnel or hop so as to fill the lowest energy localization centers first. The exponential LE PL lineshape and low power saturation behaviour have been observed in other alloy semiconductors.[13,14] A band has also been observed by other workers in their MBE SiGe[15] which, because of its appearance only at low power density and spectral position relative to the BE, appears to be LE luminescence.

The remaining spectra in Fig. 1 (b-e) show the PL for high and low excitation power density for SiGe wells of b) 5.8 nm CVD $Si_{0.8}Ge_{0.2}$, c) 3.3 nm CVD $Si_{0.8}Ge_{0.2}$, d) 1.5 nm CVD $Si_{0.65}Ge_{0.35}$, and e) 1.2 nm MBE $Si_{0.62}Ge_{0.38}$. The peaks shift as expected with alloy fraction and quantum confinement effects. However, the spectra also show a decrease in separation between the LE and BE features with decreasing SiGe well width from 20 meV for the 8.3 nm well to 7 meV in the 1.5 nm well. This is consistent with our LE model if we consider that in the thinner layers the excitons cannot move freely in the growth direction and therefore are less likely to reach the deepest alloy localization centers. Instead they become trapped on the relatively more abundant, shallower alloy fluctuations. In the thinnest sample there is essentially no change between the high and low excitation spectra, indicating either approximately identical LE and BE binding energies or the absence of one of the processes. As outlined below, the PL for this sample is in fact found to behave at low power densities in a manner in agreement with an LE process

Evidence of the LE nature of the PL in the thin quantum wells is provided by studying the dependence of the PL intensity on excitation power density. Fig. 2 shows that in the case of thick (10 nm) CVD $Si_{0.75}Ge_{0.25}$ quantum wells the LE PL (+) begins saturating with only 10 μW cm⁻² excitation, while the SiGe BE PL (■) varies as expected with a linear dependence on excitation power density. The solid curve is the power dependence of the

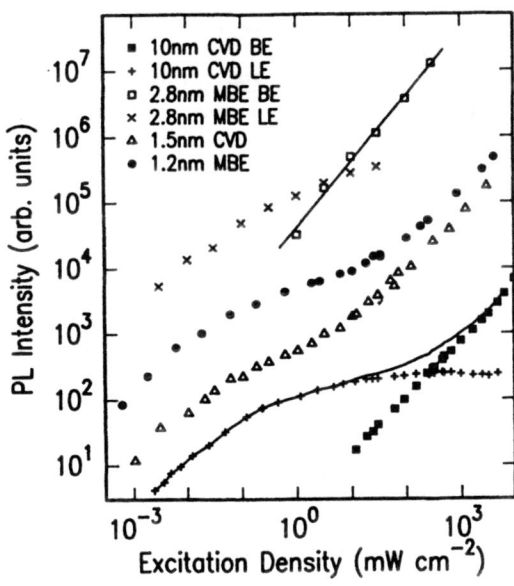

Figure 2 Dependence of PL intensity on excitation density for CVD and MBE SiGe quantum well samples at 2K. The BE PL depends linearly on power density, as indicated by the straight line fit for the 2.8nm MBE sample. The solid curve is the sum of the LE and BE signals for the 10 nm CVD sample. This curve is similar to the power dependence in the very thin (1.5nm and 1.2nm) wells. The curves do not reflect the relative intensities between different samples.

overall SiGe PL in the thick well sample, ie. the sum of the LE and BE signals. This solid curve is qualitatively in agreement with the intensity dependence of the single peak observed for the very thin (Δ 1.5 nm and ● 1.2 nm) SiGe quantum wells. The characteristic saturation at very low power levels suggests that the luminescence at low excitation density in the thin wells corresponds to the LE process seen in the thicker wells. The recovery of the linear dependence of the thin well PL intensity observed at high excitation could be due to a less efficient mechanism becoming dominant, possibly a BE or biexciton process.

Further support for the assignment of the low excitation PL in the thin quantum wells to an LE process is provided in Fig. 3 by the PL decay curves. Under very low excitation ($10^{-3} I_0$) the PL decay is a single exponential corresponding to a lifetime of 750 μs. This very slow decay is in

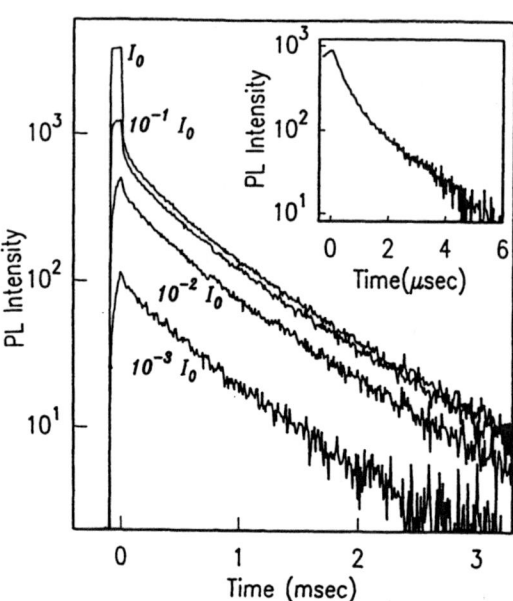

Figure 3 Time decay of the PL from the 1.2 nm MBE sample. As the excitation is increased by a factor of 1000 the contribution from the slow component ($\tau \sim 750 \mu s$) saturates, while that from the fast component increases. The inset shows the non-exponential dependence ($\tau \sim 0.35 \mu s$ to $\tau \sim 1.5 \mu s$) of the fast component at high excitation density on an expanded time scale. The four decay curves have **not** been shifted vertically.

agreement with observations for the LE process in the thick wells.[10] A long lifetime, corresponding to the free exciton <u>radiative</u> decay, is expected for the LE since the non-radiative (ie. fast) channels are eliminated. Such long lifetimes are difficult to reconcile with any other process. As the excitation power is increased by several orders of magnitude this slow component saturates, while a faster decay process becomes more obvious. As shown on an expanded time scale in the inset, the fast component is non-exponential with lifetimes of 0.35 to 1.5 μs. Again the origin for the PL observed at high excitation is not clear, however the fast decay times are consistent with processes in which non-radiative Auger decay dominates, and so would include BE or biexciton recombination. Thus even though the PL spectrum does not appear to change, the near gap PL of the very thin LE quantum well samples is dominated by the long lifetime, highly efficient process at low excitation and by a much less efficient, ~1μs lifetime process at high excitation.

Fig. 4 shows the PL spectrum of an MBE multiple quantum well sample with unusually small (~ 1.5 meV) linewidths. Luminescence from all three $Si_{0.86}Ge_{0.14}$ wells is observed, with the shifts to higher energy due to quantum confinement as expected given the nominal well widths of 2.8 nm, 4.2 nm, and 8.4 nm. The Si barriers between the wells are 30 nm, so that there should be no coupling between them. The unusual aspect of this luminescence is the sharp doublet for the 2.8 nm and 4.2 nm quantum wells. The peak separation of this doublet varies slightly with excitation power, but is typically ~2.5 meV. The doublet is shown on an expanded scale for the 2.8 nm well in Fig. 5 for excitation power densities of a) $3\times10^{-5} I_0$, b) $3\times10^{-3} I_0$, c) $10^{-2} I_0$, and d) $10^{-1} I_0$. Again, with varying excitation density the spectra evolve from one luminescence peak at low excitation to the other peak at high excitation. However, in this case the <u>higher</u> energy component persists to low power density (curve 5a), suggestive of LE luminescence. This contrasts the situation observed earlier in the thick CVD wells, where the LE peak was always deeper than the BE. However this assignment is consistent with the characteristic LE (x) saturation and linear BE (□) dependence on excitation power shown in Fig. 2. It is also in agreement with PL decay measurements which indicated a fast (0.60 μs) decay for the BE peak, while the LE decayed with a very slow component of 270 μs plus a fast component of 0.26 μs. Fig. 6 shows the decrease in LE intensity with increasing temperature from a) 1.7 K, to b) 2.8 K, c) 4.2 K, d) 6.3 K, e) 8.3 K, and finally f) 15.3 K. The excitons gradually acquire enough thermal energy to escape the very shallow alloy fluctuation potential wells, so that at 4.2 K (curve 6c) the LE peak is

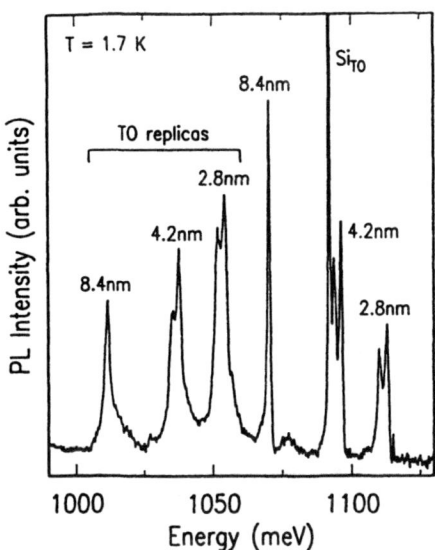

Figure 4 PL spectrum from an MBE sample with $Si_{0.86}Ge_{0.14}$ quantum wells of nominal thicknesses 2.8 nm, 4.2 nm, and 8.4 nm. The 2.8 nm and 4.2 nm quantum well PL shows a doublet structure with very small peak separations of roughly 2.5 meV.

essentially gone and BE PL dominates. An Arrhenius plot of the LE intensity between 1.7 and 4.2 K gives, as expected, a very small binding energy of ~0.4 meV. Above 4.2 K, the BE subsequently become dissociated from the impurities so that free exciton (FE) recombination dominates the spectrum at high temperatures (15.3 K).

In conclusion, we have studied SiGe quantum well samples grown by CVD or MBE techniques and found that the luminescence mechanisms are not limited to the usual FE and BE recombination found in Si and Ge PL spectra. Instead, the alloy nature of the SiGe plays an important role, with the observation of unique luminescence features at low excitation density that are consistent with a model based on excitons localized by fluctuations in alloy concentration. Clearly more work is warranted to better understand these processes and especially to determine the PL mechanism responsible for the luminescence in the thin quantum wells under high excitation conditions. We are currently working on an experiment based on the simultaneous recombination of two excitons with emission of a single photon to gain insight into these high excitation spectra (see Steiner et al., this proceedings).

REFERENCES

1. K. Terashima, M. Tajima, and T. Tatsumi, Appl. Phys. Lett. 57, 1925 (1990).
2. J.C Sturm, H. Manoharan, L.C. Lenchyshyn, M.L.W. Thewalt, N.L. Rowell, J.-P. Noël, and D.C. Houghton, Phys. Rev. Lett. 66, 1362 (1991).
3. E.R. Glaser, T.A. Kennedy, D.J. Godbey, P.E. Thompson, K.L. Wang, and C.H. Chern, Phys. Rev. B 47, 1305 (1993).

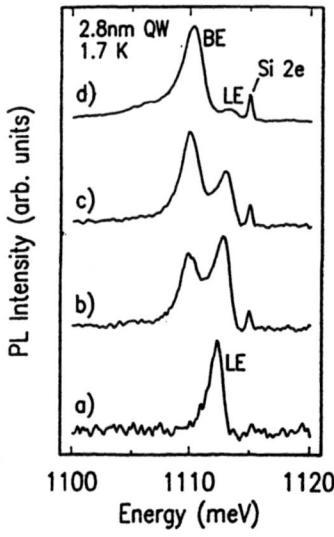

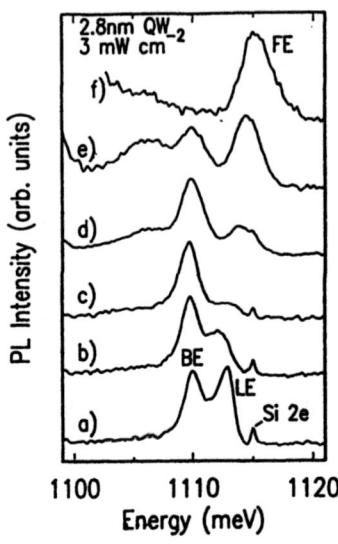

Figure 5 Evolution of the NP peak of the 2.8 nm quantum well PL, from the same sample as in Fig. 4, under increasing excitation power densities of a) 3×10^{-5} I_0, b) 3×10^{-3} I_0, c) 10^{-2} I_0, and d) 10^{-1} I_0.

Figure 6 Evolution of the NP peak of the 2.8 nm quantum well PL from the same sample as in Fig. 4, with increasing sample temperatures of a) 1.7 K, b) 2.8 K, c) 4.2 K, d) 6.3 K, e) 8.3 K, and f) 15.3 K.

4. V. Arbet-Engels, J.M.G. Tijero, A. Manissadjian, K.L. Wang, and V. Higgs, Appl. Phys. Lett., **61**, 2586 (1992).

5. J.-P. Noël, N.L. Rowell, D.C. Houghton, A. Wang, and D.D. Perovic, Appl. Phys. Lett. **61**, 690 (1992).

6. D.J. Robbins, L.T. Canham, S.J. Barnett, A.D. Pitt, and P. Calcott, J. Appl. Phys. **71**, 1407, (1992).

7. J. Spitzer, K. Thonke, R. Sauer, H. Kibbel, H.-J. Herzog, and E. Kasper, Appl. Phys. Lett. **60**, 1729 (1992).

8. T.D. Steiner, R.L. Hengehold, Y.K. Yeo, D.J. Godbey, P.E. Thompson, and G.S. Pomrenke, J. Vac. Sci. Technol. **B10**, 924 (1992).

9. N. Usami, S. Fukatsu, and Y. Shiraki, Appl. Phys. Lett. **61**, 1706 (1992).

10. L.C. Lenchyshyn, M.L.W. Thewalt, J.C. Sturm, P.V. Schwartz, E.J. Prinz, N.L. Rowell, J.-P. Noël, and D.C. Houghton, Appl. Phys. Lett. **60**, 3174 (1992).

11. L.C. Lenchyshyn, M.L.W. Thewalt, D.C. Houghton, J.-P. Noël, N.L. Rowell, J.C. Sturm, and X. Xiao, submitted to Phys. Rev. B.

12. J.C. Sturm, P.V. Schwartz, E.J. Prinz, and H. Manoharan, J. Vac. Sci. Technol. **B9**, 2011 (1991).

13. A. Fried, A. Ron, and E. Cohen, Phys. Rev. **B39**, 5913 (1989).

14. Shui T. Lai and M.V. Klein, Phys. Rev. **B29**, 3217, (1984).

15. J. Denzel, K. Thonke, J. Spitzer, R. Sauer, H. Kibbel, H.-J Herzog, and E. Kasper, to be published in Thin Solid Films.

PHOTOLUMINESCENCE INVESTIGATIONS OF HIGH PURITY MBE-GROWN Si$_{1-x}$Ge$_x$/Si SINGLE QUANTUM WELLS

M.Wachter[1], F.Schäffler[2], K.Thonke[1], R.Sauer[1], H.-J.Herzog[2], and E.Kasper[2]
[1] Dept. of Semicond. Physics, University of Ulm, P.O. Box 4066, 7900 Ulm, Germany
[2] Daimler Benz Research Center, Wilhelm-Runge-Strasse 11, 7900 Ulm, Germany

ABSTRACT

Si$_{1-x}$Ge$_x$/Si(100) single quantum well samples were grown by MBE with x varying from 18.5% to 36%, well widths L$_z$ from 1.1 nm to 175 nm, and growth temperatures T$_G$ in the range from 350°C to 750°C. We studied the photoluminescence (PL) properties in detail using Fourier transform spectroscopy. In most of the samples, the dominant and phonon-resolved SiGe band edge PL exhibits exclusively free exciton luminescence even at sample temperatures as low as 2 K. From the absence of bound excitons we conclude that the background concentration of shallow impurities is low in our SiGe layers. Variation of the growth temperature of the Si cap layer strongly influences the efficiency of the SiGe band edge PL at higher temperatures, whereas the PL efficiency is less affected at T=4 K. This indicates a reduction of the concentration of non-radiative defects in the cap layer when it is grown at higher temperatures. A broad PL band 150 meV below the SiGe bandgap, supposed to be typical for MBE-grown material, is weakly observed in only a few samples. The strength of the broad PL band does not unambiguously depend on the strain energy density in the SiGe layer as recently suggested.

INTRODUCTION

Much interest has been focused in the past years on the Si/SiGe material system prompting, e.g., optical studies by photoluminescence (PL) and electroluminescence. This situation emerges from applications for transport devices and potential use in optoelectronics in spite of the indirect bandgaps of Si and Ge. Several groups have reported on well resolved near band edge PL from SiGe single or multiple quantum wells (QW) grown pseudomorphically on Si substrates either by chemical vapor deposition (CVD) techniques[1-3] or by molecular beam epitaxy (MBE)[4-10]. In addition to this SiGe band edge luminescence a broad PL band approximately 150 meV below the SiGe bandgap is often observed in MBE-grown SiGe quantum wells. The origin of this broad luminescence is still discussed in the literature. Noël et al. recently ascribed contrast features which they observed in TEM micrographs of their MBE-grown SiGe QWs to interstitial-type Ge-rich platelets. They supposed that these defects are responsible for the broad PL band and that the defect formation is typical for MBE growth in contrast to CVD layer deposition.[11]

GROWTH AND EXPERIMENTAL DETAILS

Relevant data of the single QWs studied here are given in Table I. The samples were grown at the Daimler Benz research center in an Atomika MBE-system by electron beam evaporation for both Si and Ge.[12] The sample structures consist of p⁻ type (boron doped, $\rho \geq 1000$ Ωm) (100)-oriented Si substrates and nominally undoped epitaxial layers. The layers consist of a 100 nm thick Si buffer layer, the pseudomorphically deposited Si$_{1-x}$Ge$_x$ quantum well layer with Ge content x (18.5% $\leq$ x $\leq$ 36%), well width L$_z$ (1.1 nm $\leq$ L$_z$ $\leq$ 175 nm), and growth temperature T$_G$ (350°C $\leq$ T$_G$ $\leq$ 750°C), and a Si cap layer about 300 nm thick to reduce non-radiative recombination at the SiGe surface. The growth temperature for the Si cap was usually the same as for the SiGe QW. L_z and x were determined by high resolution x-ray diffraction (HRXRD) and for some samples also by Rutherford backscattering (RBS).

Table I. Ge content x, growth temperature T_G, and SiGe well width L_z

Serie	x(%)	T_G(°C)	L_z(nm)
A	24	350, 450, 550, 750	3.3, 3
B	18.5	600	30, 10, 5.5, 2.1, 1.1
C	24	450	175, 21, 8.5, 4.8, 2.5
D	36	450	20, 13, 6.5, 2.5
E	20.5	450	3.5, 1.8, 1.2

The spectra were measured using either a Bomem DA8 Fourier transform spectrometer or a grating monochromator with a liquid nitrogen cooled Ge detector or an InGaAs diode. The samples were mounted in a strain free manner in a helium bath cryostat, with temperature control from 2 K to 300 K. The 488 nm line of an argon ion laser with excitation power densities on the front surface of the sample between 1 μW/mm^2 and 50 mW/mm^2 was used for optical excitation, with the beam slightly focused onto a spot of about 2 mm^2 size.

RESULTS AND DISCUSSION

Fig. 1 exhibits the PL spectra taken at a sample temperature of 4.2 K and with an excitation density of 5 mW/mm^2 of the series A (see Table I). These samples were grown in a substrate temperature range between 350°C and 750°C. The spectra show the well-known multiple line structure of Si around 1.1 eV and the resolved SiGe band edge luminescence between 1.02 eV and 0.95 eV consisting of a strong no-phonon (NP) line -allowed as a consequence of alloy scattering- and four momentum conserving phonon replicas. Observed are the transverse acoustic (TA) phonon replica and the three transverse optical (TO) phonon replicas, which are associated with Si-Si, Si-Ge and Ge-Ge vibrations, as previously identified.[13] Nearly all samples exhibit only free exciton NP lines from the SiGe layer, indicating a very low background concentration of shallow impurities. In the energy range below the TO_{Si-Si} replica weak multiple phonon replicas (e.g. the $TO_{Si-Si}+O^{\Gamma}$ replica) are observed. The broad luminescence band to be discussed later is completely absent in the spectra of these samples. The linewidths and the absolute intensities of the SiGe band edge PL increase with higher growth temperatures. The sample grown at 350°C shows the sharpest NP line of this set of samples, however its intensity is reduced by a factor of five compared to the sample grown at 550°C (cf. Fig. 1). The growth temperature T_G also affects the SiGe PL intensities as a function of sample temperature T. An Arrhenius plot of the integral PL intensity at a laser excitation density of 10 mW/mm^2 is shown in Fig. 2 for the samples in Fig. 1. The sample grown at 550°C and the very similar sample grown at 750°C show only a weak change of their PL intensities in the temperature range below about 60 K, but an abrupt decrease above 77 K. The partial recovery of the intensities in Fig. 2 close to 70 K has been observed in repeated experiments, and was independently reported by Fukatsu et al..[8] In this temperature range, the lines broaden significantly changing shape gradually from exciton-like to more plasma-like. This could be related to an increased radiative transition probability, leading to the recovery peak.

The abrupt decrease is exponential with a thermal activation energy of $E_{th}=(180\pm20)$ meV, close to the expected valence band offset at the given Ge content. Hence, this strong

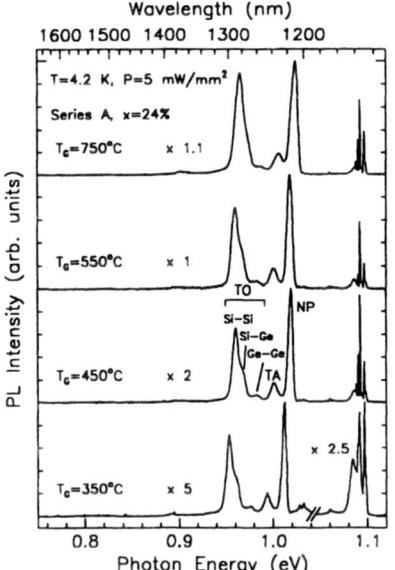

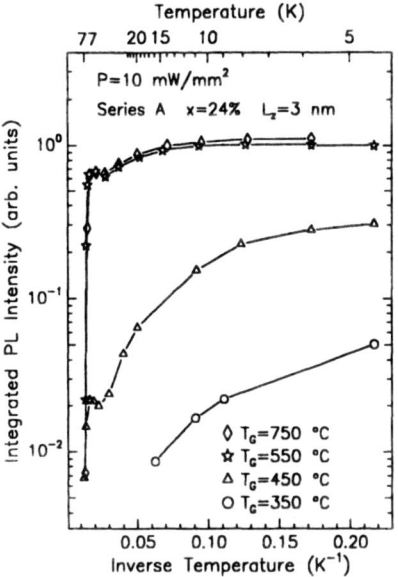

Fig.1. PL spectra of 3 nm thick Si$_{.76}$Ge$_{.24}$ quantum wells grown at different temperatures T$_G$. Indicated are multiplication factors used to obtain the depicted spectra from the measured spectra.

Fig.2. Temperature dependent SiGe band edge PL intensities of the samples in Fig.1 shown as Arrhenius plots. The curves are to scale.

intensity decay might reflect the thermal emission of holes from their confined QW states into the heterobarrier. In the whole temperature range the absolute PL intensities depend on T$_G$. The samples grown at 450°C and 350°C exhibit much lower PL intensities than those grown at 550°C and 750°C. At the same time their PL intensity loss per inverse temperature interval is much larger. These effects are evidently related to a reduced radiative quantum efficiency of the samples grown at low temperature probably through the incorporation of a larger number of non-radiative recombination centers in the QWs and the heterointerfaces. Enhanced defect densities at lower growth temperatures are indeed expected because of a reduced surface mobility of the impinging Si and Ge atoms concomitant with an easier incorporation of impurities.[15]

However, the growth temperatures do not only affect the SiGe QWs but also the Si cap layers. These were usually grown at the same temperatures as the QW in order to avoid annealing effects. The influence of the cap was directly studied in two samples, nominally identical except for the cap layers which were grown at 350°C or 600°C. These samples showed the same tradeoff in their absolute intensities as the two samples in Fig. 2 grown completely at 350°C and 550°C. We consider this strong evidence for the relatively thick Si cap and/or its interface to the QW to be a major source of non-radiative recombination when grown at lower temperatures.

Our samples show NP line positions in the wide range from 870 meV to 1120 meV. The energies are subject to the x-dependent strain and to the L$_z$-dependent quantum confinement (cf. Fig. 3). The transition energies in our single QWs calculated in a simple square well model fit to the experimental data as published elsewhere.[7]

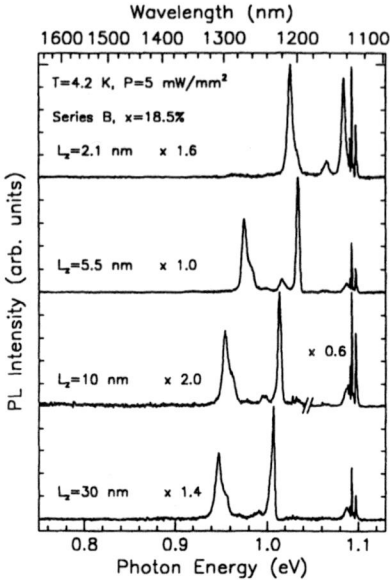

Fig.3. PL spectra of samples with varying QW widths L_z, which were grown at $T_0 = 600°C$ (series B in Table I). Measurements were performed under the same conditions as in Fig.1.

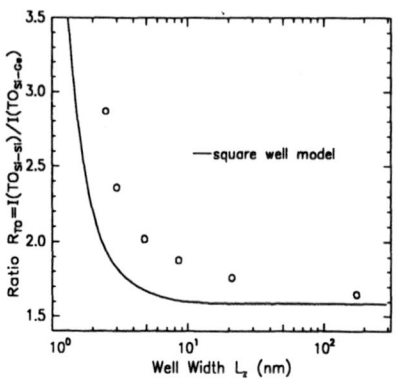

Fig.4. Intensity Ratio $R_{TO} = I(TO_{Si-Si})/I(TO_{Si-Ge})$ of the integrated intensity of the TO_{Si-Si} relative to that of the TO_{Si-Ge} phonon replica for very low excitation densities as a function of well width L_z for samples with $x = 24\%$. Circles are experimental, the solid line is a calculation using a square well model (see text).

In the following we discuss the relative intensities of the TO-phonon sidebands. The TO phonon replicas were fitted with gaussian lines to the spectra measured at very low excitation densities. From these fits the intensity ratio $R_{TO} = I(TO_{Si-Si})/I(TO_{Si-Ge})$ was determined. The ratio R_{TO} is constant in the low excitation regime for widely varying excitation densities. Therefore, band bending effects can be excluded. The ratio R_{TO} increases with smaller QW thickness L_z as shown in Fig. 4 for six samples with a Ge content of 24%. As the electronic-vibronic coupling is due to the orbital momentum of the hole the data indicates that the hole wavefunction penetrates deeper into the Si barrier for thinner QWs. This situation can be modelled. The relative probabilities of finding the hole in the Si barrier and the SiGe QW (defining an 'effective' Ge content) was calculated in the square well model. The 'effective' Ge content serves to determine the intensity ratio R_{TO} using the data of Weber and Alonso for the relative phonon strenghts in unstrained volume SiGe alloys.[13] The model values for R_{TO} thus obtained are compared to the experimental values in Fig. 4. There are large discrepancies except for very wide wells close to bulk-like layers. For the thickest QW with $L_z = 175$ nm the measured value of $R_{TO} = 1.65$ corresponds to an 'effective' Ge content of 23.3%, which is close to the value of 24% determined by RBS and HRXRD. The observed experimental-to-theoretical discrepancies for narrower wells remain to be solved.

Finally, we address the problem of the broad PL band supposed to be characteristic only of MBE-grown SiGe QWs. The spectra of the A series samples shown in Fig. 1 were entirely free of the broad PL band, independent of measurement temperature or excitation power. To investigate a possible influence of the QW thickness, as was suggested in Ref. 11, we grew three series of samples at 600°C and 450°C with L_z varying between 1.1 nm and 175 nm (series B, C and E in Table I). Four PL spectra from series B spanning a QW thickness range from 2.1 to 30 nm are plotted in Fig. 3. The spectra were recorded under the same conditions as those in Fig. 1. There is no indication of any additional PL signal far below the SiGe bandgap. Series C and E behave similarly, but the broad PL band is weakly observed, especially in the spectra of the thickest sample of series C with $L_z=175$ nm.

For a more quantitative judgement of the strength of the broad PL band, we evaluated the ratio of the integral PL intensity in the energy range below the TO_{Si-Si} replica to the integral intensity of the SiGe band edge lines. The results are plotted in Fig. 5. The PL intensities were measured at 4.2 K at a laser power density of 10 mW/mm^2. As mentioned before, only the thickest sample shows unambiguously detectable PL intensity below the SiGe near band edge PL lines. Due to the large energy range over which this PL signal was integrated, the observed intensity ratio of 0.54 corresponds to only a very small intensity of the broad PL band. For all other samples studied, the ratio remains below 0.1. The data points in Fig. 5 were derived from PL spectra excited under similar conditions as those reported by Rowell et al. in Ref. 14. In that paper, a correlation is made between the emergence of the broad PL band and the total elastic energy density E* in the strained SiGe layers. The energy density in the SiGe layers grown on a Si substrate is given by the expression

$$E^* = 2 \mu N \frac{(1+\nu)}{(1-\nu)} L_z \epsilon^2 .$$
(1)

In this formula μ is the shear modulus, N the number of SiGe layers (equal to one in our case), ν Poisson's ratio, L_z the SiGe layer thickness, and ϵ the strain in the SiGe layer. It was suggested that below a critical value E_1^* only band gap luminescence (with resolved phonon sidebands) is observed and that above a second (rather close) value E_2^* only broad band PL is detected, with a combination of both spectroscopic features seen in the

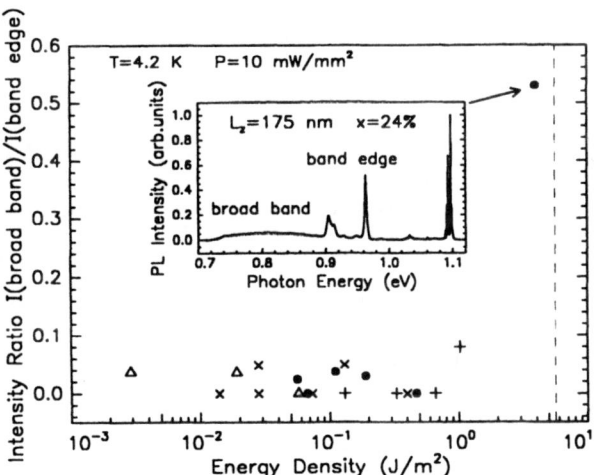

Fig.5. Ratio of the integrated PL signal in the energy range below the TO_{Si-Si} replica and of the SiGe band edge lines plotted as a function of the energy density in the SiGe layer. All samples listed in Table I were evaluated: •: series A, C; x: series B; +: series D; △: series E. The PL spectrum of the sample with the highest intensity ratio is depicted in the insert. The dashed is our calculated energy density for the critical layer thickness at x=24% and $T_0=450°C$ taken to be 250 nm.[16]

intermediate range of E*. In the light of this suggestion the presently discussed intensity ratios are plotted in Fig. 5 as a function of the elastic energy density E* following Ref. 14. The energy density which corresponds to the critical layer thickness for a Ge content of 24% and a growth temperature of 450°C is 5.7 J/m² [17] and marked in Fig. 5 with a dashed line.[18] Only one of our samples (L_z=175 nm, x=24%, T_G=450°C) is close to this critical value. On a microphotograph this sample exhibits misfit dislocations after Schimmel etching indicating partial relaxation of the SiGe layer. Hence, this sample is not relevant for the present discussion. If there is a characteristic dependence of the strength of the broad PL band on the energy density, it should -according to our data- appear in this narrow energy density range from 1 to below 5.7 J/m².

SUMMARY

The MBE-grown single QWs show excellent quality in PL measurements even for growth temperatures as low as 350°C. However, in samples grown below 550°C non-radiative recombination centers become more important leading to a significant reduction of the PL intensities at higher temperature measurements. The intensity ratio of the TO_{Si-Si} relative to the TO_{Si-Ge} phonon replica found at very low excitation densities is significantly higher than derived from a square well model. The broad PL signal about 150 meV below the SiGe bandgap has very low intensity compared to the band edge luminescence at temperatures around T=4 K and at an excitation power density of 10 mW/mm² for relatively narrow QWs. Only one sample with significantly larger L_z does not fit into this behavior exhibiting significantly stronger broad band PL. A clear correlation between the emergence of the broad PL band and the strain energy density in the SiGe layer cannot be concluded from our measurements.

The authors thank T.Baier for helpful theoretical calculations and U.Mantz for experimental assistance.

References

1. J.C.Sturm, H.Manoharan, L.C.Lenchyshyn, M.L.W.Thewalt, N.L.Rowell, J.-P.Noël, and D.C.Houghton, Phys.Rev.Lett. 66, 1362 (1991).
2. D.Dutartre, G.Brémond, A.Soufi, and T.Benyattou, Phys. Rev. B44, 11525 (1991).
3. D.J.Robbins, L.T.Canham, S.J.Barnett, A.D.Pitt, and P.Calcott, J.Appl.Phys. 71, 1407 (1992).
4. K.Terashima, M.Tajima, and T.Tatsumi, Appl.Phys.Lett. 57, 1925 (1990).
5. T.D.Steiner, R.L.Hengehold, Y.K.Yeo, D.J.Godbey, P.E.Thompson, and G.S.Pomrenke, J.Vac. Sci. Technol. B10, 924 (1992).
6. J.Spitzer, K.Thonke, R.Sauer, H.Kibbel, H.-J.Herzog, and E.Kasper, Appl.Phys.Lett. 60, 1729 (1992).
7. M.Wachter, K.Thonke, R.Sauer, F.Schäffler, H.-J.Herzog, and E.Kasper, Thin Solid Films 222, 10 (1992).
8. S.Fukatsu, H.Yoshida, N.Usami, A.Fujiwara, Y.Takahashi, Y.Shiraki, and R.Ito, Jpn.J.Appl.Phys. 31, 1319 (1992).
9. V.Arbet-Engels, J.M.G.Tijero, A.Manissadjian, K.L.Wang, and V.Higgs, Appl.Phys.Lett. 61, 2586 (1992).
10. J.Brunner, U.Menczigar, M.Gail, E.Friess, and G.Abstreiter, Thin Solid Films 222, 27 (1992).
11. J.-P.Noël, N.L.Rowell, D.C.Houghton, A.Wang, and D.D.Perovic, Appl.Phys.Lett. 61, 690 (1992).
12. E.Kasper, H.Kibbel, and F.Schäffler, J.Electrochem.Soc. 136, 1154 (1989).
13. J.Weber and M.I.Alonso, Phys.Rev. B40, 5683 (1989).
14. N.L.Rowell, J.-P. Noël, D.C.Houghton, A.Wang, L.C.Lenchyshyn, M.L.W.Thewalt, and D.D.Perovic, submitted to J.Appl.Phys. (private communication).
15. H.Jorke, H.Kibbel, F.Schäffler, and H.-J.Herzog, Thin Solid Films 183, 307 (1989).
16. H.-J.Herzog (private communication).
17. Data taken from: Landolt-Börnstein, *Numerical Data and Functional Relationships in Science and Technology*, New Series, edited by O. Madelung, (Springer, Berlin, 1982), Vol. 17a.
18. Our calculated values of E* in Fig. 5 deviate by orders of magnitude from these in Ref. 14.

SEPARATION OF NUCLEATION AND CRYSTALLIZATION IN THE SOLUTION-PHASE SYNTHESIS OF GROUP IV QUANTUM STRUCTURES

JAMES R. HEATH AND PAUL F. SEIDLER
IBM Research Division, T. J. Watson Research Center, Yorktown Heights, NY 10598.

ABSTRACT

Solution-phase chemical syntheses are presented for the fabrication of Ge quantum structures, including 0-dimensional dots, 1-dimensional wires, and 2-dimensional platelets. The syntheses, which are based on the reduction of chlorogermanes and organochlorogermanes, are carried out in two steps. Step 1, the nucleation step, is done at ambient temperature with a liquid metal reducing agent (NaK) and under the influence of sonication. Step 2, crystallization, is carried out at elevated temperatures and pressures in a sealed bomb. Separation of these two steps is shown to be critical for obtaining size and shape control over the product crystallite.

INTRODUCTION

Interest in understanding the relationship between size and electronic structure in ultrasmall semiconductor crystals has led to several beautiful solution-based chemical syntheses designed to control the product crystallite size. This chemistry, pioneered largely by Brus' group, has led to synthetic routes for the large volume (grams) production of II-VI {1,2}, and, to a lesser extent, III-V {3} nanocrystal colloids. A number of researchers have reported alternative syntheses of Group IV quantum dots, such as photolysis/pyrolysis of silanes or germanes, or recrystallization/oxidation of amorphous Si or Ge films. {4-8} These methods typically produce a small amount (milligrams or less) of crystallites characterized by a broad size distribution. Brus has somewhat circumvented the size distribution problem via the use of size-exclusion chromotography to separate sizes produced by his particle generator.{9} However, none of these alternatives match the colloid chemistries for either product volume or size control.

This lab has launched an effort to elucidate solution-phase synthetic routes for the nanofrabication of Si and Ge, with an ultimate goal of achieving rational chemical control over crystallite size. There are no Si or Ge chemistries analogous to the II-VI colloid techniques. However, the rich chemistries of organohalosilanes (germanes) suggests that these precursors may be suitable starting points for the control of crystallization reactions. The properties of Si and Ge imply that a high temperature synthesis is necessary for crystallite nucleation and annealing.

We have previously announced the success of this approach with the development of solution phase syntheses for both Si and Ge crystals and nanostructures.{10,11} For Si, the reagents are $SiCl_4$ and $RSiCl_3$, plus sodium metal as a reducing agent. The reactions are done at elevated temperatures and pressures (385 C, > 100 atm) in an alkane solvent, and are carried out for several days in a sealed bomb equipped with a stir assembly. The R-group on the trichlorosilane is either hydrogen or an alkyl ligand. For R = H, the reaction produces Si single crystals in the size range from 50 Å to 3 μm. For R = octyl, the size of the crystals is now limited to 55 +/- 25 Å. This alkyl group is intended as a strongly bound ligand which will cap the particle's surface, and thus limit growth. Hydrogen, apparently, does not perform this function. These reactions produce a large fraction of amorphous product, apparently due to the relatively low temperature (for Si) at which the reactions are carried out.

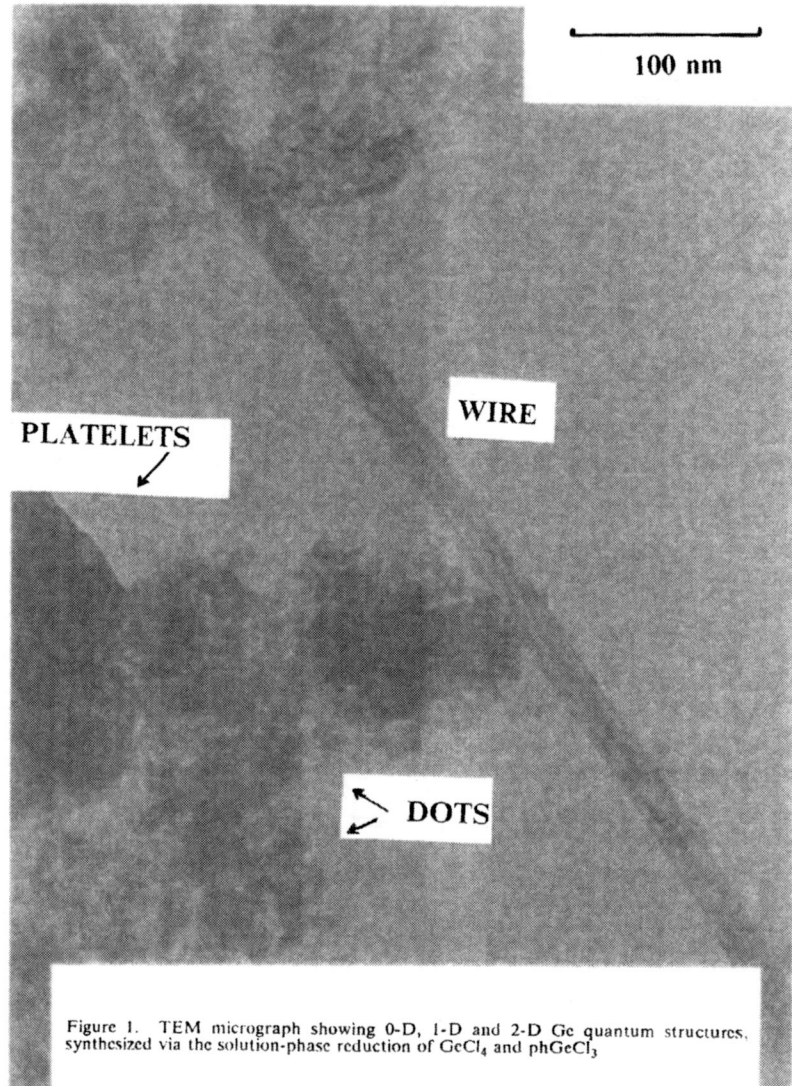

Figure 1. TEM micrograph showing 0-D, 1-D and 2-D Ge quantum structures, synthesized via the solution-phase reduction of $GeCl_4$ and $phGeCl_3$

For Ge, we have investigated similar reactions. They have produced intriguing results which imply that this chemistry may not only be used for size control, but *it may also be employed for the control of crystallite shape.* For the reduction of GeCl$_4$ at 300 C, the product is a polycrystalline powder with grain sizes near 200 Å. All products are crystalline, consistent with the fact that Ge melts ca. 500° below Si. If equivalent amounts of PhGeCl$_3$ and GeCl$_4$ are reduced, the product is Ge quantum dots of size near 100 Å, thin single crystal platelets with diameters up to a few micrometers, and single crystal Ge wires with lengths ranging from 50 nm to 10 μm and diameters ranging from 70 - 300 Å. A TEM micrograph showing all three morphologies is presented in Figure 1.

In this paper, we investigate the effect of separating nucleation and crystallization, and show that such separation is critical for crystallite shape control. The nucleation step is separated via a technique borrowed from organosilane polymer chemistry. Weidman and coworkers have utilized the ultrasonic dispersion of sodium-potassium alloy (NaK) in alkane solvents to produce small (ca. 100-Å) particles of liquid NaK evenly dispersed as a colloid.{12} If organohalosilanes are introduced into the colloid during sonication, a fast, ambient temperature reduction of the reagent is effected, leading to a polymer with a narrow molecular weight distribution. We extend this method to the synthesis of Ge crystals by first nucleating what is thought to be a polymer, and then annealing that polymer at elevated temperatures to form nanocrystals. Ge rather than Si chemistry is investigated here since the relevant temperatures make the matrix of reactions conditions more experimentally tractable. It is hoped that these results will be applicable to the chemically similar Si system. A schematic describing the technique used here is presented in Figure 2.

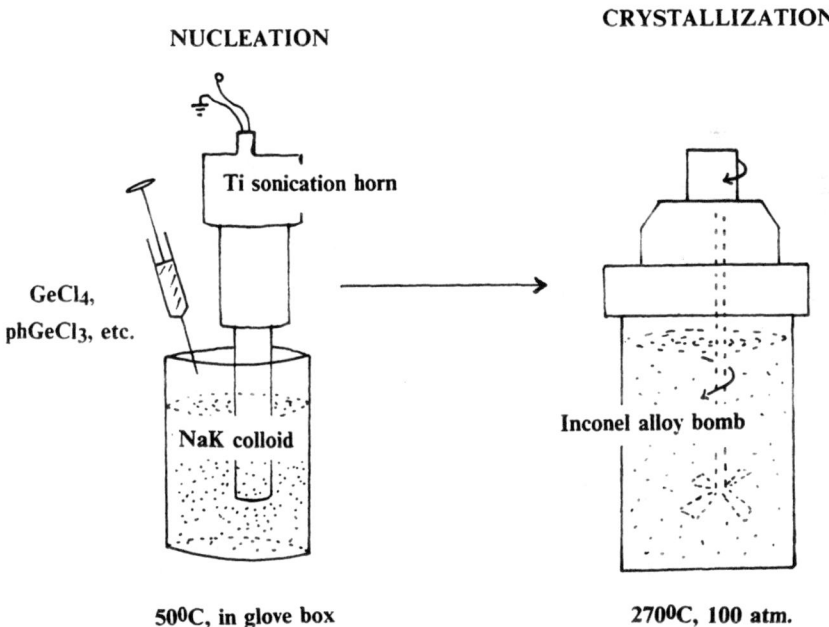

NUCLEATION

CRYSTALLIZATION

Ti sonication horn

GeCl$_4$,
phGeCl$_3$, etc.

NaK colloid

Inconel alloy bomb

50^{0}C, in glove box

270^{0}C, 100 atm.

Most reactions discussed here are not refined to the point of producing material characterized by sufficiently homogeneous and narrow size/shape distributions necessary for meaningful electronic spectroscopies. Although such refinement does appear to be forthcoming, this paper will focus exclusively on chemistry, and not on how size and shape influence the electronic structure of a given crystal. Product characterization is limited to transmission electron microscopy (TEM), x-ray powder diffraction (XRD), infrared absorption spectroscopy (IRAS), and nuclear magnetic resonance (NMR).

These reactions were carried out in two parts. Step 1, nucleation, was performed in a dry box. Unless otherwise stated, equimolar amounts of Na and K were weighed out into a beaker containing 150 mL of clean, dry heptane (HPLC grade 99 + %, dried over Na/benzophenone). The metals were fused to make a liquid 1:1 NaK alloy. A 500 W Ti sonication horn was lowered into the beaker, and the mixture was sonicated for 2-3 minutes, producing a blue-green colloid, which was ready for the nucleation step.

RESULTS AND DISCUSSION

REA-GENTS	Na/K (w/w %)	SONICATION DURING NUCLEATION?	0-D	1-D	2-D
PhGeCl₃ GeCl₄	35/65	Y	Y(100Å)	N	N
		Y (excess reagent added after sonication)	N	N	Y
	75/25	N	Y(100Å)	Y(1%)	Y
	100/0		Y(100Å)	Y(5%)	Y
	100/0	N (excess Na)	N	Y(10%)	N
GeCl₄ Me₂SiCl₂ MeSiCl₃	35/65	Y	Y(55Å)	N	N

Table 1. Germanium Chemistry Matrix

A summary of all the reactions discussed in this paper is presented above in Table 1.

R1. Ge quantum dot synthesis (size = 100 Å)

The reagents employed for R1 - R3 were phenyltrichlorogermane (PhGeCl₃) and germanium tetrachloride (GeCl₄). Aliquots of 0.5 mL PhGeCl₃ (3.1 mmole) and 0.5 mL GeCl₄ (4.4 mmole) were measured out yielding a total of 27 mmole of Cl atoms. Approximately 0.30 g Na (13 mmole) and 0.51 g K (13 mmole) were sonicated in 150 mL heptane to form a colloid, the Ge reagents were added, and sonication was continued for 2 - 3 minutes. Reaction was immediate, resulting in a black colloid which, if separated from the solvent and removed from the glove box, quickly oxidizes to a brown, amorphous powder. Instead, the colloid plus an additional 100 mL of heptane were added to the 300 mL base of an Inconel alloy bomb (Parr Instruments). The bomb was sealed and removed from the dry box. The stir mechanism on the bomb was engaged, and the bomb was heated to 270 C. Heating was accompanied by a pressure increase to ca. 100 atm. This annealing step continued for 48 hours, after which the bomb was cooled and vented, and the contents were filtered and washed with excess hexane, methanol, H₂O, and acetone. The resulting black powder was analyzed by TEM, XRD, and IRAS. TEM reveals a crystalline material with particle sizes ranging from 70 to 130 Å. XRD, shown in Figure 3, is consistent with this result. No evidence of amorphous material is observed, and the only crystals present are diamond lattice Ge. IRAS indicates that the particles are oxygen and halogen terminated, and no evidence for any remaining phenyl groups is observed. The size of the particles produced here is much greater than would be predicted if the phenyl group were a good size controlling agent (100 Å vs. 30 Å). This will be discussed later in the text.

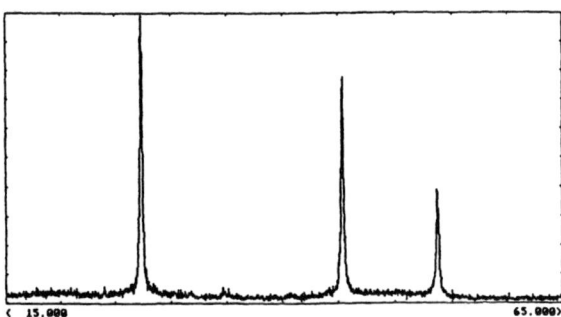

Figure 3. X-ray powder diffraction spectrum of approximately 100 Å diameter Ge crystals synthesized via the solution-phase reduction of phenyl-GeCl$_3$ + GeCl$_4$. The x-axis is in units of 2Θ, and the x-radiation is Cu-kα.

R2. Ge platelet synthesis

This synthesis was similar to R1, but with two exceptions. The reagent mixture was rich in PhGeCl$_3$ (1.0 mL, 6.2 mmole) relative to GeCl$_4$ (0.5 mL, 4.4 mmole), and only two thirds of this mixture was reduced with an equivalent amount of NaK (based on Cl). The remaining third was added into the colloid after the sonication was finished. This was intended to result in a completely reduced polymer colloid, plus an additional amount of GeCl$_4$ and PhGeCl$_3$, which were then allowed to react in the high temperature, high pressure bomb.

XRD analysis indicates that the product is crystalline Ge, with no detectable amorphous material present. TEM reveals that the major morphology present is that of platelets, similar to those pointed out in Figure 1. These platelets range from about 500 Å to 5000 Å in the two large dimensions. The thickness of a given platelet may be estimated by viewing the platelet as the plane of the TEM stage is rotated, and such measurements yield platelet aspect ratios in the range of 5 - 10.

In this reaction, annealing is performed in the presence of excess GeCl$_4$ and PhGeCl$_3$. A reasonable model for platelet growth, then, would invoke epitaxial chlorogermane addition to the surface of the particles. From Ge CVD technology, it is well known that epitaxy occurs at different rates on different crystal faces. R2 was carried out at relatively low temperatures (for Ge), suggesting that only a few chemical pathways for epitaxy are thermally accessible. This may serve to amplify the differences in various growth rates.

The discrete chemical pathways for epitaxy are of major concern here in attempting to understand the chemistry of shape and size. The results of R1 and R2 indicate that the nucleation step is of prime importance. However, still unaddressed is the role which the phenyl and excess Cl groups play in the crystal growth phase of R2. The phenyl groups have obviously not provided good size control, but their presence may serve to open up (or close) a number of epitaxy channels. For example, elimination of chlorobenzene from the surface of a Ge crystal would open up 2 sites for growth. To investigate this and other mechanisms, a couple of drops of the heptane solvent were extracted from the R2 product mixture, dissolved in acetone-d$_6$, and analyzed by

NMR. Three distinct sets of features were observed in the spectrum, at δ 7.17, 7.64, and 7.79 ppm. This signature is consistent in chemical shift, number of features, and splitting of the features with the presence of a monosubstituted phenyl group. It is *not* chlorobenzene, biphenyl or PhGeCl$_3$. Spectra of these three materials were recorded and they do not match with the unknown spectrum. However, it does seem clear that at the elevated temperatures in the bomb, the phenyl groups are relatively labile surface groups on the Ge particle, and, in some manner, do eliminate as a stable molecule. Although this mechanism does produce new sites for crystal growth, it does not yield any clues concerning the excess Cl. Even without any phenyl groups left on the surface, there is still more Cl than necessary to coordinate the available surface area.

R3. Germanium quantum wire synthesis

The synthesis of Ge wires was found to be sensitive to the nature of the reducing agent, and several variations were carried out.

R3A. Germanium quantum wires (yield < 1%).

A colloid consisting of a NaK 5.1/1 (mole/mole) alloy in heptane was prepared. This alloy has a higher melting point (near 60 C) than the 1/1 (mole/mole) dispersion (near 0 C).{13} Without further sonication, the colloid and 0.5 mL each of PhGeCl$_3$ (3.1 mmole) and GeCl$_4$ (4.4 mmole) were added into the reaction bomb. This reagent mixture was deficient in reductant (27 mmole Cl vs. 15 mole Na and K). The reaction was continued at 270 C for 2 days.

XRD indicates that the product is crystalline Ge. TEM indicates that the major product is Ge dots, with a very minor (ca. 1% or less) product of wires (100 - 300 Å diameter, 1 μm long).

R3B. Germanium quantum wires (yield ca. 5%).

For this reaction, no sonication was used. Rather, a 28% (by weight) dispersion of Na in toluene was employed as the reducing agent. Na metal has a lower reduction potential than NaK, and melts near 100 C. An approximately equimolar ratio (based on Cl) of Na was added to 0.5 mL PhGeCl$_3$ (3.1 mmole) and 0.5 mL GeCl$_4$ (4.4 mmole) in 250 mL of hexane. The reaction was run at 280 C and 150 atm for a period of 7 days. XRD indicates the product is crystalline Ge, and TEM indicates that all three morphologies (wires, dots, platelets) are made. The wires (ca. 5% yield) are 70 - 300 Å in diameter, and up to 10 μm long.

R3C. Germanium quantum wires (yield ca. 10%).

This reaction was carried out under conditions identical to those of R3B with the exception that an approximately twofold molar excess of Na was used. Although the major crystalline product here is Ge wires, the dominant product (the remaining 90%) is amorphous material, and, unlike the wires from reactions 3A or 3B, these wires are not isolated, but rather are sewn up in a web of amorphous material. The diameter of the wires is now 50 Å and below.

Apparently, the critical parameter which determines whether or not wires are produced is the reducing agent, and the yield of wires is related inversely to the efficiency of the reductant. A useful postulate to consider here comes from physical organic chemistry -- the Hammond Postulate. This states that an increased reactivity of reagents leads to decreased selectivity of products. This implies that an extremely fast nucleation/growth sequence will lead to random particle growth, with little chemical selectivity over the variety of growth sites on the particle or polymer surface. Such growth would result in a nearly spherical particle. However, for a slower nucleation/growth sequence, selectivity may be enhanced. The implication here is that when Ge is given the chance to sort itself out, it will preferentially nucleate and grow into a wire morphology.

There now exists some independent experimental results which lend some credence to this conclusion. Jarrold's group has developed an experiment for measuring the gas-phase diffusion rates of single element clusters.{14} These rates are related to shape factors for a given cluster. They find that annealed Ge clusters containing up to 50 atoms (the largest they have studied) possess prolate morphologies with length/diameter ratios near 5. The implication is, once again, that wire morphologies may be the low-energy structures for finite-sized Ge clusters and particles.

R4. Si-capped Ge quantum dots (50 - 60 Å)

In principle, the substituents bound to Ge in the reagents used in these nanostructure syntheses will determine product particle dimensions. In R1, for example, a simple calculation based on the ratio of $PhGeCl_3$ to $GeCl_4$, and assuming phenyl groups remain bound to the particle surface, predicts a surface/volume ratio suitable for an approximately 30 Å particle. The average particle size found for R1 (near 100 Å) suggests, however, that phenyl groups are poor size controlling agents. The NMR results of R2 indicate that, at the temperatures at which these reactions are carried out, the phenyl groups leave the surface of the particles. This explains the absence of phenyl features in the IR spectrum of the R1 product. A potentially better cap for the surface of these particles is methylsilyl groups. The methyl-silicon bond is extremely stable, and the Si-Ge bond is only slightly strained.

For this reaction 0.3 mL $MeSiCl_3$ (2.6 mmole), 0.1 mL Me_2SiCl_2 (0.8 mmole), and 1.0 mL $GeCl_4$ (8.8 mmole) were employed. The use here of dichlorodimethylsilane recognizes that some Si surface atoms will be bound to three interior Ge atoms, and some will be bound to only two. The ratio of methylsilanes to $GeCl_4$ yields a predicted surface/volume ratio appropriate for Ge particles in the 50 - 60 Å size range. An equivalent amount of NaK (based on Cl) colloid was prepared, and the reagents were added quickly to the reactant flask. Sonication was continued for 3-4 minutes, the mixture poured into the bomb, and the reaction continued at 280 C for 48 hours.

XRD of the product powder (Fig. 4) is notable in two ways. The peak widths of the Ge reflections indicate nanocrystals in the 50 - 60 Å size range; and, the peaks are unshifted from the bulk values. This indicates that not only do methylsilanes control particle size in a rational and predictable manner, but also that the Ge lattice (at least for this size particle) is unperturbed by Si capping agents. Further experiments aimed at extending size control, as well as experiments aimed at characterizing the surfaces of these particles are currently underway.

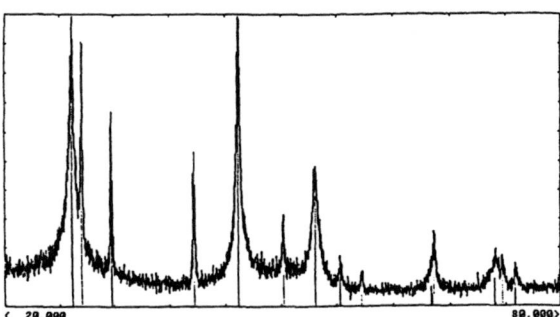

< 20.000   80.000>

Figure 4. XPD spectrum of 50 - 60 Å diam Ge crystals synthesized via the reduction of $GeCl_4$, $meSiCl_3$, and me_2SiCl_2. The sharp peaks originate from KCl and NaCl reaction byproducts. The Ge (220) feature appears artificially sharp and intense due to an accidental degeneracy with the NaCl (200) reflection.

ACKNOWLEDGEMENTS

We would like to thank T. Weidman for pointing out the utility of sonication to us. We would also like to acknowledge M. Jarrold for helpful discussions, and for disclosing his results to us prior to publication.

REFERENCES

1. R. Rossetti, S. Nakahara, and L.E. Brus, J. Chem. Phys. **79** , 1086 (1983).
2. N. Chestnoy, R. Hull, and L.E. Brus, J. Chem. Phys. **85** , 2237 (1986).
3. M.A. Olshavsky, A.N. Goldstein, and A.P. Alivasatos, J. Am. Chem. Soc. **112** , 9438 (1990).
4. S. Hayashi, H. Wakayama, T. Okada, S.S. Kim, and K. Yamamoto, J. Phys. Soc. Jap. **56** , 243 (1987).
5. H. Takagi, H. Ogawa, Y. Yamazaki, A. Ishizaki, and T. Nakagiri, Appl. Phys. Lett. **56** , 2379 (1990).
6. Y. Maeda, N. Tsukamoto, Y. Yazawa, Y. Kanemitsu, and Y. Masumoto, Appl. Phys. Lett. **59** , 3168 (1991).
7. J.M. Jasinski, and F.K. LeGoues, Chem. of Mater. **3** , 989 (1991).
8. R. Thielsch and H. Bottcher, Chem. Phys. Lett. **189** , 226 (1992).
9. K.A. Littau, P.J. Szajowski, A.J. Muller, A.R. Kortan, and L.E. Brus, J. Phys. Chem. **97** , 1224 (1993).
10. J.R. Heath, Science **258** , 1131 (1992).
11. J.R. Heath and F.K. LeGoues, submitted (1993).
12. P.A. Bianconi, F.C. Schilling and T.W. Weidman, Macromolecules **22** , 1697 (1989).
13. *Binary Alloy Phase Diagrams*, edited by T.P. Massalki, 2d ed., vol. 3, (ASM International, Ohio, 1990).
14. M.F. Jarrold (private communication) (1993).

INVESTIGATION OF COLLOIDAL Si PREPARED FROM POROUS SILICON

S. Berhane,[+] S. M. Kauzlarich, [+] K. Nishimura, [§] R. L. Smith,[§] J. E. Davis,[†] H. W. H. Lee,[‡] M. L. S. Olsen,[‡] and L.L. Chase[‡]

[+]Department of Chemistry
[§]Departments of Electrical Engineering and Computer Science
[†]Department of Applied Science
University of California
Davis, CA 95616

[‡]Lawrence Livermore National Laboratory
Livermore, CA 94550

ABSTRACT

Si nanocrystallites have been prepared by ultrasonicating thin sections of porous silicon. The materials produced from 20 and 49 wt % HF are characterized and compared. Samples were characterized by optical absorption and photoluminescence spectroscopy, and HRTEM.

INTRODUCTION

The recent production of luminescent colloidal suspensions of Si from porous Si by ultrasonic fracture[1] provides an opportunity to ascertain whether the luminescence in porous Si is due to a quantum confinement effect or to silicon compounds such as siloxene ($Si_6O_3H_6$).[2, 3, 4] In addition, characterization of particles produced in this manner may also provide more evidence for a surface layer of SiO_2 that has been proposed to be necessary for high efficiency luminescence.[5, 6] A number of methods have been used to prepare Si crystallites including thermal pyrolysis,[7] evaporation and laser ablation into an inert atmosphere,[8, 9, 10, 11] and high pressure solution phase synthesis.[12] Most methods create a wide range of sizes and structure including the technique of ultrasonic fracturing.[1] The technique that has produced the most uniform sized Si clusters reported to date involves a high temperature pyrolysis in which the Si clusters are collected as an ethylene glycol colloid.[6]

We are investigating the materials processing of porous Si in order to produce colloidal luminescing Si particles of a narrow size distribution. By starting with a well-defined microstructure,[13, 14, 15] one might be able to achieve a more narrow size distribution. This paper describes the preliminary characterization of Si colloids produced from porous Si formed in a 20 wt % and a 49 wt % HF solution. These materials have been characterized by optical absorption and photoluminescence spectroscopy, and high resolution transmission microscopy (HRTEM).

EXPERIMENTAL

Polished, single crystal p-type Si (100) wafers with a resistivity of 14 - 20 Ω-cm were used. Ohmic contacts were formed by a spring loaded metal contact touching the backside of the wafer, which is coated with aluminum deposited by electron beam evaporation. It is mounted so that only Si was exposed to the etching bath. The resulting Si electrode was used as the working electrode and Pt mesh as the counter. Porous Si was formed using a current density of 20 mA/cm^2 in a 20 wt % HF concentration solution for 15 minutes. Porous Si was also formed using a current density of 20 mA/cm^2 in a 49 wt % HF concentration solution for 15 minutes. A computer controlled PAR potentiostat-galvanostat, model 273, was used to control current and potential of formation. Electropolishing was used to remove the porous Si from the wafer. A current density of 162 mA/cm^2 in a 1-2 wt % HF solution was used to remove the surface layer of porous Si. This separation technique takes only about 30 sec. and should have minimal effect on the structure.

The porous Si was mechanically forced and removed from the Si wafer using a water gun. The samples were then removed from the bath and dried over anhydrous $CaSO_4$ for approximately 5 hours. After drying, all manipulations were carried out under a dry nitrogen atmosphere. In a typical preparation, 5 ml of distilled solvent were added to a flask containing the porous Si fragments and the mixture was placed in a Branson ultrasonic cleaning bath for 8 - 24 hours. After removal from the ultrasonic bath the solution was allowed to settle and the supernatant was removed from the settled solids. In addition, some samples were centrifuged for 5 - 10 min. and the supernatant was studied.

The samples, prepared by colloid evaporation on carbon-coated electron microscope grids, were examined in a high resolution transmission electron microscope (HRTEM) JEM 4000EX operating at 400 kv accelerating voltage. HRTEM negatives were digitized and analyzed using SEMPER VI image processing software.[16]

UV-Vis spectra were obtained with a Hewlett-Packard 8450A diode-array spectrophotometer. The photoluminescence spectra were obtained by optically exciting the colloidal suspension of Si nanoparticles with a few mW's of optical power at 355 nm or 380 nm obtained by frequency doubling the output from a CW modelocked Ti:Sapphire laser operating at a 82 MHz repetition rate. The resulting photoluminescence was imaged onto the 150 micron entrance slits of a 0.25 m monochromator and recorded with an optical multichannel analyzer. All spectra were corrected for system response using a NIST-traceable spectral calibration lamp. Contributions from solvent luminescence were also accounted for by subtracting the luminescence from a pure solvent sample.

RESULTS AND DISCUSSION

The experimental parameters for the production of porous Si were chosen in order to achieve a well-defined microstructure of narrow pore size distribution. Figure 1 shows an electron micrograph of a typical Si nanocrystallite produced from the 20 % solution. The particles in this sample were irregularly shaped and ranged in size from 2.0 - 40 nm. The majority of these particles were oriented such that visible lattice fringes measured 0.192 nm corresponding to the spacing of the (2 2 0) planes of diamond cubic Si. Figure 2 shows the Fourier filtered background subtracted periodic image of the nanocrystallite shown in Figure 1, along with the projected potential of Si along [0 0 1] at Scherzer defocus, rotated for comparison with the particle.

10 Å

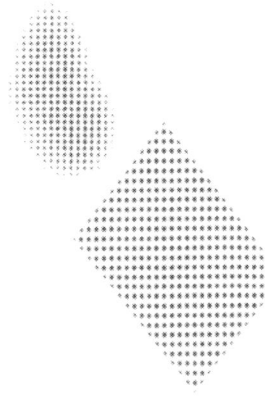

Figure 1. A HRTEM image of a silicon crystallite produced from 20 wt % HF.

Figure 2. The background subtracted image with a model of Si along [0 0 1] at Scherzer defocus.

HRTEM studies were also performed on the centrifuged sample prepared from 49 wt % HF. A large number of amorphous particles with average diameters of 10 - 50 nm were present along with a significant number of turbostratic particles with average diameters of 10 - 30 nm. The lattice fringes identifying the d-spacing between the van der Waals planes of these turbostratic particles were concentric to the center. The center region appeared amorphous or disordered. The ordering of siloxene (subsiliceous acid $Si_6H_3(OH)_3$) layers is reported to be turbostratic[17] and thus one might speculate that these samples are siloxene. However, if Si is cleaved along the (1 1 1) plane due to the ultrasonic treatment, one might obtain thin layers of Si, which could make up the turbostratic clusters. Since the diffraction patterns of these turbostratic particles provided no identifying structural information, further analysis of these particles is underway including microchemical analysis. The 49 wt % material has also been characterized by photoemission and appears to be primarily oxidized silicon.[18] It is possible that the amorphous particles observed by HRTEM are oxidized silicon which contribute to the photoemission conclusion.

The light emitting properties of the Si nanoparticles were also studied. The photoluminescence of the nanocrystallites produced from either 20 and 49 wt % HF solution were similar and the following discussion pertains to both. UV laser excitation of the Si colloid resulted in visible light emission ranging from red to blue as shown in Figure 3. The wavelength of the luminescence varied with conditions of the measurement and suggested possible luminescence mechanisms. When the cuvette containing the colloid was agitated, the photoluminescence spectra changed as the colloid settled. Immediately after agitation, the colloid emitted visible luminescence from red through blue. As the colloid settled, the red luminescence diminished in intensity, until, after a sufficiently long time period, only blue luminescence remained. Laser excitation directed near the bottom of the cuvette, where macroscopic particles were clearly observable, gave rise to the red luminescence shown in Figure 3. These particles were probably macroscopic sections of porous silicon fragmented by the ultrasonic treatment. The luminescence peak near 760 nm originates from unconverted fundamental photons from the Ti:Sapphire laser.

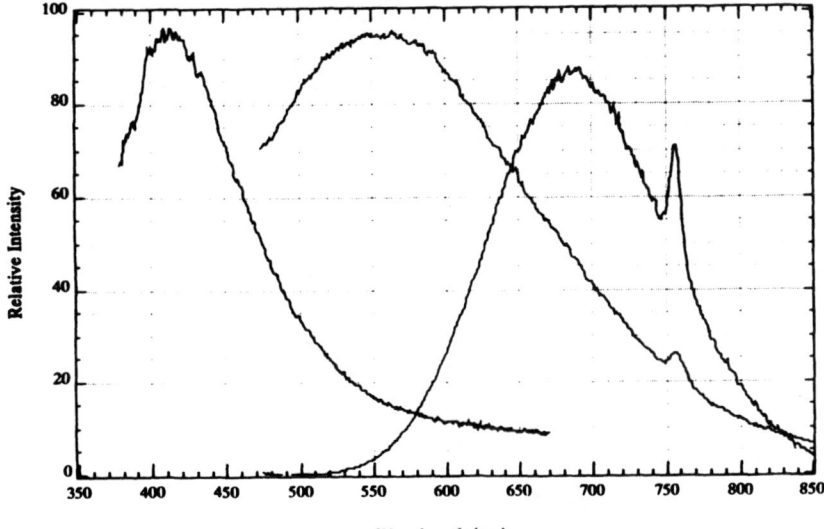

Figure 3. Room temperature photoluminescence of colloidal silicon prepared from porous silicon.

The green luminescence shown in Figure 3 was observed by placing a drop of toluene colloidal suspension onto a fused silica slide and allowing the toluene to evaporate. Laser excitation of the remaining Si nanoparticles resulted in the green luminescence. We have also observed red and blue luminescence from similarly deposited nanoparticles. When an acetonitrile colloidal suspension was centrifuged and the supernatant collected and optically excited, only blue luminescence was observed, as shown by the blue luminescence spectrum in Figure 3. The supernatant in this case contained no macroscopic particles discernible by eye.

These results may be interpreted in several ways. The smaller particles emit bluer photons, which could either indicate quantum confinement or oxidized Si. Presumably, Brownian motion suspends only the smallest particles in the solvent after gravimetric settling. Only the smallest particles remain in the supernatant after centrifugal treatment. We observed these conditions to lead to blue luminescence. However, photoemission studies[18] on these blue-emitting samples show only oxidized silicon and defect states SiO_2 are known to luminesce in the blue. Interfacial interactions and surface states, which may be more important in smaller particles,[6] could also play a role in the blue luminescence. These possibilities are presently under investigation.

In summary, we have investigated silicon particles produced from ultrasonicating thin sections of porous Si produced from 20 and 49 wt % HF. Although nanocrystallites that can be identified as silicon are produced from the 20 wt % HF, the particles produced from the 49 wt % HF show significant oxidation. It is probable that the thin sections of porous silicon are very oxygen sensitive and are oxidized before they are ultrasonicated. Since 49 wt % HF produces smaller pore sizes and interpore spacings compared with 20 wt %,[13] it is possible that the fraction of oxide to silicon is significantly higher. Experiments are underway to minize oxidation of the thin sections of porous silicon.

ACKNOWLEDGMENT

We thank Dr. George J. Thomas for permission to use the HRTEM at Sandia National Laboratories, Livermore and Dr. Doug L. Medlin for his assistance. We thank Dr. Michael. A. O'Keefe at the National Center for Electron Microscopy, Lawrence Berkeley Laboratory for assistance in the analysis of the HRTEM image. This work was supported by the National Science Foundation (DMR-9201041), and performed under the auspices of the U. S. Department of Energy by Lawrence Livermore National Laboratory under contract No. W-7405-ENG-48.

REFERENCES

1.   J. L. Heinrich, C. L. Curtis, G. M. Credo, K. L. Kavanagh, M. J. Sailor, Science **255**, 66 (1992).
2.   S. M. Prokes, J. Appl. Phys. **73**, 407 (1993).
3.   H. D. Fuchs, M. S. Brandt, M. Stutzmann, J. Weber, Mater. Res. Soc. Symp. Proc. **256**, 159 (1992).
4.   M. J. Sailor, K. L. Kavanagh, Advanced Materials **4**, 432 (1992).
5.   S. M. Pillai, Z. Y. Xu, M. Gal, R. Glaisher, M. Phillips, D. Cockayne, Jpn. J. Appl. Phys. **31**, L1702 (1992).
6.   K. A. Littau, P. J. Szajowshki, A. J. Muller, A. R. Kortan, L. E. Brus, J. Phys. Chem. **97**, 1224 (1993).
7.   J. J. Wu, R. C. Flagan, J. Appl. Phys. **61**, 1365 (1987).
8.   S. Hayashi, S. Tanimoto, K. Yamamoto, J. Appl. Phys. **68**, 5300 (1990).
9.   R. Okada, S. Ijima, Appl. Phys. Lett. **58**, 1662 (1991).
10.   S. Ijima, Jpn. J. Appl. Phys. **26**, 357 (1987).
11.   Y. Saito, J. Cryst. Growth **47**, 61 (1979).
12.   J. R. Heath, Science **258**, 1131 (1992).
13.   R. L. Smith, S. D. Collins, J. Appl. Phys. **71**, R1 (1992).
14.   R. Herino, G. Bomchil, K. Barla, C. Bertrand, J. L. Ginous, J. Electrochem. Soc. **134**, 1994 (1987).
15.   G. Bomchil, R. Herino, K. Barla, J. C. Pfister, J. Electrochem. Soc. **130**, 1611 (1983).
16.   W. O. Saxton, T. J. Pitt, M. Horner, Ultramicroscopy **4**, 343 (1979).
17.   A. Weiss, G. Beil, H. Meyer, Z. Naturforsch **34b**, 25 (1979).
18.   L. J. Terminello, L.L. Chase, M. Balooch, S. Berhane, S. M. Kauzlarich unpublished, (1993).

Si AND Ge NANOCRYSTALLITES EMBEDDED IN CaF$_2$ BY MOLECULAR BEAM EPITAXY (MBE)

A. P. Taylor, B. M. Kim, P. D. Persans, and L. J. Schowalter
Physics Department and Center for Integrated Electronics, Rensselaer Polytechnic Institute, Troy, New York 12180

ABSTRACT

Thin films of CaF$_2$ containing layers of Si and Ge nanocrystallites were grown epitaxially on Si(111) substrates by MBE while varying the substrate temperature. The sticking coefficient of both Si and Ge to CaF$_2$ are less than unity at the temperatures studied. Evidence that chemical reactions are minimal between Si and CaF$_2$ leading to the formation of volatile species such as SiF$_x$ is presented and it is surmised that the dominant mechanism for re-evaporation is thermal desorption. Both Si and Ge sticking coefficients vary exponentially with 1/T and activation energies are determined. A cluster growth model describing the evolution of the amount of Si in clusters over time is given. Solving in the limit of low cluster coverage, a solution that varies linearly with time and exponentially with 1/T is obtained. Weak room temperature photoluminescence from Si nanocrystallites in CaF$_2$ is seen, however, it is unclear whether the luminescence is coming from the Si nanocrystallites or the CaF$_2$. Second harmonic generation is observed from samples containing single layers of Ge in CaF$_2$.

INTRODUCTION

Semiconductor nanocrystallites have received much attention recently due to their novel optical properties[1,2]. Currently there is a heated debate as to the origins of photoluminescence from porous Si[3]. Besides linear optical properties, nonlinear effects are expected to be large in nanocrystallite systems[2]. By fabricating nanocrystallites of Si and Ge embedded in CaF$_2$ by MBE, we will have the ability to probe the optical properties of the nanocrystallites existing in a clean environment.

The heteroepitaxial systems of Si and Ge nanocrystallites embedded in CaF$_2$ are attractive for a number of reasons. CaF$_2$ grows two dimensionally on Si(111) whereas both Si and Ge form clusters when grown epitaxially on CaF$_2$(111) at elevated temperatures. A simplistic surface free energy argument can be used to explain this phenomena[4]. By varying the growth temperature, time, and flux rates the cluster size distribution can be varied. Since the nanocrystallites are embedded in CaF$_2$, they are protected against oxidization in atmosphere. Within the CaF$_2$ host crystal the nanocrystallites may be F terminated which could be important for their optical properties. Also, CaF$_2$ has a 12 eV band gap making it a suitable medium for waveguiding over a broad range of wavelengths. From a technological point of view, the system is compatible with Si VLSI.

FABRICATION METHOD

A Fisons VG90S MBE system equipped with e-beam evaporators for Si and Ge and a high temperature Knudsen cell[5] for CaF$_2$ evaporation were used to grow Si and Ge nanocrystallites on CaF$_2$. Base pressures were in the mid 10^{-11} mb range and in the 10^{-9} mb range during growth.

Layers of Si nanocrystallites interleaved with CaF$_2$ were grown epitaxially on CaF$_2$ buffer layers on Si(111) substrates. The composite region containing 10 or 20 Si layers was capped with additional CaF$_2$ to protect the nanocrystallites. The CaF$_2$ buffer layer ranged in thickness from less than 10nm to 500nm and the Si content in the composite region ranged from 1% to 25%. The Si flux rate was set at 0.10nm/s or 0.20nm/s and the substrate temperature was varied between 600°C and 850°C.

Single layers of Ge nanocrystallites were grown epitaxially on CaF$_2$ buffer layers on Si(111) substrates and capped with CaF$_2$. A 220nm thick CaF$_2$ layer was grown at 700°C on a 4 inch

Si(111) wafer and then cleaved into pieces. A series of single Ge layers were grown on the pieces after a 10min. thermal cleaning at 850°C and a thin (<10nm) CaF_2 layer was grown at 700°C. A single buffer layer thickness was chosen for consistency. The Ge flux rate was fixed at 0.10nm/s and growth times were set at 10s or 40s. Growth temperatures were varied between 400°C and 750°C. Within a few seconds after Ge growth, a thin (<10nm) CaF_2 layer was grown to protect the Ge nanocrystallites.

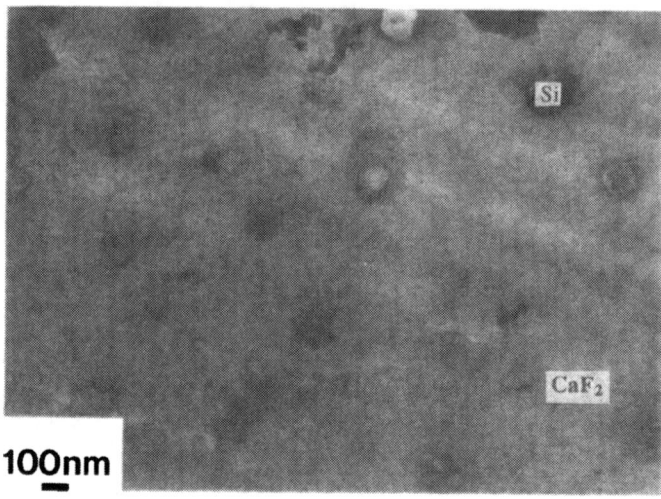

Fig.1 Planar TEM micrograph of Si nanocrystallites embedded in CaF_2.

RESULTS AND DISCUSSION

While working on reducing the size of the nanocrystallites by varying the growth conditions we have gained valuable information on how the nanocrystallite growth proceeds. To date, the smallest nanocrystallites of Si that we have observed by planar TEM have dimensions of about 20nm. As for the Ge nanocrystallites, we have observed sizes on the order of 10nm by planar TEM. Fig. 1 is a planar TEM micrograph of Si nanocrystallites embedded in CaF_2. The average particle surface area in the micrograph is 110nmX110nm and we estimate the particle thickness to be about 50nm. The nanocrystallites cover about 4% of the area. The facetted appearance of the nanocrystallites is evidence of their crystallinity.

A large discrepancy was found between the total amount of either Si or Ge deposited and the amount measured in the composites by Rutherford backscattering spectroscopy(RBS). This implies sticking coefficients of both Si and Ge to CaF_2 are less than unity over the range of growth temperatures studied. Re-evaporation of Si and Ge from a CaF_2 surface could occur by either thermal desorption of the elemental species or by the formation of a volatile compound such as SiF_x or GeF_x through reactions with the CaF_2. Evidence suggesting that the amount of material lost to reactions is minimal for Si on CaF_2 is presented and we conclude that the majority of re-evaporation is due to thermal desorption. This is in disagreement with earlier work[6]. As for the case of Ge on CaF_2, we do not have strong evidence suggesting that reactions are not important but speculate that they probably are not. Also, a cluster growth model is developed that describes the evolution of the total amount of Si in the nanocrystallites over time that takes into account re-evaporation.

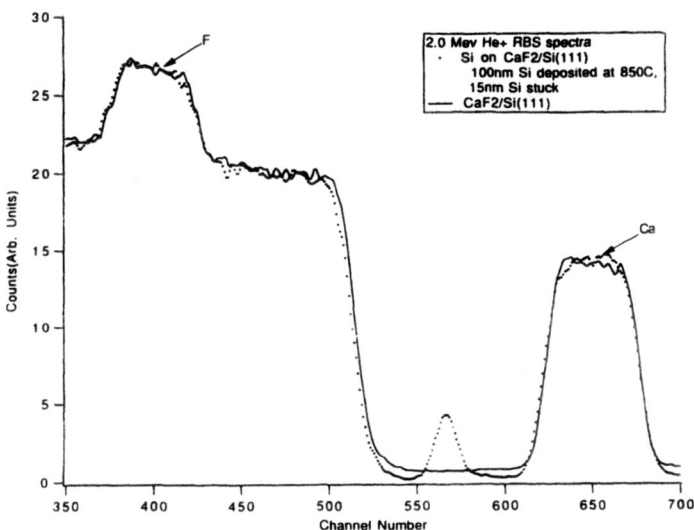

Fig.2 RBS spectra of an 180nm CaF$_2$ layer on a Si(111) substrate with and without Si grown on top.

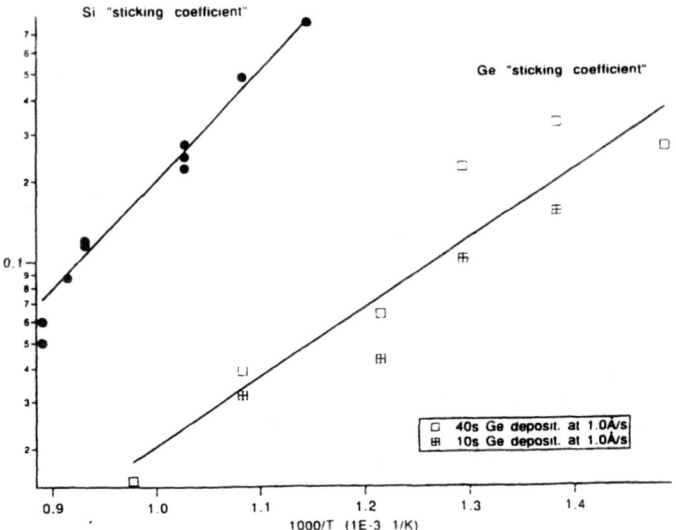

Fig.3 Plots of Si and Ge "sticking coefficients" versus 1000/T (10^{-3} K^{-1}). The lines are best fits to the data points obtained from experiments.

A Si layer 100nm thick was deposited on a piece of a 180nm thick CaF_2 layer on a Si(111) substrate at 850°C. Only 15nm of Si stuck to the CaF_2 as determined by RBS and optical microscopy revealed a smooth Si surface. The edges of the piece were masked from the Si flux. Fig. 2 is a superposition of the RBS spectra taken from the areas of the sample with and without Si grown on the CaF_2. The F and Ca peaks are nearly identical implying that little Ca and more importantly, F, had been removed after Si growth. In Fig. 2 the Ca edges have been aligned since the one from the area with Si on top is at lower energy. Also, one of the spectra has been slightly shifted on the counts axis for comparison since a lower number of counts were collected for that spectra. Ca:F ratios calculated by integrating areas under the Ca and F peaks also indicate no significant amount of F had been removed after Si growth. A ratio of 1:1.99 was calculated from dotted line spectrum in Fig. 2. A comparable ratio of 1:1.96 was calculated from the solid line spectrum in Fig. 2. A similar experiment to the one described for Si was carried out for Ge at a substrate temperature of 750°C, however, the results are inconclusive due to the nature of the RBS spectra.

Arrhenius plots of the "sticking coefficient" versus $1000/T$ (10^{-3} K^{-1}) for the samples containing Si and Ge nanocrystallites are presented in Fig. 3. The "sticking coefficient" was calculated by dividing the areal density of Si and Ge atoms obtained from RBS measurements by the total areal density deposited. Total areal density was determined by multiplying the flux rate by the time of deposition. Best fits to the data yield the equations

$$S_{Si} = 1.76 \times 10^{-5} \exp(0.81 eV/kT), \text{ and} \tag{1}$$

$$S_{Ge} = 5.33 \times 10^{-5} \exp(0.52 eV/kT). \tag{2}$$

Activation energies of 0.81eV and 0.52eV for Si and Ge, respectively are obtained from the fits. The activation energy for Ge is smaller than that for Si as expected since the vapor pressure of Ge is higher than that of Si.

A cluster growth model was developed to describe the evolution of the total amount of Si in clusters with time. The rate equations,

$$dR_{Si}/dt = JC - r_1 + r_2, \text{ and} \tag{3}$$

$$dR_{CaF2}/dt = J(1-C) + r_1 - r_2 - DR_{CaF2}, \tag{4}$$

are proposed where R_{Si} is the number of Si atoms per unit area in clusters, R_{CaF2} is the number of Si atoms per unit area on the CaF_2 surface not in clusters, J is the Si flux, C is the Si cluster coverage, r_1 is the rate of Si atoms per unit area leaving R_{Si} to R_{CaF2}, r_2 is the rate of Si atoms per unit area leaving R_{CaF2} to R_{Si}, and DR_{CaF2} is the rate of thermal desorption per unit area of Si from CaF_2 where D is equal to thermal desorption constant. The equations are coupled through the term r_2 which depends upon R_{CaF2}, the perimeter of a cluster, the number of clusters, and the surface diffusion constant, D_2.

A number of assumptions are made in order to solve the equations: clusters are treated as hemispheres and the critical radius of the hemisphere has a power law dependence on time[7]; the sticking coefficient of Si to Si is unity; R_{CaF2} is in equilibrium; R_{Si} is much greater than R_{CaF2}; C is much less than unity (i.e. at early times in the deposition); the rate of thermal desorption dominates over r_2 and r_1 is negligible; and the clusters stop growing once CaF_2 is deposited on them. For the given assumptions, we find that

$$R_{Si} \propto const. JtD_2/D$$

where the surface diffusion constant, $D_2 = D_{20} \exp(-E_2/kT)$, and the thermal desorption constant $D = D_0 \exp(-E/kT)$. The model may also hold for Ge cluster growth. The quantity R_{Si} is directly proportional to the Si sticking coefficient. By taking the activation energy for thermal desorption equal to the Ca-Si bond strength of 1.2 eV, a value obtained from the literature[8], we can estimate the activation energy for surface diffusion to be about 0.4 eV.

The optical properties of the nanocrystallites embedded in CaF_2 have been examined with various techniques. Raman scattering and photoluminescence experiments have been conducted on samples consisting of multilayers of Si nanocrystallites interleaved with CaF_2 on $CaF_2/Si(111)$. The Raman scattering shows evidence for small Si particles[4]. We observe weak luminescence in the visible under 488nm Ar ion laser excitation at room temperature, however, it is unclear whether the origin of the luminescence is from the Si particles or from the CaF_2 host crystal. Second harmonic generation signals were measured from single layers of Ge nanocrystallites on $CaF_2/Si(111)$, $CaF_2/Si(111)$, and from bulk semi-insulating $GaAs(111)$ under identical experimental conditions. The signals from the single layers of Ge nanocrystallites were from 40% to 60% weaker than the bulk $GaAs(111)$ signal and the signal from the $CaF_2/Si(111)$ sample was about 100 times smaller than the $GaAs(111)$ signal. Since samples containing Ge nanoparticles were only comprised of single layers of Ge it is implied that by going to many layers the second harmonic signal could possibly be 2 or 3 orders of magnitude more intense than that from bulk $GaAs(111)$.

CONCLUSION

From growth studies of Si and Ge nanocrystallites embedded in CaF_2 the sticking coefficients of both Si and Ge to CaF_2 are less than unity over the temperature ranges examined. Evidence that the dominant mechanism for re-evaporation of Si from CaF_2 is thermal desorption and no significant amount of reactions occur between the Si and CaF_2 has been presented. We speculate this is also the case for Ge on CaF_2, however, we do not have strong confirming evidence. Arrhenius plots of "sticking coefficient" versus $1/T$ yield activation energies for the re-evaporation process. A cluster growth model has been proposed and with given assumptions yields an approximate solution that conforms to our data.

We observe weak photoluminescence come from samples containing Si nanocrystallites embedded in CaF_2, however, it is unclear whether the luminescence is coming from the Si particles or the CaF_2 host crystal. Second harmonic generation has been observed in samples containing single layers of Ge nanocrystallites in CaF_2 and the signal is comparable to that from bulk semi-insulating $GaAs(111)$ under identical experimental conditions. By going to multilayers of Ge nanocrystallites embedded in CaF_2 it could be possible to increase the second harmonic signal by 2 or 3 orders of magnitude.

ACKNOWLEDGEMENTS

This work was supported in part by the NSF contract No. DMR-9203183. We also wish to thank Prof. X.-C. Zhang and T. Hewitt for making preliminary second harmonic measurements on the samples containing Ge nanocrystallites.

REFERENCES

[1] Y. Wang and N. Herron, J. Phys. Chem., **95**, 525, 1991.
[2] L. Brus, Appl. Phys. A **53**, 465, 1991.
[3] See for example, Mat. Res. Soc. Symp. vol. 256, editors, S. S. Iyer, R. T. Collins, and L. T. Canham, 1992.
[4] A. P. Taylor, K. Stokes, Z. C. Wu, P. D. Persans, L. J. Schowalter, and F. K. LeGoues, in Mat. Res. Soc. Symp. vol. 283, 1993.
[5] A. P. Taylor, K. Yang, and L. J. Schowalter, J. Vac. Sci. Technol. A **9**, 3181, 1991.
[6] T. Asano and H. Ishiwara, J. Appl. Phys., **55**, 3566, 1984.
[7] M. Zinke-Allmang, Scan. Microscopy, **4**, 523, 1990.
[8] F. J. Himpsel, U. O. Karlsson, J. F. Morar, D. Rieger, and J. A. Yarnoff, Phys. Rev. Lett., **56**, 1497, 1986.

CENTRIFUGALLY-ASSISTED SIZE CLASSIFICATION AND IMMOBILIZATION OF SILICON CRYSTALLITES IN GELS

D.J. DUVAL*, S.H. RISBUD*, Z.A. MUNIR*, B.J. McCOY**
*University of California, Division of Materials Science and Engineering, Davis, CA 95616
**University of California, Department of Chemical Engineering, Davis, CA 95616

ABSTRACT

Stokes settling in viscous gels was used as a method for classifying mixed powders of silicon crystallites and amorphous silicon; Quantum-size materials derived from porous silicon were encapsulated and trapped in the viscous gels. After complete gelation the matrix retained size gradations developed during Stokes settling, thus yielding a new variant of functionally gradient materials (FGM). Production and rheological properties of the sol and preparation of specimens for centrifuging are discussed. The relative position of particles with various sizes are quantitatively interpreted in relation to the increase in gel viscosity with time.

INTRODUCTION

Previous work in our laboratory has shown that curtailing the reaction of silicon powders with glass melts can create materials that exhibit luminescence, possibly associated with quantum confinement of nanosize silicon clusters [1]. These materials generally exhibit non-uniformly distributed silicon crystallites with a large size distribution. In the present work we have suspended silicon particles in sol-gels to provide a stable matrix for classifying particles of specific sizes by Stokes settling in a high gravity environment. This type of processing is one variant of the rapidly evolving FGM technology and could lead to homogeneous distributions of silicon crystallites in a glass matrix. After FGM processing it will be possible to thermally etch the silicon crystallites during glass consolidation so that they are further reduced to quantum dot size (≈ 10 nm).

BACKGROUND AND THEORY

The creation of a composition or particle-size gradient in a sample of material results, expectedly, in parallel gradients useful properties of this material. Such FGMs have recently been the focus of intense investigations, primarily in Japan. Although the initial emphasis was on the synthesis and processing of thermal barrier material for space applications [2] (the National Space Plane and shuttle engines), subsequent investigations have focused on other areas in which the application of FGM provides novel and effective solutions to existing materials problems. These latter areas include the use of FGM in nuclear fusion and fast breeder reactors (as first-wall composite materials) [3,4], in electronic and magnetic applications (electro-ceramics, sensors), in optical applications (high-performance laser rods, optical disks), in chemical applications (membranes, catalysts), in biomedical materials (tooth implants, artificial bones), and in joining applications (ceramic engines, heat and corrosion resistance coatings) [5-7].

The prospect of inducing deliberately-designed composition gradients or controlled particle-size distributions in optical and electronic materials is extremely attractive in the synthesis of special materials. For instance, gradient index (GRIN) optical elements are now a well-accepted part of modern photonic and communication devices. These materials have a well-controlled and continuous change in the refractive index, and find applications in fiber-optic couplers, photocopiers, miniaturized optical systems, and medical endoscopes. The effectiveness of such GRIN elements is strongly determined by the extent of the radial or axial refractive index change (Δn) that can be obtained in bulk disks or cylindrical preforms of the optical component. GRIN lenses with large refractive index variations ($\Delta n > 0.1$) and low dispersion are sought as optical blanks for processing components with a variety of profiles and symmetry.

Another important area in which FGM processing can be desirable is quantum dot materials of narrow size distribution. In quantum dot materials, the band gap of a bulk semiconductor is shifted significantly by reducing its particle size to a value smaller than the exciton (electron-hole pair) Bohr radius. The practical possibility of centrifugally assisted size classification has the potential for tailoring quantum devices by size-selection of the semiconductor and subsequent heat treatment to produce particles in nanometer dimensions. Bulk quantum dot materials are composites of semiconductor particles (e.g., Si, CdSe, GaAs) suspended in a ceramic or glass medium. The quality of quantum confinement, i.e., the optical density, is related to the concentration of semiconductor particles. It is here that FGM processing can result in unique materials with a gradient in semiconductor quantum dot concentration and/or size. The optical absorption spectra of such samples have never been studied before, and can be expected to give useful insights into the band structure transitions in these materials.

PROCEDURES AND RESULTS

Preparation of the Sol

The procedure for sol production was modified from that described in reference 8 and is outlined in Figure 1. For a given batch, the sol was prepared first by mixing 16 ml water, 32 ml ethanol, and 0.4 ml phosphoric acid. The pH was adjusted to 2, if necessary, with additional acid after vigorous mixing. Separately, 32 ml ethanol was mixed with 20 g TEOS. These solutions were then mixed. Phosphoric acid was used rather than hydrochloric acid because of its lower vapor pressure, and because it eliminates chlorine evolution during glass consolidation and may lower glass consolidation temperature by including phosphate complexes into the glass network. The sol was pre-reacted in a sealed environment for 20 hours under constant agitation by magnetic stirrer in a closed flask. Variations of the above procedure were performed, and the effect on change in viscosity with time is illustrated in Figures 2 and 3.

The sol was then subjected to a vacuum while being mildly heated to enhance polymerization and reduce ethanol concentration, resulting in an increased viscosity. The initial viscosity was similar to that of water, but after vacuum hydrolysis the viscosity resembled that of a thin lacquer.

Particle Incorporation

Two types of silicon particles were incorporated into the sol-gel matrix. A mixture of amorphous Si (Johnson Matthey −325 mesh) and crystalline Si (NBS 640b) was used in most cases. This broad and uneven distribution of powders was chosen to enhance the

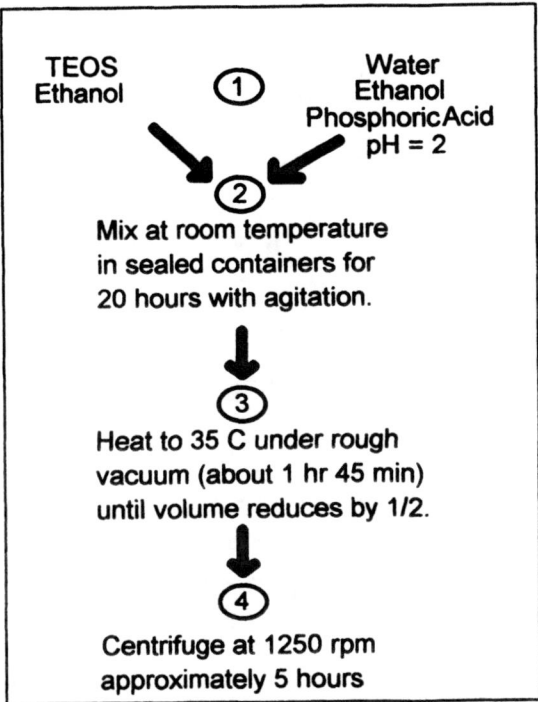

Fig. 1. Schedule for sol-gel synthesis

ability to observe particle size gradations due to Stokes settling. In another set of experiments, particles derived from porous Si (particles 10 – 200 Å, agglomerates 30 nm – 1000 nm) were used. The goal here was to separate large agglomerates from the more finely dispersed particles resulting from ultrasonic fragmentation of the porous silicon architecture.

Silicon particles were introduced into the sol at stages 1, 3, or 4 of the sol-gel schedule shown in Figure 1. The stage where the silicon powders were added to the gel appears to have no observable effect on the viscosity or optical properties of the sols. In all cases, various gradations in optical density of the gels were observed after centrifuging. Porous silicon was added to the water/acid/ethanol solution before mixing at stage 1, and in another case at stage 4. In the previous case, the sol exhibited a marked orange hue under UV irradiation throughout the process, changing to a dark red as the gel slowly dried. Only a bluish hue was observed when porous Si was added just prior to stage 4. Again, various gradations in optical density of the gels were observed after centrifuging.

Calculations of Relative Particle Position

The distribution of particles due to enhanced Stokes settling by centrifuging can be determined by calculating the velocity of solid particles in the liquid sols. The relationship is expressed in terms of velocity v, time t, viscosity n, particle density D_p, liquid density D_L,

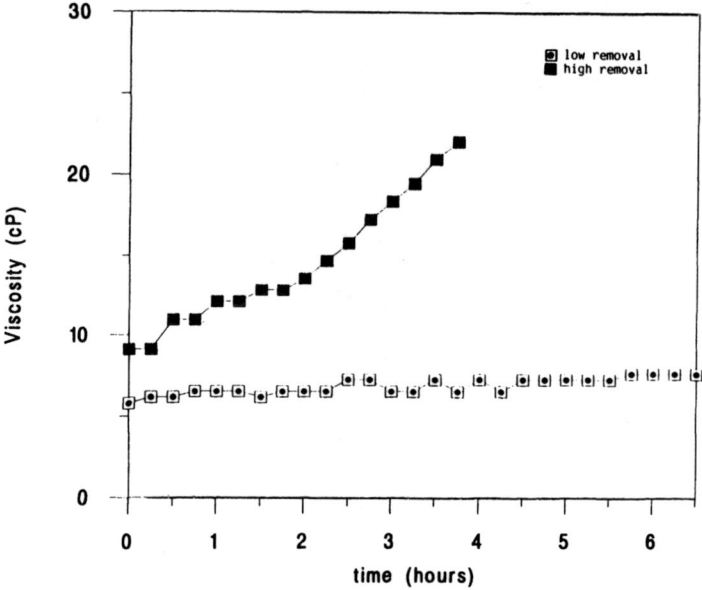

Fig. 2. Change in viscosity with time at 25 C for two different amounts of solvent removal.

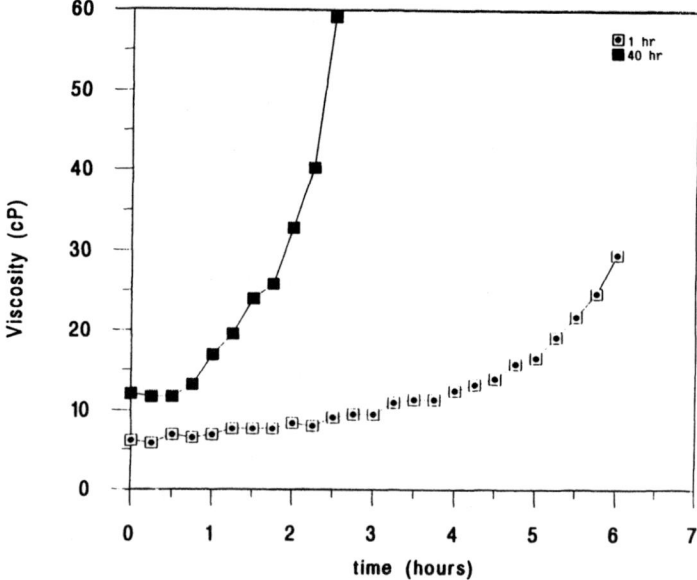

Fig. 3. Change in viscosity with time at 35 C for two different gel "incubation" periods.

particle radius r, and the centrifugal force $g = Rw^2$ where R is the average length of the centrifuge arm and w is the angular velocity,

$$\frac{dR}{dt} = \frac{2}{9} \left(D_P - D_L\right) r^2 Rw^2 \frac{1}{n} \tag{1}$$

The calculation for the experiments is complicated by one feature: in all cases over the viscosity range of interest, the change in viscosity with time $n(t)$ was experimentally determined to be of the form,

$$n(t) - n_o = A \exp(bt) \tag{2}$$

where A is the viscosity of the sol when mixed at stage 1 ($t = 0$), n_o is the change in viscosity due to vacuum hydrolysis, and b is an empirically derived constant related to factors which affect the change of the sol's viscosity with time. Two such fits are depicted in Figures 4 and 5. Upon solving for the change in distance a particle travels in time dt by substituting $n(t)$ for n in equation (1),

$$\int_{R_o}^{R_o + x} \frac{dR}{R} = \frac{2}{9} (D_P - D_L) r^2 w^2 \int_{t_o}^{t_1} \frac{dt}{n_o + A \exp(bt)} \tag{3}$$

integration yields,

$$\ln \left(1 + \frac{x}{R_o}\right) = \frac{2}{9} (D_P - D_L) r^2 w^2 \left[\frac{t_1 - t_o - \frac{1}{b} \ln \left[\frac{n_o + A \exp(bt_1)}{n_o + A \exp(bt_0)}\right]}{n_o} \right] \tag{4}$$

where x is the distance a particle has drifted due to settling, and R_o is the radius of the centrifuge to the silicon layer on top of the sol. Equation (4) was used to analyze the particle positions for settling of the silicon particles in the sol-gel glass precursor whose viscosity characteristics are shown in Figure 4. Centrifuging was begun when the viscosity reached about 50 cP at a time of about 2.75 hours. Calculations of x for a variety of particle sizes and centrifuging times are shown in Table I.

Table I. Drift Distances for Various Particles at 1250 rpm

Particle size (μm)	Drift length (cm)	Centrifuging time (hr)
1.0	0.085	0.25
1.0	0.21	1.0
1.0	0.25	2.0
1.0	0.26	8.0
1.0	0.26	16.0
0.1	0.0025	8.0
5.0	9.72	8.0
10.0	185.0	8.0

114

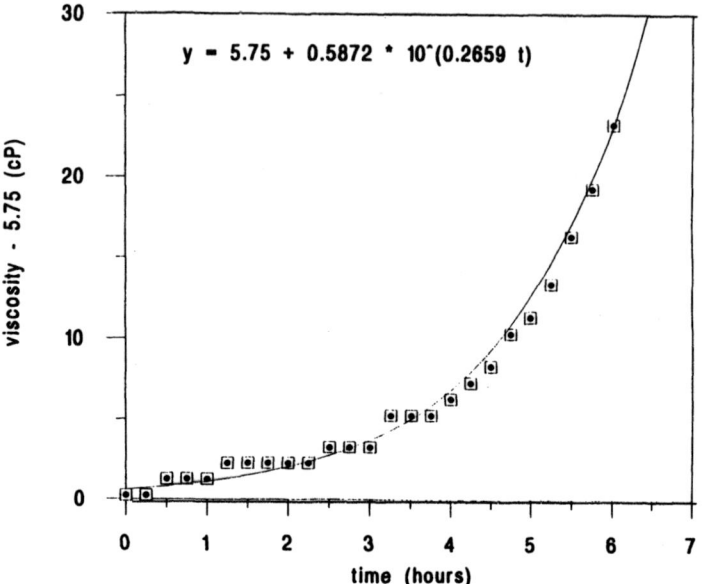

Fig. 4. Change in viscosity with time at 35 C.

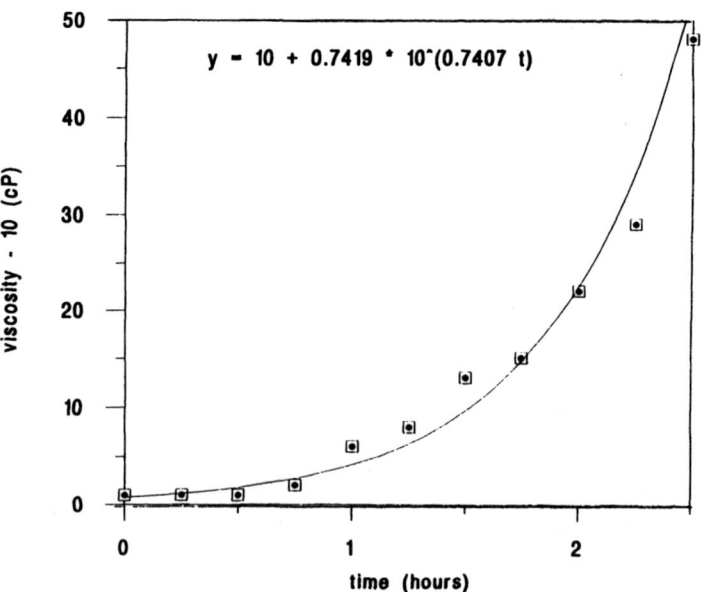

Fig. 5. Change in viscosity with time at 35 C.

Listed below are values used in the calculations.

n_o = 0.1 g/cm s (i.e., 10 cP; the change in viscosity due to vacuum hydrolysis)
A = 7.419 X 10^{-3} g/cm s (initial viscosity of the sol and fit from data)
b = 4.738 X 10^{-4} /s (derived from best fit of the data)
t_o = 2.75 hours (time after vacuum hydrolysis when centrifuging began)
t_1 = 3, 3.75, 4.75, 10.75, 118.75 hours (times during centrifuging plus t_o)
R_o = 8 cm (radius of the centrifuge to location of silicon particles)
D_P = 2.33 g/cm^3 (density of silicon)
D_L = 0.954 g/cm^3 (measured after vacuum hydrolysis)
w = 130 /s (measured to be 1250 rpm)
r = 5 μm, 2.5 μm, 0.5 μm, 0.05 μm (a proposed range of sizes)

Note that the gelation time is less than 8 hours for this particular sol. Assuming that the viscosity does not change with centrifugal force or particle concentration, rearrangement of the viscosity term in Eq. (4) shows that particle drift stops at sufficiently long times. The time when no further settling occurs is thus independent of centrifuge speed, particle size, and liquid or particle density. The total drift after gelation can be determined from

$$\ln\left[1 + \frac{x}{R_o}\right] = \frac{2}{9}(D_P - D_L)r^2w^2 \left[\frac{\frac{1}{b}\ln\left[\frac{n_o}{A} + \exp(bt_o)\right] - t_o}{n_o}\right] \qquad (5)$$

CONCLUSIONS

Viscosity characteristics of sol-gels were modeled. This enabled the calculation of particle size distributions for silicon in gels after high gravity settling. Gradients in optical density of silicon in the gels were observed. These gradients are attributed to population density differences among particles of different size. Porous silicon was incorporated in gels. Large agglomerates were separated from finely dispersed particles. The stage of sol-gel development when porous silicon is introduced seems to affect the luminescence characteristics of the silicon.

ACKNOWLEDGMENTS

NSF support through grant DMR 92-01685 (Materials Synthesis & Processing Initiative) is acknowledged. Professor Susan Kauzlarich and Samson Berhane supplied porous silicon materials for our gel experiments. Alex Nickel helped in the sol preparation.

REFERENCES

1. L.C. Liu, S.H. Risbud, in MRS Proceedings, Vol. 272 (1992) pp. 35–39.
2. T. Hirano, Second Symposium for Functionally Gradient Materials, July 1, 1988, Tokyo, The FGM Research Society, *Kino Zairyo*, Vol. 8 (1988) p. 15.

3. M. Seki, *Kino Zairyo,* Vol. 8 (1988) p. 7.
4. T. Igari, A. Notomi, H. Tsunoda, K. Hida, T. Kotoh, and S. Kunishima, in Proceedings of the First International Symposium on FGM, M. Yamanouchi, M. Koizumi, T. Hirai, and I. Shiota, editors, 1990, p. 11.
5. M. Niino, *Kino Zairyo,* Vol. 7 (1987) p. 31.
6. T. Kawai, S. Miyazaki, and M. Araragi, in Proceedings of the First International Symposium on FGM, M. Yamanouchi, M. Koizumi, T. Hirai, and I. Shiota, editors, 1990, p. 191.
7. M. Yuki, T. Murayama, T. Irisawa, A. Kawasaki, and R. Watanabe, in Proceedings of the First International Symposium on FGM, M. Yamanouchi, M. Koizumi, T. Hirai, and I. Shiota, editors, 1990, p. 2032.
8. W.C. LaCourse, *Sol-Gel Technology for Thin Films, Fibers, Preforms, Electronics, and Specialty Shapes,* L.C. Klein, editor, Noyes Publications (1988) pp. 184–98.

PHOTOLUMINESCENCE FROM NANOCRYSTALLINE SILICON
PREPARED BY PLASMA CVD AND OXIDATION

S. Vepřek, M. Rückschloß, B. Landkammer and O. Ambacher
Institute for Chemistry of Information Recording, Technical University Munich, Lichtenbergstraße 4,
D-W 8046 Garching/Munich, Germany

ABSTRACT

Light emitting nanocrystalline silicon has been prepared by a completely dry processing which uses standard silicon technology. This enables us to prepare compact films on various substrates and to control the crystallite size. Dependence of the photoluminescence intensity and its peak energy on the crystallite size is reported and compared with current theoretical models.

INTRODUCTION

The majority of papers on light emitting silicon deals with the "porous silicon", PS, prepared by anodic etching in a HF-ethanol-water solution [1, 2], and there are only few alternative preparation techniques available so far. These include HF-etching followed by wet anodic oxidation [3] and spark erosion [4]. An increasing number of papers is devoted to the fundamental studies of optical properties of nanosized clusters prepared either by gas phase reactions followed by a posttreatment (typically oxidation) [5-7] or by chemical precipitation in solution [8, 9].

The aim of our work is the preparation of compact thin films of light emitting silicon by conventional silicon technology with a controllable crystallite size in the relevant range of < 40 Å and their maximum possible content in the film. In such a way we hope to be able to control also the electrical transport properties of the material which are the necessary pre-requisites for obtaining an efficient and stable electroluminescence (EL). The control of the crystallite size and of their fraction allows us to study also the effect of the grain size on some fundamental properties and to compare the results with the suggested theoretical models. This is the main goal of the present paper.

EXPERIMENTAL PROCEDURE AND RESULTS

The films have been prepared by a combination of low temperature plasma CVD of a-Si:H or nc-Si followed by high temperature oxidation and annealing in forming gas. The experimental details can be found in our recent papers on this subject [10, 11] and in the references to our earlier work on the a-Si and nc-Si deposition given there.

The crucial step in the preparation of light emitting nc-Si-films is the control of the crystallite size and of the quality of the nc-Si/SiO_2-interface during the oxidation. The oxidation rate laws of a planar Si/SiO_2-structure, as obtained by the oxidation in dry or wet oxygen and well documented in the literature [12], do not apply to the oxidation of small nc-Si crystallites in the SiO_2-matrix. The nanosized particles show much lower oxidation rate than the planar Si/SiO_2-interface. The origin of this difference is currently being studied by several groups.

The free Gibbs energy of formation of small (spherical) particles of a radius R, $\Delta F_f(R)$, is usually larger than that of a bulk crystal, ΔF_f^0 (R→∞), due to the destabilizing contribution of the surface energy $\sigma > 0$:

$$\Delta F_f(R) = 4/3 \; \pi \; R^3 \; \Delta F_f^0 + 4\pi \; R^2 \; \sigma, \qquad \sigma > 0.$$

Thus, isolated small particles have lower melting point, higher vapour pressure and higher reactivity. The opposite effect, i.e. the significantly smaller reactivity of nc-Si/SiO$_2$ towards oxygen can be explained by lowering of the surface energy of the crystallites due to the formation of strong Si-O bonds. An alternative explanation could be a compressive stress which builds up in the films due to their oxidation, and slows down the oxygen diffusion. If this were the case, then the oxidation rate should sharply increase above the transformation temperature of SiO$_2$ (ca. 1040°C). As no such effect has been seen in our experiments we conclude that the lowering of the surface free energy of the nanocrystals due to oxidation on (σ(Si-O) ≪ 0) is mainly responsible for their smaller reactivity.

The slow oxidation rate of the nc-Si enables us to control their size even when using a conventional oven. The as-oxidized Si/SiO$_2$-interface contains a relatively large density of defects (e. g. dangling bonds as seen by ESR and lattice strain revealed in high resolution transmission electron microscopy). These can be annealed out in forming gas (3 - 5 mol% of H$_2$ in N$_2$) at high temperature.

Figure 1 shows the effect of the oxidation and annealing in forming gas on the intensity of the photoluminescence (PL). The PL-intensity of as-deposited nc-Si is very weak, but it increases upon oxidation. This has been reported in our previous paper [10, 11] and it can be seen in Fig. 1 when comparing the PL-intensity of the film after various oxidation times. Notice, that the total fraction of nc-Si in the film decreases upon prolonged oxidation. Thus, the increase of the PL-intensity is due to a strong increase of the transition probability for the optical transition (absorption or emission?) with decreasing crystallite size (see below).

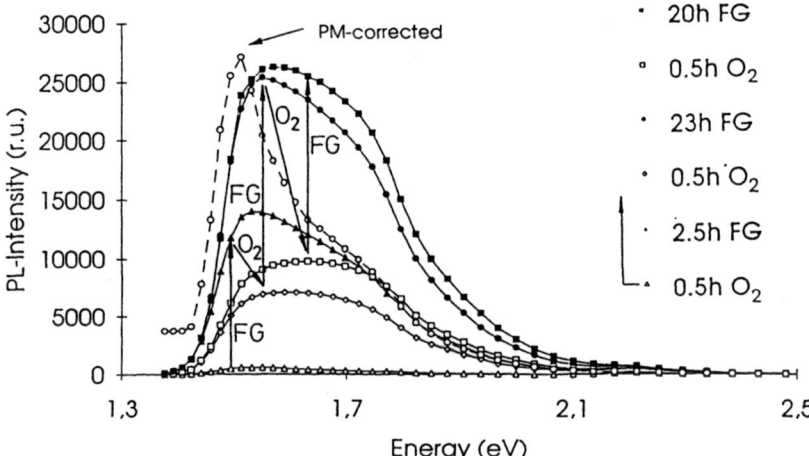

Fig.1: Effect of the oxidation in dry oxygen (1250°C, 1 at.) and annealing in forming gas (1250°C, 1 at., 5 mol% H$_2$ in N$_2$) on the photoluminescence of nc-Si. The broken line shows the spectral dependence corrected for the spectral response of the photomultiplier.

Figure 1 shows that the annealing of the films in forming gas strongly increases the PL-intensity although the crystallite size and their fraction in the film do not change. The relative scale of the PL-intensity in Fig. 1 is the same as in our previous paper (see Fig. 3 in [10]). A comparison of the present and 5 months older data shows that we were able to improve the PL-intensity by a factor of 10. The PL is clearly visible by naked eye under relatively weak UV-illumination and its intensity is fairly comparable to that of anodically etched PS. No degradation of the PL-intensity during several months of exposure of the samples to air and to the light has been noticed so far.

In order to compare the relative PL-efficiency (for financial limitation we could not perform the absolute calibration of our system yet) of different films we determined the average crystallite size and the fraction of nc-Si in the films using X-ray diffraction. Because of the small crystallite size and their fraction in the SiO_2-matrix, the Bragg reflections of the nc-Si are buried in the strong background of the X-rays scattered from the amorphous SiO_2. Therefore, about one week repetitive scans per one sample and signal averaging are needed to obtain the data.

The average crystallite size is obtained from the measured integral half width of the Bragg reflections using the Scherrer formula [13]. Unlike the full width at half maximum (which, in the case of a broad distribution of the crystallite sizes, overestimates the contribution of the large crystallites), the integral width yields values of the average crystallite size which are in a fair agreement with those obtained from the Warren-Averbach analysis [14]. The relative PL-efficiency is obtained from the measured PL-intensity when dividing it by the measured amount of nc-Si. The results are plotted in Fig.2 using the average crystallite size.

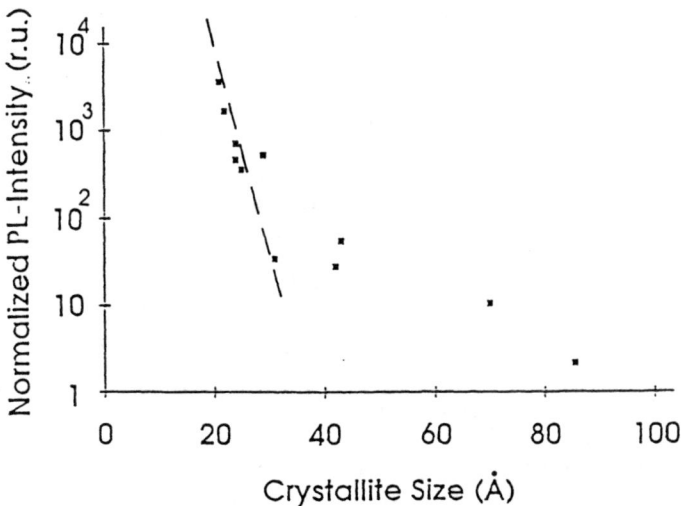

Fig.2: The points show the relative PL-efficiency as a function of the average crystallite size. The broken line corresponds to the average trend of the calculated oscillator strength of various optical transitions [15].

The data in Fig.2 show two different regimes: A relative small increase of the PL-efficiency with decreasing crystallite size from ~80 to 40 Å and a very strong increase below about 30 Å. The apparent dependence between 40 and 80 Å has little physical meaning for the following reasons: Due to the strong increase of the PL-efficiency below 30 Å a volume fraction of only 0.1% of 20 Å small crystallites in the disribution yielding an average size of 80 Å is sufficient to dominate the measured PL-intensity. Such a small fraction in total ≤ 6 vol % of nc-Si in the SiO_2-matrix cannot be seen in the XRD data. On the other hand, even 50% of 60-80 Å large crystallites (if they were present) in the sample with average size of 21-25 Å would not contribute noticeably to the measured PL-intensity but they were clearly visible in the XRD. Therefore, only the data below 30 Å show a meaningful dependence of the PL-efficiency on the average crystallite size. To best of our knowledge, these are the first experimentel data showing the dependence of relative PL-efficiency on the average crystallite size.

How does it compare with theoretical predictions? The radiative recombination of an electron from the lowest energy state of the conduction band (near the X-point in the Brillouin zone) with a hole at the top of the valence band (lowest energy at the central, Γ-point) is symmetry (and momentum) forbidden. The selection rule can be relaxed in a phonon assisted process. Because of a low probability of this event to occur the light absorption and emission at the band gap energy is small. Decreasing the crystallite size below 100 Å results in Brillouin zone folding which allows a mixing of the k-space representation of the bulk crystal states. The band gap transitions become increasingly allowed [15,16]. Simultaneously, the band gap should increase with decreasing grain size.

Several recent theoretical papers discuss these questions. Probably the most complete first principle calculations (LDF-approximation) have been done by Delley and Steigmeier [15] who calculated the optical transition matrix elements (or oscillator strength)

$$f_{ij} = 4\pi m/h \ \omega_{ij} \left(\int \Phi_i^*(r) x \Phi_i(r) \ d^3r \right)^2$$

for the dipole transitions between various quantum states of small Si-crystallites with two different configurations (either a Si-atom or a tetrahedral interstitial in the centre of the clusters) and surfaces terminated with bonded hydrogen. For a given cluster size the values of the oscillator strength depend on the nature of the transitions and of the clusters, and they can differ by almost two orders of magnitude (Fig.1 in [15]). Nevertheless, a strong increase of the oscillator strength with decreasing cluster size below about 30 Å (clusters containing about 1000 Si-atoms) has been found for all types of transitions [15]. Simultaneously, the calculated band gap increases proportionally to the reciprocal value of the cluster size (see Fig.2 in [15]).

For a fair comparison of the theoretical data of the oscillator strength with the measured dependence of the relative PL-efficiency on the crystallite size one would have to calculate (for each crystallite size) a weighed average value of the contributions of the various matrix elements in the given spectral region. As this is beyond the scope of this paper we plotted in Fig.2 as broken line the mean trend of the data from Delley and Steigmeier (Fig.1 in [15]). One can see that this trend is in reasonable agreement with our experimental data for crystallite sizes ≤ 30 Å.

If the PL were due to the recombination of an exciton confined within the small crystallite (i.e. electron near the bottom of the conductance band and hole at the top of the valence band), it should show a blue shift corresponding to the band gap increase with decreasing crystallite size D ($E_g \propto 1/D$, [15]). Our data (as well as those of some other researchers) however show that the main PL-peak at 1.5 eV does not change its position within the whole range of the average crystallite sizes covered here (20 to 90 Å). According to the data of Dalley and Steigmeier it should increase from about 1.7 eV for D ≈ 60 Å to more than 3 eV for D $\leq$ 20 Å. Interestingly, the data of Takagi et al. [5] show a smaller but clearly visible increase of the PL peak energy of the quasi-isolated Si-clusters with oxidized surfaces from about 1.4 eV for D ≈ 50 Å to about 1.65 eV for D ≈ 30 Å.

DISCUSSION AND CONCLUSIONS

The experimentally found increase of the relative PL-efficiency with decreasing crystallite size below 30 Å (Fig.2) is in reasonable qualitative agreement with the quantum confinement model and with the theoretical calculations by Delley and Steigmeier [15] of the oscillator strenght for optical transitions in small Si-crystallites. The absence of the blue shift, however, contradicts to the model according to which the PL originates in a radiative recombination of an excition confined within the crystallite. Thus, our data lend strong support for the surface-state recombination model of Koch et al. [17] which is pictured in Fig.3.

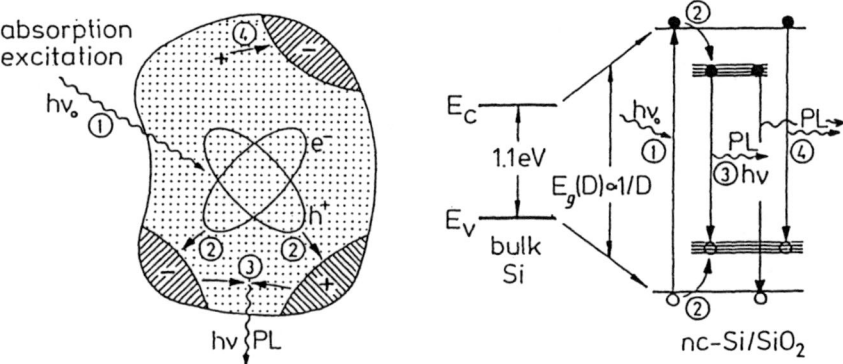

Fig. 3: The basic idea of the surface-state recombination model [17] as applied to the present data.

The absorption of the excitation radiation in the small nc-Si crystallites occurs at energies $h\nu \geq E_g(D) \propto 1/D$. The exciton is confined within the crystallite. There is a certain, but relatively small probability that it can recombine emitting a PL-photon with an energy of $h\nu \approx E_g(D)$. However, due to the large density of surface states which are localized below E_c for electrons and above E_v for holes, there is a much larger probability that the electron will be trapped (step 2 in Fig.3). The PL occurs by radiative recombination of electron and hole from these states (step 3). The reported long radiative time for the red luminescence of ≈ 100 μs (which has been also confirmed in our measurements) is in agreement with this picture.

Recent calculations of Gavrilenko et al. [18] for a thin Si-slab terminated with hydrogen have shown that the LDOS states which are associated with the Si-H and Si-Si bonds of the surface Si atoms directed into the slab dominate the DOS of the low laying states at E_c. Although no systematic set of such calculations is available for various slab thicknesses, one would expect the states to move somehow parallel with E_g when D decreases. The absence of any blue shift with 1/D in our PL-dates indicates that the states from wich the red luminescence originates may be of a "chemical nature" which fixes their energy to a more or less constant value. The fact, that we did not observe so far any measurable energy shift also for the PL-peak at $h\nu \approx 1.7$ and 2.2 eV suggests that there may be more such states. Of course, also the model of a hole at E_v recombining with an electron in the surface state (see 4 in Fig.3 and [17]) (or a free electron with trapped hole) is basically in accord with our data. Optical absorption data measurements on our samples clearly show the blue shift with decreasing D as expected from the model (Fig.3).

ACKNOWLEDGEMENT

We should like to thank Prof. F. Koch and his coworkers for many stimulating discussions. This work has been supported in part by the Deutsche Forschungsgemeinschaft.

REFERENCES

[1]   V. Lehmann and U. Gösek, Appl. Phys. Lett. **58**, 856 (1991).

[2]   L. T. Cunham, Appl. Phys. Lett. **57**, 1046 (1990).

[3]   E. Bustarret, M. Ligeon, J. C. Bruyere, F. Muller, R. Heerino, F. Gaspard, L. Ortega and M. Stutzmann, Appl. Phys. Lett. **61**, 1552 (1992).

[4]   R. E. Hummel and Sung-Sig Chang, Appl. Phys. Lett. **61**, 1965 (1992), MRS Symp. Proc. 283 (Fall Meeting, Boston, Dec. 1992, in press).

[5]   H. Takagi, H. Ogawa, Y. Yamazahi, A. Ishizaki and T. Nakagiri, Appl. Phys. Lett. **56**, 2379 (1990).

[6]   L. Brus, Appl. Phys. **A 53**, 465 (1991).

[7]   Y. Maeda, H. Uto, Y. Kanemitsu and Y. Masumoto, Int. Conf. on Solid State Devices and Materials, Tsukuba 1992, MRS Symp. Proc. 283 (Fall Meeting, Boston, Dec. 1992, in press).

[8]   J. R. Heath, this volume, paper B **4.1**.

[9]   S. Berhane, S. M. Kauzlarich, K. Nishimura, R. L. Smith, M. L. S. Olson, H. W. H. Lee and L. L. Chase, this volume, paper B **4.2**.

[10]   M. Rückschloß, B. Landkammer, O. Ambacher and S. Vepřek, MRS Symp. Proc. 283 (Fall Meeting, Boston, Dec. 1992, in press).

[11]   M. Rückschloß, B. Landkammer and S. Vepřek, Appl. Phys. Lett., submitted.

[12]   S. M. Sze, Physics of Semiconductor Devices, J. Wiley & Sons Inc., Singapore 1981.

[13]   H. P. Klug, L. E. Alexander, X-Ray Diffraction Procedures, John Wiley & Sons, New York 1974.

[14]   M. Rückschloß, Ph. D. Thesis, Tech. University Munich 1993.

[15]   B. Delley and E. F. Steigmeier, Phys. Rev. **B 47**, 1397 (1993).

[16]   N. F. Mott, E. A. Davis, Electronic Processes in Non-Crystalline Materials, Clarendon Press, Oxford 1979.

[17]   F. Koch, V. Petrova-Koch, T. Muschik, A. Nikolov and V. Gavrilenko, MRS Symp. Proc. 283 (Fall Meeting, Boston, Dec. 1992, in press).

[18]   V. Gavrilenko, P. Vogl and F. Koch, MRS Symp. Proc. 283 (Fall meeting, Boston Dec. 1992, in press).

PHOLUMINESCENCE STUDIES
ON POROUS SILICON QUANTUM CONFINEMENT MECHANISM

SHULIN ZHANG[*], KUOKSAN HE[*], YANGTIAN HOU[*], XIN WANG[*], JINGJIAN LI[**], PENG DIAO[**], BIDONG QIAN[**], SHENGMIN CAI[**] AND AKIRA FUJISHIJMA[***]
*Dept of Physics, Peking University, Beijing 100871, China
**Dept. of Chemistry, Peking University, Beijing 100871, China
***Dept. of Synthetic Chemistry, University of Tokyo, Japan

ABSTRACT

A novel step—like and pinning behavior of photoluminescence peak energy connected with changes in the concentration of HF and current density were for the first time observed for p— type porous silicon. Based on a theoretical calculation of the electron structure of the silicon quantum wire it is argued that these behaviors can be explained in terms of a novel formation mechanism model of porous silicon.

Introduction

Porous silicon (PS) has become a very interesting subject because of the discovery of its high efficiency of visible light emission. Based on the assumption that the remaining silicon walls between pores could be considered as a quantum wire array, the quantum confinement model was proposed to explain the light emission and formation mechanism [1, 2]; however, more experimental work is necessary to confirm this model. We measured the photoluminescence (PL) spectra of samples prepared by different anodizing conditions. The quantum confinement effects were used with success to analyses the above experimental results.

Experimental

PS samples were formed on n^- and p^- type silicon wafers with a resistivity of $10-15$ Ω cm by the anodic oxidation method similar to that used in Reference 3, but for n^- type samples the formation was carried out under light illumination. The PL spectra were measured by a Spex—1403 double monochromator with a cooled photomultiplier tube (PMT) RCA 31034 and a conventional photocounting electronics at room temperature. The excitation was provided by an Argon laser.

Experimental Results

(1) The step—like behavior of the PL peak energy of PS

In Fig.1(a) and (b) the PL spectra are shown for samples formed on a p^- type Si substrate in various HF concentrations, C_{HF}, at two fixed current densities J=50 and 20 mA cm^{-2} respectively for the forming time t= 8 min. We see the same characteristics in Fig 1 (a)and (b),

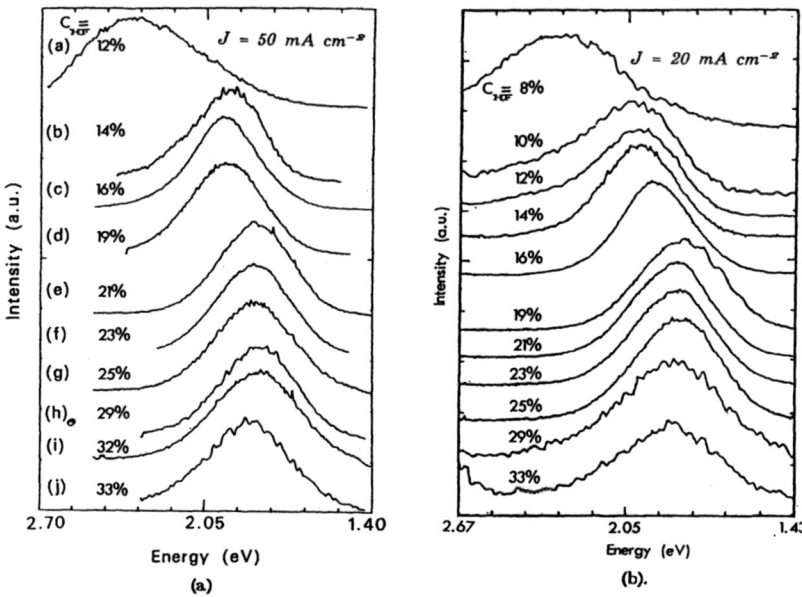

Fig.1 Photoluminescence spectra of p$^-$ type porous silicon samples formed in various HF concentrations C$_{HF}$ for the forming time t= 8 min at fixed current density J=50 mAcm^{-2} (a) and J= 20mAcm^{-2} (b).

where the PL peak positions do not obviously shift in the larger range of C$_{HF}$ but do shift in the smaller C$_{HF}$ change when there is a concentration of only 2–3 wt%. For clarity we reorganized the results of Fig.1(a) and (b) to give Fig.2 (b) and (c) respectively, where the PL peak energies E$_{PL}$ are given as a function of the HF concentration C$_{HF}$. In Figs 2(b) and (c) a step–like behavior appears for the variation of the PL peak energy versus HF concentration C$_{HF}$. After lapse of one year, the measurements were repeated on the samples in Fig.1 (a) and the step–like behavior could still be observed although the intensities of the PL spectra had considerably increased and the peak wavelengths had shifted a little towards high energy.

For the sake of contrast, the experimental results for n$^-$ type PS samples formed by the same forming conditions as those for the samples in Fig.1 (a) are shown in Fig2(a). In Fig.2 (a) no step– like behavior could be seen.

(2) The pinning behavior of the PL peak energies of PS

For PS samples grown on p$^-$type silicon substrate and formed at a fixed C_{HF} for a fixed time t but at various current densities J , we did not observe obvious variation of the PL peak energies E_{PL} with forming current densities J, as can be seen in Figs.3 (a) and (b) where the fixed values of C_{HF} are 35 wt% and 24 wt% respectively. We refer this behavior as the pinning behavior.

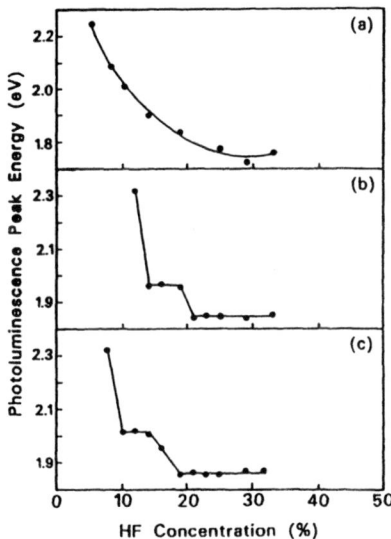

Fig.2 Variation of observed photoluminescence peak energies E_{PL} as a function of the HF concentration. (a) n − type PS samples formed in the various HF concentrations at fixed current density J=50 mAcm^{-2} for forming time t=8 min; (b) p− type PS samples formed in the same condition with that of Fig.2(a); (c) p− type PS samples formed in various HF concentrations at fixed current density J=20 mAcm^{-2} for the forming time t=8 min.

Discussion

To discuss the experimental results of PS on the basis of the quantum confinement model we have calculated the electron structure of Si quantum wires by the empirical pseudopotential method[4]. The main results are consistant with those calculated by Sanders and Chang using the tight bond method[5]. We understand that the microstructure of p− type PS is similar to nanometer crystalline silicon [6] and that the surface states play an important role in PS light emission. But for a qualitative discussion on the mechanism we think the simple quantum confinement model (QCM) is suitable.

(1) The calculation of the electron structure of quantum wires

In Fig.4 the variation of Si quantum wire energy gaps Egs is plotted against size L of the Si quantum wire based on calculations using the QCM. From Fig.4 we see that while the PS size L is less than 2.5 nm the energy band gap increases rapidly and when L is larger than 10 nm Eg decrease very slowly and gradually approaches the energy gap of bulk Si (~ 1.1 eV).

If we consider that the PL peak energy could be corresponded approximately to Eg and the quantum confinement model is true, for the step—like behavior of PL peak energies in P^- type PS,it suggests that the variation of the size L of a silicon quantum wire with a change in the HF concentrations C_{HF} is not continuous but step— like. This phenomenon has not been reported before.

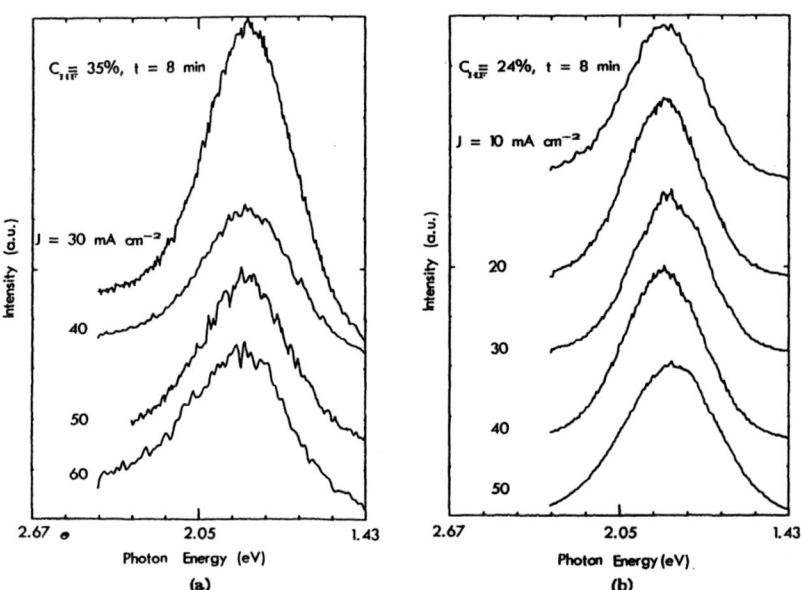

Fig.3 Photoluminescence spectra of p^- type PS samples formed in various current densities J for the forming time t=8 min at fixed HF concentrations C_{HF}=35 wt% (a) and C_{HF}=24 wt% (b).

(2) Formation Mechanism

The step—like change of silicon quantum wire sizes with the variation of HF concentration cannot be explained by the conventional electrochemical theory. According to the Schottky barrier model of anodization, using the conventional electrochemical theory [7], with decrease in the HF concentration the Schottky barrier V_e at the Si/electrolyte interface is reduced, which induces the dissolution of Si and then leads to a reduction in the size of Si quantum wire, hence the HF concentration is directly proportional to the size of the silicon quantum wire. So the size of the silicon quantum wire must be continuously reduced and can not exhibit a step— like change as the concentration of HF is decreased as predicted in the frame of the conventional electrochemical theory.

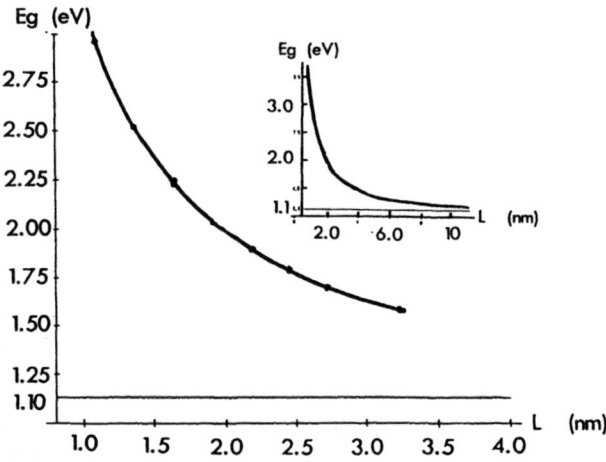

Fig.4 Calculation current of Si quantum wire energy gaps Eg's against size L of the Si quantum wire by the empirical pseudo— potential method.

In order to explain the above behavior, we have proposed a Schottky formation model combined with the quantum effect[8]. In this new model, besides the Schottky barrier V_e at the Si/electrolyte interface, an additional quantum potential V_q due to the quantum size effect of silicon quantum wires should be involved. Only if the holes pass over the new barrier $(V_e + V_q)$, can the dissolution reaction take place. At the stage where the concentration of HF gradually decreases but the dissolution of Si does not take place, it is the quantum confinement barrier V_q which prevents holes passing over the barrier $(V_e + V_q)$ and silicon being dissolved. As soon as there is a decrease in the electrochemical barrier V_e so that $(V_e + V_q)$ reduces to the value at which the holes can pass over the barrier and reach the surface of the silicon quantum wire, dissolution at once takes place and the size of the silicon quantum wire begins to decrease. Because of the quantum confinement effect, a decrease in the size of Si quantum wires must induce an increase of the quantum barrier V_q. When the increase of V_q exceeds the decrease of V_e the barrier $(V_e + V_q)$, under this new condition, would again block the holes arriving at the interface of silicon /HF solution and the dissolution would stop. In the case of small size of Si quantum wires, here about 2 nm ,from the experimental and the theoretical results, only when the decrease in the average size of PS is one or two silicon layers, is the increase of barrier V_q enough to exceed the decrease of the electrochemical barrier V_e due to the gradual decrease of HF concentration in the before and behind dissolution stage [3]. Therefore a new plateau of a fixed new size would appear until the more decrease of V_e due to the farther decrease of HF concentration makes the holes pass over the new barrier (V_e+V_q). The process mentioned above may explain the step−like behavior of size changes in the silicon quantum wire size. The appearance of the new plateau must correspond to an increase of V_q larger than the decrease of V_e. The values estimated from experiments satisfy the above requirement, and demonstrate that the Schottky formation model combined with the quantum confinement effect can correctly explain the formation mechanism of p−type PS.

In term of the above model, a explanation of the pinning behavior can be obtained .When the HF concentration and the resistivity of the electrolyte are kept to constant, variations in the anodizing current were obtained by varying the voltage of the applied power. Although any variation of the applied voltage will produce a variation in the hyper electrical potential, the varying of hyper electrical potential only influences the pore bottoms and does not influence the barrier at the pore wall/electrolyte interface because the direction of applied field is parallel to the pore wall in the quantum wire model. As a result, the forward etching speed should change, which agrees with other experiments , but the etching of the pore wall should not be influnced by the hyper electrical potential and hence the size of the silicon quantum wire does not change.

Thanks are due to the support from the NSFC.

References

1 L.T.Canham , Appl. Phys. Lett. 57, 1046(1990).

2 V.Lehmannand and H.Foll, J.Electrochem. Soc., 137, 653(1990).

3 Shu−lin Zhang, Yongtian hou, and Huok−San Ho, Bidong Qian and Shengmin Cai, J. Appl. Phys., 72, 4469(1992).

4 Meng−Yan Shen and Shu−Lin Zhang to be published

5 G.D.Sanders and Y.C.Chang, Phys. Rev. B45 9202(1992).

6 Shu−lin Zhang, Xin Wang, Huok−San Ho, Jinjian Li,Peng Diao,and Shengmin Cai, Submitted to Appl. Phys. Lett.

7 H.Gerisher, in Physical Chemistry, an Advanced Treatise, edited by H. Egring (Academic, New York, 1970), Vol.9A, Chap.5.

8 Shu−lin Zhang, Huok−San Ho, Yongtian hou, Bidong, Qian Peng Diao and Shengmin Cai, Appl. Phys. Lett., 62, 642(1993)

CHARACTERIZATION OF Si_nGe_m STRAINED LAYER SUPERLATTICE P-N JUNCTIONS

Jesper Engvall, Janos Olajos, Hermann G. Grimmeiss
Dept. of Solid State Physics, University of Lund, Box 118, S-221 00 Lund, Sweden

Hartmut Presting, Horst Kibbel and Erich Kasper
Daimler-Benz AG, Research center, Wilhelm-Runge-Str. 11, D-7900 Ulm, Germany

Abstract

A series of three p-n junctions consisting of ten monolayer strain adjusted Si_nGe_m superlattices (Si_6Ge_4, Si_5Ge_5, Si_4Ge_6) has been characterized with respect to Current-Voltage, Capacitance-Voltage and Short Circuit Photocurrent measurements. The samples were grown by MBE with an Antimony surfactant resulting in improved interface sharpness. Superlattice interband transitions were measured by Short Circuit Current spectroscopy and bandgap energies were determined by a fitting procedure. The temperature dependence of the junction capacitance and of the Short Circuit Current indicate a potential barrier for electrons at the superlattice-buffer interface that impedes electron transport from the p-n junction at lower temperatures.

Introduction

The large interest in strained layer superlattices (SLS) of Si and Ge originates largely in theoretical calculations. Predictions have been made that although the constituent materials Si and Ge have indirect bandgaps, such superlattices can be made quasi-direct through the combined effects of strain and zone-folding[1]. The difficulty of producing strained layers of sufficient quality for optical applications has recently been overcome in growth of Si-rich $Si_{1-x}Ge_x$ Quantum Wells (QW) when it was shown that high growth temperatures were needed for the photoluminescence properties[2]. The high growth temperatures used in growth of $Si_{1-x}Ge_x$ QW's cannot directly be transferred to growth of Si_nGe_m Superlattices since the high growth temperatures also result in smearing of the interfaces due to Ge segregation. The introduction of Sb as a surfactant has however been shown to reduce interface smearing and therefore made it possible to use the higher growth temperatures during SLS growth.

We report measurements performed on a series of strain adjusted Si_nGe_m SLS p-n junctions grown by Molecular Beam Epitaxy (MBE) where the SLS was grown at 500°C utilizing Sb as a surfactant. Capacitance-Voltage and Current-Voltage measurements were performed and measurements of Short Circuit Photocurrent (I_{SC}) was used to determine the fundamental bandgap of the superlattices for different temperatures.

Sample Structure

The layer structure is shown in the inset in fig. 1. The substrate material is n^+ Si. On the substrate was first grown a 50nm-thick Si layer, doped with Sb by Secondary implantation (DSI) to a concentration of more than 1×10^{17} cm^{-3}. On this was grown a 100 nm $Si_{1-x}Ge_x$ buffer layer, grown at 575°C, followed by a 145 period Si_nGe_m superlattice (200nm), grown at 500°C. The growth of both the alloy buffer and the superlattice was preceded by the deposition of a monolayer (ML) of Sb. On top of the superlattice the p-layer was grown at 500°C as a $Si_{1-x}Ge_x$ alloy with the same Ge concentration as the SLS, doped by boron coevaporation in excess of $1 \times 10^{19} cm^{-3}$. The alloy was then capped with 10 nm of p^+ Si.

The p-n junctions were fabricated into mesa diodes of different dimensions, and the p-layer was contacted with Au/Ti contacts with a window to allow the optical studies.

Sample	buffer	SLS	p-layer
B2460	$Si_{0.50}Ge_{0.50}$	Si_6Ge_4	$Si_{0.60}Ge_{0.40}$
B2517	$Si_{0.40}Ge_{0.60}$	Si_5Ge_5	$Si_{0.50}Ge_{0.50}$
B2518	$Si_{0.25}Ge_{0.75}$	Si_4Ge_6	$Si_{0.40}Ge_{0.60}$

Table 1 Nominal Values of Layer Structure

The nominal growth data are summarized in table 1. The samples have been studied by Transmission Electron Microscopy (TEM) and X-Ray Diffraction (XRD) to investigate the morphology, strain and layer composition. The TEM investigations show that the morphology is good for the Si_6Ge_4 sample and that the SLS period and strain is close to achieving the nominal values. However, the TEM studies also show that the morphology is significantly worse for the Si_5Ge_5 and Si_4Ge_6 samples, which makes it difficult to determine the strain and SLS period in these samples. In the following we will mainly concentrate on the Si_6Ge_4 sample and less on the more Ge-rich samples.

Characterization

The diodes were characterized by Capacitance-Voltage (C-U) measurements to determine the depletion layer width and doping concentration. The C-U for the Si_6Ge_4 sample is shown in fig.1. As seen, the doping in the superlattice and buffer are different, 1.5×10^{17} for the SLS, but less than 6×10^{16} for the buffer. The doping in the superlattice is constant enough to determine the built in Voltage (V_{bi}) as an extrapolation of $1/C^2$ to zero. This was done for a series of temperatures and gives $V_{bi}=0.699, 0.746, 0.769, 0.782$ V at T=293K, 242K, 211K, 194K, respectively, where a correction of 2kT/e has been made to account for the carrier distribution tails[3].

The zero bias capacitance translates to a space-charge region at room temperature of 90nm for Si_6Ge_4 sample and 100 nm for the Si_5Ge_5 sample. For the Si_4Ge_6 sample the

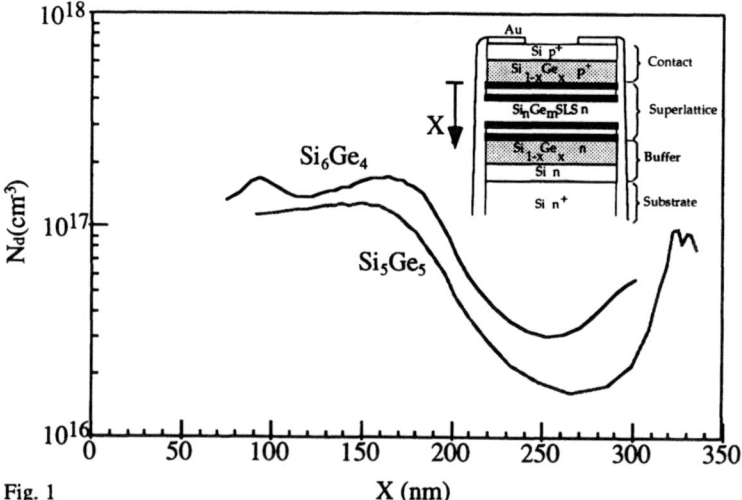

Fig. 1

Doping concentrations obtained from Capacitance-Voltage measurements for the Si_6Ge_4 and Si_5Ge_5 samples. The inset shows the layer structure.

capacitance is significantly less than expected and the depletion layer width is less certain. For the Si_6Ge_4 sample and the Si_5Ge_5 sample this means that the entire Space charge region is contained in the superlattice layer. Since the I_{SC} is only generated in the space charge region[4] the I_{SC} will be a measurement of the absorption in the superlattice and we therefore use the I_{SC} to determine the fundamental energy gap of the superlattices.

In fig. 2 is the square root of the I_{SC} plotted vs photon energy at 229K for the three samples studied. As can be seen there is a linear relationship in this graph and the I_{SC} is therefore investigated by fitting the current to an expression of the type

$$I_{SC} = qA\Phi\left[f_{tail}(h\vartheta) + C_1\left(h\vartheta - E_{G,SLS}\right)^2 \Theta\left(E_{G,SLS}\right) + C_2\left(h\vartheta - E_2\right)^2 \Theta\left(E_2\right)\right] \tag{1}$$

where A is the diode window area, ϕ is the incident photon flux, $h\upsilon$ is the photon energy, C_1 and C_2 are constants. The first term in the parentheses, $f_{tail}(h\upsilon)$, is an exponential tail for photon energies below the energy $E_{g,SLS}$ which has the shape

$$f_{tail}(h\vartheta) = \begin{cases} C_0 \exp\left(\dfrac{h\vartheta - E_0}{E_1}\right) & ; h\vartheta < E_0 \\ C_0 & ; h\vartheta > E_0 \end{cases} \tag{2}$$

The second term in the parentheses is used to identify the bandgap of the SLS. The third term is used to identify any higher lying onsets due to absorption in the Si substrate or the SiGe buffer layer. There is a second onset (not shown in fig. 2) at 1.07 eV for the Si_4Ge_6 sample, that is interpreted as contributions to the I_{SC} from absorption in the buffer. The function Θ is the Heaviside stepfunction.

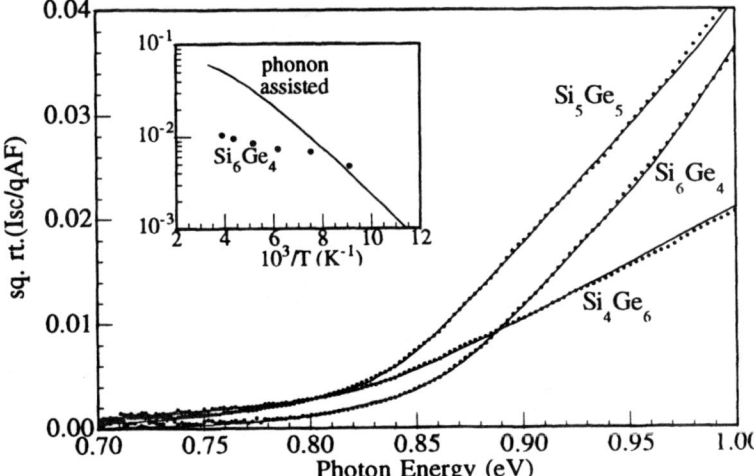

Fig. 2
The square root of I_{SC} for the three samples. The points are measured, the solid line is the square root of the fitted expression.

There is no a priori reason that this should be the functional dependence of the absorption in a Si_nGe_m SLS. The square of the photon energy is a good approximation for describing the absorption in indirect semiconductors. For these materials it can be justified theoretically when assuming parabolic bands, a constant matrix element and an assumption that only one phonon branch contributes significantly. Then one ends up with the Macfarlane-Roberts expression[5]:

$$\alpha(h\vartheta,T) = K\left[\left(h\vartheta - E_G + \hbar\omega_q\right)^2 n_q + \left(h\vartheta - E_G - \hbar\omega_q\right)^2 \left(n_q + 1\right)\right]$$

(3)

where n_q is the Bose-Einstein distribution function of the participating phonon of energy $\hbar\omega_q$ and K is a temperature independent constant. Neither the assumption of constant matrix element nor parabolic bands are justified in a Si_nGe_m superlattice[6]. Expression 3 also implies a change in the shape of the absorption edge with temperature, due to the temperature dependence of the phonon population. The steepness of the absorption edge would change according to:

$$\frac{\alpha(E_G,T)}{\alpha\left(E_G + 2\hbar\omega_q,T\right)} = \frac{1}{9 + \exp\left(\dfrac{\hbar\omega_q}{kT}\right)}$$

(4)

This ratio is plotted semilog vs inverse temperature in the inset of fig.2. for a phonon energy of 52meV (This seems to be an appropriate phonon energy. The Transverse Optical (TO) phonon energy in Si is 63 meV and fitting absorption in Si to (3) gives a phonon energy of about 50 meV[7]. The luminescence of a Si_6Ge_4 superlattice[8] was interpreted as a no-phonon and TO-phonon line and the phonon energy was about 52meV.). Also plotted is the ratio of the photocurrent at $h\upsilon = E_{G,SLS} - 52meV$ to the photocurrent at

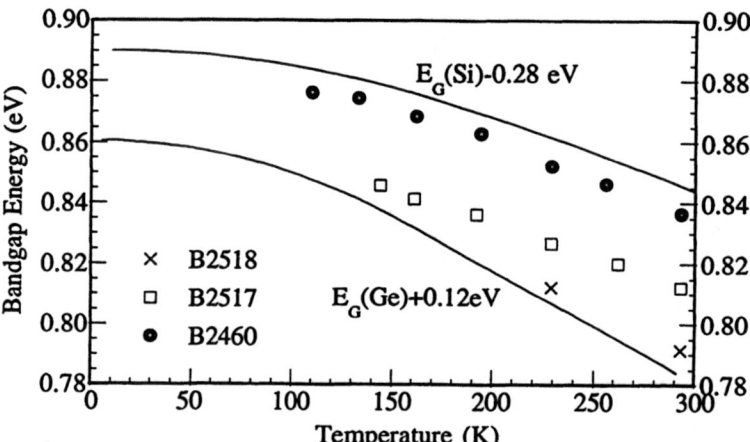

Fig. 3
Bandgap energies for the three samples determined from measurements of I_{SC}.

$h\upsilon = E_{G,SLS} + 52\,\text{meV}$. As can be seen the current ratio remains essentially constant. There is nothing to suggest that the absorption is phonon assisted, and the parabolic dependence should be seen as an experimental result, not as evidence of an indirect transition.

The fitted energy $E_{G,SLS}$ is plotted as a function of temperature in fig.3 for the three samples. The temperature dependence is compared to the temperature dependence of the Si and Ge bandgaps[9]. Assuming the same temperature dependence for the Si_6Ge_4 sample as for Si at lower temperatures, the bandgap for this sample at 4K is extrapolated to 0.88 eV, in close agreement with the NP PL peak position of a Si_6Ge_4 SLS at 0.877 eV[7]. This luminescence was suggested to originate from weakly bound excitons. For the Si_5Ge_5 sample the measured bandgap of 0.84 eV at 150K is significantly higher than the value 0.76 eV at 150K that we have reported earlier[10]. The discrepancy is probably partly due to the larger than nominal in-plane strain in the Si layers of the SLS in that sample, which is expected to lower the bandgap. For a Si_4Ge_6 SLS we have found no previous experimental data to compare with.

The measured junction Capacitance deviates from the expected values at low temperatures. The capacitance drops rapidly when the temperature is lowered, and the transition temperature is different for the different samples. Capacitance measurements at lower frequencies give higher values, as can be seen in the inset of fig. 4, indicating that it is the RC-time-constant of the junction that is increasing with decreasing temperature. At even lower temperatures the large RC-time-constant is also affecting the I_{SC}-measurement, resulting in a slow response to a modulation of the light, and eventually a large reduction in the photocurrent. This is what limits the temperature range in which reliable I_{SC}-measurements can be performed (fig.3).

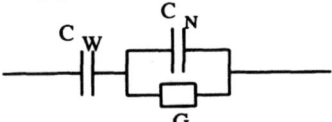

We model this RC-constant with a small signal equivalent circuit as above, where the capacitance C_W is the depletion layer capacitance, and C_W in series with C_N gives the high

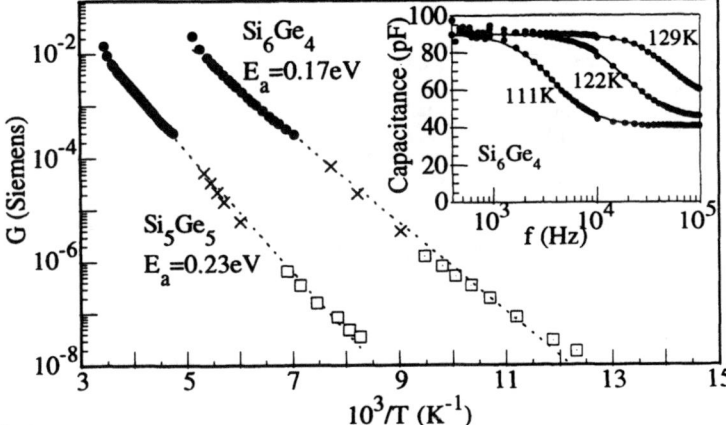

Fig.4
Conductance calculated from measurements of junction RC-time-constant. The solid lcircles, crosses and open squares are obtained from capacitance measurements at 1 MHz, capacitance measurements as a function of frequency and measurements of I_{SC} falltime respectively. The inset shows the capacitance vs frequency for the Si_6Ge_4 sample.

frequency (low temperature) capacitance. For each measurement the value of the conductance, G=1/R can then be calculated, taking into account the expected temperature dependence of the depletion layer hrough a linear extrapolation of V_{bi}, but assuming C_N to be constant. The calculated conductance is displayed in the Arrhenius plot of fig.4. It shows a thermally activated behaviour, with an activation energy of 0.17 eV for the Si_6Ge_4 sample and 0.23 eV for the Si_5Ge_5 sample. Even if the exact value of the Sb shallow donor level in these materials is unknown, this is much larger than expected from a freezeout of the dopant in the SLS. The thermally activated behaviour indicates that the buffer layer is acting as a potential barrier between substrate and superlattice for the electrons. Two factors can contribute to this:

The net n-doping in the buffer is much lower than in the SLS as seen by the C-U measurements. A lower degree of incorporation of Sb in Ge than in Si will make the Sb-concentration less in the buffer than in the SLS. The n-doping is also reduced by a background Boron doping in the layers from the Ge-source. The low doping will also make the resulting free electron concentration very sensitive to any other unintentional deep acceptor levels, due to the crystal defects.

A bandoffset between Superlattice and buffer or between buffer and substrate can produce this effect. Thermionic emission theory used to describe conductance over SIScapacitors[11] and n-n heterojunctions[12] predict a thermally activated conductance. From Kronig-Penny calculations, the bandoffset between a Si_5Ge_5 SLS and an unstrained $Si_{0.50}Ge_{0.50}$ alloy[13] is expected to be small (32 meV). These buffers are more Ge-rich than the SLS and strained, which will change the Kronig-Penny-value, but not by 200meV.

Summary

By measurements of I_{SC} on p-n junction diodes, we have determined bandgap energies for a series of strain adjusted Si_nGe_m superlattices, grown with an Sb surfactant at 500°C. The temperature dependence of the diode capacitance and photocurrent suggest that these measurements are affected by potential barriers at the $Si_{1-x}Ge_x$ buffer layer.

This work was partially financed by the European Basic Research Programme (ESPRIT) under contract P7128.

[1] R. J. Turton and M. Jaros, Mat. Sci. Eng. B **7**, 37 (1990)

[2] N.Usami, S. Fukatsu and Y. Shiraki, Appl. Phys. Lett. **61**, 1706, (1992)

[3] C.G. B. Garrett and W.H. Brattain, Phys. Rev. **99**, 376 (1955)

[4] S. Braun and H.G. Grimmeiss, J. Appl. Phys **44**, 2789 (1973)

[5] G.G. Macfarlane and V. Roberts, Phys. Rev. **97**, 1714, (1955)

[6] R.J. Turton, M. Jaros and I. Morrison, Phys. Rev. B **38**, 8397 (1988)

[7] R.Braunstein, A. R. Moore and F. Herman, Phys. Rev., **109**, 695, (1958)

[8] U. Menczigar, G. Abstreiter, J. Olajos, and H.G. Grimmeiss, Phys. Rev. B, **47**, 4099, (1993)

[9] O. Madelung (ed.), Landolt-Börnstein, *Numerical Data and Functional Relationships in Science and Technology*, Group III, Vol 17a, Springer, Berlin, 1982, pp 66-67

[10] J. Olajos, J. Engvall, H.G. Grimmeiss, U. Menczigar, G. Abstreiter, H. Kibbel, E. Kasper and H. Presting, Phys. Rev. B, **46**, 12 857, (1992)

[11] T.W. Hickmott, P.M. Solomon, R. Fischer and H. Morkoc, J. Appl. Phys **57**, 2844 (1985)

[12] L. L. Chang, Solid State Electronics, **8**, 721, (1965)

[13] U.Menczigar, J. Brunner, E. Friess, M. Gail, G. Abstreiter, H. Kibbel, H. Presting and E. Kasper, Thin Solid Films, **222**, 227 (1992)E-MRS

CHARACTERIZATION OF
Ge AND C IMPLANTED $Si_{1-x}Ge_x$ AND $Si_{1-y-z}Ge_yC_z$ LAYERS

ASHAWANT GUPTA*, YAO-WU CHENG*, JIANMIN QIAO*, M. MAHMUDUR RAHMAN*, CARY Y. YANG*, SEONGIL IM**, NATHAN W. CHEUNG**, AND PAUL K.L. YU***
*Microelectronics Laboratory, Santa Clara University, Santa Clara, CA 95053
**Department of Electrical Engineering and Computer Sciences, University of California, Berkeley, CA 94720
***Department of Electrical and Computer Engineering, University of California, San Diego, La Jolla, CA 92093.

ABSTRACT

In an attempt to substantiate our previous findings of boron deactivation and/or donor complex formation due to high-dose Ge and C implantation, SiGe and SiGeC layers were fabricated and characterized. Cross-sectional transmission electron microscopy indicated that the SiGe layer with peak Ge concentration of 5 at% was strained; whereas, for higher concentrations, stacking faults were observed from the surface to the projected range of Ge as a result of strain relaxation. Results of spreading resistance profiling were found to be consistent with the model of dopant deactivation due to Ge implantation and subsequent solid phase epitaxial growth of the amorphous layer. Furthermore, for unstrained SiGe layers (Ge peak concentration ≥ 7 at%), formation of donor complexes is indicated. Preliminary photoluminescence results correlate with the spreading resistance profiling results and indicate shallow donor complex formation.

INTRODUCTION

Alloy of silicon and germanium is a promising material for use in the fabrication of high-performance devices for the future. Fabrication of Si-based heterojunction bipolar transistors (HBTs) and p-channel metal oxide semiconductor field effect transistors (MOSFETs) using ultra-high vacuum chemical vapor deposition (UHV-CVD) and molecular beam epitaxy (MBE) for the epitaxially grown SiGe layer have been reported [1]. These devices are promising as they show improved performance over existing Si transistors. SiGe devices have also been fabricated in layers formed by high-dose Ge implantation followed by solid phase epitaxy [2]. The potential advantage of implantation over UHV-CVD, MBE, and other epitaxial growth processes is in its compatibility with the standard silicon fabrication line and convenience in selective area growth. However, SiGe layers formed by high-dose Ge implantation present some problems. Extrinsic dislocation loops are formed during implantation due to excess recoiled Si interstitials. Additionally, high Ge dose induces a lattice strain in the regrown SiGe layer, resulting in the formation of surface defects (stacking faults) [2,3]. These defects degrade the performance of devices formed in this layer. Compared with room-temperature (RT) implantation, Ge implantation performed at a low-temperature (LT) is reported to result in the reduction of the dislocation loops. Also, sequential implantation of C following Ge results in the reduction of surface defects due to strain compensation [2,3]. In this paper, in an attempt to substantiate our previous findings of boron deactivation and/or donor complex formation due to high-dose Ge implantation, we report the characterization of Ge implanted SiGe and Ge and C implanted SiGeC layers. Some preliminary photoluminescence (PL) results are correlated with cross-sectional transmission electron microscopy (XTEM) and spreading resistance profiling (SRP) results. The observations are found to be consistent with previously reported electrical characteristics of SiGe devices [2].

GROWTH OF SiGe AND SiGeC LAYERS

SiGe and SiGeC layers were fabricated by performing high-dose Ge and C implantation into 1 ohm-cm n-type Si <100> substrates. The Ge implantation was performed at liquid-nitrogen temperature. A range of Ge dose from 2×10^{16} /cm^2 to 5×10^{16} /cm^2 was used. An ion beam energy of 120 keV was used to obtain a 170 nm thick amorphous layer with peak Ge concentration ranging from 5 at% to 12 at%. Carbon implantation was subsequently performed in one sample at room temperature with a dose of 2×10^{15} /cm^2 and an energy of 20 keV to obtain a peak concentration of 0.5 at% and a projected range (R_p) of about 65 nm. Rutherford backscattering spectrometry (RBS) results indicate that the R_p for Ge is about 70 nm. All samples were annealed at 800 °C for 1 hour in nitrogen ambient to regrow the amorphous layer. Table 1 shows the implant conditions for the samples used in this study. Sample 1 is the Si control.

Table I. Implant conditions for samples

Sample	Ge Implant Temperature (K)	Dose (/cm^2)	Energy (keV)	Peak Ge (at%)	Peak C (at%)
1	NA	0	0	0	0
2	77	2×10^{16}	120	5	0
3	77	3×10^{16}	120	7	0
4	77	5×10^{16}	120	12	0
5	77	5×10^{16}	120	12	0.5

ELECTRICAL AND OPTICAL CHARACTERIZATION

In a previous study, we suggested boron deactivation and/or donor complex formation due to Ge implantation into Si [2]. In an effort to confirm that model which was based on capacitance-voltage characteristics of n^{++}-p^+ diodes formed in the SiGe layer, SRP was performed on all 5 samples. The results are shown in figure 1. The profile for sample 1 is not shown as it is the Si control and had a constant resistivity. For the SiGe and SiGeC samples, the majority electron concentration was estimated from the resistivity profile under the assumption that the free carrier mobility is the same as that in Si. Hot point probe was used to ascertain that all samples were still n-type after Ge and C implantation. The resistivity for sample 2 around the projected range of Ge ions is higher than the bulk resistivity by about an order of magnitude. Since this change is larger than the expected change in mobility for this alloy [1], it may be attributed to dopant deactivation due to Ge implantation. One expects that the magnitude of dopant deactivation would somewhat relate to the Ge dose, and an increase in Ge dose would result in an increase in resistivity around the projected range. Contrary to this reasoning, as can be seen for sample 3, the resistivity decreases with an increase in Ge dose. XTEM results of samples 2 and 3 indicate that there are no surface defects for sample 2 where as the strained SiGe layer in sample 3 relaxes to produce stacking faults from the surface to R_p (for XTEM micrographs for all samples except Si control see ref. 3). Thus, the decrease in resistivity for sample 3 may be attributed to compensation of phosphorus deactivation by formation of donor complexes resulting from strain relaxation. For samples 4 and 5, the resistivity decreases by about two orders of magnitude, supporting the fact that once the critical Ge dose for strain relaxation is exceeded (~3×10^{16} /cm^2), any further increase in Ge dose results in a significant increase in net donor concentration.

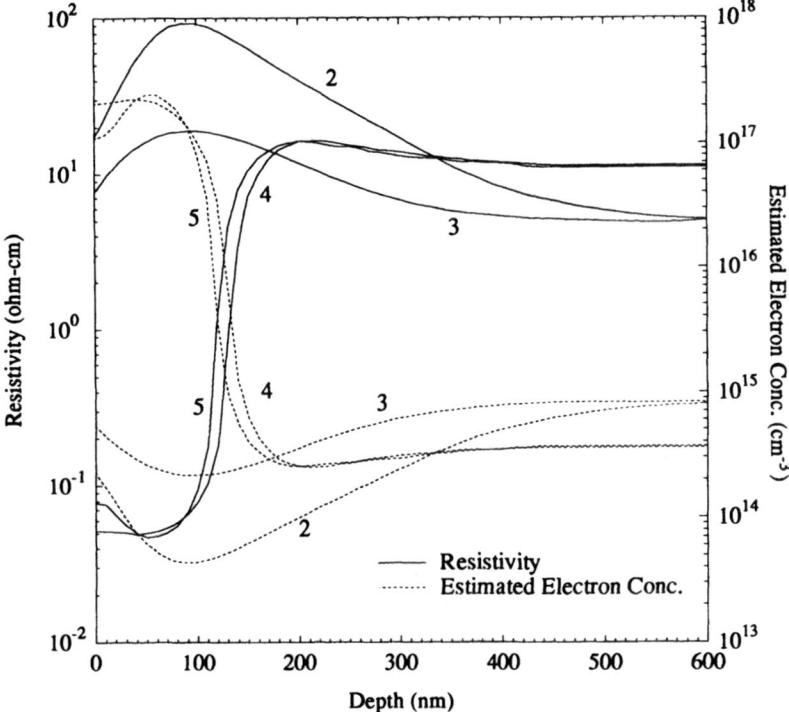

Figure 1: Resistivity profiles obtained from SRP. Estimated electron concentration profiles calculated from the resistivity profiles assuming free carrier mobility to be same as in Si.

Preliminary results of photoluminescence measurements performed on the samples are shown in figure 2. The excitation was provided by an Argon ion laser with a wavelength of 514.5 nm. The laser beam diameter was 2 mm and the incident power was kept constant at 500 mW. The temperature was maintained at ~10 K. The results once again seem to correlate with the hypothesis that Ge implantation results in phosphorus deactivation as well as donor complex formation. It has been reported that boron and phosphorus concentration in silicon can be estimated by computing the ratio of the dopant peak intensity to the intrinsic peak intensity for the bound and free excitons, respectively [4]. The emission lines for boron, phosphorus, as well as intrinsic silicon lie very close together, around 1.09-1.10 eV. Due to the higher temperature (~10 K) as well as higher ΔE (0.62 meV) used in this study compared with Tajima's study [4], the individual peaks could not be resolved. Thus, we were unable to estimate the phosphorus concentration from our PL results even for the Si control sample. However, as indicated by Tajima and other authors [5,6], the dopant peak intensity dominates over the intrinsic peak intensity for doping concentrations greater than ~10^{14} /cm^3. Thus, under the assumption that the emission line due to the so called donor complexes also lies in the same energy range (shallow donor complex), the maximum peak intensities were compared for our samples. Peak intensities for all samples correlate very well with the increase and decrease in the carrier concentration as shown by SRP results.

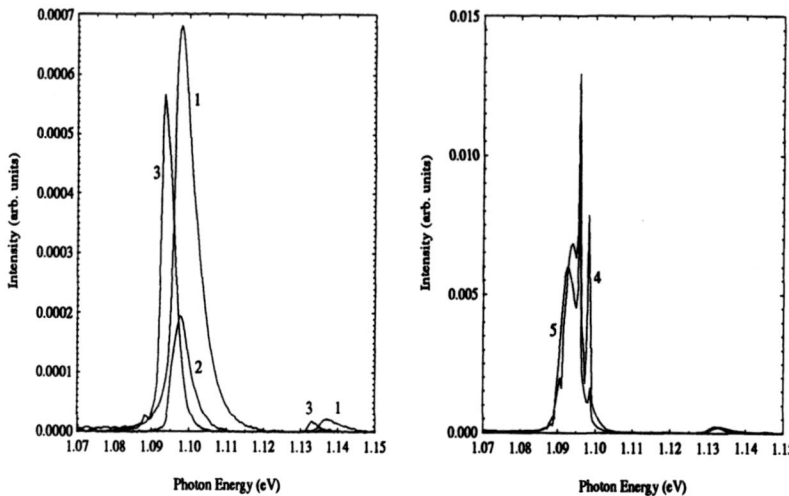

Figure 2: PL spectra for Si, SiGe, and SiGeC samples.

SUMMARY

Results of SRP and PL measurements performed on SiGe and SiGeC specimens support the earlier investigation using C-V measurements that indicated dopant deactivation due to high-dose Ge implantation into Si. Furthermore, once the critical Ge dose for strain relaxation is exceeded, shallow donor complexes seem to be formed in the SiGe layer from the surface to the projected range of Ge. Any increase in the Ge dose beyond the critical dose appears to result in a significant increase in the net donor concentration.

The authors are grateful to Mr. Richard Williams of Siliconix for his assistance in SRP measurements. Thanks are also due to Mr. X.S. Jiang and Mr. M. Markarian of University of California, San Diego, for performing the PL measurements.

This work was partially supported by the National Science Foundation under contract No. 442427-21153 and by the Division of Materials Science, Office of Basic Energy Sciences of the U.S. Department of Energy under contract No. DE-AC03-76SF00098.

References

[1] S.C. Jain and W. Hayes, Semicond. Sci. Technol. **6**, 547 (1991).
[2] A. Gupta, C. Cook, L. Toyoshiba, J. Qiao, C.Y. Yang, K. Shoji, A. Fukami, T. Nagano, and T. Tokuyama, J. Electronic Mater. **22** (1), 125 (1993).
[3] S. Im, J. Washburn, R. Gronsky, N.W. Cheung, and K.M. Yu, Proceedings of the MRS Fall 1992 Meeting, in press.
[4] M. Tajima, Appl. Phys. Lett. **32** (11), 719 (1978).
[5] P.J. Dean, J.R. Haynes, and W.F. Flood, Phys. Rev. **161** (3), 711 (1967).
[6] K. Kosai and M. Gershenzon, Phys. Rev. B **9** (2), 723 (1974).

REDUCING DISLOCATION DENSITY BY SEQUENTIAL IMPLANTATION of Ge AND C IN Si

Seongil Im*, Jack Washburn, Ronald Gronsky, Department of Materials Science and Engineering, University of California, Berkeley, CA 94720

Nathan W. Cheung, Department of Electrical Engineering and Computer Sciences,University of California, Berkeley, CA 94720

Kin Man Yu and Joel W. Ager, Center for Advanced Materials, Materials Sciences Division, Lawrence Berkeley Laboratory, 1 Cyclotron Road, Berkeley,CA 94720

ABSTRACT

Carbon implantation was performed after high dose(5×10^{16}/cm^2) Ge implantation into [100] oriented Si substrates to study the effect of sequential implantation on dislocation nucleation. When the nominal peak concentration of implanted C is over 0.55 at%, Dislocations in the SiGe layer containing C are considerably reduced in density after solid phase epitaxial(SPE) annealing at 800°C for 1 hour, compared to the SiGe layer without C. These results suggest that during annealing, C atoms compensate the Ge-induced misfit strain which causes dislocation generation in the region of peak Ge concentration. Channeling spectra obtained by RBS analysis show only 5% to 6% minimum back scattering yield as C atoms suppress the dislocation generation.

INTRODUCTION

Strained SiGe layer is of technological interest because of its narrower band gap compared to Si, high mobility of carriers and possible optoelectronic properties[1,2]. Strain-relaxed SiGe alloy layer by ion beam synthesis(IBS) has also been studied extensively because IBS is a very simple and economic technique that can achieve the band gap narrowing effect[3,4,5]. Heterojunction bipolar transistor is a potential IBS application[6].

However, due to Ge-induced dislocations and elongated end-of-range(EOR) dislocation loops formed during SPE regrowth, it is of concern that pipe-diffusion of dopants may occur through the dislocations during device processing leading to device failures. In the present IBS study, a sequential implant of C following liquid nitrogen temperature(LNT) implant of Ge has been performed in [100] oriented Si, to minimize the density of dislocations in SiGe layers.

EXPERIMENTAL PROCEDURES

High dose Ge ions(Ge$^+$) were implanted into [100]-oriented Si to generate amorphous SiGe alloy layers with a Ge peak concentration of 12 at%.(The corresponding dose was 5×10^{16}/cm^2.) A Ge ion beam energy of 120keV was chosen and two implantation temperatures of 298K(RT) and 77K(LNT) were used. The projected range(R$_p$) of Ge ions was about 70nm and the amorphized layer thickness was about 170nm. The 12 at% Ge peak concentration alloy layer was then subjected to a second implantation of C ions(C^+) at RT, using a C^+ beam energy of 20keV to position the implanted C ions at the same depth into the alloy layer as the 12 at% Ge peak. Three different nominal C doses were used to yield three anticipated C peak concentrations as 0.3 at%, 0.55 at%, 0.9 at%. The corresponding doses are 10^{15}/cm^2, 2x10^{15}/cm^2 and 3.5x10^{15}/cm^2 respectively. For all the implanted SiGe layers, SPE annealing was performed in a nitrogen ambient at 800°C for 1 hour. The redistribution of Ge and C during SPE regrowth at 800°C is negligible due to their low diffusivities at the temperature[7,8].

In order to measure the Ge dose and depth profile, Rutherford Backscattering Spectrometry(RBS) were used. To characterize the structure and crystallinity of the regrown SiGe alloy layers, cross-sectional transmission electron microscopy(XTEM) was performed using both symmetrical and <022> two beam conditions in a <110> zone axis orientation. <110> oriented channeling in RBS was also done to assist the XTEM results. Raman spectroscopy was used to observe the local vibration mode(LVM) of C on the substitutional Si site.

RESULTS AND DISCUSSION

XTEM micrograph in Fig.1 shows two defect bands formed during SPE annealing after RT implant of Ge : EOR loops and Ge-induced stacking faults.

According to previous reports, EOR dislocation loops form due to excess recoiled Si interstitials in the end range of implantation[9]. Since the migration energy of Si interstitials is only 0.1~0.4eV, a thermal budget at RT is sufficient to cause a dynamic recombination of vacancies and interstitials in the end of range during implantation. The excess Si interstitials remaining after the recombination form extrinsic fault loops which are the origin of EOR loops. These EOR loops has already been minimized by using LNT implant of Ge as shown in Fig.2 and previous works[9,10].

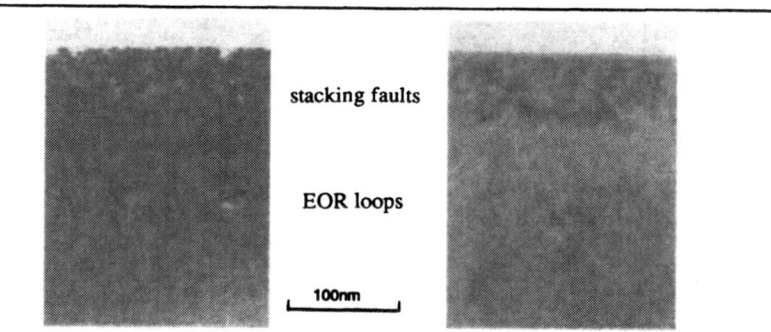

stacking faults

EOR loops

100nm

Fig.1 SiGe alloy layer obtained by RT implant of Ge and SPE anneal. EOR loops and strain-induced stacking faults exist in high density.

Fig.2 SiGe alloy layer obtained by LNT implant of Ge and SPE anneal. EOR loops are minimized in density.

However, misfit-induced stacking faults resulted from high dose of Ge have never been eliminated for Ge peak concentration above 6 at%[9]. In the present study, C sequential implantation shows a dramatic reduction in the density of the stacking faults at Ge peak concentration of 12 at% when a critical dose of C atoms is implanted. XTEM micrograph in Fig.3 shows that both EOR loops and stacking faults are considerably reduced in density in a SiGe layer with 0.55 at% C peak. Because C atoms are much smaller than Ge or Si atoms in volume, it has been speculated that a small dose of C atoms in Si lattice can compensate the misfit-strain induced by high dose of Ge atoms during SPE regrowth. Observation of the LVM of Si-C at a Raman shift of 608cm^{-1}in Fig.4 shows that at least part of the C atoms sit on substitutional sites strongly supporting the possibility of strain-compensation by C atoms[8,11]. No evidence of SiC precipitate formation was found in the present study.

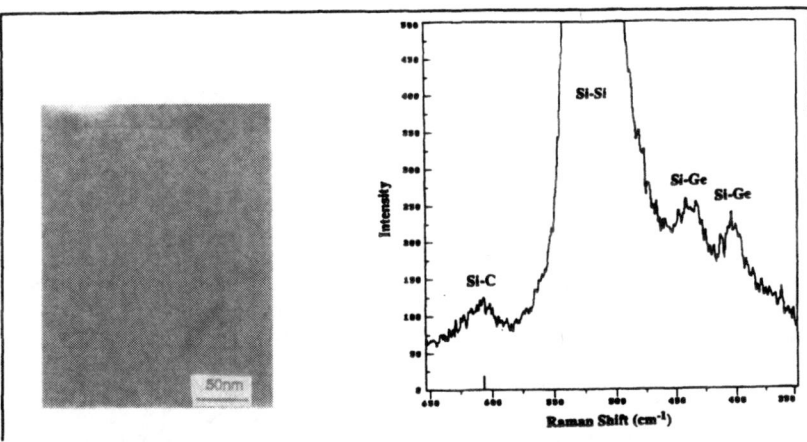

Fig.3 SiGe alloy layer obtained by C sequential implant after Ge LNT implant. (Nominal C peak concentration = 0.55 at%) Both EOR loops and stacking faults are minimized in density.

Fig.4 Raman spectra obtained from SiGe layer with C.(nominal C peak =0.55 at%), LVM of Si-C is detected at about 608cm^{-1}, showing substitutionality of C in Si lattice.

In Fig.5, the minimum back scattering yields(χ_{min}) measured from SiGe layers with C or without C are plotted by using <110> ion channeling with RBS. Note that the minimum yield suddenly decreases in the middle between 0.3 at% and 0.55 at% of C peak concentration implying a sudden reduction of defect density. Such an effect of C on SiGe crystallinity or strain-compensation has slightly been improved from 0.55 at% of C peak to 0.9 at% peak of C. The smallest minimum back scattering yield is obtained from a SiGe layer with 0.9 at% of C peak(χ_{min}=5%).

142

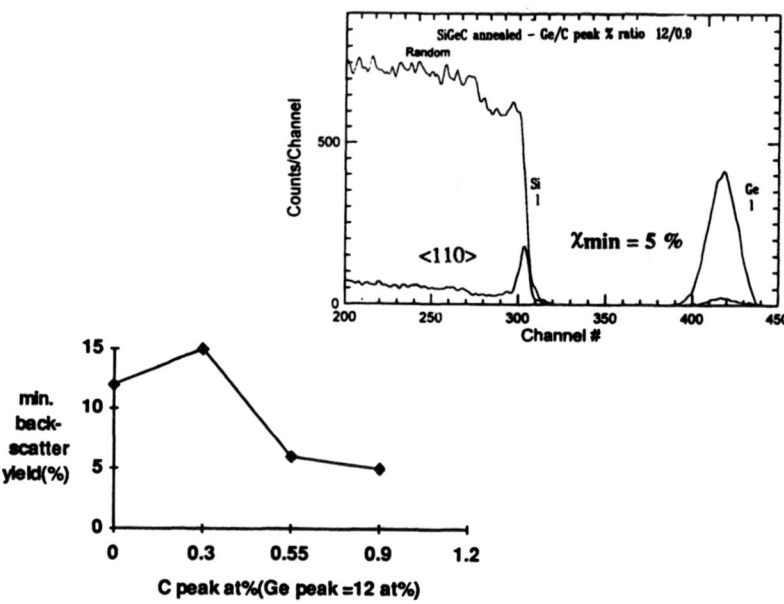

Fig.5 <110> ion channeling spectra of SiGe alloy layer with 0.9 at% C peak and a plot of minimum backscattering yields obtained from each SiGe alloy layer. Note a sudden improvement of cyrstallinity at C peak concentration between 0.3 at% and 0.55 at%.

CONCLUSION

Dislocations and stacking faults in SiGe alloy layer formed by IBS were considerably reduced in density by using C sequential implant after Ge LNT implant. During SPE regrowth, very small dose of C atoms in the SiGe layers effectively compensate the misfit strain induced by high dose of Ge atoms. This defect-reduction by C atoms abruptly begins at a C peak concentration between 0.3 at% and 0.55 at%, saturating with higher C content. The smallest minimum backscattering yield obtained from a SiGe layer with 0.9 at% C peak is 5%.

ACKNOWLEDGMENT

The authors gratefully acknowledge the assistance of Dr. Xiang Yao and Dr. Ian Brown for C implantation at the Lawrence Berkeley Laboratory.

This work was supported by the National Science Foundation under contract No. 442427-21153 and by the Division of Materials Science, Office of Basic Energy Sciences of the U.S. Department of Energy under contract No. DE-AC03-76SF00098.

REFERENCES

1. S. C. Jain and W. Hayes, Semicon. Sci. Technol. **6**, 547-576, (1991)
2. **Strained-Layer Superlattices : Materials Science and Technology**, Semiconductors and Semimetals, Vol.33, edited by Thomas P. Pearsall, ACADEMIC PRESS, INC., 223-304 (1991)
3. K. M. Yu, I. G. Brown and Seongil Im, Mat. Res. Soc. Symp. Proc., **235**, 293(1992)
4. D.C. Paine, D. J. Howard, and N. G. Stoffel, J. Electronic Materials, **20**(10), 735(1991)
5. Akira Fukami, Ken-ichi Shoji, Takahiro Nagano and Cary Y. Yang, Microelectronic Engineering, **15**, 15 (1991)
6. Akira Fukami, Ken-ichi Shoji, Takahiro Nagano and Cary Y. Yang,Appl. Phys. Lett., **5**(22), 2345(1990)
7. Properties of Silicon, EMIS Data reviews series No.4, published by INSPEC, The Institution of Electrical Engineers, New York, (1988) p. 285
8. S. Im, J Washburn, R. Gronsky, N. W. Cheung and K. M. Yu, "Defect Control During Solid Phase Epitaxial Growth of SiGe Alloy Layers", submitted to Appl. Phys. Lett. 1993
9. S. Im, J Washburn, R. Gronsky, N. W. Cheung and K. M. Yu, MRS Symp. Proc. 1992 (in press)
10. S. Prussin and K. S. Jones, J. Electrochem. Soc., **137**(6), 1912 (1990)
11. K. Eberl, S. S. Iyer, S. Zollner, J.C. Tsang and F. K. LeGoues, Appl. Phys. Lett., **60**(24), 3033(1992)

Raman Scattering Spectroscopy of GeSi/Si Strained Layer Superlattice

Zhang Rong, Zheng Youdou, Gu Shulin and Hu Liqun
Department of Physics, Nanjing University, Nanjing 210008, CHINA
Fax: 86-25-302728

ABSTRACT

Raman scattering measurements have been carried out on $Si_{1-x}Ge_x$/Si SLS. It is found that the Ge-Ge optic phonon frequency shift is proportional to strain in the SiGe film, and the Ge-Ge strain shift coefficient is $408cm^{-1}$. Based on these study a new method for analyzing the Raman spectra of SiGe/Si SLS has been proposed. Using the new method we can obtain the composition of the alloy sublayers as well as the strain in SLS. The strain distribution in the SiGe/Si SLS has been discussed, and strain in both SiGe and Si sublayers of the SLS have been calculated.

I. INTRODUCTION

The growth of $Si_{1-x}Ge_x$/Si strained layer superlattice(SLS) on silicon substrate has recently stimulated a lot of interest because of its potential application in microelectronics and optoelectronics, and its compatibility with current Si technology. The great lattice mismatch(maximum with 4.2%) between the $Si_{1-x}Ge_x$ sublayer and Si sublayer has important influence on the structure and property of this kind of material and device. Generally, the lattice misfit can be accommodated in two ways depending on the thickness and the condition of the process. In low temperature epitaxy, thin films are strained to match the substrate, the in-plane lattice parameter is determined by the substrate or thicker buffer layer and along the growth direction the lattice is strained correlated to in-plane lattice with Poisson's ratio. If the thicknesses of sublayer is greater than a critical value, misfit dislocations appear and result in strain relief[1]. The critical thickness changes at different Ge composition x. In the case of $Si_{1-x}Ge_x$/Si superlattice, the strain distribution is determined by both the average composition which determines the critical thickness of total SLS, and the composition of each sublayer which determines the critical thickness of that sublayer[2]. Usually the strain in the SLS may be partially relaxed resulted from thermal process[3] or structure destruction[4].

Many techniques have been employed to measure the strain and strain relaxation in the $Si_{1-x}Ge_x$/Si strained layer superlattice[5-8]. Among them Raman scattering spectroscopy has proved to be a versatile and nondestructive tool for measuring the physical properties of semiconductor superlattices[9]. Results from Raman measurements of the acoustic and optic phonons in the $Si_{1-x}Ge_x$/Si SLS can provide information on the layer thicknesses, composition and strain[10]. In measuring the strain or strain relaxation attention is given to the shift of the optical-phonon frequencies away from their values in the bulk material with the same composition. In Si sublayer only one Raman peak is observed due to the optic phonon around $520cm^{-1}$, whereas in $Si_{1-x}Ge_x$ sublayer three peaks are found near 300, 400 and $500cm^{-1}$ due to optic modes labeled as Ge-Ge, Si-Ge and Si-Si respectively[11]. Many investigators[11-13] have reported their Raman measurements of $Si_{1-x}Ge_x$/Si SLS and demonstrated that the Si-Si and Ge-Si phonon frequencies vary linearly with strain. On the other hand there is ambiguity on the relation of the Ge-Ge phonon frequency to strain in the $Si_{1-x}Ge_x$ alloy[14]. Such information, however, is essential for analyzing Raman spectra of $Si_{1-x}Ge_x$/Si SLS.

We have carried out measurements of strain, using X-ray diffraction(XRD), X-ray photoelectron spectroscopy(XPS) and the optic-phonon frequencies in a variety of $Si_{1-x}Ge_x$/Si SLS and heterostructures grown on Si for Ge concentration in the range 0.3<x<0.6. The Raman measurements confirmed that the Ge-Ge phonon frequency varied linearly with strain. Based on these results we propose in this paper a new method for analyzing the Raman spectra of $Si_{1-x}Ge_x$/Si SLS. With the new method we can obtain the composition of the $Si_{1-x}Ge_x$ sublayer as well as the strain in SLS. The strain distribution in the $Si_{1-x}Ge_x$/Si SLS will also be presented. The strain distribution in the SiGe/Si SLS has been discussed, and strain in both SiGe and Si sublayers of the SLS have been calculated and explained depending on the thickness of sublayers.

II. EXPERIMENT

Table I   x data obtained from XPS and Raman

No.	Type[a]	ω_{Ge-Ge}	ω_{Si-Si}	x_{XPS}	$x_{(8)}$	$x_{(9)}$	$x_{(10)}$	λ
A	SL	295	510	0.35		0.35	0.36	100%
B	SL	291	504	0.40		0.37	0.36	63%
C	SL	293	498	0.45	0.43	0.47	0.45	52%
D	SL	292	497	0.45	0.43	0.47	0.45	40%
E	SL	290	495	0.45	0.43		0.44	17%
F	SL	294	487	0.60	0.57		0.58	23%
G	SL	293	484	0.60	0.60		0.60	10%
H	H	292	504	0.35		0.37	0.36	63%
I[11]	SL	296	493	0.65		0.58	0.55	53%
J[19]	SL	294	507	0.33		0.37	0.37	100%

[a] SL means superlattice, H means heterostructure.

The $Si_{1-x}Ge_x$/Si SLS were grown by RTP/VLP-CVD under optimized growth conditions[15-17]. All samples were grown at ~600°C under 10-100mTorr on 3-5 Ωcm n-type 50-mm Si wafers. SiH_4 and GeH_4 were used as reactant source gases. The thickness of the various SLS sublayers was chosen cross the critical curve at the growth temperature so that samples had different strain and strain relaxation. The uniformity in thickness and composition from the center to the edge of the wafer is estimated to be better than 2%. Each sample was capped with a thin Si layer.

XRD and XPS were used to determine the thickness and germanium fraction x of the SLS. A representive small-angle XRD spectrum is shown in Fig.1. The compositions obtained from XRD and XPS are given in Table I.

The Raman spectra of the $Si_{1-x}Ge_x$/Si SLS were recorded at 300K on SPEX Raman spectrometer in a quasibackscattering geometry using 300mW of argon laser light at 514.5nm for excitation. The measurements were performed on the same specimens used for the

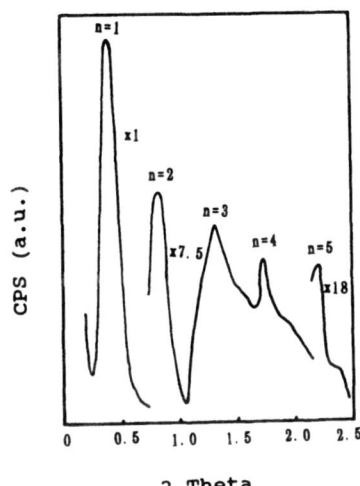

CPS (a.u.)

n=1

x1

n=2

n=3

x7.5

n=4 n=5

x18

0   0.5   1.0   1.5   2.0   2.5

2 Theta

Fig.1 A typical small-angle XRD spectrum of SiGe/Si SLS.

above-mentioned experiments, and in those experiments care was taken to position the probe within 1mm from the center of specimens. Wavenumber resolution was $0.5cm^{-1}$. The standard Si sample was used to calibrate the wavenumber of monochromator. The wavenumber of Si-Si and Ge-Ge line were also listed in Table I.

III. RESULTS AND DISCUSSION

A typical Raman spectrum is shown in Fig.2. The sample is a $Si/Si_{0.5}Ge_{0.5}/Si$ double heterostructure. The thickness of $Si_{0.5}Ge_{0.5}$ alloy is 1500Å while the Si caplayer is 500Å thick. The spectrum exhibits three main peaks attributed to scattering from longitudinal-optical phonons corresponding to Ge-Ge ($290cm^{-1}$), Ge-Si($408cm^{-1}$) and Si-Si($490cm^{-1}$) modes in the alloy layer, and a weak peak due to the signal from the Si caplayer and substrate.

A. Optic Phonons

It is well known that when a strain is applied along the major axes of the diamond-type crystal, the triply degenerate optical mode will be split into a singlet with wave vector parallel to the strain, and a doublet with wave vectors perpendicular to the strain directions. In our experiment under backscattering geometry only the singlet can be observed, and the phonon frequency shift $\delta\omega$ is[18]

$$\delta\omega=1/2\omega_0[p\epsilon_{zz}+q(\epsilon_{xx}+\epsilon_{yy})] \quad (1)$$

where ω_0 is the optic phonon frequency of bulk material, p and q are phenomenological parameters, ϵ_{ij} represents the components of the strain tensor and $\epsilon_{xx}=\epsilon_{yy}=-\epsilon$, $\epsilon_{zz}=-2S_{12}\epsilon/(S_{11}+S_{12})=\alpha\epsilon$. Here the S_{ij}'s are the elastic compliances of the material. Available values for S_{ij}'s give $\alpha=0.77$ for pure Si and $\alpha=0.75$ for pure Ge. Hence, for the alloy we take $\alpha=0.76$, resulting in a frequency shift as

$$\delta\omega=1/2\omega_0(0.76p-2q)\epsilon=b\epsilon \quad (2)$$

Assuming that p and q are not composition or strain dependent, values for b and ω_0 can be obtained from the slope and from an extrapolation at $\epsilon=o$ by curve fitting using Eqn(2). This result is illustrated in Fig.3. The X-ray axis of lattice mismatch

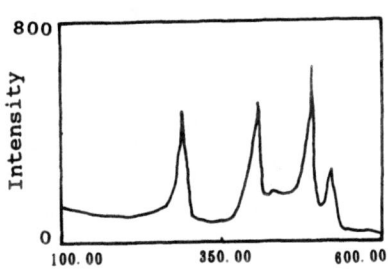

800

Intensity

0

100.00   350.00   600.00

Raman Shift (cm^{-1})

Fig.2 Raman spectrum of a $Si/Si_{0.5}Ge_{0.5}/Si$ double heterostructure.

strain in the $Si_{1-x}Ge_x$ alloy is deduced from the Si-Si vibration mode frequency shift. The Y-axis is the frequency shift of Ge-Ge optic phonon in the $Si_{1-x}Ge_x$ film. It is obvious that the Ge-Ge optic phonon frequency decreases linearly with strain in the film. The strain-shift coefficient b of the Ge-Ge optic phonon can be determined from Fig.3 and is $408cm^{-1}$ which is very agreement with the value deduced from [11].

B. Composition and Strain

Raman scattering has been proposed to measure the composition of $Si_{1-x}Ge_x$ alloy. The three main peaks maned Si-Si, Ge-Si and Ge-Ge are the Raman frequencies of the vibrations between Si-Si, Si-Ge and Ge-Ge first neighbor atoms, respectively. In a random alloy, the number of Si-Si, Si-Ge and Ge-Ge bonds is proportional to $(1-x)^2$, $2x(1-x)$ and x^2, respectively. Assuming that the intensity of a Raman peak is proportional to the corresponding number of bonds, the film composition can be evaluated by measuring the intensity ratio between the two most intense peaks.

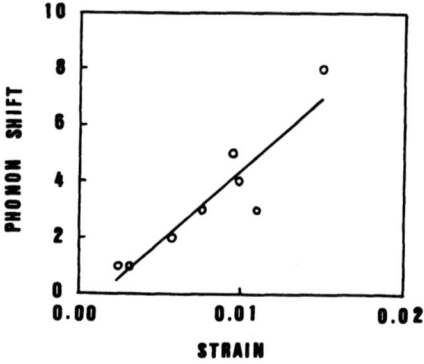

Fig.3   Ge-Ge optic phonon frequency shift vs strain

There are several factors which could affect the peak amplitude of a Raman spectrum. First, the Raman peak intensity is very sensitive to sample surface, so same composition with different surface condition will lead to different peak intensity. Second, the peak intensity is also direction sensitive. Under different scattering geometry, the Raman spectra will include contribution from different vibration modes. If the scattering direction deviates from the backscattering geometry, the phonons other than longitudinal optical modes will mix into the backscattering spectrum. Third, the peak intensity is sensitive to order structure. If there is long-range order structure in the alloy film, the number of Si-Si, Si-Ge and Ge-Ge bonds will not be proportional to $(1-x)^2$, $2x(1-x)$ and x^2.

Based on results of Raman spectroscopy mentioned in last section, here we propose a new method to obtain the composition and strain simultaneously. The frequencies of the Si-Si and Ge-Ge lines in the alloy film vary almost linearly with both concentration and strain and can be written as

$$\delta\omega_{Si-Si}=-66x+b_{Si-Si}\epsilon, \quad \delta\omega_{Ge-Ge}=20x+b_{Ge-Ge}\epsilon \qquad (3)$$

So we can obtain Ge composition x and strain ϵ from Eqn(3). The $Si_{1-x}Ge_x$ alloy film in SLS is often partially strained. Define ß as lattice mismatch between the $Si_{1-x}Ge_x$ sublayer and the Si sublayer, and λ as percentage relaxation, i,e., λ representing the part that the strain accommodates the lattice mismatch, and $1-\lambda$ representing the other part that the misfit dislocation or Si sublayers accommodate the lattice mismatch. Then

$$\lambda=[a(Si_{1-x}Ge_x)-a(Si)]/a(Si), \quad \epsilon=\lambda ß$$

According to the Vegard's law, $a(Si_{1-x}Ge_x)=(1-x)a(Si)+xa(Ge)$, $a(Si)=0.54309nm$ and $a(Ge)=0.56575nm$, therefore

$$ß=0.0416 x \qquad (4)$$

Using the data deduced from [11] and (4), we can rewritten (3) as:
$$\delta\omega_{Si\text{-}Si}=-66x+25\lambda x, \quad \delta\omega_{Ge\text{-}Ge}=20x+17\lambda x \quad (5)$$
For fully relaxed film (sublayer) or bulk alloy of GeSi, $\lambda=0$, we have
$$\omega_{Si\text{-}Si}=522-66x, \quad \omega_{Ge\text{-}Ge}=280+20x \quad (6)$$
and for fully strained GeSi film (sublayer), we have
$$\omega_{Si\text{-}Si}=522-41x, \quad \omega_{Ge\text{-}Ge}=280+37x \quad (7)$$
Hence we can deduced from eqn(6) and (7)
$$x=[242-(\omega_{Si\text{-}Si}-\omega_{Ge\text{-}Ge})]/86 \quad (8)$$
for unstrained film, and
$$x=[242-(\omega_{Si\text{-}Si}-\omega_{Ge\text{-}Ge})]/78 \quad (9)$$
for fully strain film. For partially strained film, $0<\lambda<1$, x and λ can be solved from Eqn(5) as
$$x=[1874-(17\omega_{Si\text{-}Si}-25\omega_{Ge\text{-}Ge})]/1622,$$
$$\lambda=[(10\omega_{Si\text{-}Si}+33\omega_{Ge\text{-}Ge})-14460]/811x \quad (10)$$
Compositions and percentage relaxations obtained from Eqn(10) have been listed in TableI. In the table, if calculated λ is greater than 100%, it will be set to 100%. From the table, we can see that the compositions determined from Raman measurements by Eqn(10) agree well with the data obtained from other technique. Fig. 4 plots the composition gotten from Raman spectroscopy versus value obtained from other technique.

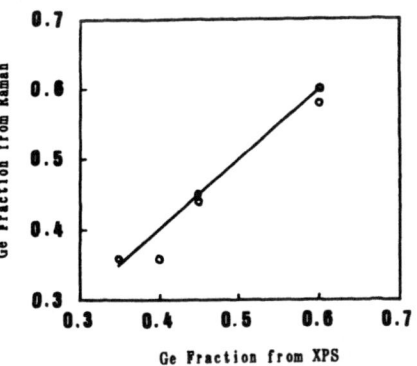

Fig.4 Ge fraction calculated from Raman spectra vs from XPS

C. Strain Distribution in $Si_{1-x}Ge_x/Si$ SLS

In the $Si_{1-x}Ge_x/Si$ SLS, when the SiGe alloy sublayer is much thinner than both the Si sublayer and the critical thickness, and the total thickness of SLS is less than the critical thickness of average Ge fraction at growth temperature, the lattice mismatch between $Si_{1-x}Ge_x$ sublayer and Si sublayer will be accommodated by strain in the $Si_{1-x}Ge_x$ sublayer. If the total thickness is greater than the average-composition-related critical thickness at growth temperature, it will be different. Fig.5 gives a Raman spectrum of sample F listed in TableI. The sample consists of 20 periods of $Si_{0.4}Ge_{0.6}$(8.0nm) sublayer and Si(2.4nm) sublayer, and 10nm Si caplayer. In Fig.5, five Raman peaks can be observed at 294cm^{-1}, 416cm^{-1}, 487cm^{-1}, 509cm^{-1} and 522cm^{-1} respectively. Three vibration modes under 500cm^{-1} are from the $Si_{1-x}Ge_x$ sublayer, hence the strain and composition can determined using Eqn(10). The calculated Ge concentration is 0.58 and is very close to the 0.60 determined by other technique. The percentage relaxation given by Eqn(10) is 23%, and the strain in the SiGe sublayers is 0.55%.

It is very interesting that between SiGe-related peaks under 500cm^{-1} and bulk Si optical phonon at 522cm^{-1}, there is a weak peak located at 509cm^{-1}. Because the total thickness of this sample is greater than the "effective" critical value, and the Si sublayer is thinner than the alloy sublayer, we consider that some misfit dislocations may exist near the SLS/Si buffer interface and the

150

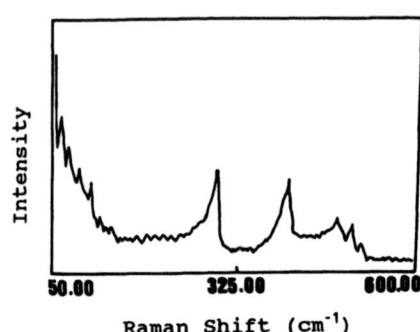

Raman Shift (cm⁻¹)

Fig.5 Raman spectrum of sample F
listed in Table I

Si sublayer is also strained. For the Si sublayer, if we neglect frequency shift due to confinement effect in the Si sublayers, the Raman peak red-shift corresponding to tensile strain is $\delta\omega=6.76\times10^2\epsilon_{Si}$, in this sample $\delta\omega=13$cm⁻¹, therefore $\epsilon_{Si}=1.9\%$. The lattice mismatch λ between the $Si_{0.4}Ge_{0.6}$ sublayer and the Si sublayer is 2.4%, hence $\beta=\epsilon_{SiGe}+\epsilon_{Si}$, that means the $Si_{0.4}Ge_{0.6}/Si$ interface in SLS is fully strained and dislocation free, and all misfit dislocation exist outside the SiGe/Si SLS.

REFERENCES

1. A.T.Fiory, J.C.Bean, L.C.Feldman and I.K.Robinson,
   J.Appl.Phys., 56, 1227(1984)
2. S.C.Jain and W.Hayes, Semicon.Sci.Technol., 6,547(1991)
3. S.M.Jang, H.W.Kim and R.Reif, Appl.Phys.Lett., 61, 315(1992)
4. W.Freiman and R.Beserman, K.Dettmer and F.R.Kessler,
   Appl.Phys Lett., 60, 1673(1992)
5. R.Hull, J.C.Bean, L.J.Peticolas, D.Bahnck, B.E.Weir and
   L.C.Feldman, Appl.Phys.Lett., 61, 2802(1992)
6. X.F.Duan, Appl.Phys.Lett., 61, 324(1992)
7. L. Qin, Y. Zheng, R. Zhang and D. Feng, Superlattices and
   Microstructures, 12, 517(1992)
8. M.A.G.Halliwell, M.H.Lyons, S.T.Davey, M.Hockly, C.G.Tuppen
   and C.J.Gibbings, Semicond. Sci. Technol. 4, 10(1989)
9. F.Meyer, M.Zafrany, M.Eizenberg, R. Beserman, C. Schwebel
   and C.Pellet, J. Appl. Phys., 70, 4268(1991)
10. J.Sapriel and B.D.Rouhani, Surf.Sci.Reps., 10, 189(1989)
11. F.Cerdeira, A.Pinczuk, J.C.Bean, B.Batlogg and B. A. Wilson,
    Appl. Phys. Lett., 45, 1138(1984)
12. Z.Sui, I.P.Herman and J.Bevk, Appl.Phys.Lett., 58,2352(1991)
13. D.J.Lockwood and J.M.Baribeau, Phys.Rev., B45, 8565(1992)
14. J.Xu, J.Wang, C.Sheng, H.Sun, S.Zheng and W.Yao, Chinese
    J.Semicond, 11, 822(1990) (in Chinese)
15. Y.Zheng, R.Zhang, L.Hu, S.Gu, R. Wang, P. Han and R. Jiang,
    in Mechanisms of Heteroepitaxial Growth, edited by
    M.F.Chisholm, R.Hull, L.J.Schowalter and B.J.Garrison(Mater.
    Res. Soc. Proc. 263, Pittsburgh, PA, 1992), pp227-232
16. R.Zhang, Y.Zheng, R.Jiang, L.Hu, P.Zhong, S.Yu, Q.Li, D.Feng
    Appl. Surf. Sci., 48/49, 356(1991)
17. S.Gu, R.Wang, P.Han, L.Hu, R.Zhang and Y.Zheng,
    Superlattices and Microstructures, 12, 4, 513(1992)
18. E.Anastassakis, A.Pinczuk, E.Burstein, F.H.Pollack and
    M.Cardona, Solid State Commun., 8, 133(1970)
19. D.C.Houghton et al, J. Cryst. Growth, 81, 434(1987)

MULTI-WAVELENGTH ELLIPSOMETRY FOR EFFECTIVE CHARACTERIZATION OF THIN EPITAXIAL Si$_{1-x}$Ge$_x$ LAYERS ON SILICON SUBSTRATE

MICHAEL EICHLER, MARITA WEIDNER AND THOMAS MORGENSTERN
Institute of Semiconductor Physics, Walter Korsing Straße 2, O-1200 Frankfurt (Oder), Germany

ABSTRACT

The range of composition (x) is one of the parameters we often have to measure if Si$_{1-x}$Ge$_x$ layers are generated by chemical vapour deposition (CVD). It is important in this case, in which way the optical properties of Si$_{1-x}$Ge$_x$ layers depend on the range of composition. We are interested in using multi-wavelength ellipsometry as a technique for rapid, nondestructive characterization of these samples, without large preparations, especially for series of measurements (2D profiles or wafer-lots). The number of unknown parameters and the multiple solutions are reduced by using several wavelengths during the measurement. The calculation is prepared by the help of parameter-correlation based on results of spectroscopical ellipsometry. To examine the results, thickness and composition were controlled for selected samples by cross-sectional transmission electron microscopy (XTEM) and X-ray double crystal diffractometry (DCD).

INTRODUCTION

Epitaxial Si$_{1-x}$Ge$_x$ layers are of famous interest for development of new devices in silicon based semiconductor technologies. Measurements with optical methods have an important significance as nondestructive kind of control, especially inside the production process. In this case ellipsometry with fixed wavelength is able to achieve a lot of automatic thickness measurements of thin films on substrate during a short time. Because ellipsometry with fixed wavelength for process control mostly use very stable light sources (LASER) and a mainly fixed equipment, it will offer high precision.

In the set up of CVD processes the main problems are often the unknown parameters thickness and range of composition of the Si$_{1-x}$Ge$_x$ layer. Because of the inhomogeneity of these parameters across the wafer and the influence of the wafer position in the reactor, it is necessary to get profiles and maps by the help of large procedures with a lot of measured points for several wafers. Fortunately it is possible to measure the range of composition in Si$_{1-x}$Ge$_x$ layers by optical methods, because the optical properties depend on the range of composition. So the ellipsometry with fixed wavelength should be a preferred method to solve this problem. On the other hand the quantity of unknown parameters requires more measurements for one sample with several conditions to reduce the multitude of solutions. To modify the conditions for example, it is possible to change the angle of incidence or to change the wavelength. In this case it is an effective technique to switch the wavelength during the measurement for one sample.

The ellipsometrical measures Ψ and Δ are the intensity ratio (tanΨ) and the phase difference (Δ) between the parallel (R$_p$) and perpendicular (R$_s$) components of the electrical field vector to the plane of incidence of linear polarized light after reflection on the sample surface:

$$\tan\Psi \; \exp(i\Delta) \; = \; R_p/R_s \; = \; f(\lambda,\Phi,N_s,N_f,N_{ox},t_f,t_{ox}). \tag{1}$$

This so-called ellipsometer equation [1] connects the ellipsometrical measures for the wavelength (λ) and the incidence angle (Φ) with the model parameters of the test specimen, there are N$_s$, N$_f$ and N$_{ox}$ the refractive index of substrate, SiGe film and native oxide, respectively, and t$_f$, t$_{ox}$ the thickness of the film and native oxide. From two measured quantities Ψ and Δ at one wavelength two model parameters are obtained, as a rule via an iteration procedure. By using two wavelengths there are two of such equations. A wavelength correlated iteration procedure [2] allows to estimate the real part of the refractive index for both wavelengths and the thickness of the $Si_{1-x}Ge_x$ layer. Modern personal computers with appropriate software are able to calculate the iteration procedure as the inverse problem of ellipsometry and the correlation between two ore more wavelengths in acceptable time for the measurements.

The optical constants of $Si_{1-x}Ge_x$ alloys are a complex problem. Primarily the refractive index and the absorption constant depend on the wavelength and the range of composition. But for the measurement there are a lot of other influences, like inhomogeneity of the layer, defects on the interfaces and effects from lattice mismatch.

For many years several studies of the optical constants of $Si_{1-x}Ge_x$ alloys have been made. Lukes [3] investigated $Si_{1-x}Ge_x$ alloys from bulk material with Ge content from 0 to 1 at. % only. Humlicek [4] offered experimental data of fully relaxed material, where the lattice parameter are different from strained epitaxial material. Figure 1 illustrates this problem by the help of the prepared parameter $n_{SiGe}-n_{Si}$.

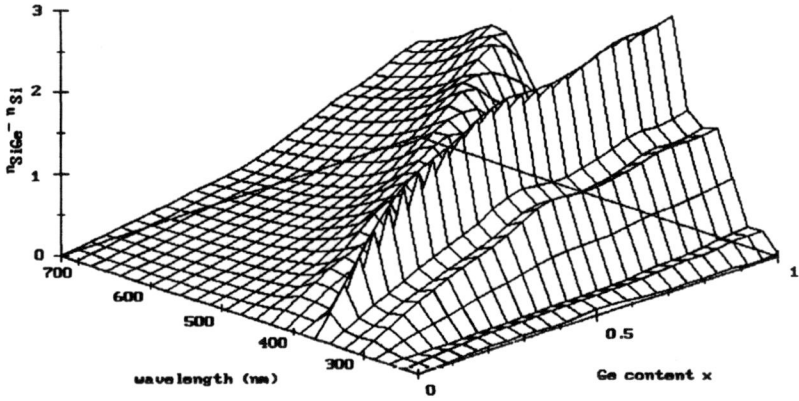

Fig. 1 Difference of refractive index n_{SiGe} and n_{Si} as a function of wavelength and Ge content

This parameter consists of the difference between the real parts of refractive index of $Si_{1-x}Ge_x$ and Silicon and it is able to demonstrate the difficult relations of the contrast between substrate and layer.

In the last 3 years there are also experimental results about the complex refractive index from strained epitaxial films of $Si_{1-x}Ge_x$ on silicon substrate [5-7], obtained from ellipsometrical investigations at 632.8 nm.

EXPERIMENTAL

One ellipsometrical measurement provides information to calculate 2 parameters for the given sample layer system from the measured ellipsometrical angles Ψ and Δ. If there are more than 2 unknown parameters for the sample, it will be necessary to acquire additional experimental data. Just in this case $Si_{1-x}Ge_x$ layers with unknown optical properties and unknown thickness of these layers and unknown parameters of natural oxide film on the top, have to be investigated.

Fortunately the used option of ellipsometer PLASMOS SD2000 contains two wavelengths of 632.8nm and 785 nm and a supplementary suitable software module. The connection of these hardware and software elements offers the possibility for a correlated calculation with the use of more than one (with other hardware options up to 4) wavelengths to solve the ellipsometrical numerical-inversion problem for one sample. The numerical iteration process mainly includes several iteration algorithms, like algorithm after Levenberg-Marquardt [8], with a determined search for the minimum of an error function. The error function is calculated from the root mean square values (RMS) of measured and model values and has to be tested also for its regular behaviour in a random environment. It is critical to set reasonable start conditions to take the iteration process to an expected progress and a regular result. That's why a completed automatic measurement and calculation mode, without the handling of an experienced operator, his declaration of the well-known parameters and his selection of the correct result out of the amount of multiple numerical iteration solutions, is impossible for these problems.

This automatic ellipsometer with two wavelengths according to the principle of rotating analyzer was applied to ellipsometrical measurement of Ψ and Δ and an incidence angle of 70°. Because of the computer controlled scanning xy-stage the measurement time is below 1 sec. per point for one wavelength in a rasterscanning analysis mode. Table 1 shows the complex refractive index of silicon substrate for the used wavelengths.

Table 1 Complex refractive index of the silicon substrate

wavelength λ (nm)	complex refractive index, real part n	complex refractive index, imaginary part k
632.8	3.858	-0.018
785	3.660	-0.006

The strained epitaxial $Si_{1-x}Ge_x$ layers were deposited in a CVD reactor at atmospheric pressure at 700°C from SiH_4, GeH_4 in HCl and H_2 with a thickness from 20 nm up to 80 nm. The processing conditions were choosen in such a way as to obtain a thickness variation across the 4" silicon wafer. This makes it possible to spread an ellipsometer curve for different layer thicknesses under otherwise comparable conditions by measuring along the diameter of the wafer perpendicular to the wafer flat. Furthermore the Ge content of different wafers was varied to get the dependence of the refractive index on the x-value.

Figure 2 shows in a plot a small but representative selection of measured and calculated Ψ-Δ couples for $Si_{1-x}Ge_x$ layers on Si substrate at 785 nm and an incidence angle of 70°. In the plot every measured curve (identification number 5 to 9) consists of an amount of measured points from one scan across the wafer in the direction of the wafer flat. The calculated curves (identification number 1 to 4) are computed in a model for several values of the real part of refractive index. Everyone of them is calculated for a thickness of the $Si_{1-x}Ge_x$ layer from 0 nm up to 95 nm with nearly a complete circulation. All of them are fitted to the measured curves with the same parameters for the native oxide (layer 1, thickness t=1.3 nm, n=1.46, k=0), for the substrate and of course for the ambient. The imaginary part of the refractive index of the $Si_{1-x}Ge_x$ layer was assumed unchanged to the value of silicon for these wavelengths, too. So the different real parts of the refractive index of SiGe layers will present the Ge content in a reasonable kind of presentation.

In this case the measured curves are selected for SiGe layers with almost constant Ge content across the wafer and a variable thickness. Because the different parameters of the CVD process, there also are samples with variable Ge content across the scan of measurement. In the Ψ-Δ-plot the curves of these samples cross the calculated curves for an "iso-content" distinct. Now it is obvious, that there is no possibility to get information about the thickness and the Ge content in parallel, or to detect the influence of these parameters on the measuring result separately.

In order to get this information it is necessary to change the conditions of measurement by switching the wavelength and achieve a correlated calculation with two sets of parameters. Because there is an acceptable difference between the real parts of refractive index for both wavelengths, the solution is mainly multiple continuously, but if most of the parameters are well-known for a secure scope, the correct result will be distinct.

For the first steps parallel XTEM and DCD inspections realize the reference to get the secure scope of these parameters, to reduce the amount of solution and to prove the results. These mainly expensive, destructive methods respectively present a good agreement with the ellipsometrical conclusion.

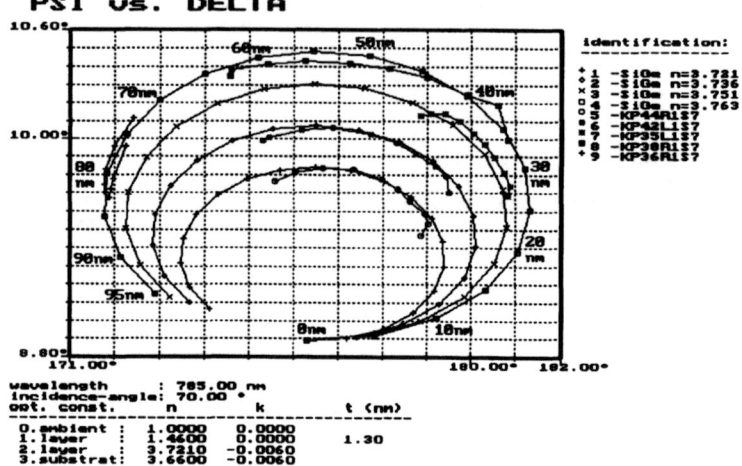

Fig. 2 Ψ-Δ-plot for a wavelength of 785 nm with measured curves (identification number 5-9) and fitted model curves (identification number 1-4)

RESULTS AND DISCUSSION

The fundamental task of this investigation is to find out a practical formula to express the dependence of the optical properties of $Si_{1-x}Ge_x$ alloys on the range of the Ge content x for the used wavelength and in a useful scope of Ge content for conventional measurements by ellipsometry. The standard wavelength 632.8 nm has already been examined in this case. Racanelli [5] described this dependence for the real part of the complex refractive index n=n(x) approximately with following square-law rectification:

$$n(x) = 3.857 + 0.502x + 0.521x^2 \quad k=-0.018 \quad \lambda=632.8nm \quad 0<x<0.30 \quad (2)$$

The imaginary part is assumed unchanged at k=-0.018. Own measurements of samples with a Ge content from 10 up to 23 at. % confirm these parameters in comparison with X-ray double crystal diffractometry and offer the possibility to describe the dependence at a wavelength of 785 nm in a corresponding form by using the same points on the samples for the measurements. The best fit for n(x) from the measured Ψ-Δ-values is realised with following linear rectification:

$$n(x) = 3.674 + 0.395x \quad k=-0.006 \quad \lambda=785nm \quad 0.10<x<0.23 \quad (3)$$

For the used range of 0.1 < x < 0.23 and with the parameters from figure 2 the error in n(x) is less than 0.005. The imaginary part of the complex refractive index is unchanged too and its value has been found out to k=-0.006 during the fitting procedure.

A special preparation provided a reference to estimate the errors for the ellipsometrical measurements and calculations and to check the parameters of the CVD process. Some wafers are prepared in an analogous CVD process without Ge. So on the top of the wafer there is only a Si layer and its natural oxide. Figure 3 shows the Ψ-Δ-plot at a wavelength of 632.8 nm from these samples across the wafers. To compare it there is an additional model curve for the Ge content x=0.225 and the resulting refractive index n=3.996. Because the strong zoomed scope for the measured angles in the plot there is only visible the start value of the model curve at 0 nm. The next calculated value of the model curve for the thickness (2. layer) t=5 nm is out of range at the values Ψ=10.48 and Δ=177.61°. But the ellipse in the plot shows the relation of the error range for the ellipsometer (SD 2000 error specification: ±0.1° in Ψ and ±0.3° in Δ).

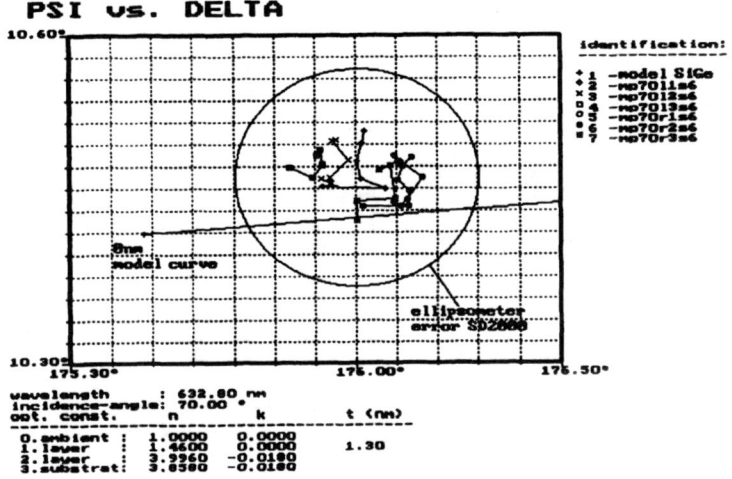

Fig. 3 Ψ-Δ-plot at a wavelength of 632.8 nm for Si buffer on Si substrate

All of the measured values are around the beginning of the model curve for the $Si_{1-x}Ge_x$ layer with a thickness of 0 nm and a refractive index at n=3.996. There is a good agreement between the model and the measurement with only a small error in the correlated calculation of both wavelengths. The Si buffer has almost got the optical properties of the substrate.

In consideration of the random error in measurement figure 4 present a test with 1000 samples during a time of 6 hours in stable environmental conditions of a clean room in an automatic repetition mode. This test shows a sample (KP44) taken from figure 2 at a wavelength of 632.8 nm and an incidence angle of 70° and offers an error distribution almost similar to the normal distribution with σ=0.05 nm for a mean value for thickness of the $Si_{1-x}Ge_x$ layer of 58 nm.

SUMMARY AND CONCLUSIONS

It has been shown that the parallel determination of thickness and range of composition of strained thin epitaxial $Si_{1-x}Ge_x$ layers on silicon substrate is possible by using fixed wavelengths ellipsometry. Because the optical properties at the used wavelengths at λ=632.8 nm and λ=785 nm have a substantial difference for the examined range of composition, this effective method can be used for rapid nondestructive characterization especially inside the production process.

For the wavelength at 632.8 nm the results from [5] have got a verification. In comparison with X-ray double crystal diffractometry a linear relation has been found between the real part of the refractive index and the Ge content for the wavelength 785 nm in the range $0.10 < x < 0.23$.

XTEM inspections are used to characterize the quality of the layers and interfaces. In this connection it has been found out, that ellipsometry reacts sensible to defects in the layers and interfaces, therefore it is possible for example, to use ellipsometrical maesurements to detect defects in homogenous surfaces.

To obtain best agreement of calculated ellipsometrical curves and measured data a native oxide of 1.3 nm thickness on the top of the SiGe layer has to be assumed. This sensible parameter should be determined anew for any set of samples prepared under similar conditions since it cannot be considered to be a constant.

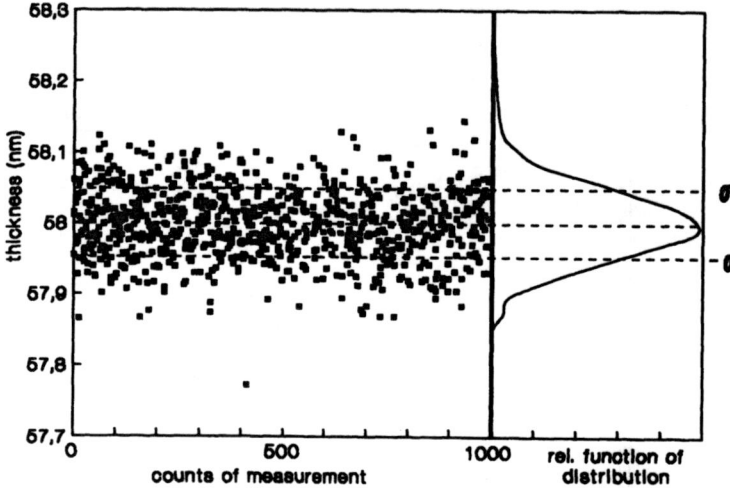

Fig. 4 Random error in measurement for 1000 samples

ACKNOWLEDGEMENTS

The authors would like to thank P. Paduschek from PLASMOS Munich (Germany) for the provided support and the useful discussions, G. Morgenstern and P. Zaumseil from the Institute of Semiconductor Physics for XTEM inspections respectively for the DCD examination. The majority of this work, including the preparation of the $Si_{1-x}Ge_x$ layers, was carried out at the laboratories of the Institute of Semiconductor Physics, Frankfurt (Oder), Germany.

This work was supported by the Ministry of Research and Technology (BMFT Germany, indication: JESSI E69 (B) FKZ 01 M 2918 A). The authors take the responsibilities for the content of this publication.

REFERENCES

1.   R. M. A. Azzam, N. M. Bashara Ellipsomery and Polarized Light , 3rd ed. (North-Holland Physics Publishing, Amsterdam, 1986), p. 539.
2.   PLASMOS GmbH, Ellipsometrical Software for Wavelength Correlated Iteration Procedure, Munich (1992)
3.   F. Lukes, W. Schmidt, J. Humlicek, M. K. Kekoua, E. Khoutschvili, Phys. Stat. Sol. 53, 321-325 (1979).
4.   J. Humlicek, M. Garriga, M. I. Alonso, M. Cardona, J. Appl. Phys., 65, 2827-2832 (1993).
5.   M. Racanelli, C. I. Drowley, N. D. Theodore, R. B. Gregory, H. G. Tompkins, Appl. Phys. Lett. 60, 2225-2227 (1992).
6.   G. M. W. Kroesen, G. S. Oehrlein, E. de Fresart, G. J. Scilla, Appl.Phys.Lett. 60, 1351-1353 (1992).
7.   T. I. Kamins, Electronics Letters, 27, 451 (1991).
8.   Marquardt, D. W.: J. Soc. Ind. Appl. Math., 11, 431 (1963).

SELECTIVE REMOVAL OF SILICON-GERMANIUM: CHEMICAL AND REACTIVE ION ETCHING

F. SCOTT JOHNSON, VEENA MISRA, AND J. J. WORTMAN
North Carolina State University, Department of Electrical and Computer Engineering, Raleigh, NC 27695-7911
LEANNE R. MARTIN, GARI A. HARRIS, AND DENNIS M. MAHER
North Carolina State University, Department of Materials Science and Engineering, Raleigh, NC 27695-7916

ABSTRACT

The use of both chemical and reactive ion etching for the selective removal of Si_xGe_{1-x} alloys with respect to both silicon and silicon dioxide has been investigated. We have found that a solution of NH_4OH:H_2O_2:H_2O is effective in selectively etching the Si_xGe_{1-x} films with respect to both of these materials. The chemical composition of the substrate surface after removal of insitu doped Si_xGe_{1-x} films was evaluated using EDS and SIMS. Diffusion from insitu doped $Si_{0.7}Ge_{0.3}$, followed by selective removal, was used to demonstrate self-aligned npn dopant profiles with narrow base widths. Reactive ion etching of Si_xGe_{1-x} alloys was investigated using SF_6, CF_4, and Cl_2 based chemistries. Pressure, power, and gas flow ratios were optimized to provide the least isotropic etch possible for Si_xGe_{1-x} films containing approximately 40% Ge. Selectivity and degree of anisotropic etching were determined as a function of Ge content for samples with 0% to 100% Ge. Samples were analyzed using SEM and ellipsometry. Highest selectivities were achieved using SF_6 and O_2.

INTRODUCTION

The use of Si_xGe_{1-x} alloys in silicon-based electron devices can provide advantages both in device performance, and fabrication. Bandgap tailoring in these alloys allow majority and minority carriers to be separately influenced, adding an additional degree of freedom in device design. In polycrystalline Si_xGe_{1-x} films, the work function, mobility, and dopant activation energies are functions of the germanium content. These properties make polycrystalline Si_xGe_{1-x} an attractive alternative to polysilicon in applications such as CMOS gates. From a fabrication perspective, Si_xGe_{1-x} has the advantage of a low deposition temperature and the potential of selective deposition and removal. Polycrystalline Si_xGe_{1-x} has recently been proposed as a diffusion source for the formation of ultra shallow junctions [1]. In this paper, we evaluate the use of both chemical and reactive ion etching techniques for the selective removal of Si_xGe_{1-x} from silicon and silicon dioxide. The effect of Si_xGe_{1-x} removal using a NH_4OH:H_2O_2:H_2O solution on silicon/silicon dioxide device structures is presented. The selective dry etching of Si_xGe_{1-x} using SF_6 and O_2 is also presented.

SAMPLE PREPARATION

Polycrystalline Si_xGe_{1-x} films used in this work were deposited using a Leisk^{TM} rapid thermal processor, modified for low pressure deposition [2]. Si_xGe_{1-x} films were deposited using SiH_2Cl_2 and GeH_4 diluted to 8% in a hydrogen carrier gas [3]. Films were deposited on bare silicon and silicon dioxide covered substrates. Plan view and cross sectional Scanning Electron Microscopy (SEM) were used to verify that comparable grain structure and surface roughness were maintained for these samples for all germanium contents used. Secondary Ion Mass Spectroscopy (SIMS) and auger electron spectroscopy analysis indicate oxygen to be present in the deposited films in quantities of approximately one atomic percent.

SELECTIVE CHEMICAL ETCHING

A solution of $NH_4OH:H_2O_2:H_2O$ has previously been shown to selectivity etch Si_xGe_{1-x} with respect to both silicon and silicon dioxide [4]. The Si_xGe_{1-x} etching rates (and etching selectivities) have been determined to be approximately exponentially proportional to the Ge content of the alloy. In Figure 1a, the etching rate of polycrystalline Si_xGe_{1-x} films are given versus the silicon content, for two temperatures.

To determine the effect of this solution on both single crystal silicon and silicon dioxide device structures during over-etching, a patterned sample was prepared. Undoped Si_xGe_{1-x}, containing approximately 40% Ge, was deposited on single crystalline silicon. A 400nm low temperature oxide was then patterned as an etching mask. No annealing steps were performed to densify the deposited oxide. The alloy was then etched using a 1:1:5 solution of $NH_4OH:H_2O_2:H_2O$ at 75°C. The sample was etched for over 4 times as long as the time needed to remove the $Si_{0.6}Ge_{0.4}$ film. The result of this demonstration is shown in the SEM image of Figure 1b. While it is clear from this figure that the $Si_{0.6}Ge_{0.4}$ layer has been over-etched by many times, there is no visible etching or damage to either the deposited oxide mask or the silicon substrate.

To test the effectiveness of the solution in removing all of the germanium over a large area of the surface, light element Energy Dispersive X-ray Spectroscopy (EDS) was performed on samples while in a SEM. To maximize information gathered from the surface of the sample, the excitation volume of the incident electrons was reduced by using the minimum accelerating voltage for excitation of the germanium L-α line (1.18KeV), while providing an adequate signal for detection. The relative intensity of the L-α peak versus acceleration voltage can be seen in Figure 2a. From this graph, it was determined that an accelerating voltage of 1.6KeV could be used to provide an ample germanium signal while maximizing the surface information. Four samples were then prepared for analysis. Sample A was a bare silicon wafer. Samples B and C were prepared by depositing $Si_{0.6}Ge_{0.4}$ on a bare silicon wafer, followed by selective chemical removal of the alloy. Prior to selective removal, sample B was given a 10 hour anneal in nitrogen at 850°C. Sample D was prepared by depositing a thin $Si_{0.6}Ge_{0.4}$ film on a bare silicon wafer. EDS was performed on each sample with an accelerating voltage of 1.4KeV and 1.6 KeV. EDS results showed that no detectable level of germanium remained on any of the etched samples. This can be seen by comparing the 1.18 KeV peak of the four samples in Figure 2b.

A final study of the silicon substrate surface after the removal of a $Si_{0.7}Ge_{0.3}$ film was accomplished using SIMS. Bipolar junction transistor test structures were fabricated as samples for this analysis. Samples were prepared using the base emitter formation steps of a self aligned polysilicon emitter bipolar junction transistor process. The typical base ion implantation was replaced by a diffusion of boron from insitu doped $Si_{0.7}Ge_{0.3}$. For comparison, a control sample was fabricated without the $Si_{0.7}Ge_{0.3}$ deposition, diffusion, and selective etching steps. A SIMS depth profile of this structure is shown in Figure 3. The Ge signal plotted in this figure is at the background level throughout the structure. The slight increase of the Ge signal at the polysilicon silicon interface is due to an oxygen enhanced background signal resulting from the native oxide at the polysilicon silicon interface. This signal is well below that which would result from an incomplete removal of the $Si_{0.7}Ge_{0.3}$ alloy and is identical to that of the control sample.

SELECTIVE REACTIVE ION ETCHING

Reactive ion etching of Si_xGe_{1-x} alloys using SF_6 and CF_4 was performed in a Semi Group RIE System 1000TP. Reactive ion etching using Cl_2 was accomplished using a Drytek Quad System. Wafers prepared for RIE were masked using a low temperature oxide so that 40 cm^2 of material was exposed for all samples. A comparison of reactive ion etching of polycrystalline Si_xGe_{1-x} was performed using Cl_2, $SF_6:O_2$, and $CF_4:O_2$ chemistries. Pressure, power, and O_2 flow ratio were optimized to provide the least isotropic etch possible for intrinsic Si_xGe_{1-x} films containing 40% Ge. These conditions were also found to be approximately the same conditions determined to provide the least isotropic etching of intrinsic polycrystalline silicon films. Etching using Cl_2 was found to result in highly anisotropic etchwalls, while etching using CF_4 resulted in less anisotropic etching. Etchwall profiles were independent of the

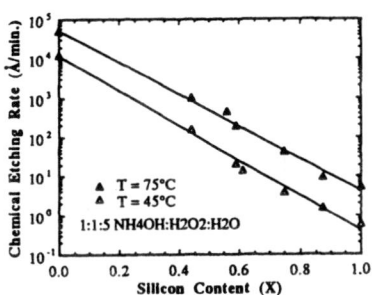

Figure 1. a) Chemical etching rates of SixGe1-x in a solution of NH4OH:H2O2:H2O versus silicon content at temperatures of 45°C and 75°C b) SEM micrograph shows selectively etched SiGe, over etched by more than 3 times, demonstrating high selectivity to both Si and SiO2

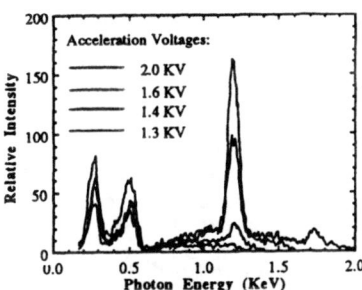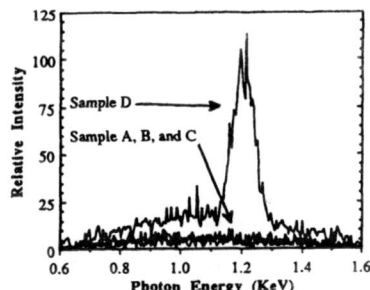

Figure 2. a) Relative intensity of 1.18 KeV peak as accelerating voltage is increased from 1.3KeV to 2.0 KeV. b) Peak intensity for silicon standard, selectively etched SiGe samples, and SiGe standard.

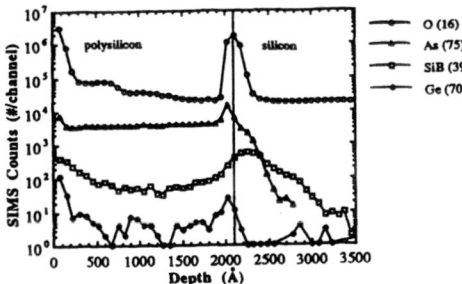

Figure 3. SIMS analysis of Ge, O, and dopant profiles in a polysislicon emitter BJT structure indicating complete removal of germanium following shallow base diffusion.

germanium content of the alloys for both cases. The etch selectivity to silicon of $SF_6:O_2$, and $CF_4:O_2$ chemistries is compared in Figure 4 for polycrystalline Si_xGe_{1-x} alloys as a function of Ge content. These etches were conducted using a power of 100 Watts, a pressure of 80 mTorr, and an oxygen flow of 40% and 25%, for $SF_6:O_2$ and $CF_4:O_2$, respectively. In Figure 4, the etch rate of the alloys is shown to increase with germanium content. It can be seen that, for these reactant gas ratios, a much higher selectivity can be obtained using SF_6 for films containing more than 30% germanium.

For the SF_6 and O_2 chemistry, sidewall slopes were found to be a strong function of germanium content. In Figure 5a, the slope of the etchwalls is plotted versus the silicon content of the Si_xGe_{1-x} films for reactive ion etching using SF_6 and O_2. As can be seen from this figure, the anisotropic nature of this etch is severely effected for films containing 30% or more germanium. It is revealing to compare the silicon concentration dependence of the anisotropic nature of the etch in Figure 5a, with the silicon concentration dependence of the etch rate, as indicated in Figure 5b. The correlation of these two parameters indicates that the increase in etch rate of the Si_xGe_{1-x} samples is due primarily to an increase in the spontaneous (non-ion assisted) etching component with increasing Ge content. The increase of etching selectivity with pressure for reactive ion etching using CF_4 and O_2 has been attributed to this high reactivity of Ge and F [5]. This result suggests that, by using higher etching pressures, the selectivity to silicon can be increased even further. Increasing the pressure increases the available fluorine, while reducing the ion assisted components of the etch rate [6]. This effect can be seen in Figure 6. With an increase in chamber pressure from 100mTorr to 200mTorr, the etch rate of a 40% germanium alloy sample as well as that of a pure germanium sample remains relatively unchanged, while the silicon etch rate decreases as the pressure increases. Selectivity of $Si_{0.6}Ge_{0.4}$ to Si increases to 12:1 at 200mTorr. The presence of germanium has been determined to increase the rate of volatization of both silicon and silicon dioxide [7, 8]. This effect could explain the similarity of the pressure dependence of the 40% Ge and pure Ge samples as shown in Figure 6.

CONCLUSIONS

A solution of $NH_4OH:H_2O_2:H_2O$ has been shown to be effective as a selective etch for Si_xGe_{1-x} alloys. The etch does not effect silicon/silicon dioxide device structures. EDS and SIMS indicate that all measurable traces of germanium can be removed, even after annealing of the Si_xGe_{1-x} film. The use of $Si_{0.7}Ge_{0.3}$ as a selectively removed diffusion source was demonstrated in a NPN bipolar structure. Selective reactive ion etching of Si_xGe_{1-x} has been investigated using $SF_6:O_2$ and $CF_4:O_2$ chemistries. For the gas ratios used, SF_6 was found to provide a higher degree of selectivity. The primary mechanism for selectivity was found to be due to the spontaneous reaction of germanium with fluorine. Increasing RIE pressure to 200mTorr increases selectivity of $Si_{0.6}Ge_{0.4}$ to 12:1. These findings suggest that either chemical or dry etching methods may be used to selectively remove polycrystalline Si_xGe_{1-x} films from silicon dioxide and single crystal silicon surfaces in semiconductor device fabrication.

ACKNOWLEDGEMENTS

This work has been partially supported by the NSF Engineering Research Centers Program through the Center for Advanced Electronic Materials Processing (AEMP) (Grant CDR-8731505) and SRC Microstructures Sciences Program (Grant 90-SJ-081). The authors would like to thank Dale Bachelor of the Analytical Instrumentation Facility at North Carolina State University for scanning electron microscopy and Ray Hamaker and the rest of the staff of the AEMP Microelectronics Fabrication Facility for their assistance and helpful discussion.

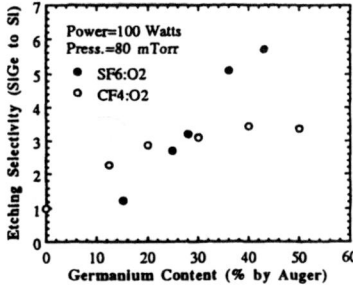

Figure 4. RIE etch rates of Si_xGe_{1-x} alloys versus silicon content for $SF_6:O_2$ and $CF_4:O_2$ chemistries. Power=100Watts, Pressure=80mTorr.

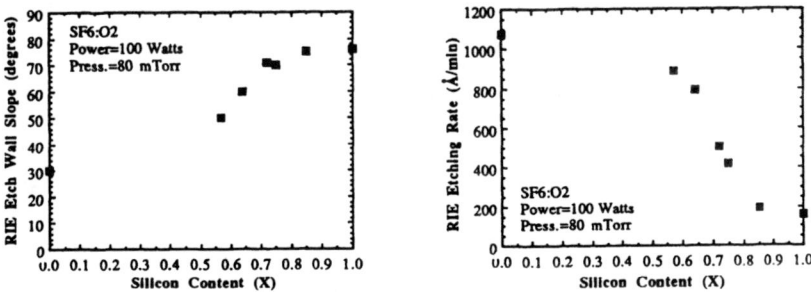

Figure 5. a) RIE etchwall slope versus silicon content for $SF_6:O_2$. b) Increase in RIE etch rate of Si_xGe_{1-x} versus silicon content is inversely proportional to etchwall slope.

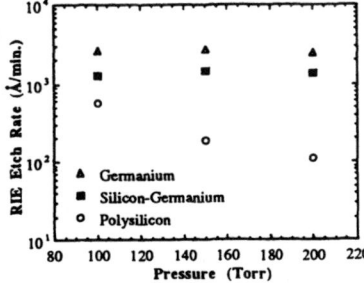

Figure 6. RIE etch rates for Si, Ge, and a $Si_{0.6}Ge_{0.4}$ alloy versus pressure. Increase of pressure to 200mTorr increases selectivity of $Si_{0.6}Ge_{0.4}$ to Si of 1:12.

REFERENCES

1.   D. T. Grider, M. C. Öztürk, and J. J. Wortman, submitted to Transactions on Electron Devices, Nov. 1992.

2.   F. S. Johnson, R. M. Miller, M. C. Ozturk, and J. J. Wortman, Materials Research Symposium Proceedings, 146, p. 345 , (1989).

3.   M. Sanganeria, M. C. Öztürk, G. Harris, D. M. Maher, D. Batchelor, J. J. Wortman, B. Zhang, and Y. L. Zhong, Presented at 1991 Electrochemical Society Third International Symposium on Ultra Large Scale Integration, Washington DC (1991).

4.   F. S. Johnson, D. S. Miles, D. T. Grider, and J. J. Wortman, J. Elect. Mat., 21, p. 805 , (1992).

5.   G. S. Oehrlein, Y. Zhang, G. M. W. Kroesen, E. d. Fresart, and T. D. Bestwick, Appl. Phys. Lett., 58, p. 2252 , (1991).

6.   G. S. Oehrlein, T. D. Bestwick, P. L. Jones, M. A. Jaso, and J. L. Lindstrom, J. Electrochem. Soc., 138, p. 1443 , (1991).

7.   G. A. Porkolab and E. D. Wolf, Appl. Phys. Lett., 56, p. 2319 , (1990).

8.   G. S. Oehrlein, G. M. W. Kroesen, E. d. Fresart, Y. Zhang and T. D. Bestwick, J. Vac. Sci. Technol., 9, p. 768 , (1991).

HYDROGEN RADICAL ANNEALING EFFECT
ON THE GROWTH OF MICROCRYSTALLINE SILICON

Jung Mok JUN, Kyu Chang PARK, Sung Ki KIM, Kyung Ha LEE, Mi Kyung CHU, Min Koo HAN*, Young Hee LEE**, and Jin JANG
Department of Physics, Kyung Hee University, Dongdaemoon-ku, Seoul 130-701, Korea
*Seoul National University, Seoul 151-742, Korea
**Jeonbuk National University, Jeonjoo 556-756, Korea

ABSTRACT

We have studied the growth of microcrystalline silicon (μc-Si) and amorphous silicon (a-Si:H) by layer by layer deposition technique, where the deposition and the radical exposure are done alternatively. He or hydrogen plasma exposure gives rise to the etching effect of both μc-Si and a-Si:H even though the etch rate by He plasma is much smaller. The long exposure of hydrogen radical on a-Si:H gives rise to the formation of μc-Si at low substrate temperature (T_s), whereas the hydrogen content decreases at high T_s. The growth mechanism of the crystallite is proposed on the basis of experimental results.

INTRODUCTION

Doped microcrystal silicon (μc-Si) films have been received considerable attention recently because of their important properties such as high electrical conductivity and high optical band gap compared to the hydrogenated amorphous silicon (a-Si:H). High hydrogen dilution in a glow discharge silane plasma is known to produce μc-Si films at relatively low substrate temperatures (around 250°C). Three possible roles of H atoms have been proposed to explain the growth of μc-Si from the SiH$_4$ and H$_2$ plasma [1]; (1) the hydrogen coverage of the growing surface enhances the diffusion of the absorbed radicals such as SiH$_3$ or SiH$_2$, (2) hydrogen radicals act as etchant of weak Si-Si bonds, thus promoting the growth of stable crystalline phase, (3) hydrogen atoms soak into several layer beneath the top surface and promote the network relaxation.
The layer by layer deposition technique, i.e., the deposition of thin a-Si:H layer and the hydrogen radical exposure (HRE), has been studied recently [2-6]. This technique is used to make both low hydrogen content a-Si:H and μc-Si. However, the role of hydrogen radicals on the growing surface is still controversial.
In the present work, we have studied the repeated deposition of μc-Si (or a-Si) and the exposure of hydrogen radicals (or He plasma) on the surface of the thin layer with varying exposure time and temperature. The role of HRE on the growing surface is discussed on the basis of the experimental results.

EXPERIMENTAL

We used a remote plasma CVD technique, where the plasma generating region is connected to a remote deposition chamber. The base pressure of the reaction chamber is less than 2×10^{-7} torr, pumped by turbomolecular pump. 20 % silane in hydrogen is introduced into the downstream reactor. While helium passes through the plasma generating region inside of the cylindrical quartz tube of diameter 3.8 cm, some of them are ionized and excited to metastable state, and used to dissociate infeeding SiH$_4$.
Films of about 0.4 μm in thickness were deposited on both Corning 7059 glass plates for optical absorption and Raman scattering measurements, and high resistivity Si wafer for Fourier-Transform infrared absorptions. The details of the preparation and HRE conditions for μc-Si are listed in Table I. Table II shows the deposition and He plasma exposure conditions for μc-Si.

Mat. Res. Soc. Symp. Proc. Vol. 298. ©1993 Materials Research Society

We fixed the deposition conditions and the HRE or He plasma exposure conditions, as shown in Tables I and II, and varied only the exposure time for each layer. The substrate temperature was fixed at 330℃.

Table III shows the conditions for the deposition of a-Si:H, HRE and He plasma exposure. The substrate temperature was fixed at 350℃ or 250℃.

Table I   The conditions for the deposition of μc-Si and HRE.

	Deposition	HRE
RF power (W)	100	100
Flow rate (sccm)		
He	160	160
H₂	0	50
SiH₄	0.15	0
Substrate temperature (℃)	330	330
Time (s)	30(14 Å)	variable

Table II   The conditions for the deposition of μc-Si and He plasma exposure.

	Deposition	He plasma
RF power (W)	100	100
Flow rate (sccm)		
He	250	250
SiH₄	0.09	0
Substrate temperature (℃)	330	330
Time (s)	30(12 Å)	variable

Table III   The conditions for the deposition of a-Si:H, HRE and He plasma exposure.

	Deposition	HRE	He plasma
RF power (W)	30	30	30
Flow rate (sccm)			
He	100	100	100
H₂	0	50	0
SiH₄	1.0	0	0
Substrate temperature (℃)	350(250)	350(250)	350
Time (s)	60(40 Å)	variable	variable

RESULTS AND DISCUSSION

In the previous work, we have deposited μc-Si by RPCVD using silane and He without hydrogen dilution [7]. The optimum growth temperature is 330℃ and the grain size is 200Å. In the present work we used this μc-Si as an element layer for HRE treatment.

Figure 1 shows the average deposition rate of μc-Si films plotted against the HRE time. The deposition rate was obtained from the total thickness divided by the deposition time, where the HRE time is not included. The each layer thickness is 14Å. The deposition rate increases with the HRE time until 10s, but starts to decrease after that. The increasing part is due to the deposition of the remaining silane in the reactor and the decreasing part is the etching of Si-Si bonds by hydrogen radicals. If the HRE time is 30s, the deposition rate becomes zero, indicating that the etching is dominant compared with the deposition.

Figure 2 shows the deposition rate of μc-Si films plotted against the He plasma exposure time. Each layer thickness is 12Å. The deposition and the helium plasma exposure conditions are shown in Table II. Compared to Fig. 1, the etch rate is much smaller, but it is clear that the μc-Si layer can be etched by He plasma exposure on the surface. The top surface appears to be reconstructed during the exposure of He metastables and/or He ions, even though the detailed mechanism is not clear now. One possibility is that the soaking of metastable He atoms beneath the surface region gives rise to enhance the lattice relaxation. Another possibility is the etching of Si by He ion sputtering. The detailed etching mechanism by He plasma exposure will be studied.

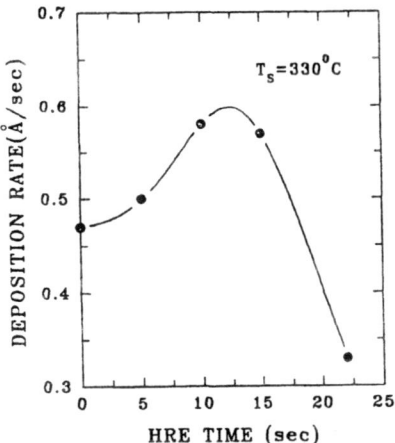

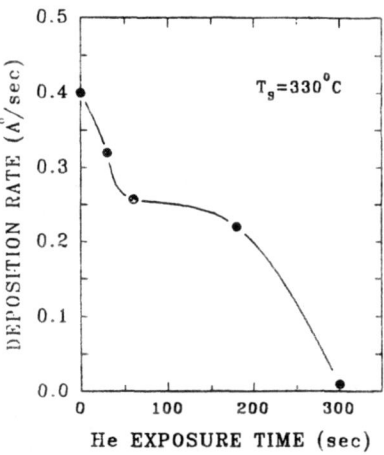

Fig. 1 The deposition rate of μc-Si films prepared by the repeated deposition and HRE process.

Fig. 2 The deposition rate of μc-Si films prepared by the repeated deposition and He plasma exposure.

Figure 3 shows the dark- and photo- conductivities of μc-Si films deposited by layer by layer deposition technique using HRE. The photoconductivity increases with the HRE time, indicating that the grain boundary regions are saturated with the hydrogen atoms during the HRE. The sample using HRE time of 22s shows the photoconductivity of 6×10^{-4}S/cm and dark conductivity $\sim 10^{-6}$S/cm, resulting the photosensitivity of 600 under 100mW/cm^2

Figure 4 shows the average deposition rate of a-Si:H by layer by layer deposition technique. The substrate temperature was fixed at 350°C. The deposition rate decreases with increasing the HRE time because of the etching of a-Si:H by HRE.

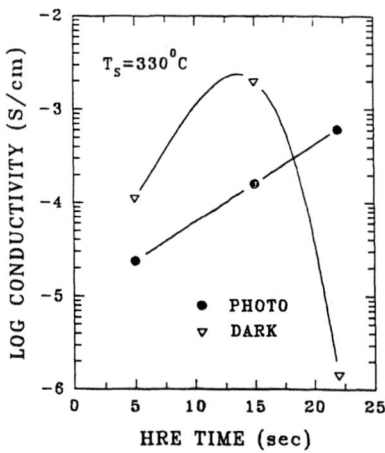

Fig. 3 The dark- and photo-conductivities of μc-Si films deposited by layer by layer deposition technique using HRE.

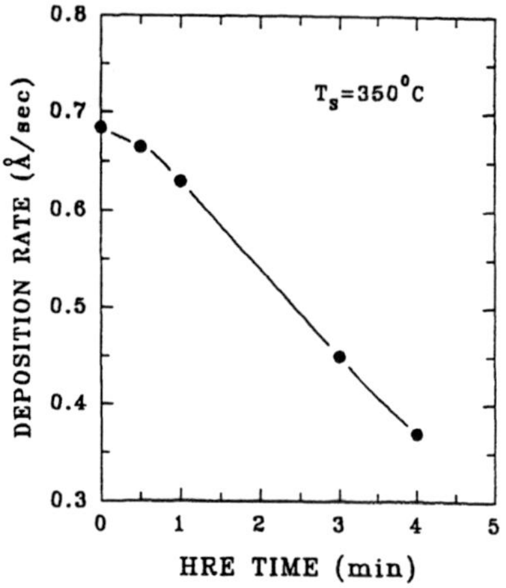

Fig. 4 The deposition rate of a-Si:H films prepared by the repeated deposition and HRE.

Figure 5 shows the hydrogen content in the a-Si:H films treated by HRE. The hydrogen content decreases with the increase of the HRE time, indicating the existence of the chemical annealing by HRE if the T_s is 350°C. The hydrogen atoms attached to weakly bonded Si atoms appear to be driven out through the lattice relaxation by the soaked hydrogen atoms beneath the surface. However, the hydrogen content increases with the HRE time until 1 min and then is out-diffused when the T_s is 250°C

Figure 6 shows the deposition rate of a-Si:H films treated by periodic He plasma exposure. The deposition rate decreases slightly with increasing the exposure time. It is clear, from this figure, that the He plasma can etch the Si layer whether it is μc-Si or a-Si:H.

Figure 7 shows the optical band gap of a-Si:H films treated by HRE. The optical band gap decreases with the HRE time because the hydrogen content decreases with increasing HRE time. It is noted that the hydrogen content increases with the optical band gap in a-Si:H [8].

It is generally accepted that the high hydrogen dilution to silane gives rise to the formation of microcrystal Si. However, the mechanism of the growth of microcrystal Si is still controversial. It is clear that the hydrogen soaks into the amorphous silicon network and increases the hydrogen content when the T_s is 250°C, however, the hydrogen atoms bonded to Si diffuse out from the beginning of the HRE process when the substrate temperature is high, in this experiment, 350℃.

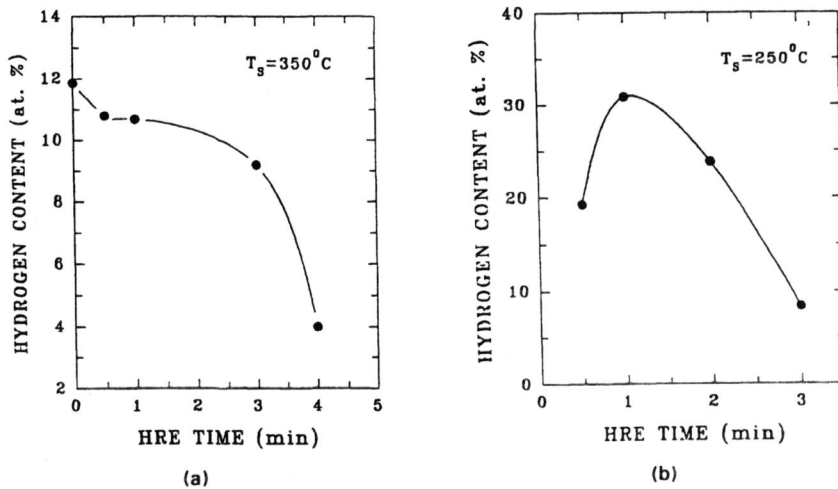

Fig. 5 The hydrogen content in a-Si:H films prepared
by the repeated deposition and HRE. The substrate
temperature was 350℃ (a) and 250℃ (b).

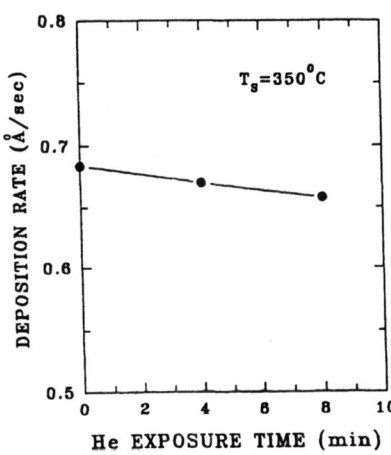

Fig. 6 The deposition rate of a-Si:H
films prepared by the repeated
deposition and He plasma exposure.

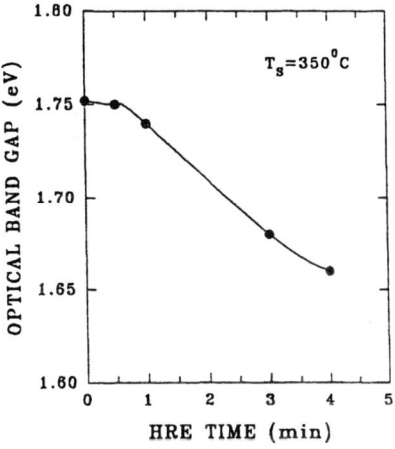

Fig. 7 The optical band gap of
a-Si:H films deposited by the
repeated deposition and HRE
treatment.

The hydrogen atoms soak into the network during the HRE at 250°C. When the hydrogen content is very high in the surface region of Si, the in-diffused hydrogen atoms are able to bond with H atom in Si-H_n bonds, forming hydrogen molecule and diffuse out from the network. In this process the Si atoms are relaxed and form crystallites. The energy released from the formation of hydrogen molecules can enhance the lattice relaxation, resulting in the formation of crystallite or the growth of crystallite. The substrate temperature is one of the most important deposition variable in a-Si:H as well as in μc-Si. When the substrate temperature is 350°C, the hydrogen content in the surface layer is not large because the out-diffusion of hydrogen from the network is dominant. In this process, the silicon atoms can be relaxed and the hydrogen content in the film decreases.

Otobe et al. [9,10] proposed that the incoming atomic hydrogen in the a-Si network let the crystallite grow if there is a nucleus in the network by the separation of the nucleation and the grain growth of the crystallite. It can be presumed that the hydrogen atoms are saturated to the Si atoms in the surface of a nucleus, and then the incoming atomic hydrogen bonds with these hydrogen atoms, resulting in the formation of hydrogen molecule and the concomitant structural relaxation of the network. The nucleus appears to grow by this process. It is expected, therefore, the nucleus can not be grown at high T_s, 350°C, by hydrogen radical exposure.

CONCLUSIONS

The exposure of He plasma as well as HRE on the surface of both μc-Si and a-Si:H gives rise to the etching effect. The HRE on a-Si:H surface gives rise to the increase of hydrogen content(C_H) when the T_s is low (250°C), but to the reduction of C_H through the chemical annealing effect when the T_s is high (350°C). The crystallite can be grown by the formation of molecular hydrogen through the bonding of in-diffusion hydrogen atom and H in Si-H_n in the surface of a crystallite.

ACKNOWLEDGEMENTS

This work was supported by Korea Science and Engineering Foundation through the Semiconductor Physics Research Center.

References

1. A. Matsuda and T. Goto, Mat. Res. Soc. Symp. Proc. 164, 3 (1990).
2. A. Asano, Appl. Phys. Lett. 56, 533 (1990).
3. D. Das, H. Shirai, J. Hanna and I. Shimizu, Jpn. J. Appl. Phys. 30, L239 (1991).
4. H. Shirai, J. Hanna and I. Shimizu, Jpn. J. Appl. Phys. 30, L679 (1991).
5. H. Shirai, J. Hanna and I. Shimizu, Jpn. J. Appl. Phys. 30, L881 (1991).
6. H. Shirai, D. Das, J. Hanna and I. Shimizu, Tech. Digest of 5th International Photovoltaic Science and Engineering Conf. (Kyoto, Japan, 1990), p. 59.
7. S.C. Kim, K.C. Park, S.K. Kim, J.M. Jun and J. Jang, Mat. Res. Soc. Symp. Proc. (Fall, 1992), to be published.
8. H.Fritzshe, Solar Energy Mater. 3, 447 (1980)
9. M. Otobe and S. Oda, Jpn. J. Appl. Phys. 31, L1388 (1992).
10. M. Otobe and S. Oda, Jpn. J. Appl. Phys. 31, L1443 (1992).

A Study on Mechanism of Visible Luminescence From Porous Silicon

Zhou Yongdong, Jin Yixin, Ning Yongqiang, Yuan Jinshan.
Changchun Institute of Physics, Academia Sinica, Changchun 130021, China.

ABSTRACT

The efficient visible light emitting porous silicon was made using the standard method of anodic oxidation. The characteristic photoluminescence spectra and Raman Scattering spectra of the porous silicon were obtained. A kind of stable yellow powder (not the fragments of the porous silicon thin film, the porous structure was still on the crystalline silicon substrate) was taken off from the visible light emitting porous silicon with nonchemical method. The powder can not dissolve in water, alcohol, acetone and some other common solvents, and it still emits efficient visible light after further grinding. The PL spectrum of the powder shows the same peak position, the same shape and the same FWHM as that of the porous silicon wafer. The microstructure of the porous silicon wafer and the microscopic shape of the powder were studied using the scanning electron microscopy (SEM). The X-ray photoelectron spectroscopy (XPS) shows that the fluorescent powder in the surface layer of the porous silicon is composed of many kinds of elements, such as Si, C, O, N and so on. The Si content is only 50% or less in the surface layer of porous silicon. The above-mentioned experiments were performed again on the unpolished surface of the single crystal silicon substrate. So we suggest that the visible luminescence of the porous silicon is from the fluorescent powder maybe not due to the quantum confinement effect in the nm-scale crystalline silicon pillars.

Key words: porous silicon, fluorescent powder, photoluminescence, SEM, XPS.

INTRODUCTION

Since Canham reported that the porous silicon can emit efficient visible light, many groups have been concentrating their research work on the visible luminescence from the porous silicon. As silicon has an indirect bandgap of only 1.1eV, so it's difficult to produce efficient infrared light from this system. Now, people want to use the efficient visible light emission of the porous silicon to push the frontier of the silicon-based material into the opto-electronics territory which is dominated by III-V compound semiconductors now. The efficient cheap display instruments may also be made of porous silicon. The current-induced light emitting device of porous silicon[1] and the highly sensitive photodetector of porous silicon[2] were fabricated in the laboratory. But the origin of the visible light from the porous silicon is still in researching. One main current point of view is the quantum confinement of the recombining electrons and holes in the one-dimensional quantum wires[3]. But some other groups believe that the visible light is from some kinds of moleculars such as the siloxene[4]. And there are some other points of view too. The exact mechanism of the visible luminescence from porous silicon is still contentious now. This paper reports the experimental study on the visible luminescence from porous silicon.

EXPERIMENT

1. Preparation of samples:

Silicon wafers used in the experiments were n-type single crystals (2. 4~3. 6Ω·cm in the resistivity) with a mirror surface and a unpolished reverse side. In order to make the current uniform, the ohmic contact on the back of the silicon wafer was achieved by evaporating a thin film of Al or introducing the solution of KCl touching the wafer's back side when the silicon wafer's front surface was in anodic oxidation. During anodization the sample was illuminated uniformly by a 250W infra-red lamp. The porous silicon layers studied here were all produced by electrochemical etching in HF (49wt%) −C$_2$H$_5$OH solution (1:1 by vlume) at a constant current density of 20mA/cm^2 or 80mA/cm^2 using Pt as a counter-electrode. After anodization, the wafers were rinsed in pure alcohol and then blown dry in the air for test. The powder-like fluorescent material was brushed down directly from the porous silicon surface. The porous silicon and the fluorescent powder are just stored in the air. The manual grinding to the fluorescent powder lasted for about ten minutes in an agate mortar.

2. Spectral measurement:

PL measurements: The samples were excited by the 337nm emission of a pulse width of about 10ns from a N$_2$ laser. The peak power of the laser is several hundred kilowtt (kW). The luminescence of the sample was detected by the R-928 type photomultiplier. All the data were inputed into a computer and the PL spectra were drawn out.

Raman spectrum measurements: The samples were excited by the 488nm line of a Ar$^+$ laser. The Raman Scatting spectra were carried out using the JY-T800 type laser Raman spectrometer.

All the spectra were carried out at room temperature.

3. Analyses of the microstructures:

The analyses of the microshape of the porous silicon wafer, the cross section microstructure of the wafer and the powder morphology were performed using the 1000B type SEM.

4. Analyses of the compositions:

The exact compositions of the fluorescent powder in the surface of the porous silicon wafer were carried out through the ESCALAB-MKII type XPS apparatus.

RESULTS AND DISCUSSION

Spectrum measurements:

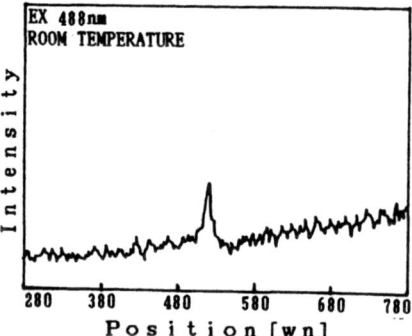

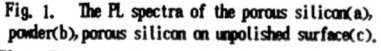

Fig. 1. The PL spectra of the porous silicon(a), powder(b), porous silicon on unpolished surface(c).

Fig. 2. The Raman Scattering spectru of porous silicon

The PL spectra from the porous silicon wafer is shown in Fig. 1a. The orange luminescence spectrum peaks at 17260cm^{-1} (580nm), a FWHM of

3600cm^{-1} (122nm) . The Raman Scatting spectrum of the porous silicon sample is shown in Fig. 2. The spectrum has a peak near 520cm^{-1}, with a width of 15cm^{-1}.

The PL spectrum from the fluorescent powder is shown in Fig. 1b. The orange luminescence spectrum centred at 17000cm^{-1} (588nm) with the FWHM of 3480cm^{-1} (122nm) . Fig. 1c shows the PL spectrum from the porous silicon etched on an unpolished surface of the single crystal silicon wafer. The spectrum peaks at 16540cm^{-1} (604nm) with the FWHM of 3200cm^{-1} (118nm).

SEM measurements:

Fig. 3a shows the SEM micromorphology of the porous silicon wafer. There are large number of micropores at the place where the surface layer has been dropped out. The morphology of the surface layer is similar to that in literature[5]. Fig. 3b presents the SEM photograph of

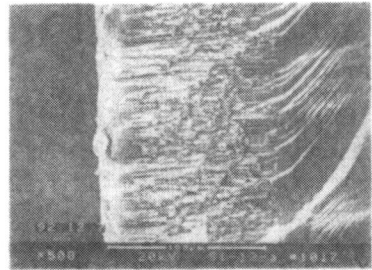

Fig. 3a. The SEM morphology of the surface of porous silicon

Fig. 3b. The cross-sectional SEM morphology of porous silicon

cross-section microstructure of the porous silicon wafer which has three separate layers: surface layer, porous layer, crystalline silicon substrate. The clear μm-scale micropores with the length nearly 100μm are observed.

Fig. 4 shows the SEM morphology of the fluorescent powder. Fig. 4a is about the fluorescent powder before grinding. Fig. 4b shows the fluorescent powder after grinding. Both the figures show no deducible situation of the micro pillars structure. The micropowder will reunite after trituration.

Fig. 4a. The SEM morphology of the powder before trituration

Fig. 4b. The SEM morphology of the powder after trituration

XPS measurements:

Fig. 5 shows the result of the XPS study of the fluorescent powder in the surface layer of the prous silicon. The spectrum shows that the fluorescent powder is composed of many kinds of elements, such as Si, C, O, N and so on. The further study of XPS illustrated that the Si content varies from approximately 30% to 50% (hydrogen was not counted in).

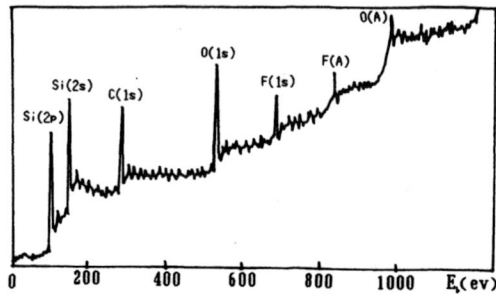

Fig.5. The XPS composition study on the powder

Considering the similar peak positions, the FWHMs and the shapes of the spectra shown in Fig. 1a and 1b, we suggest that the efficient visible luminescence is from the yellow fluorescent powder in the surface layer of porous silicon formed during the process of porous silicon fabrication. Considering the PL spectrum of porous silicon fabricated on the unpolished surface of the crystalline silicon wafer, shown in Fig. 1c, it is obvious that the surface treatment (polishde or not) has no effect on the PL of the porous silicon. The fact that the fluorescent powder can also emit efficient visible light after further grinding proves that the powder is unlikely composed of the one–dimensional crystalline silicon quantum wires (nm–scale in diameter). Considering the SEM microstructure study of the porous silicon, it can be agreed that the fluorescent powder was contained in the surface layer over the porous layer. The study of the XPS also support the point of view above and indicate that the composition of the powder is complicated.

CONCLUSIONS

The visible light is from the fluorescent powder–like material. (The composition and structure of the powder are complicated, and are being studied now.)

The efficient visible light emitting porous silicon can be divided into three separate layers: surface layer, porous layer, crystalline silicon substrate. The fluorescent powder is in the surface layer over the porous layer of the porous silicon

The visible luminescent of the porous silicon shows no relationship with that whether the silicon wafer had been polished before anodic oxidation.

ACKNOWLEDGMENTS

The auhtors thank professor Shihua Huang and Baogui Yu for the help on the PL spectra measurements, Jinxiu Jiang and Naikang Liu's help on the SEM measurements.

REFFERENCES

1. Axel Richter, Member, IEE, P, Steiner, F. Kozlowski, and W. Lang, IEEE ELECTRON DEVICE LETERS 12, 691 (1991).
2. J. P. Zheng, K. L. Jiao, W. P. Shen, W. A. Anderson, and H. S. Kwok, Appl. Phys. Lett. 61, 459 (1992).
3. L. T. Canham, Appl. Phys. Lett., 57, 1046 (1990).
4. M. S. Brandt, H. D. Fuchs, M. Stutzmann, J. Weber and M. Cardona, Solid state Communication, 81, 307 (1992).
5. Nobuaki NOGUCHI, Ikuo SUEMUNE, Masamichi YAMANISHI, G. C. HUA and Nobuo OTSUKA, Jpn. J. Appl. Phys., 31, 490 (1992).

THE RELATIONSHIP BETWEEN SURFACE CHEMISTRY AND PHOTOLUMINESCENCE OF POROUS SILICON

Kun-Hsi Li*, Chaochieh Tsai*, Joe C. Campbell*, Milan Kovar** and John M White**
*Microelectronics Research Center, Department of Electrical and Computer Engineering
**Department of Chemistry and Biochemistry
The University of Texas at Austin, Austin, TX 78712

ABSTRACT

Green photoluminescence (PL) is observed from an as-anodized porous Si wafer immersed in the anodization electrolyte and the PL turns red after the sample is removed from the electrolyte and is blown dry. The PL of porous Si immersed in alcohol exhibits a blue shift and a marked decrease in intensity relatively to dry, as-anodized wafers. However, when the immersed samples are treated with UV for a few minutes, the PL peak shifts to a longer wavelength. Fourier-transform infrared spectroscopy reveals that alkoxy surface species and silicon hydride species backbonded to oxygen atoms appear on the UV-treated samples. Furthermore, the PL characteristics and surface species of the UV-treated samples can be recovered to those of as-anodized wafers by dipping in HF. These results point out the importance of surface chemistry in the luminescence process of porous Si.

INTRODUCTION

The discovery of efficient room-temperature, visible photoluminescence (PL) from porous Si has stimulated extensive research and extensive debate on the physical mechanisms responsible for this luminescence. The above-bandgap emission was initially modeled as two-dimensional confinement in quantum wire-like structures of highly porous Si[1]. Transmission electron microscopy (TEM)[2][3], Raman spectroscopy[4][5], and X-ray diffraction[6] measurements have confirmed the presence of Si crystallites with column or particle dimensions of 3 to 5 nm. Alternatively, it has also been suggested that the luminescence originates in the amorphous regions in porous Si[7][8]. Other hypotheses for the luminescence of porous Si include (1) the formation of hydride complexes[9] or siloxene[10], and (2) recombination through the energy states at the disordered surface[11] or the localized states[12], which may be conduction and valence sublevels in the Si microstructures.

Hydrogen thermal-desorption studies[9][13] revealed the importance of hydrogen passivation for light emission from as-anodized wafers. It has also been demonstrated that surface passivation with good quality oxide grown by electro-oxidation[14] or rapid thermal oxidation[15] produces efficient luminescence. The relatively low dangling bond density on the hydrogen or oxygen passivated porous Si is thought to enhance radiative recombination. In addition to passivating the nonradiative recombination centers, those surface atoms may contribute to the recombination process more actively, such as formation of hydride complexes or siloxene, and the energy shift of surface states. In this paper, we will describe the changes of PL characteristics achieved by changing the surface chemistry of porous Si.

EXPERIMENTAL RESULTS AND DISCUSSIONS

The porous Si layers were formed on boron-doped (100) Si wafers (resistivity $\approx$ 7-10 Ω-cm). Aluminum was deposited on the backside to provide uniform current distribution. The backside aluminum was protected by wax during anodization. The samples were anodized in H_2O:acetic acid:49% HF (2:1:1) electrolyte at 10 mA/cm^2 for 10-20 minutes under illumination by a tungsten lamp. Acetic acid is added to the electrolyte to reduce the surface tension and to aid in removing bubbles from the surface. This improves the uniformity of the PL across the wafer. After anodization, no further chemical etching in HF was needed to obtain visible photoluminescence.

Prior to being removed from the electrolyte, the porous Si samples were illuminated with a UV lamp and green luminescence was observed. After the samples were dipped into DI water and blown dry with nitrogen, the PL changed from green to red. Room temperature PL measurements were performed using an unfocused Ar-ion laser (λ= 457.9 nm; 20 mW) as the

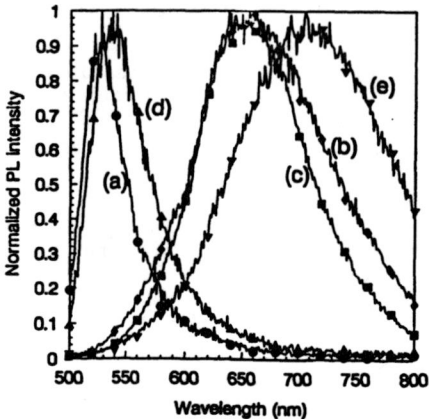

Figure 1. PL spectra of (a) an as-anodized sample which was still kept in H_2O:acetic acid: 49% HF = 2:1:1 electrolyte, (b) the sample which was removed from the electrolyte, dipped into DI water and blown dry, (c) the sample which was returned to H_2O : acetic acid : 49% HF = 3 :1 solution, (d) the sample which was returned to H_2O : acetic acid : 49% HF = 2:1:1 solution, and (e) the sample which was again blown dry.

excitation source. During the PL measurements the samples were positioned in a quartz ampule that had been filled with an electrolyte. Curve(a) in Fig. 1 shows the PL spectrum of an as-anodized sample before being removed from the elctrolyte. Only the long-wavelength portion of the PL spectrum is seen, owing to the presence of the long-pass filter (3 dB cutoff wavelength @ 505 nm) which was used to block the laser light. We presume that the peak wavelength of the PL is somewhat shorter and the spectral width is broader than shown in curve (a). Green luminescence was observed from porous Si samples that were anodized in either the H_2O:49% HF (3:1) or the H_2O:acetic acid:49% HF (2:1:1) electrolytes. After the samples were removed from the electrolytes, dipped into DI water and blown dry, the PL changed from green to red (curve (b)). Returning the samples to a H_2O:49% HF (3:1) solution did not change result in significant PL spectral change (curve (c)). However, returning the samples to a 2:1:1 solution of water, 49% HF and acetic acid resulted in green luminescence immediately (curve (d)). Finally, when the sample was again blown dry in air the spectrum shifted back to the red (curve (e)). Samples prepared without illumination during anodization exhibited the similar behavior.

Possible explanations for the observed shifts in the PL spectra include (1) sequential etching of the top surface of the porous layer and (2) modification of the energies of the surface states. If porous Si oxidizes in air, the size of the Si features would be expected to decrease when the samples were removed from the electrolyte. As a result, one might expect a blue shift instead of the observed red shift. We have noted that the luminescence is not uniform over the thickness of the porous Si layer. The surface emits at a shorter wavelength than the lower part of the porous layer. It is possible that the small structures responsible for the green luminescence are totally oxidized when removed from the HF solution and the lower portion of the porous Si layer with larger features exhibits red luminescence. Under this scenario, when the samples are returned to the electrolyte, etching reduces the feature size resulting in a blue shift. There are, however, several problems with this explanation. It is known that the porous Si layer is hydrogen-passivated and does not oxidize quickly[16]. Furthermore, the red shift was still observed even though the sample was removed in a dry nitrogen glove box and loaded immediately into a vacuum chamber. The fact that the spectral shift between green and red can be repeatedly cycled more than one hundred times would also seem to argue against sequential etching.

The second explanation is the modification of the energies of the surface states. It is possible that the spectral shifts reported here may be attributed to electrolytically-induced changes in the energies of the surface states relative to the band edges. In support of this interpretation, we have also observed reversible spectral changes in several "non-etching" solutions such as ethanol, methanol, and propanol. Exposure to Br_2 vapor also resulted in reversible quenching of the photoluminescence. Lauerhaas et al.[17] related the change of PL spectra to the dipole moment of solvents. But H_2O and KCl which have larger dipole moments do not result in the observed spectral shifts. We also provide a further evidence of the importance of surface chemistry by investigating the changes in the PL and the surface

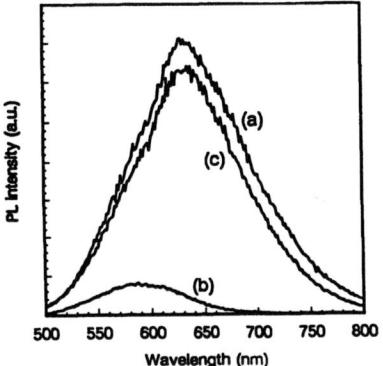

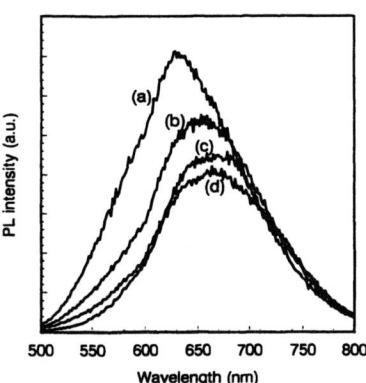

Figure 2. PL spectra of (a) a blown-dry as-anodized porous Si sample, (b) the sample while still immersed in propanol, and (c) the sample that was blown dry after a 15 min immersion in propanol.

Figure 3. PL spectra of the samples which were immersed in propanol with UV treatment for (a) 0 min, (b) 1 min, (c) 3 min, (d) 15 min. These samples were removed from propanol and were blown dry before measurement.

species of porous Si treated in alcohol. In this paper, the "UV-treated" and the "non-UV-treated" designations refer to samples that were immersed in alcohol with and without exposure to UV, respectively. The UV treatment was achieved by illumination from a 6 W UV lamp that was located approximately 2 cm away from samples. Room temperature PL measurements were obtained using the same UV lamp as the excitation source. For the PL measurements the irradiance was approximately five times less than that used for the UV treatment.

Curve (a) in Fig. 2 is the PL spectrum of an as-anodized porous Si wafer measured in air. As soon as an as-anodized sample was immersed in alcohol, dim green photoluminescence could be observed. Curve (b) shows the PL of a similar sample placed in a quartz ampule filled with propanol. It is clear that immersion of non-UV-treated samples in propanol results in a blue shift of the PL peak and a marked decrease in intensity relative to dry, as-anodized wafers. The UV excitation during the PL measurement appears to have negligible effect on the PL spectrum owing to its low level and short duration. Curve (c) shows the PL spectrum of the sample which was blown dry with nitrogen after a 15 min immersion in propanol without UV treatment. There is no significant difference between the PL spectra of the samples before and after immersion in propanol. Similar results were observed from samples treated in methanol and ethanol. Like the as-anodized sample, dim green luminescence was observed when the blown dry, non-UV-treated samples were returned to alcohol again. These reversible PL changes may be due to alcohol physisorption on the surface of porous Si immersed in alcohol. When the samples were removed from alcohol and blown dry, the PL spectra recovered their initial characteristics, as alcohol was desorbed. Reversible changes of the silicon surface in a methanol atmosphere have previously been observed in resistance measurements[18]. The decrease in resistance when exposed to methanol was ascribed to a downward curvature of the semiconductor band due to induced surface donor states that were created by methanol physisorption.

After the sample immersed in alcohol had been treated with UV light for several minutes, the PL peak shifted to a longer wavelength. Contrary to the reversible PL change observed from the non-UV-treated samples, this red shift in the PL of the UV-treated sample was irreversible. After the UV-treated sample was blown dry, it still exhibited red PL. Unlike an as-anodized sample and the non-UV-treated sample, however, the UV-treated sample exhibited a memory-effect; i.e., no significant blue shift of the PL peak and intensity drop were observed when the blown dry, UV-treated samples were returned to alcohol again.

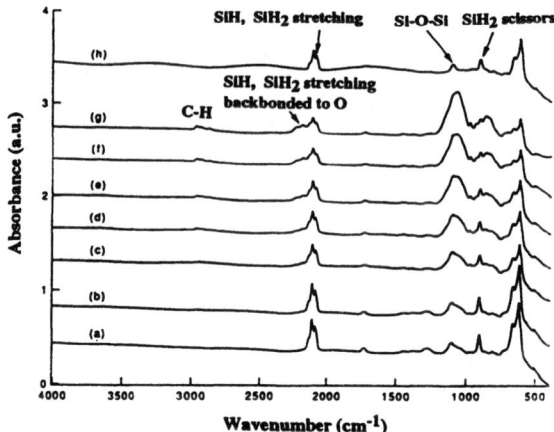

Figure 4. FTIR spectra of (a) an as-anodized porous Si sample, (b) the sample that was immersed in propanol for 15 min without UV treatment, (c) the sample that was UV treated during immersion in propanol for 1 min, (d) for 3 min, (e) for 7 min, (f) for 15 min, (g) for 30 min and (h) the 15-min UV-treated sample after dipping in HF. These samples were blown dry before measurement.

Figure 3 shows the PL spectra of samples that were immersed in propanol with UV treatment for different times and then blown dry with nitrogen. Increasing the treatment time caused larger spectral red shifts with ever decreasing PL intensity. The PL spectra of samples prepared in this way exhibited a tendency toward saturation for longer treatments. Similar results were also observed for samples treated in methanol and ethanol.

Transmission Fourier-transform infrared (FTIR) spectroscopy with a resolution of 4 cm^{-1} was used to investigate the surface species on the porous Si sample with different treatments. All FTIR spectra shown in Fig. 4 were taken from blown-dry samples in a nitrogen ambient. Curve (a) in Fig. 4 shows the FTIR spectrum of an as-anodized wafer which has peaks at 2115-2089 cm^{-1}, 1105 cm^{-1}, 907 cm^{-1}, 665 cm^{-1}, and 625 cm^{-1}. The peaks at 2115-2089 cm^{-1} are from the stretching modes of SiH and SiH$_2$. The 907 cm^{-1} peak is the scissors mode of SiH$_2$. The peak at 1105 cm^{-1} corresponds to the bulk Si-O-Si asymmetric stretching mode. The peaks at 665 cm^{-1} and 625 cm^{-1} are the SiH and SiH$_2$ deformation modes. Curve (b) shows the FTIR spectrum of the non-UV-treated sample which was immersed in propanol for 15 min, removed and blown dry. The similarity between curves (a) and (b) indicates that 15 min immersion in propanol without UV treatment does not result in a permanent change of surface species. Curves (c)-(g) show the FTIR spectra of the UV-treated samples for different treatment times. After UV-treatment, new peaks appear at 2900 cm^{-1}, 2253 cm^{-1}, 2202 cm^{-1}, and 870 cm^{-1}. These new peaks and Si-O-Si stretching at 1100 cm^{-1} increase with increasing treatment time. The absorption at 2900 cm^{-1} which is associated with C-H stretching vibration appeared after 1 min of UV treatment. The blue-shifted infrared absorption in the Si-H stretching region observed at 2253 cm^{-1} and 2202 cm^{-1} is associated with silicon surface atoms backbonded to oxygen. A similar blue shift of silicon hydride stretching peaks was observed from an as-anodized porous Si exposed to visible illumination in air[19]. In addition, a shoulder at 1070 cm^{-1} appears and grows into a peak after 30 min of treatment. No prominent Si-C stretching mode[20] near 730 cm^{-1} was observed in the FTIR spectra of UV-treated samples. These FTIR results may imply that the alkoxy surface species, Si-O-R (R: C_nH_{2n+1}), and silicon hydride species backbonded to oxygen have formed on the surface of the UV-treated sample. It has been shown that methoxy species can attach to a Si surface exposed to methanol vapor[18]. Though no new surface species were formed on the blown-dry non-UV-treated porous Si samples after a 15-min immersion in alcohol, the new surface species formed on UV-treated samples could be found on non-UV-treated samples that had

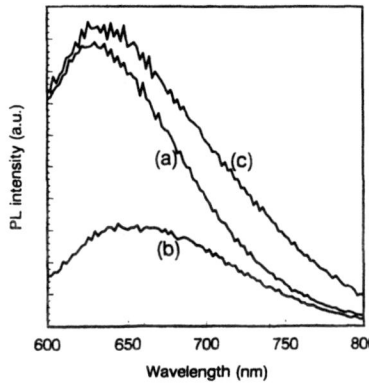

Figure 5. PL spectra of (a) an as-anodized porous Si sample, (b) the sample that was immersed in propanol with UV treatment for 15 min, and (c) the 15-min UV-treated sample after dipping in HF. These samples were blown dry before measurement.

been immersed in alcohol for, typically, 5-10 days. This indicates that UV treatment can enhance the formation of alkoxy surface species and silicon hydride species backbonded to oxygen.

It has recently been suggested that localized surface states play a central role in the luminescence process[11]. This suggests that the photoluminescent characteristics of porous Si can be changed with a modification of porous Si surface. A blue shift of the PL peak and a drop in the PL intensity observed from an as-anodized sample immersed in alcohol can be explained to be due to the modification of the porous Si surface by alcohol physisorption. The UV-treated samples with the alkoxy surface species and silicon hydride species backbonded to oxygen exhibited longer PL peak wavelengths than an as-anodized wafer. Furthermore, hydrogen passivation can be totally recovered by dipping the UV-treated sample in HF as shown by curve (h) in Fig. 4. Figure 5 shows the PL spectra of (a) an as-anodized porous Si, (b) the UV-treated sample, and (c) the UV-treated sample after dipping in HF. All of the samples were blown dry before PL measurement. The PL spectrum of the UV-treated sample after dipping in HF is similar to that of an as-anodized sample. The recovery of the PL spectrum after HF treatment is probably related to the recovery of surface species.

CONCLUSIONS

We have demonstrated that the PL of porous Si is surface sensitive. Completely reversible shifts in the PL spectra between green luminescence from a sample immersed in electrolyte and red luminescence from the sample after blow-drying were obtained repetitively. A reversible spectral blue-shift and a drop in the PL intensity were observed when porous Si samples were immersed in alcohol without UV treatment. The spectral changes are thought to be due to alcohol physisorption on the porous Si surface. After the samples were immersed in alcohol with UV treatment, alkoxy surface species and silicon hydride species backbonded to oxygen were found to form on the surface. These samples also exhibited different luminescent characteristics from an as-anodized sample. Dipping the UV-treated samples in HF recovered the hydrogen coverage on the surface and restored the PL spectrum to that of an as-anodized sample. Changes in the PL spectra concurrently with changes of surface species on porous Si suggest the importance of surface chemistry on the luminescence process of porous Si.

ACKNOWLEDGEMENTS

We are grateful to A. F. Tasch, D.-L. Kwong, S. Shih, and K. H. Jung for encouragement and helpful discussions. This work has been supported by grants from office of Naval Research, ONR-N00014-92-J-1085, the Texas Advanced Research Program-178, and the NSF-supported Science and Technology Center at the University of Texas at Austin (CHE-8920120).

REFERENCES

[1] L. T. Canham, Appl. Phys. Lett. **57**, 1047 (1990).

[2] A. G. Cullis and L. T. Canham, Nature, **353**, 335 (1991).

[3] M. W. Cole, J. F. Harvey, R. A. Lux, D. W. Eckart, and R. Tsu, Appl. Phys. Lett. **60**, 2800 (1992).

[4] R. Tsu, H. Shen, and M. Dutta, Appl. Phys. Lett. **60**,112 (1992).

[5] Z. Sui, P. P. Leong, I. P. Herman, G. S. Higashi, and H. Temkin, Appl. Phys. Lett. **60**, 2086 (1992).

[6] A. Bensaid, G. Patrad, M. Brunel, F. de Bergevin, and R. Herino, Solid State Comm. **79**, 923 (1991).

[7] T. George, M. S. Anderson, W. T. Pike, T. L. Lin, R. W. Fathauer, K. H. Jung, and D. L. Kwong, Appl. Phys. Lett. **60**, 2359 (1992).

[8] J. M. Perez, J. Vilalobos, P. McNeill, J. Prasad, R. Cheek, J. Kelber, J. P. Estrera, P. D. Stevens, and R. Glosser, Appl. Phys. Lett. **61**, 563 (1992).

[9] S. M. Prokes, W. E. Carlos, and V. M. Bermudez, Appl. Phys. Lett. **61**, 1447 (1992).

[10] H. D. Fuchs, M. S. Brandt, M. Stutzmann, and J. Weber, Mat. Res. Soc. Šymp. Proc. **256**, 159 (1992).

[11] T. Muschik, V. Petrova-Koch, A. Kux, and F. Koch, Mat. Res. Soc. Fall, 1992.

[12] T. Miyoshi, K.-S. Lee, and Y. Aoyagi, Jpn. J. Appl. Phys. **31**, Pt. 1 , 2470 (1992).

[13] C. Tsai, K.-H. Li, J. Sarathy, S. Shih, J. C. Campbell, B. K. Hance, and J. M. White, Appl. Phys. Lett. **59**, 2814 (1991).

[14] J. C. Vial, A. Bsiesy, F. Gaspard, R. Herino, M. Ligeon, F. Muller, R. Romestain, and R. M. Macfariane, Phys. Rev. **B45**, 14171 (1992).

[15] V. Petrova-Koch, T. Muschik, A. Kux, B. K. Meyer, F. Koch, and V. Lehmann, Appl. Phys. Lett. **61**, 943 (1992).

[16] D. B. Fenner, D. K. Biegelsen, and R. D. Bringans, J. Appl. Phys. **66**, 419 (1989).

[17] J. M. Lauerhaas, G. M. Credo, J. L. Heinrich, and M. J. Sailor, J. Am. Chem. Soc. **114**, 1911 (1992).

[18] J. N. Chazalviel, J. Electroanal. Chem. **233**, 37 (1987).

[19] C. Tsai, K.-H. Li, J. C. Campbell, B. K. Hance, J. Mf. White, J. of Elect. Mat. **21**, 589 (1992).

[20] V. M. Bermudez, J. Appl. Phys. **71**, 5450 (1992).

LIGHT-EMITTING POROUS SILICON AFTER STANDARD MICROELECTRONIC PROCESSING

C. PENG, L. TSYBESKOV AND P.M. FAUCHET
Department of Electrical Engineering, University of Rochester, Rochester, NY 14627

F. SEIFERTH AND S.K. KURINEC
Department of Microelectronic Engineering, Rochester Institute of Technology, Rochester, NY 14623

J.M. REHM AND G.L. MCLENDON
Department of Chemistry, University of Rochester, Rochester, NY 14627

ABSTRACT

We have investigated the properties of light-emitting porous silicon (LEpSi) after standard microelectronic processing steps such as annealing, thermal and chemical oxidation, ion implantation, and reactive ion etching. The nature of the physical and chemical changes induced by these processing steps is studied. After thermal or chemical oxidation, the photoluminescence (PL) from LEpSi is blue shifted and more stable. Low dose dopant implantation essentially keeps the PL spectrum unchanged. Thermal annealing after ion implantation affects the PL intensity differently, depending on the type of ions. Reactive ion etching changes the surface morphology and shifts the PL peak to blue.

INTRODUCTION

Due to its natural compatibility with crystalline silicon, LEpSi is believed to be a promising building block for a new generation of optoelectronic integrated circuits (OEICs) fully compatible with silicon-based microelectronics. Some simple devices including light-emitting diodes [1,2] and photodetectors [3] have been successfully fabricated in LEpSi, suggesting that OEICs made of LEpSi might become reality in the not too distant future.

Before attempting to demonstrate OEICs purely based on silicon, it is essential to test the compatibility of LEpSi with the present Si microelectronic technology. Examples of processes widely used in today's IC production lines are thermal oxidation, thermal diffusion or ion implantation, and chemical or ion etching. In this paper, we present a study of the effects of thermal and chemical oxidation, ion implantation and reactive ion etching (RIE) on the light emitting properties of LEpSi. These studies also provide useful new information on the origin of the PL in LEpSi.

BACKGROUND

Because of its mechanical and chemical stability, silicon dioxide (SiO_2) is widely used as a mask against implantation or diffusion of dopants, a means of surface passivation and an electrical isolator to separate devices in IC fabrication technology. Wet thermal oxidation is a method of fabricating thick and high quality SiO_2 layers by mixing O_2 with water vapor before entering the high temperature reaction region. The thickness d_0 of the SiO_2 layer after a wet oxidation time t can be expressed as [4]:

$$d_0 = \frac{A}{B}t \qquad\qquad (1)$$

where A, B are temperature-dependent rate constants.

Ion implantation [5] is a technique in which the dopant atoms are vaporized, accelerated, and directed to a silicon substrate. The average doping depth and concentration can be easily controlled by the acceleration energy and the beam current. As each ion impinges on the target, it undergoes a series of nuclear collisions causing damage to the silicon lattice. The disruption of the silicon lattice caused during implantation can be reduced by high temperature annealing. Annealing after ion implantation also activates the implanted dopants.

Reactive ion etching (RIE) [6] is a powerful tool in modern semiconductor microelectronic technology. RIE is more directional, controllable and precise than chemical etching. The basic idea of RIE is to produce a discharge under a high electric field in which gases become ionized and produce chemically active species (plasmas). When such species interact with the exposed substrate, a chemical reaction takes place and the atoms of the substrate are removed or 'etched'. In addition to the chemical process, a physical process is important especially at low pressure, when the energetic ions directly sputter the substrate atoms.

EXPERIMENTS AND RESULTS

LEpSi layers were fabricated by anodizing n-type (100) silicon substrates in a 50%HF:CH_3OH =1:1 (volume) solution for 40 min. under white light assistance [7]. The anodization current density was 6 mA/cm^2. The porous layer thickness was approximately 80 μm as revealed by cross-sectional microscopic measurement.

The photoluminescence (PL) was excited by the 4579Å Ar^+ laser line. The spectrum was recorded by a grating spectrometer attached to an optical multichannel analyzer (OMA). Time resolved photoluminescence (TRPL) experiments were conducted using a frequency doubled Q-switched Nd:YAG laser producing 7 ns long pulses at 532 nm. A photomultiplier attached to a monochromator was used to record the PL decay at different wavelengths. At room temperature, the PL decay is not exponential. With our signal to noise ratio, the decay can be fitted by a distribution of lifetimes ranging from μsec to msec, or more simply by a nearly exponential, fast component, followed by a much slower component. The decay times quoted below correspond to this fast component.

1. Thermal and chemical oxidation

Thermal oxidation of LEpSi was performed by wet oxidation with the bubble technique at a temperature T_{ox} from 400°C to 1000°C for 2 min. Assuming that there is no difference in oxidation rates between crystalline silicon and porous silicon, we obtain from Eq.1 that d_0 is about 150Å after 2 min. for T_{ox}=1000°C. Chemical oxidation was performed at room temperature by dipping the porous silicon in 70% HNO_3 for 2 sec. followed by thermal drying at 150°C.

XPS measurements showed that SiO_2 layers were formed after both high temperature thermal oxidation and chemical oxidation. The changes of PL intensity and peak wavelength with the thermal oxidation temperature have been described elsewhere [8]. The PL intensity dropped dramatically at T_{ox}=400°C; it recovered and became more stable starting from T_{ox}=800°C. The PL peak wavelength also shifted to higher energies. After chemical oxidation, the PL intensity increased by less than a factor of two and the peak shifted to the blue. These results support the quantum size effect hypothesis and establish that oxidation of LEpSi can maintain its useful properties.

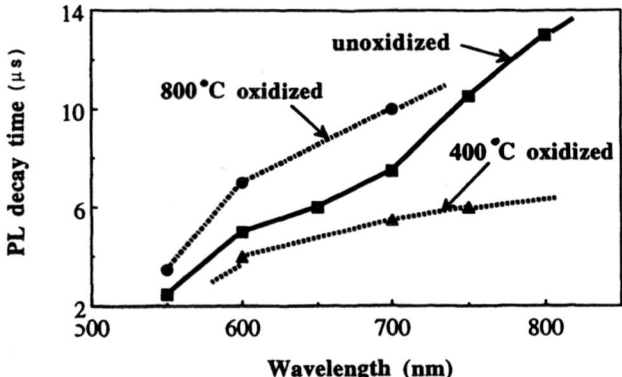

Fig. 1 PL decay time spectra at three stages of oxidation (un-oxidized, T_{ox}=400°C and T_{ox}=800°C).

Figure 1 shows the PL decay time measured at different wavelengths and three stages of oxidation. Low temperature oxidation (T_{ox}=400°C) leads to shorter PL decay times and high temperature oxidation (T_{ox}=800°C) to longer PL decay times. This results from the poor surface passivation at T_{ox}=400°C and the good surface passivation at T_{ox}=800°C and is in agreement with the change of spin densities measured at these temperatures [9].

2. Ion implantation

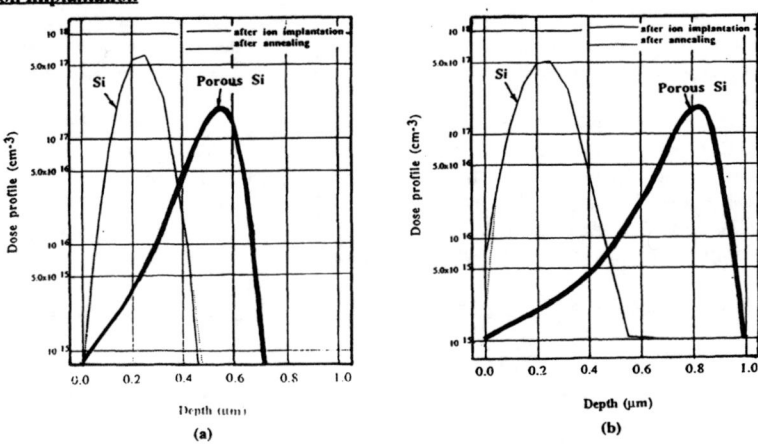

Fig.2 Computer simulation of ion implantation doping profiles with:
(a) boron (energy: 75keV, dose: 1×10^{13} cm^{-2})
(b) phosphorus (energy: 170keV, dose: 1×10^{13} cm^{-2})
for crystalline silicon (thin line) and porous silicon with a porosity of 70% (thick line). The simulation for porous silicon assumes a substrate density reduction to 30% of its original value. The dotted lines show the effect of annealing at 850°C for 30 min. for implanted crystalline silicon.

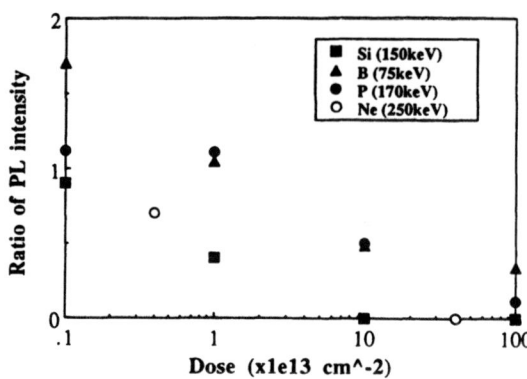

Fig.3 Ratio of PL intensity (average of several measurements) after different types of ion implantation and different doses. Results of Ne implantation from Ref.[11] are given for comparison.

Ion implantation was performed using silicon, boron and phosphorus at an energy of 150keV, 75keV and 170keV respectively. The implantation dose varied from 1×10^{12} to 1×10^{15} cm^{-2} and the ions were incident at 7° off the normal to the substrate. Some of the samples were then annealed in N_2 at a temperature of 850 °C for 30 min.

Computer simulations (by SUPEREM IV) of the dopant profile after ion implantation and after annealing were carried out (Fig.2). Simulations for porous silicon of 70% porosity were performed by simply reducing the density of the substrate to 30% of its original value. The junction depth for porous silicon is twice as deep as for crystalline silicon. Annealing does not increase the junction depth appreciably.

The excitation laser line employed in our PL measurements has a penetration depth of ~ 0.5 μm in LEpSi [10]. This guarantees that the PL comes from the implanted region only. Changes of PL intensity and peak position after ion implantation were carefully recorded with respect to a reference sample. All PL peaks were slightly red shifted by about 40 meV after ion implantation. It is the implantation dose, and not the energy, that had the largest effect on the PL intensity and spectrum. As shown in Fig.3, low dose implantation does not change the PL intensity much and even enhances it in some samples. Higher doses quench the PL and cause a larger red shift in the PL peak. Results of Ne implantation [11] are also shown for comparison.

Annealing generally caused a reduction of PL intensity in all samples implanted with dopants (P or B). However, for high dose Si self-implantation, annealing restored partially the PL intensity (Table I). We interpret this behavior as a competition between (i) removal of the damage introduced by ion implantation and (ii) elimination of the surface hydrogen which passivates the surface. As a result, for heavily damaged samples, the damage removal process is more important, causing a partial recovery of PL intensity, while for the less damaged

Table I. Changes of PL Intensity and Decay Time after Treatments

Type of treatment	PL Intensity (a.u.)			PL Decay Time (μs)	
	Before Implantation	After Implantation	After Annealing	Before Implantation	After Implantation
Si, 150keV, 1×10^{15} cm^{-2}	5400	~ 0	69	3.0	2.7
B, 75keV, 1×10^{13} cm^{-2}	5700	5900	63	3.0	9.8
P, 170keV, 1×10^{13} cm^{-2}	6200	6851	1238	3.0	9.5

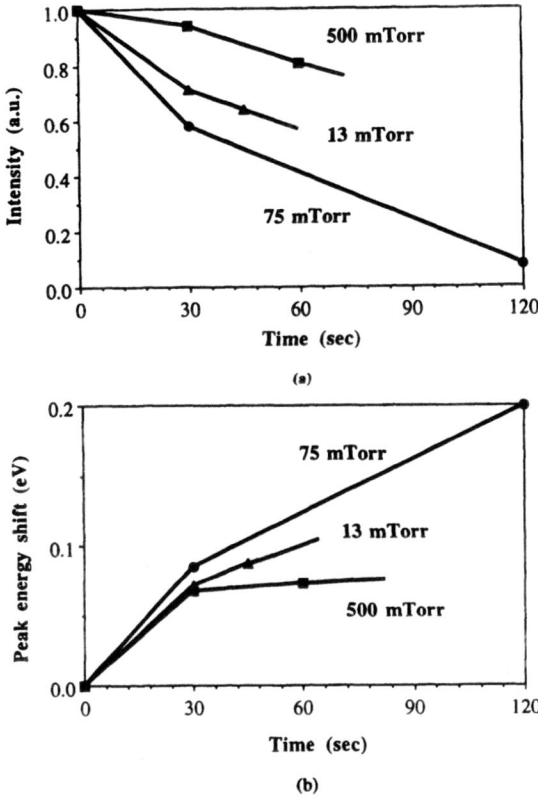

Fig.4 Changes of (a) PL intensities and (b) peak
energies after RIE of different time duration. The
numbers indicate the pressure used in RIE.

samples, the PL intensity is reduced due to the loss of hydrogen.

TRPL were recorded on the samples before and after ion implantation (Table I). At room temperature, the fast component of decay time for this series of samples was approximately 3 μs. Low dose implantation increased the decay time, as it did increase the PL intensity. The decay time of the silicon implanted samples did not change appreciably. Since the decay time at room temperature is dominated by nonradiative processes [12], the increase of the PL decay time suggests that passivation or elimination of nonradiative centers has taken place. However, it should be remembered that the penetration depth at the excitation wavelength of 532 nm is about 2.5 μm [9]. This is deeper than the junction depth, and thus the measured PL decay time may also be affected by unimplanted LEpSi.

It appears that LEpSi survives ion implantation at low to medium doses. Implantation with a dose of 10^{13} cm^{-2} is enough to compensate the substrate doping of 3×10^{15} cm^{-3}. Thus, a p-n junction can be formed without changing appreciably the light emitting properties.

3. Reactive ion etching

RIE was performed at room temperature on LEpSi in 30 standard cm^3 per minute (sccm) SF$_6$ and 3 sccm O$_2$ with 75W forward power under 500 mTorr (high pressure), 75 mTorr or 13 mTorr (low pressure) for a time duration from 30 sec to 2 min. At low pressure, the physical process dominates due to the lack of scattering between the ions. At high pressure, the chemical etching dominates and the accumulation of reaction products (SiF$_4$) slows the etching speed. Thus, the RIE etching rates are slow under either high pressure or low pressure. A maximum etching rate of about 1000Å/min was achieved at 75 mTorr in crystalline Si.

RIE treatment of porous Si caused a significant blue shift in the PL peak and reduced the PL intensity (Fig. 4). These changes were the largest at 75 mTorr. The PL intensities of the samples immediately after RIE were very low but recovered by a factor of ~ 100 after two days of exposure to laboratory air. This is due to partial surface passivation by ambient oxygen which reduces the dangling bond density believed to be the nonradiative recombination centers. Spatially resolved reflectance (SRR) and spatially resolved PL (SRPL) maps [13] show little difference between as-prepared samples and RIE treated samples. SEM with sub-micron resolution suggests that the surface structural periodicity becomes smaller after longer RIE treatment.

These results indicate that LEpSi can survive RIE. The blue shift of the PL after RIE may be caused by a size reduction of the silicon nanocrystalline structures which are thought to be the origin of light emission from LEpSi.

CONCLUSIONS

We have shown that LEpSi can survive several important processes required for the development of silicon-based optoelectronic devices. We have studied the modification of the LEpSi properties induced by these treatments and established the range of parameters that lead to acceptable changes. Our results are consistent with the hypothesis that light originates from nano-scale crystalline silicon objects.

The authors thank G. Carver from AT&T Bell Laboratories for SRR and SRPL measurements. The work is supported by Rochester Gas & Electric, the New York State Energy Research and Development Authority and Xerox Corporation (C.P, L.T, P.M.F) and the National Science Foundation Center on Photo-induced Charge Transfer (J.M.R., G.L.M). C. Peng is a Link Fellow.

REFERENCES

[1]N. Koshida and H. Koyama, Appl. Phys. Lett. 60, 347 (1992)
[2]P. Steiner, F. Kozlowski, H. Sandmaier and W. Lang, Mat. Res. Soc. Symp. Proc. 283, 343 (1993)
[3]J.P. Zheng, K.L. Jiao, W.P. Shen, W.A. Anderson and H.S. Kowk, Appl. Phys. Lett. 61, 459 (1992)
[4]B.E. Deal and A.S. Grove, J. Appl. Phys. 36, 3770 (1965)
[5]J.F. Gibbons, Proc. IEEE 56, 295 (1968) and Proc. IEEE 60, 1062 (1970)
[6]J.W. Coburn, Vacuum 34, 157 (1984)
[7]L. Tsybeskov, C. Peng, S.P. Duttagupta, E. Ettedgui, Y. Gao, P.M. Fauchet and G.E. Carver, 'Comparative study of light-emitting porous silicon anodized with light assistance and in the dark', 1993 MRS spring meeting, this volume
[8]C. Peng, L. Tsybeskov and P.M. Fauchet, Mat. Res. Soc. Symp. Proc. 283,121 (1993)
[9]V. Petrova-Koch, T. Muschik, A. Kux, B.K. Meyer, F. Koch and V. Lehmann, Appl. Phys. Lett. 61,943 (1992)
[10] F. Koch, V. Petrova-Koch, T. Musehik, A. Nikolov and V. Gavrilenko, Mat. Res. Soc. Symp. Proc. 283,197 (1993)
[11] J.C. Barbour, D. Dimos, T.R. Guilinger and M.J. Kelly, Nanotechnology, 3, 202 (1992)
[12] J.C. Vial, A. Bsiesy, F. Gaspard, R. Herino, M. Ligeon, F. Muller, R. Romestain and R.M. Macfarlane, Phys. Rev. B 45, 14171 (1992)
[13] E. Ettdgui, C. Peng, L. Tsybeskov, Y. Gao and P.M. Fauchet, G.E. Carver, H.A. Mizes, Mat. Res. Soc. Symp. Proc. 283,173 (1993)

ENHANCED LUMINESCENCE AND OPTICAL CAVITY MODES FROM UNIFORMLY ETCHED POROUS SILICON

Vincent V. Doan, C. L. Curtis, G. M. Credo and M. J. Sailor
Department of Chemistry, University of California at San Diego,
La Jolla, CA 92093-0506

R. M. Penner
Institute for Surface and Interface Science, Department of Chemistry,
University of California at Irvine, Irvine, CA 92717

ABSTRACT

Uniform layers of porous silicon have been produced in a photoelectrochemical etch that show intensity enhancements of up to 100 fold, relative to samples etched in the dark. These films can also show fine structure in their photoluminescence (PL) spectra characteristic of longitudinal optical cavity modes. The highly luminescent, uniform porous layer is generated by illumination with blue or green light during the electrochemical etch of single-crystal (B-doped) Si, and the enhancement is attributed to a localized photochemical etch process. The relevance of the increased PL intensity and interference-induced spectral changes to measurements of the intrinsic emission spectrum of porous Si are discussed.

INTRODUCTION

Visibly photoluminescent porous silicon is produced by electrochemical and/or chemical etches of either n- or p-type crystalline silicon.[1-3] It has recently been shown that photoluminescent porous Si can also be produced by a photoelectrochemical etch.[4, 5] The photoelectrochemical etch takes advantage of the rectifying nature of the semiconductor/liquid junction; illumination reduces the net etch rate (hole current) at p-type, while it increases the etch rate at n-type Si. In this work, we report that if p-Si is irradiated with low levels of short-wavelength light during the galvanostatic electrochemical etch, the PL of the resulting porous Si is 1-2 orders of magnitude more intense than p-Si etched in the dark. In addition, the luminescent films are optically uniform enough to display well-resolved Fabry-Perot fringes in their PL spectra.

EXPERIMENTAL

Single-crystal polished wafers of p-type (boron-doped) (100) Si of either 9.5 or 3.0 Ω-cm resistivity were cut into rectangles with areas of ca. 0.2 to 0.5 cm^2. These were ohmically contacted on the back by scratching with Ga/In eutectic and

Mat. Res. Soc. Symp. Proc. Vol. 298. ©1993 Materials Research Society

affixing a Cu wire with conductive Ag paint. The contact method was checked for ohmic behavior by running an I-V curve of a wafer contacted on both sides in this manner. The entire contact and edge were coated with epoxy, and the resulting device was used as the working electrode in a two-electrode electrochemical cell. A Pt flag with a press-contacted Pt wire attached was used as the counter electrode. The etching bath was a 50:50 (by volume) solution of aqueous 49% HF and 95% ethanol. Photoelectrochemical etching was carried out in optical quality polystyrene cuvettes. A 300-W ELH-type (tungsten) lamp fitted with an appropriate 20-nm-bandpass interference filter (either 700-, 500- or 450-nm wavelength) was used as the etching light source. Luminescent porous Si resulted after etching at a constant current density of either 3 or 5 mA/cm^2 for 30 min. The samples were then removed from the bath, rinsed with ethanol, and dried under a stream of nitrogen. The PL spectra were always obtained (in air) immediately after drying.

DISCUSSION

Enhanced Photoemission   Single-crystal Si electrodes immersed in HF etching solutions display rectification properties analogous to Schottky junctions.[6-8] The anodic current necessary for the corrosion reaction that forms porous Si corresponds to the forward-bias current for p-Si, and the reverse-bias current for n-Si. Similar to Schottky solar cells, illumination of a p-Si/solution interface generates a photocurrent that is opposite in sign to the forward-bias current, and thus illumination of p-Si electrodes is expected to inhibit the corrosion process.[4]

Surprisingly, illumination of p-Si with low intensity, short-wavelength light during the electrochemical etch produces porous Si that shows a PL blue-shift and dramatic enhancement in PL intensity (Fig. 1). In addition, roughening of the

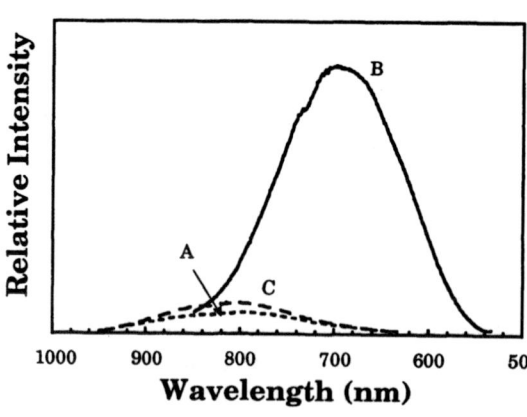

Figure 1. Emission spectra of samples etched in the dark (A), with 1.75 x 10^{15} photon-s^{-1}cm^{-2} of 450-nm (B) and 700-nm light (C).

sample surfaces is more noticeable, as measured by atomic force microscopy (Table 1). Features with lateral dimensions of *ca.* 800Å are observed on samples etched with 450-nm light (see Fig. 1B for representative PL spectrum), while samples etched in the dark (see Fig. 1A for representative PL spectrum) appear much smoother. The photoelectrochemical etch was always performed in a constant current mode (5 mA/cm^2), and so the net anodic (corrosion) current was the same on all samples, regardless of illumination intensity. Similar observation of increased PL intensity from anodization of p-Si under white light illumination has been reported by several groups.[9-12]

Our interpretation of the observed increase in PL intensity is that the electrochemical etch can only occur in regions with good electrical contact to the substrate. Thus, as the regions within the porous Si layer become thinner, the electrically injected holes required to etch the material are less likely to make it from the substrate into these thinner regions. In contrast, the photoetching process is more effective at etching the thinner regions. The short penetration depth of blue (450 nm) light in Si (0.43 μm)[13] ensures that photo-generated holes are produced close to the Si/solution interface, Scheme 1. It is also known that in bulk Si crystals, electronic holes generated close enough to the surface are capable of overcoming the built-in potential of the junction and crossing the interface.[14] In the case of porous Si, this effect should become even more pronounced as the size of the Si material becomes smaller than the width of the space-charge region. Thus the close proximity of photogenerated holes to the surface and the small dimensions of the Si fibers combine to enhance the photochemical oxidative etching process for short illumination wavelengths. The enhanced etch rate results in an intensity enhancement and blue-shift of the peak in the PL spectrum relative to a sample etched in the dark. It has been shown that chemical etches that reduce the physical dimensions of porous silicon's nanostructures will lead to such a blue shift in the PL spectrum, and this has been interpreted within the context of a quantum confinement mechanism.[3] Presumably the photoelectrochemical etch used in the present work generates a larger number of the smaller fibers.

Table 1: Tabulated data for samples in Fig. 1.

Scheme 1

	Intensity	Roughness (Å)
Dark	1	10
Red	1.3	15
Blue	10	43
Unetch	--	1

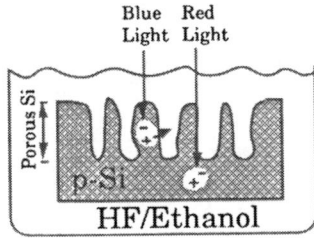

Alternatively, red (700 nm) light illumination is expected to result primarily in excitation of the substrate (Scheme 1) because of its deep penetration depth in silicon (8.4 μm),[13] which is greater than the thickness of the porous Si layer (0-5 μm). Since the experiments in this work were performed under constant current conditions, red light illumination has a negligible effect on the net etch current density. The data for illumination of Si during the electrochemical etch support this interpretation.

Figure 2 shows the PL enhancement (with respect to a sample etched in the dark) as a function of the etch light intensity (He/Cd 442 nm). The PL intensity reaches a maximum for electrodes etched under 4×10^{15} photon-s^{-1}-cm^{-2} irradiation. The PL enhancement drops off and the surfaces of the electrodes appear cracked at higher irradiation intensities. The dependence of PL intensity on photoetch intensity is interpreted as a balance between the rate of production of new luminescent material (via the electrochemical or photochemical etch) and the rate of destruction of this material (via photochemical dissolution). The initial rise of the curve in Fig. 2 corresponds to an increase in production of highly photoluminescent material, driven by the blue photocorrosion process. At some point, the rate of photoetching will exceed the rate of production of new porous Si from the electrochemical process. When this happens, the total amount of luminescent material present will decrease, and this corresponds to the drop off in PL intensity at higher photocorrosion light intensities. A similar interpretation has been proposed for p-Si etched with various intensities of white (tungsten-halogen) light.[10] Under the conditions of our experiments, there is no correlation between the PL emission maximum and the light intensity used in the photoelectrochemical etch for photon fluxes above 10^{15} photon-s^{-1}-cm^{-2} (Fig. 3). At lower light intensities, the general trend is a blue shift in the PL wavelength maximum with increasing illumination intensity, as has been observed in the white light photocorrosion experiments.[10]

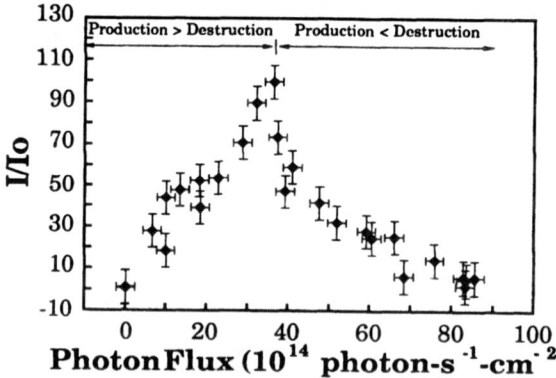

Figure 2 . Correlation of PL enhancement to photoetch light intensity. All samples were etched for 30 min, at 5 mA/cm^2, and uniformly irradiated with 442-nm light during the electrochemical etch. The abscissa (I/I_0) was measured as the ratio of integrated PL intensity of the samples (I) to dark-etch sample (I_0).

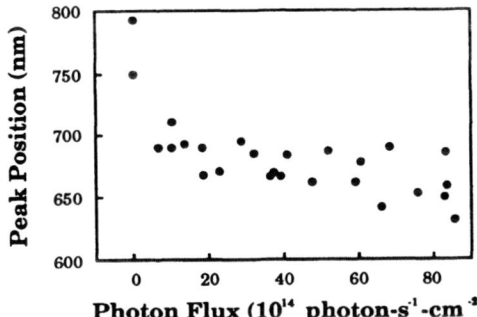

Figure 3. Correlation of the photoluminescence emission maximum to the intensity of light used during photo-electrochemical etch.

It should be pointed out that the data presented here do not indicate a specific luminescence mechanism for porous Si.[15] The enhanced corrosion with short wavelength light may produce a greater number of smaller quantum structures, or it may induce a different type of amorphous Si or other chemical species formation in the film. Raman data of Asano *et al.*[12] have shown that photoelectrochemically etched p-type porous Si is still crystalline, and infrared measurements on the samples in the present study showed no detectable surface oxide species (and the same ratio of Si mono- and di-hydride stretching modes) on either light- or dark-etched samples. However, XPS and Infrared data of Tsai *et al.*[10] indicate that the concentration of surface SiH_2 species and silicon oxides correlates with the light intensity used during white light photoelectrochemical etches. The data are consistent with either the quantum-confined Si model, with specific surface species acting as nonradiative traps, or with the luminescent chemical species model. Further studies into the chemical and structural nature of porous Si are required to elucidate the mechanism of luminescence.

<u>Observation of Optical Cavity Modes</u>   The porous Si films grown in the above manner are very uniform. The air/porous Si and porous Si/substrate interfaces are flat, and the porous Si layer itself displays no features larger than 200 nm (limit of SEM measurement). Characteristic Fabry-Perot fringes are seen in the PL spectrum (Fig. 4) of a 5 μm-thick porous Si layer, if the spectrum is observed from a 1-mm spot on the film. Current densities larger than 5 mA/cm² or etch times longer than 30 min. lead to less uniform porous Si layers, in which no fine structure in the PL spectra could be observed. A plot of the inverse of the wavelength of fringe maxima vs. n, where n is an integer (Fig. 5-A), is linear, as expected from the interference relationship:[16]

$$n\lambda = 2\mu L \tag{1}$$

where λ is the wavelength, μ is the refractive index of the film medium, and L is the film thickness. Assuming that the porous Si layer consists only of pure Si (with refractive index 3.49), the value of μ should be the porosity-weighted average refractive index of Si and air (effective medium model).

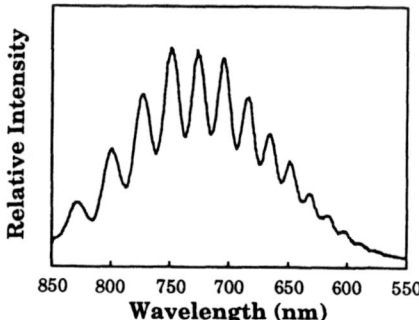

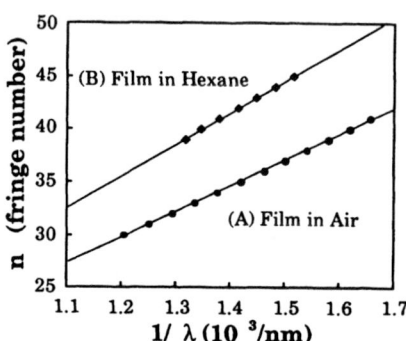

Figure 4. PL spectrum from a 1 mm-diameter spot of porous Si wafer etched at 3mA/cm^2 for 30 min. under 1.31x10^{14} photon-s^{-1}-cm^{-2} of 500-nm light.

Figure 5. Plot of the inverse of the wavelength of fringe maxima vs. n (A) in air; (B) with hexane. The effective optical thickness (μ x L) of this layer in air is 12.2 ± 0.2 μm and in hexane it is 15.0 ± 0.8 μm.

The porosity and thickness of a porous Si layer that exhibits Fabry-Perot fringes can be calculated by replacing the air in the pores with hexane (refractive index = 1.38). The interference fringes shift because of the difference in refractive index between air and hexane. Solution of the simultaneous equations found from the two different PL spectra yielded thickness values that tended to overestimate the film thicknesses as measured by SEM. Optical absorption is not taken into account with this simple index of refraction model and is probably the cause of this inaccuracy. This conclusion is consistent with the results of detailed studies of the refractive index of non-luminescent porous Si films by Pickering et al.[1]

CONCLUSIONS

Porous silicon prepared via a photoelectrochemical etch of p-Si with short-wavelength illumination displays enhanced intensity as much as 100-fold. We interpret the enhancement as arising from a localized photochemical etch process. Short-wavelength light is effective at modifying the corrosion rate of electrically isolated regions because of its small penetration depth. In addition, these low current density etches produce porous Si layers uniform enough to display Fabry-Perot fringes in their PL spectra. The reversible change of fringe spacing on immersion of the porous Si layer in hexane which has larger refractive index (1.38) than air shows that the fine structure in the PL spectrum arises from optical interference.

ACKNOWLEDGMENTS

The work at UCSD was supported in part by the U.S. Office of Naval Research, through grant #N00014-92-J-1810, and at UCI by the NSF Young Investigator Award through grant #DMR-9257000.

REFERENCES

1.	C. Pickering, M.I.J. Beale, D.J. Robbins, P.J. Pearson and R. Greef, J. Phys. Chem. **17**, 6535 (1984).

2.	V. Lehmann and U. Gosele, Appl. Phys. Lett. **58**, 856-858 (1990).

3.	L.T. Canham, Appl. Phys. Lett. **57**, 1046-1048 (1990).

4.	V.V. Doan and M.J. Sailor, Appl. Phys. Lett. **60**, 619-620 (1992).

5.	V.V. Doan and M.J. Sailor, Science **256**, 1791-1792 (1992).

6.	R.L. Smith, S.-F. Chuang and S.D. Collins, J. Electron. Mater. **17**, 533-541 (1988).

7.	F. Gaspard, A. Bsiesy, M. Ligeon, F. Muller and R. Herino, J. Electrochem. Soc. **136**, 3043-3046 (1989).

8.	R.L. Smith, S.-F. Chuang and S.D. Collins, Sens. Actuat. **A21-A23**, 825-829 (1990).

9.	C. Tsai, K.-H. Li, D.S. Kinosky, R.-Z. Qian, T.-C. Hsu, J.T. Irby, S.K. Banerjee, A.F. Tasch, J.C. Campbell, B.K. Hance and J.M. White, Appl. Phys. Lett. **60**, 1700-1702 (1992).

10.	C. Tsai, K.-H. Li, J.C. Campbell, B.K. Hance, M.F. Arendt, J.M. White, S.-L. Yau and A.J. Bard, J. Electron. Mater. **21**, 995-1000 (1992).

11.	I. Suemune, N. Noguchi and M. Yamanishi, Jpn. J. Appl. Phys. **31**, L233-L236 (1992).

12.	T. Asano, K. Higa, S. Aoki, M. Tonouchi and T. Miyasato, Jpn. J. Appl. Phys. **31**, L373-L375 (1992).

13.	D.E. Aspnes and A.A. Studna, Phys. Rev. B **27**, 985 (1983).

14.	A. Kumar and N.S. Lewis, J. Phys. Chem. **94**, 6002-6009 (1990).

15.	M.J. Sailor and K.L. Kavanagh, Adv. Mater. **4**, 432-434 (1992).

16.	B. Rosi. *Optics* (Addison-Wesley, Reading, MA, 1957), p. 125.

THE RELATIONSHIP OF POROUS SILICON FILM
MORPHOLOGY TO THE PHOTOLUMINESCENCE SPECTRA

JOHN PENCZEK and R. L. SMITH
Department of Electrical and Computer Engineering, University of California, Davis
Davis, CA 95616

ABSTRACT

Photoluminescence (PL) spectra are presented for porous silicon samples formed under various formation conditions in aqueous HF solutions. Formation conditions are chosen that effect maximal changes in morphology of resultant films and that correspond to varying formation electrochemistry. The trends in PL center wavelength and full width at half maximum (FWHM) with formation conditions are examined and compared to the resulting morphology. The PL spectra were observed to be most affected by changes in formation conditions when porous films are formed where the electrochemical process of silicon dissolution changes from a $2e^-$ to $4e^-$ (oxide production) reaction. Under these conditions, decreasing HF concentration and/or increasing current density produces a spectral blue-shift which is proportional to the narrowing of the FWHM. This behavior corresponds to morphology changes which are consistent with the quantum confinement model.

INTRODUCTION

Following the initial report of visible PL from porous silicon [1], a great deal of attention has been directed toward understanding the physical mechanism responsible for PL. Early reports of PL from porous silicon have shown that the pore formation conditions, substrate type, substrate doping level, and even exposure time in air can have an affect on the PL spectra. The quantum confinement model [1] for porous silicon PL argues that the changes in PL spectra are caused by changes in the remaining silicon skeletal geometry and dimensions. Given that varying the formation conditions alters the porous silicon morphology in a presumed manner, the quantum model should be able to account for the corresponding PL spectral shifts.

Porous silicon layers are often described in terms of their porosity, i.e. the percent of void space. Porous silicon porosity has been measured using gravimetric or gas adsorption techniques. For n-type and heavily doped p-type (p^+) silicon, these techniques have been employed to extract the porosity, pore size, and the pore size distribution [2]. TEM measurements can also be used to verify pore sizes, but the complex network of interconnecting pores make it difficult to obtain an accurate size distribution. For lightly doped p-type (p^-) silicon, pore size and pore size distribution data is not readily available since gas adsorption techniques are not considered reliable due to the very small pore size and tortuosity [2]. However, gravimetric measurements [3] show that the dependence of porosity on HF concentration and current density is consistent with the p^+ data. Therefore, the p^+ data is extrapolated to p^- silicon in order to determine the dependence of morphology on the pore formation conditions.

Based on the trends observed for p^+ and the limited p^- data, porosity, pore size, and pore size distribution tend to increase with increasing formation current density and decreasing HF concentration. However, as HF concentration increases, porosity has been observed to become less dependent on changes in formation conditions. This is demonstrated in Figure 1, which plots porosity versus formation conditions with data points taken from previously published results on p+, 0.01 Ωcm samples [2,3]. Although there is no straightforward correspondence between porosity and silicon nanocrystal size, such a correlation is often made when interpreting the PL spectra with the quantum model. In order to determine the consistency of the quantum model with respect to porosity trends in p^- silicon, PL spectra were taken on samples made using formation conditions that are known to produce varying morphologies and electrochemical processes. Specifically, pore formation conditions were selected based on the work of Zhang, Collins, and Smith [4]. Their work showed that the morphology produced by anodic oxidation of silicon in aqueous HF can be represented by

Mat. Res. Soc. Symp. Proc. Vol. 298. ©1993 Materials Research Society

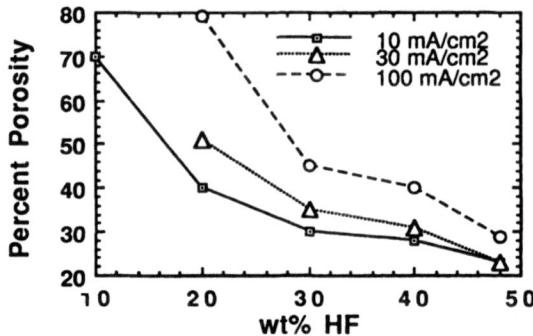

Figure 1: Porosity dependence on formation conditions. Data is taken from Herino [2] and Beale [3] on 0.01 Ωcm p-type (100) silicon.

three regions on the current density vs. HF concentration curve (Figure 2). The top dark line in Figure 2 defines the boundary between the electropolishing and transition regions. The bottom dark line is the boundary between the transition and pore formation regions. In the electropolishing region, silicon is dissolved by a 4e⁻ oxidation to SiO_2, followed by oxide dissolution via F⁻. In the pore formation region, silicon is believed to undergo direct dissolution by a 2e⁻ oxidation and proceeds locally to form pores. The transition region corresponds to conditions where both reactions are believed to take place.

The dashed lines in Figure 2 identify the position of the selected formation conditions relative to the formation boundaries. For a given HF concentration, the current densities were chosen so that the formation conditions would produce films that lie in (a) the transition region near the porous boundary, (b) in the porous region near the porous boundary (near porous), and (c) far into the porous region (far porous). The measured PL spectra are then interpreted in terms of the quantum model and compared to the corresponding morphology.

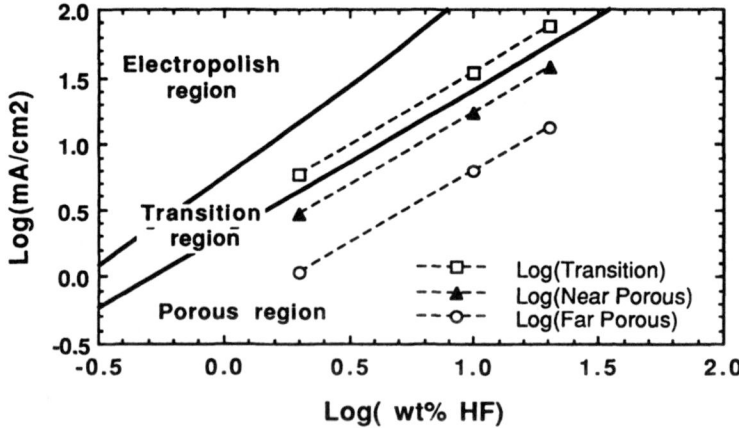

Figure 2: Dependence of silicon morphology on the electrochemical formation conditions [4]. Data symbols indicate the location of formation conditions used in this study.

SAMPLE PREPARATION

All of the porous silicon films used in this study were formed on (100) p-type silicon substrates (14-21 Ωcm). An ohmic backside contact to the substrate was produced by depositing 0.4 microns of aluminum and annealing in nitrogen atmosphere at 450 C. A 1.8 cm diameter polyethylene bottle, sealed to the polished side of the wafer by Apiezon (type W) wax, was used to define the pore formation area. Anodic oxidation was performed by establishing a constant current between a platinum wire mesh in the aqueous hydrofluoric (HF) acid and the wafer back contact. The porous film uniformity is improved by using a wire mesh to maintain uniform current lines, and by mechanical agitation of the electrolyte to dislodge bubbles formed on the surface.

The sample formation conditions are given in Table I. The anodization times were selected to produce porous films that were approximately 10 microns thick [5]. However, the etch times for some of the samples formed in low HF concentrations were limited to 1200 seconds to avoid excessive chemical dissolution [5,6].

Table I: Sample formation conditions used and their corresponding locations in Figure 1.

HF Concentration (wt%)	Currenty Density (mA/cm^2)	Formation Region
20	75	Transition
20	38	Near Porous
20	14	Far Porous
10	35	Transition
10	17	Near Porous
10	6.3	Far Porous
2	5.9	Transition
2	2.9	Near Porous
2	1.1	Far Porous

In order to determine the repeatability of the pore formation process, the group of samples with formation conditions listed in the Table I were formed three times, with about 3 weeks separating each group. All samples were stored in ambient air at room temperature and the data presented here was taken approximately 5 months after formation.

PHOTOLUMINESCENCE SPECTRA

PL measurements were performed at room temperature on a Spex Fluorolog 2 Spectrofluorometer using a GaAs photomultiplier tube. The center of the porous samples were excited with a 6x10 mm incoherent beam at a wavelength of 355 nm. The PL emission was corrected for instrument response. The samples were measured periodically over several months to examine aging effects. A blue-shift was observed for samples formed in the far porous region, presumably due to the thinning of the silicon skeleton by oxidation [7]. However, there was also a red-shift for the 20 wt% HF and 10 wt% HF samples formed in the transition region. After an initial 1 month exposure to air, the PL FWHM of most samples narrowed slightly (<10 nm) over a subsequent 4 month period.

The PL spectra exhibit several interesting trends when correlated to the formation conditions. Figure 3a illustrates the effect of current density on the center PL wavelength for the three selected HF concentrations. Samples formed in 2 wt% HF concentration had the most dramatic blue-shift with increasing current density, and this effect decreased with increasing HF concentration. The effect of current density on FWHM is plotted in Figure 3b. Again, the 2 wt% HF data has the most dramatic decrease in FWHM with increasing current

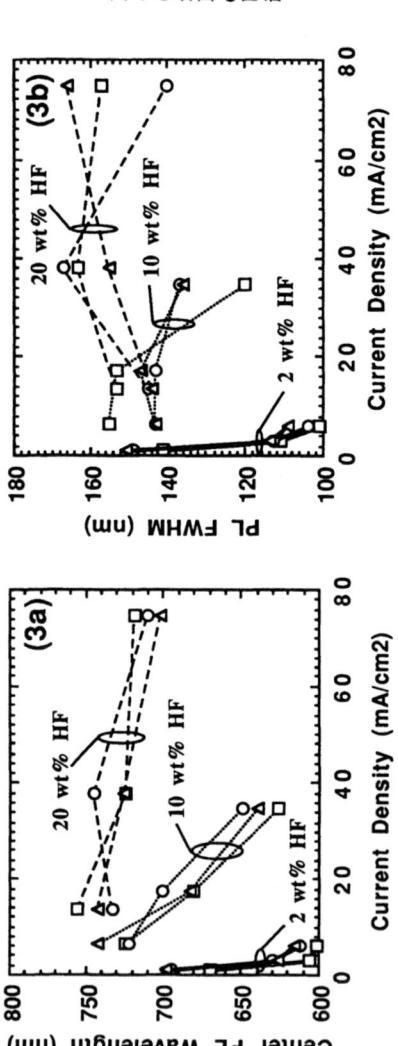

Figure 3: Spectral shifts in (a) PL center wavelength and (b) FWHM as a function of formation current density. Encircled lines represent the degree of repeatability between sample sets at the indicated HF concentration.

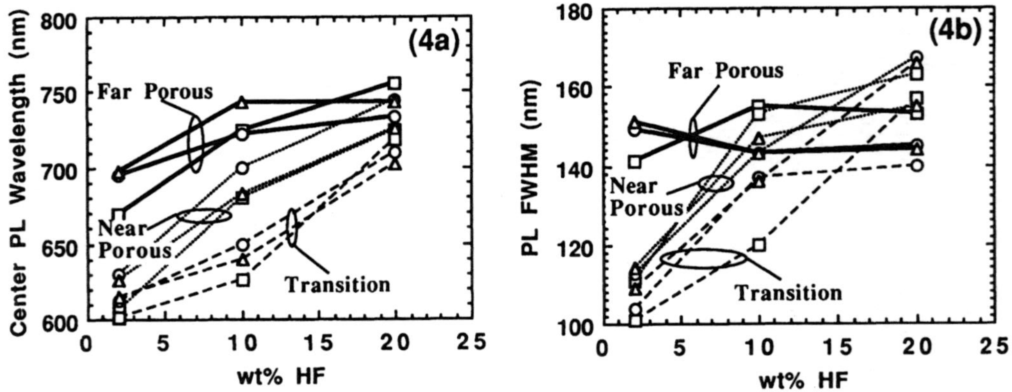

(3b)

20 wt% HF

10 wt% HF

2 wt% HF

PL FWHM (nm)

Current Density (mA/cm2)

(3a)

20 wt% HF

10 wt% HF

2 wt% HF

Center PL Wavelength (nm)

Current Density (mA/cm2)

(4b)

Far Porous

Near Porous

Transition

PL FWHM (nm)

wt% HF

(4a)

Far Porous

Near Porous

Transition

Center PL Wavelength (nm)

wt% HF

Figure 4: Spectral shifts in (a) PL center wavelength and (b) FWHM as a function of HF concentration. Encircled lines represent the degree of repeatability between sample sets in the indicated formation region.

density. The 20 wt% HF data has less repeatability and exhibits little change in the FWHM with formation current density. When the center PL wavelength and the FWHM are plotted with respect to the HF concentration (Figures 4a and 4b), both are seen to increase with the concentration. The effect tends to diminish for the higher concentrations. Figure 5 gives a plot of center PL wavelength versus FWHM for all of the PL spectra taken from these samples. It suggests that there is a proportional relationship between the center PL wavelength and the FWHM for samples formed in the transition region and near the boundary between the transition and pore formation regions. The figure also indicates that a larger range of center wavelengths and FWHMs are attainable for samples formed in or near the transition region.

DISCUSSION

As discussed earlier, published data on p-type porous silicon [2,3,5] shows that the pore size and porosity increase with decreasing HF and/or increasing current density. The remaining silicon would then have to be decreasing in dimension as well. According to the quantum confinement model, a blue shift in the PL spectrum would result from a decrease in silicon nanocrystal size. As the HF concentration increases, however, the porosity of porous silicon films does not change as rapidly with formation conditions, yet pore diameters still follow the same trends. Therefore, at higher HF concentrations, the spectral shift should be less sensitive to current density variations. These trend are clearly observed in our PL center wavelength results (Figure 3a and 4a).

If we assume that the PL spectra are inhomogeneously broadened by the silicon nanocrystal size distribution, then changes in the FWHM should follow the variations in the distribution. If a narrowing of the pore size distribution is taken to imply a narrowing of the nanocrystal size distribution, then p^+ pore size distribution data suggests that the FWHM of p^- samples would narrow for decreasing current density and increasing HF concentration. However, the 2 wt% HF and 10 wt% HF samples were observed (Figure 3b) to have the opposite behavior. This contradiction may be reconciled if the FWHM broadening is interpreted as a consequence of coupling effects between silicon nanocrystals. When the distance between the nanocrystals is reduced, the wavefunctions would begin to overlap and split degenerate energy levels. This effect would manifest itself in a broader PL spectrum. However, the broadening is only expected to be significant if the homogeneous linewidth is relatively broad to begin with. The concept of coupling effects between silicon nanocrystals was also used by Vial et al. [8] to explain PL decay rates of porous silicon.

At a given HF concentration, increasing the current density is known to give larger pore sizes. This would presumably increase the distance between nanocrystals. Based on the proposed coupling model, the FWHM should narrow since there would be less coupling between nanocrystals. Figure 3b follows this trend for samples formed at 2 wt% HF and 10 wt% HF. The 20 wt% HF data indicates that the FWHM is relatively unaffected for increasing current density. This may be accounted for the reduced sensitivity to formation conditions of the higher HF concentrations.

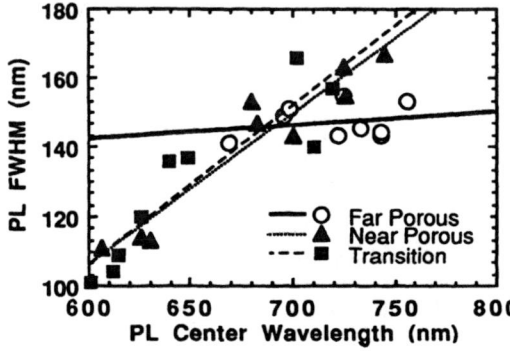

Figure 5: Correspondence of PL center wavelength with the PL linewidth. The lines are a least squares fit to the corresponding formation region data.

Qualitative TEM observations of porous silicon morphology indicate that the pore size and inter-pore distance decreases with increasing HF concentration. Thus, we assume that increasing the HF concentration will lead to a decrease in the distance between silicon nanocrystals. The coupling model would then predict that the increased coupling between nanocrystals should broaden the FWHM. This was the trend observed in Figure 4b for samples formed in the near porous and transition regions. The samples formed in the far porous region showed little change in FWHM with increasing HF concentration. However, when samples with similar current densities were compared, all three formation regions were found to have narrower FWHM with increasing HF concentration.

From Figure 5, one can see that to effect the maximum variation in both PL center wavelength and FWHM, porous silicon should be formed near or inside of the transition region. This corresponds to a current density to HF concentration logarithmic ratio of ≥ 1.2 [log(mA/ cm^2)/log(wt%)]. This also corresponds to the formation conditions which produce the maximum change in porosity with HF concentration and/or current density (Figure 1). According to electrochemical analysis, this is also the region where the rate of silicon anodic oxidation to SiO_2 increases with respect to the direct dissolution of silicon [4,11]. Unfortunately, these results do not exclude or lend more credence to either quantum confinement, siloxene, or oxide trap explanations for porous silicon PL. Further experiments with higher HF concentrations are needed before one can say if PL spectra behavior can be linked more convincingly to one model or the other.

SUMMARY

In order to maximally vary the PL center wavelength and FWHM with formation conditions, the optimal current densities and HF concentrations should lie in or near the transition between pore formation and electropolishing. This also corresponds to the formation conditions that produce the maximum change in porosity and maximum anodic oxide formation. Within this region of formation, a proportional relationship results between PL center wavelength and FWHM such that a blue-shift is associated with narrowing of the PL spectra. We were able to produce a wavelength shift from 757 to 601 nm and FWHM of 167 to 101 nm using only aqueous HF solutions. Better reproducibility and uniformity of films can be produced with the addition of alcohols. However, one then introduces an additional chemical parameter which may play a role in the PL characteristics, especially if the mechanism is not exclusively due to quantum size effects. Extensions of this work will include higher HF concentrations and examination of nearest neighbor distances in order to examine more closely the possibility of wavefunction coupling and its effect on the PL spectra.

ACKNOWLEDGMENT

The authors would like to thank M. J. F. Talbot for her valued assistance with the PL measurements and Prof. Ken Sauer for the use of his PL facility at the Dept. of Chemistry, UC Berkeley and the Division of Structural Biology, LBL.

REFERENCES

[1] L. T. Canham, Appl. Phys. Lett. **57**, 1046 (1990)
[2] R. Herino, G. Bomchil, et al., J. Electrochem. Soc. **134**, 1994 (1987)
[3] M. I. J. Beale, J. D. Benjamin, et al., J. Crystal Growth **73**, 622 (1985)
[4] X. G. Zhang, S. D. Collins, and R. L. Smith, J. Electrochem. Soc. **136**, 1561 (1989)
[5] H. Unno, K. Imai, and S. Muramoto, J. Electrochem. Soc. **134**, 645 (1987)
[6] D. Brumhead, L. T. Canham, et al., Electrochimica Acta **38**, 191 (1993)
[7] L. T. Canham, M. R. Houlton, et al., J. Appl. Phys. **70**, 422 (1991)
[8] J. C. Vial, A. Bsiesy, et al., Phys. Rev. B **45**, 14171 (1992)
[9] G.D. Sanders and Yia-Chung Chang, Appl. Phys. Lett. **60**, 2525 (1992)
[10] T. Ohno, K. Shiraishi, and T. Ogawa, Phys. Rev. Lett. **69**, 2400 (1992)
[11] R. L. Smith, S.-F. Chuang, and S. D. Collins, J. Electron. Mat. **17**, 533 (1988)

INITIAL STAGE CARRIER DYNAMICS IN POROUS SILICON USING ULTRAFAST SPECTROSCOPY

T. Matsumoto*, O. B. Wright*, T. Futagi*, H. Mimura*, and Y. Kanemitsu**
*Electronics Research Laboratories, Nippon Steel Corporation,
5 − 10 − 1 Fuchinobe, Sagamihara, Kanagawa 229, Japan
**Institute of Physics, University of Tsukuba, Tsukuba, Ibaraki 305, Japan

ABSTRACT

We have studied the electronic relaxation processes in porous silicon using a femtosecond pump and probe pulse−correlation technique at 440−nm wavelength. We have observed photoinduced absorption with a response time on the order of 5 ps. From picosecond luminescence decay measurements and the femtosecond pump and probe experiments, the carrier dynamics corresponding to our excitation conditions in porous Si is clarified as follows : carriers are excited in Si microcrystals and then rapidly thermalize to the surface state within 5 ps. After this, strong luminescence occurs from this state with decay components on the order of 1 ns.

INTRODUCTION

Porous silicon (Si) fabricated by electrochemical anodization has attracted much interest because it exhibits strong luminescence at room temperature [1]. The structure of this material is similar to a condensed state of nanometer−order Si microcrystals[2, 3], and both the luminescence peak [1] and the absorption edge [4] shift to higher − energies with increasing chemical dissolution. For these reasons the quantum−confinement effect is suggested as the origin of the strong luminescence. However, other mechanisms, such as an amorphous effect, siloxene cluster effect[5] or surface effect [6 − 8] are proposed, and the luminescent mechanism still remains unknown. The difficulty of this problem mainly arises from the lack of information about the carrier dynamics in the initial decay stage.

Generally the electronic state of the excited carriers is very sensitive to the surrounding environment. The relaxation which reflects the interaction between the excited carriers and the surrounding field (photons, phonons, and carriers) is very fast especially for nanometer−order semiconductor microcrystals. Picosecond and femtosecond spectroscopy have therefore often been used to probe the transient carrier dynamics [9 − 13]. To clarify the luminescent mechanism in porous Si, it is necessary to understand the relaxation dynamics in the initial decay stage through ultrafast spectroscopy.

In this paper, we have studied electronic relaxation processes in porous Si using a femtosecond pump and probe pulse−correlation technique at 440−nm wavelength. We have observed photoinduced absorption with a response time on the order of 5 ps. This result suggests that carriers are generated in nanometer − order microcrystalline Si cores and rapidly thermalize to the luminescent surface states.

EXPERIMENTAL

Free – standing porous Si layers were used in this experiment. Anodization was carried out for the Si substrates with 3.5–4.5 Ω cm resisitivity in HF–ethanol solution (HF : H_2O : C_2H_5OH = 1 : 1 : 2) at a current density of 10 mA∕cm² for ten minutes under light irradiation. After anodization, the porous Si layer was removed by increasing the current density abruptly to 700 mA∕cm².

Figure 1 shows the experimental arrangement. Our experiments were performed with the second harmonic (SH) of a cw mode–locked titanium doped sapphire laser. The excitation wavelength was 440 nm, the excitation power 10 mW, the pulse duration 200 fs, and the repetition rate 81 MHz. This laser beam was split into a pump and probe by a beam splitter. The pump beam was modulated at MHz frequencies to improve the signal–to–noise ratio using synchronous detection. The probe beam passes through a polarization rotator and is rotated 90° with respect to the pump–polarization to avoid the coherent artifact. Both the pump and the delayed–probe beams are focused onto the same spot (diameter ≒ 20 μ m) of the sample. The modulated probe intensity induced by the pump is detected by a Si photodiode and a lock–in amplifier.

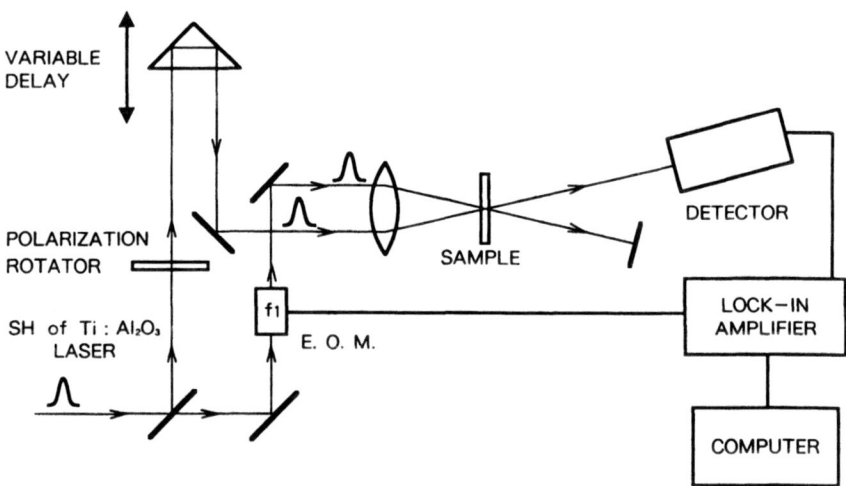

Figure 1. Experimental arrangement for studying electronic relaxation processes in porous Si.

PRINCIPLE OF PHOTOINDUCED ABSORPTION

Porous Si can be considered as a random system in which k−vector conservation is absent. In this case the absorption cross section becomes smaller for deeper states and induced absorption can be measured [14].

Our experimental results previously obtained [6−8] suggest that there are two distinct excited states in porous Si, the microcrystalline state and the surface state, and that the microcrystalline state is responsible for the absorption process whereas the surface state is responsible for the luminescence process.

Figure 2 shows the electronic relaxation processes in porous Si. Carriers are excited to the microcrystalline state $|M\rangle$ by the intense pump beam. These carriers then relax to the ground state $|G\rangle$ or to the surface state $|S\rangle$. The delayed pulse with much smaller intensity can probe the decay rates α, β, and γ shown in Fig. 2. To analyze these parameters we have to perform a third−order perturbation calculation using a density matrix formalism. Details of these calculations are given in reference 15.

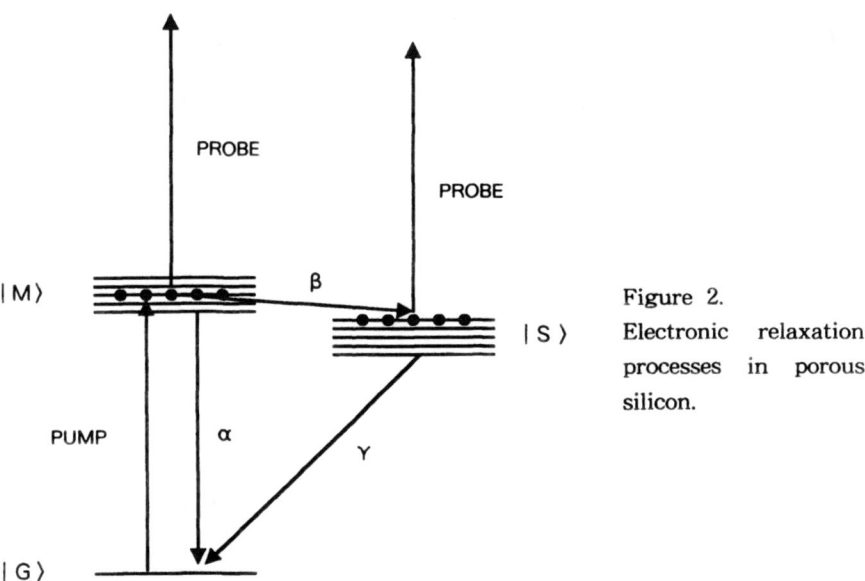

Figure 2. Electronic relaxation processes in porous silicon.

RESULTS AND DISCUSSION

Figure 3 shows the fast decay of the photoinduced absorption in the free standing porous Si film in the vicinity of time delay $\tau = 0$. The initial response time of this induced absorption shows two exponential decay components with relaxation times on the order of 5 ps and 15 ps. By modulating the light at frequencies on the order of 200 Hz we have also observed a very slow induced−absorption change on these time scales. This slow induced—absorption change may be attributed to the redshift of the band gap due to the sample temperature changes. The amplitude of this slow induced—absorption is very large (it is possible to observe it with naked eye at low frequencies). This suggests the possibility not only to fabricate luminescent materials but also to make an optically switched optical modulator with porous Si.

From the study of absorption and Raman spectrum measurements [8], a correlation between the blueshift of the absorption edge and the Si microcrystal size is evident. However, the luminescence peak does not shift in spite of the blueshift of the absorption. This suggests that there are two distinct states : the absorption state and the luminescence state. We have previously performed picosecond luminescence decay measurements [6, 7] and also observed decay components on two different time scales. The decay component of the microcrystalline state is on the order of 100 ps (denoted α in Fig. 1) and that of the surface state (denoted γ in Fig. 1) is subnanosecond. The transfer rate (denoted β in Fig. 1) from the microcrystalline state to the surface state has not been clarified due to insufficient time resolution. In the present work relaxation time β^{-1} is approximately equal to 5 ps. Therefore the carrier dynamics in porous Si can be described as follows : carriers are excited in Si microcrystals and rapidly thermalized to the surface states. After the thermalization strong luminescence occurs from this state. Slow decay components on the order of μs to ms may be attributed to relaxation associated with the hopping and/or diffusion processes [16].

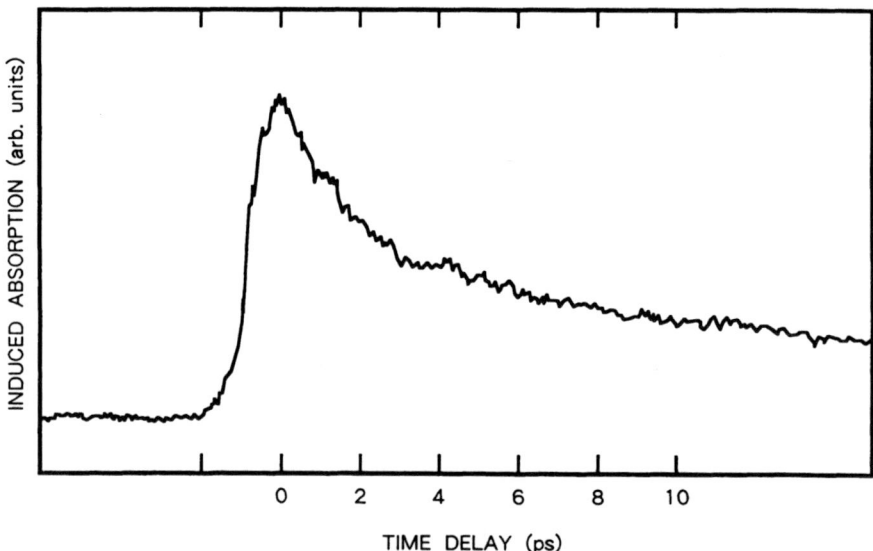

Figure 3. Photoinduced absorption in porous Si film in the vicinity of time delay $\tau = 0$.

CONCLUSION

We have studied electronic relaxation processes in porous silicon using a femtosecond pump and probe pulse – correlation technique at 440 −nm wavelength and observed the photoinduced absorption with a response time on the order of 5 ps. This is the fastest decay component observed until now in porous Si and can be explained from the thermalization process of microcrystalline states to surface states. Much faster relaxation processes should be investigated to probe the coherence time of the excited carriers in Si microcrystals at low temperature.

REFERENCES

[1] L. T. Canham, Appl. Phys. Lett. 57, 1046 (1990).
[2] A. G. Cullins and L. T. Canham, Nature 353, 335 (1991).
[3] M. W. Cole, J. F. Harvey, R. A. Lux, D. W. Eckart, and R. Tsu, Appl. Phys. Lett. 60, 2800 (1992).
[4] V. Lehmann and U. Gosele, Appl. Phys. Lett. 58, 856 (1991).
[5] M. S. Brandt, H. D. Fuchs, M. Stutzmann, J. Weber, and M. Cardona, Solid State Commun. 81, 307 (1992).
[6] T. Matsumoto, T. Futagi, H. Mimura, and Y. Kanemitsu, Extended Abstracts of the 1992 International Conference on Solid State Devices and Materials (Tsukuba, August 26 – 28, 1992) p. 478.
[7] T. Matsumoto, T. Futagi, H. Mimura, and Y. Kanemitsu, Phys. Rev. B47 (1993) in press.
[8] Y. Kanemitsu, H. Uto, T. Matsumoto, T. Futagi, and H. Mimura (unpublished).
[9] Y. Kanemitsu, K. Suzuki, H. Uto, Y. Masumoto, T. Matsumoto, and H. Matsumoto, Appl. Phys. Lett. 61, 2446 (1992).
[10] A. Nakamura, H. Yamada, and T. Tokizaki, Phys. Rev. B40, 8585 (1989).
[11] T. Itoh, M. Furumiya, T. Ikehara, and C. Gourdon, Solid State Commun. 73, 271 (1990).
[12] N. Peyghambarian, B. Fluegel, D. Hulin, A. Migus, M. Joffre, A. Antonetti, S. W. Koch, and M. Lindberg, IEEE J. Quantum Electron. QE−25, 2516 (1989).
[13] M. G. Bawendi, W. L. Wilson, L. Rothberg, P. J. Carroll, T. M. Jedju, M. L. Steigerwald, and L. E. Brus, Phys. Rev. Lett. 65, 1623 (1990).
[14] J. Tauc : Semiconductors and Semimetals, ed. J. I. Pankove (Academic, London,1984) Vol. 21, part B, p. 299.
[15] T. Matsumoto, K. Ueda, and M. Tomita, Chem. Phys. Letters, 191, 627 (1992).
[16] R. A. Street : Semiconductors and Semimetals, ed. J. I. Pankove (Academic, London, 1984) Vol. 21, Part B, p. 197.

VISIBLE PHOTOLUMINESCENCE FROM
RAPID-THERMAL-OXIDIZED POROUS SILICON

Y. KANEMITSU[*], T. MATSUMOTO[**], T. FUTAGI[**], and H. MIMURA[**]

[*]Institute of Physics, University of Tsukuba, Tsukuba, Ibaraki 305, Japan
[**]Electronics Research Laboratory, Nippon Steel Corporation, Kanagawa 229, Japan

ABSTRACT

We have studied the origin of the visible photoluminescence (PL) from oxidized porous Si. The hydrogen-passivated surface of porous Si prepared by electrochemical etching is converted to stable silicon oxides by rapid-thermal-oxidization processes. At low oxidation temperature (T_{ox}), the PL spectrum with a peak near 700 nm is observed. At high T_{ox} above 800 °C, a strong blue PL is observed near 400 nm. We discuss the origin of blue and red PL by employing the results of *ab initio* electronic structure calculations of silicon-oxygen compounds.

INTRODUCTION

Optical and electronic properties of nanometer-size semiconductor crystallites have attracted much attention, because they exhibit new quantum phenomena and have potential for novel and future photonic devices. Recently, a great deal of research effort is focused on nanometer-size Si [1,2] and Ge [3,4] crystallites exhibiting strong visible photoluminescence (PL) even at room temperature. In particular, many researchers are trying to clarify the origin of the strong visible PL of Si nanostructures fabricated by electrochemical etching (often called porous Si). However, the mechanism of the strong visible PL from porous Si still remains unclear. Recent models are based on either quantum confinement effects or surface effects in nanometer-size Si crystallites.

The surface of luminescent Si nanocrystallites is usually covered by silicon hydride SiH_x or silicon dioxide SiO_2 and the efficient PL from of porous Si is closely related to a passivation of nanocrystallite surfaces [5]. H-terminated [6,7] and oxidized porous Si [8,9] exhibit strong visible PL. Unfortunately, the strong visible PL from H-terminated porous Si is unstable against oxidation at room temperature, and the PL degradation in porous Si is also discussed [6]. For optoelectronic device applications and discussions on the mechanism of visible PL form porous Si, we need to control the structure of the internal surface of porous Si and to preserve the crystalline Si (c-Si) cores with diameters of several nanometers. It is reported that a stable high-quality oxide is formed on the surface of the c-Si core by rapid thermal oxidization (RTO) processes [9,10]. In this work, we have studied visible PL from oxidized porous Si fabricated by RTO processes. We discuss the origin of blue and red PL by employing the results of *ab initio* electronic structure calculations of silicon-oxygen compounds.

EXPERIMENT

Porous Si layers were formed on p-type (100) Si wafers with ~4 Ωcm resistivity. Thin

Al films were evaporated on the back of the wafers to form a good ohmic contact. The anodization was carried out in HF–ethanol solution (HF:H_2O:C_2H_5OH=1:1:2) at a constant current density of 10 mA/cm^2 for 5 min. Porous Si samples were processed in a RTO apparatus. The heating rate was 200 °C/s and the samples were kept for 35 s at the oxidation temperature T_{ox} ranging from 480 to 1200 °C. The cooling rate was <100 °C/s.

PL spectra from oxidized porous Si were measured in a vacuum by 325 nm excitation light from a He–Cd laser or 193 nm excitation from an ArF laser. The PL lifetimes were measured using a streak camera and a 220–nm and 200–fs pulse from the fourth harmonic of a cw–mode–locked Ti:Al_2O_3 laser.

OPTICAL PROPERTIES OF OXIDIZED POROUS Si

A broad Raman signal in oxidized porous Si was observed at low Raman frequency regions below 500 cm^{-1}. The Raman signal from the samples was divided into three components due to the crystalline Si (c–Si) substrate, c–Si spheres with the average diameter of about 3 nm, and amorphous silicon–based materials. Even in oxidized porous Si with high T_{ox}, RTO processes (times less than one minute) preserve the c–Si core, although the signal due to amorphous materials increases [11].

Figure 1 shows PL spectra from oxidized porous Si under 325 nm excitation at room temperature as a function of T_{ox}. In porous Si oxidized at low T_{ox} (<800 °C), the PL peak is around 700 nm. At higher T_{ox} above 800 °C, the strong blue PL near 400 nm appears, while the red PL near 700 nm disappears. To clarify the PL peak near 400 nm in oxidized

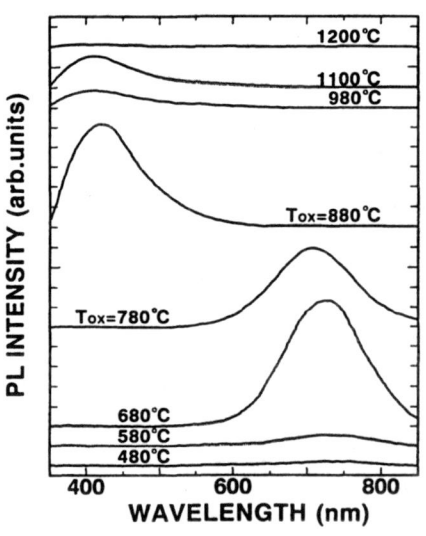
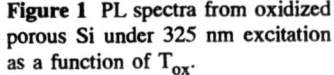
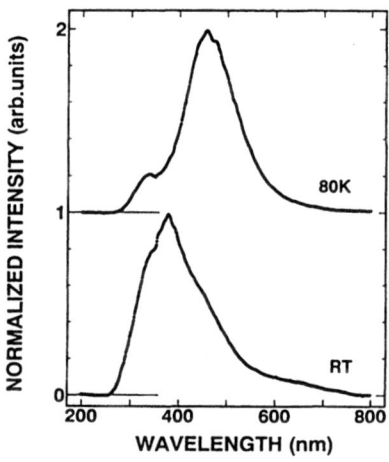

Figure 1 PL spectra from oxidized porous Si under 325 nm excitation as a function of T_{ox}.

Figure 2 PL spectra from oxidized porous Si (T_{ox} =980 °C) under 193 nm excitation measured at room temperature and 80 K.

porous Si with high T_{ox}, we also measured the PL spectrum under 193 nm excitation. Figure 2 shows PL spectra oxidized porous Si of T_{ox}=980 °C. The temperature dependence of the PL spectrum clearly shows that the blue PL spectrum near 400 nm consists of two peaks of 390 nm and 460 nm.

Fourier–transform infrared (FTIR) spectroscopy provides information about the internal surface of porous Si. In as–etched porous Si samples (before RTO processes), the absorption peaks due to Si–H and Si–H$_2$ are observed near 900 and 2100 cm^{-1}. After oxidization, these two peaks disappear. Figure 3 shows the dependence of FTIR spectra on T_{ox}. At low T_{ox} below 800 °C, an absorption band due to silicon oxyhydride (Si:O:H) is observed near 3400 cm^{-1}. At T_{ox} higher than 800 °C, this broad absorption band also disappears. With rising T_{ox}, the amount of silicon oxide monotonically increases. These FTIR spectra strongly suggest that at T_{ox} lower than 800 °C, the surfaces of Si nanocrystallites are terminated by Si–O–H or SiO$_2$. At T_{ox} higher than 800 °C, the surfaces of Si crystallites are covered by SiO$_2$. The PL spectrum is sensitive to these surface–termination conditions. From Figs.1 and 2, it is found that both the surface termination and the PL spectrum change at T_{ox} higher than 800 °C.

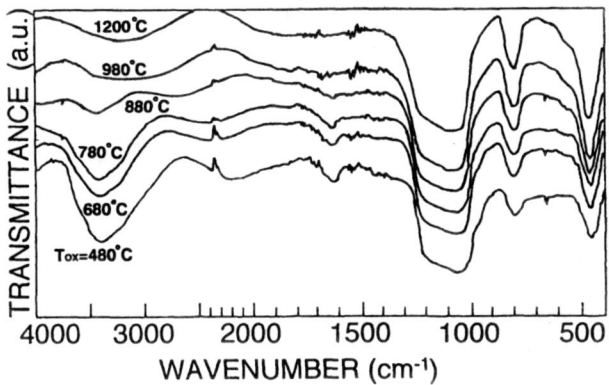

Figure 3 FTIR spectra of oxidized porous Si as a function of T_{ox}.

Figure 4 shows picosecond initial decay of PL. The picosecond PL decay is approximately expressed by a stretched exponential:

$$I(t)=I_0 exp[-(t/\tau)^\beta]. \qquad (1)$$

The picosecond decay time constant τ of blue PL at 400 nm is larger than that of red PL at 700 nm. The value of β at 400 nm ($\beta\sim0.55$) is different from that at 700 nm ($\beta\sim0.47$). The decay time constant in microsecond time region at 400 nm is smaller than that at 700 nm. These results suggest that the origin of blue PL is different from that of red PL. Moreover, Fig.5 shows the time–resolved blue PL spectrum in the oxidized sample with T_{ox}=880 °C. In H–terminated porous Si, we observed the redshift of the PL peak in the initial decay stage [5]. However, the redshift of the PL peak in the initial decay stage is not observed in oxidized porous Si. It is considered that the origin of visible PL in oxidized porous Si is different from that in H–terminated porous Si.

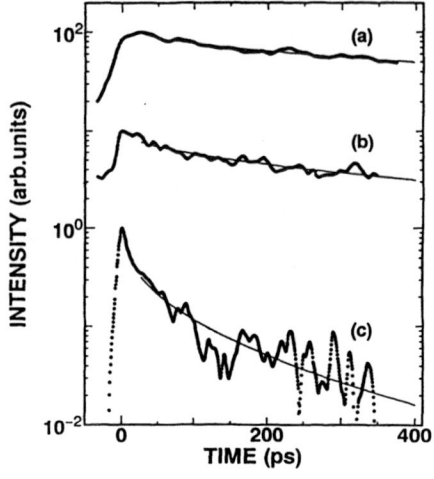

Figure 4 Picosecond Pl decay in oxidized porous Si:(a) T_{ox}= 980 °C, τ=680 ps at 400 nm, (b) T_{ox}=880 °C, τ=300 ps at 400 nm, and (c) T_{ox}=580 °C, τ=80 ps at 700 nm. With increasing T_{ox}, τ at 400 nm increases.

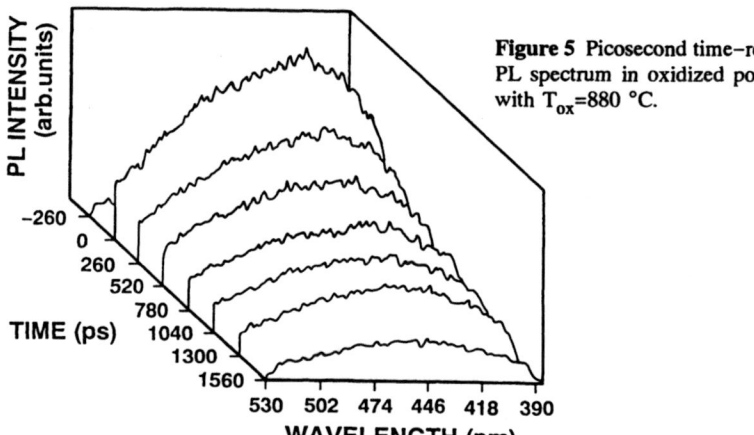

Figure 5 Picosecond time–resolved PL spectrum in oxidized porous Si with T_{ox}=880 °C.

EFFECTS OF OXYGEN ATOMS ON ELECTRONIC STRUCTURES OF Si NANOSTRUCTURES

Let us now discuss the effect of oxygen atoms on electronic properties of Si nanostructures in order to clarify the origin of blue PL. During RTO processes, the composition of the internal surface changes from silicon hydride Si:H to silicon oxyhydride Si:O:H and SiO_2 for T_{ox} < 800 °C or to SiO_2 for T_{ox} > 800°C. The surface of the Si nanocrystallites is mainly terminated by oxygen atoms. The oxidized surface layer of nanocrystallites plays two important roles:(i) the reduction of surface dangling bonds as a nonradiative recombination center of the c–Si core and (ii) the modification of the electronic structure of the near–surface region due to oxygen atoms. The PL peak energy also depends on T_{ox} and the PL wavelength suddenly changes from about 700 nm to 400 nm with rising T_{ox} above 800 °C. Here, we stress the importance of the oxidized surface layer in visible PL processes.

It is known that quasi–two–dimensional (2D) Si sheets [12] and oxygen–terminated Si sheets (siloxene) [13] exhibit visible PL and the oxygen atoms strongly affects the electronic structures of 2D Si sheets [14,15]. Therefore, it is considered that the electronic structure of the oxidized surface layer is different from that of the c–Si core. We employ the results of the *ab initio* electronic structure calculations to discuss the band of oxidized surface monolayer. The details of the calculation results are reported in Ref.[14–16]. Here we consider three types of Si:O compounds as the model surface structure and these compound structures are illustrated in Fig.6.

In Fig.6 (a), the oxygen atoms are located out–of–plane in the Si monolayer. We consider that at low T_{ox}, H atoms at the surface are only replaced by OH groups or O atoms and O atoms cannot migrate into Si network. The most characteristic feature of this layer is a direct–gap structure of 1.7 eV at Γ point [15]. Oxygen atoms do not cut the electron and hole delocalization in the Si layer and the orbital mixing of the Si σ electrons with the O non–bonding n electrons (σ–n mixing) effectively reduces the band gap at Γ point in the two–dimensional Si structure. Therefore, we propose the O–terminated Si surface layer having the direct–gap as a possible origin of strong red PL in oxidized Si spheres [16].

At higher temperature, we believe that oxygen atoms migrate into the Si–skeleton layer and oxygen atoms form the bridges or interconnect Si atoms. Two surface structures are illustrated in Fig.6 (b) and (c). The band structure of sheetlike Si:O compound [Fig.6 (b)] is very complicated. The optical allowed band gaps are 1.5, 2.6, and 4.5 eV [14]. In the other Si:O compound [Fig.6(c)], linear Si chains are interconnected by oxygen atoms. This structure has a direct gap of 3.2 eV at X point. Therefore, these Si:O compounds as the surface of nanocrystallites are possible origin of the blue PL.

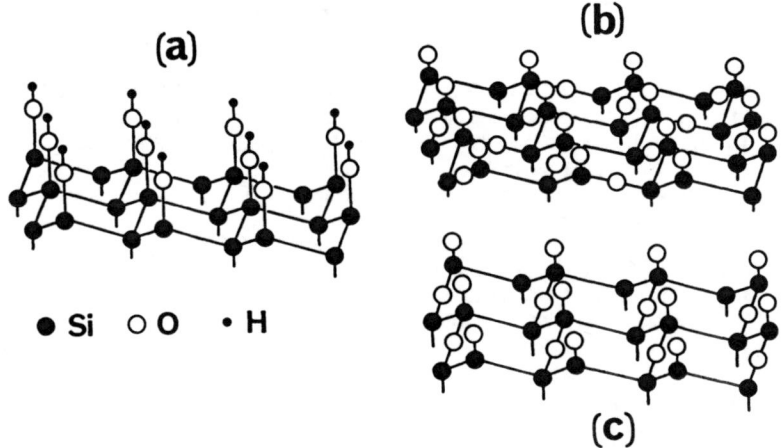

Figure 6 Model surface structures of oxidized porous Si.

Finally, we discuss defects in SiO_2 glass. During high temperature RTO processes, the surface is covered by the SiO_2 layer and this surface layer contains defects probably. It is known that the point defects in SiO_2 such as nonbridging–oxygen hole centers (≡Si–O•) and neutral oxygen vacancies (≡Si–Si≡) play a key role in visible PL process (1.9 and 2.7 eV) in SiO_2 glass [16]. Defect states in SiO_2 as well as oxidized surface monolayer can act as the luminescent centers in oxidized porous Si. The electronic properties of oxidized Si nanostructures are sensitive to the position of O atoms in Si:O compounds.

CONCLUSION

In conclusion, we have studied the origin of the strong PL of oxidized porous Si. We stress that the oxidized surface layer plays the most essential role in strong visible PL. Our study shows that visible PL from Si nanostructures is sensitive to the surface chemical structure.

ACKNOWLEDGMENT

The authors would like to thank Drs. T. Ogawa, K. Takeda, and K. Shiraishi for discussions.

REFERENCES

[1]   L.T.Canham, Appl. Phys. Lett. **57**, 1046 (1990).
[2]   A.G.Cullis and L.T.Canham, Nature **353**, 335 (1991).
[3]   Y.Kanemitsu, H.Uto, Y.Masumoto, and Y.Maeda, Appl. Phys. Lett. **61**, 2187 (1992).
[4]   Y.Maeda, N.Tsukamoto, Y.Yazawa, Y.Kanemitsu, and Y.Masumoto, Appl. Phys. Lett. **59**, 3168 (1991)
[5]   T.Matsumoto, T.Futagi, H.Mimura, and Y.Kanemitsu, Phys. Rev. B **47** (1993) in press.
[6]   M.A.Tischler, R.T.Collins, J.H.Stathis, and J.C.Tsang, Appl. Phys. Lett. **60**, 639 (1992).
[7]   C.Tai, K.H.Li, D.S.Kinosky, R.Z.Qian, T.C.Hsu, J.T.Irby, S.K.Banerjee, A.F.Tasch, J.C.Campbell, B.K.Hance, and J.M.White, Appl. Phys. Lett. **60**, 1700 (1992).
[8]   J.C.Vial, A.Bsiesy, F.Gaspard, R.Herino, M.Ligeon, F.Muller, R.Romestain, and R.M.Macfarlane, Phys. Rev. B **45**, 14171 (1992).
[9]   V.Petrova–Koch. T.Muschik, A.Kux, B.K.Meyer, F.Koch, and V.Lehmann, Appl. Phys. Lett. **61**, 943 (1992).
[10] H. Linke, P. Omling, B.K.Meyer, V.Petrova–Koch. T.Muschik, and V.Lehmann, Mater. Res. Soc. Symp. Proc. **283**, to be published.
[11] A. Yamamoto and Y. Kanemitsu, unpublished.
[12] Y.Kanemitsu. K.Suzuki, Y.Masumoto, T.Komatsu, K.Sato, S.Kyushin, and H.Matsumoto, Solid State Commun. (1993) in press.
[13] M.S.Brandt, H.D.Fuchs, M.Stutzmann, J.Weber, and M.Cardona, Solid State Commun. **81**, 307 (1992).
[14] K.Takeda and K.Shiraishi, Solid State Commun. **85**, 301 (1993).
[15] K.Takeda and K.Shiraishi, Phys. Rev. B **39**, 11028 (1989).
[16] Y. Kanemitsu, K. Takeda, K. Shiraishi, and T. Ogawa, (unpublished).
[17] D. L. Griscom. J. Ceramic Soc. Jpn. **99**, 923 (1991).

PHOTOLUMINESCENCE OF CHEMICALLY ETCHED
POLYCRYSTALLINE AND AMORPHOUS Si THIN FILMS

A. J. Steckl, J. Xu and H . C. Mogul
Nanoelectronics Laboratory, Department of Electrical and Computer Engineering, University of Cincinnati, Cincinnati, OH 45221-0030

ABSTRACT

Si thin films were deposited on quartz at temperatures (T_D) ranging from 540 to 640°C. X-ray diffraction indicates that films deposited at $T_D < 580$°C are amorphous, while those deposited above 600°C are poly-crystalline with a <220> texture. The Si films were made porous by stain-etching in $HF:HNO_3:H_2O$. Only Si films deposited at 590°C and above show photoluminescence (PL), centered at ~650-670 nm under UV excitation. Films deposited at $T_D < 580$°C do not luminesce even after very long etch times, which produce a highly porous structure. The PL intensity and the x-ray signal follow a very similar trend with T_D. It appears that a minimum level of crystallinity is required for photoemission in porous Si and that a strong relationship exists between them.

INTRODUCTION

Three major mechanisms have been proposed to explain the visible room temperature photoluminescence (PL) from porous silicon (PoSi) prepared by anodization and stain etching of crystalline silicon (c-Si): (a) quantum confinement of charge carriers [1-6] in a low-dimensional crystalline Si skeleton; (b) localized emission due to Si polysilanes or hydrides [7-9] formed by the bonding of Si surface atoms to hydrogen; (c) formation of a specific class of Si-O-H compounds (siloxene) [10-11], whose optical properties are due to chemical confinement of Si in linear chains and six-fold rings by oxygen atoms. To date, the explanation for photoluminescence in PoSi still remains an open topic due to the lack of unambiguous experiments.

Recently, strong PL has been reported from anodized [12] and from chemically stained [13, 14] poly-crystalline Si (poly-Si) films with a spectrum similar to that of PoSi prepared from c-Si, which indicates that long range order is not necessary for photoluminescence. This is an exciting prospect for many optoelectronic applications of PoSi, wherein one could incorporate photoemissive poly-Si thin films into selected device structures using a process requiring only a relatively low temperature. Two important examples are poly-Si photoemitters on VLSI circuits for inter- and intra-chip optical communications and on glass for flat panel display devices. Interestingly, amorphous Si films which were obtained by molecular beam epitaxy (MBE) at 25°C annealed at temperature below 700°C [13] and by low pressure chemical vapor deposition (LPCVD) at 560°C [14] did not exhibit a PL response upon stain-etching. In this paper, we report on a more systematic study of the effect of deposition temperature (T_D) on the relationship between the crystallinity and the photoluminescence of Si thin films. This relationship can hopefully provide the key to understanding the PL mechanism in PoSi, as well as establishing some important guidelines for the fabrication of related devices.

EXPERIMENTAL PROCEDURE

Si thin films were deposited on quartz substrates by LPCVD with silane at temperatures ranging from 540°C to 640°C. The deposition process was carried out at 0.5 Torr using pure silane at a flow rate of 25 sccm. The thickness of the Si films was around 1 μm. Companion samples prepared on oxidized Si substrates yielded results identical to those described below. The crystallinity of the deposited thin films was studied by X-ray diffraction spectroscopy (XRD).

Following deposition, the films were rendered porous by stain-etching, which is a significantly simpler process than anodization. The stain-etching [15,16] was performed by immersion of samples into a HF:HNO$_3$:H$_2$O solution with 1:3:5 composition at 27°C for 2 min in all cases, and for additional time periods in some cases. The surface morphology of as-deposited and etched films was studied by scanning electron microscopy (SEM). PL spectra were obtained at room temperature with ultraviolet (UV) excitation at 370-380 nm from a filtered Hg source.

RESULTS AND DISCUSSION

X-ray diffraction (XRD) was performed on the as-deposited Si films using a diffractometer with a Cu K$_\alpha$ line. Films deposited at T$_D$ < 580°C exhibited no x-ray peaks, indicating an amorphous structure. For deposition at T$_D$ > 580°C, the films displayed XRD peaks corresponding to the <111>, <220> and <311> directions, as shown in Fig. 1. For 580 < T$_D$ < 600°C, the Si films undergo a transition from amorphous to poly-crystalline structure. Finally, for T$_D$ > 600°C, the films exhibit a pronounced <220> texture. The grain size, calculated from the <220> line width using the Scherrer formula [17], ranges between 680 Å for T$_D$ = 600°C to 600 Å for T$_D$ = 640°C. These results are similar to those previously reported [18,19] for the structural properties of LPCVD Si films.

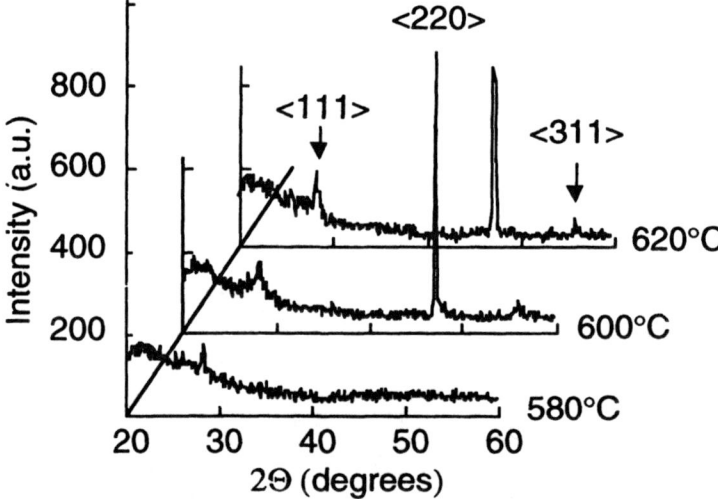

Fig. 1   X-ray diffraction spectra for Si films deposited at 580 , 600 and 620°C.

As shown in the SEM micrographs of Fig. 2, films with T$_D$ ≤ 580°C have a very smooth surface, while a much rougher surface was observed for T$_D$ > 580°C. The T$_D$ dependence of the surface morphology for as-deposited films is very similar to that reported [20] by H. Watanabe et al. for Si films deposited under similar conditions. After a 2 min etch, the films became porous to varying degrees . For 540 ≤ T$_D$ ≤ 580°C, the porosity was very low and did not vary greatly, as seen from Figs. 2(b) and (d). Above 580°C, the porosity was observed to vary significantly with T$_D$. The highest level of porosity was observed (see Fig. 2(f)) at T$_D$=600°C. For T$_D$ > 600°C, the porosity decreases somewhat but does not reach the very low level observed in films with T$_D$ < 580°C. Furthermore, the morphology of the PoSi films

as-deposited ◄─┼─► 2 min etched

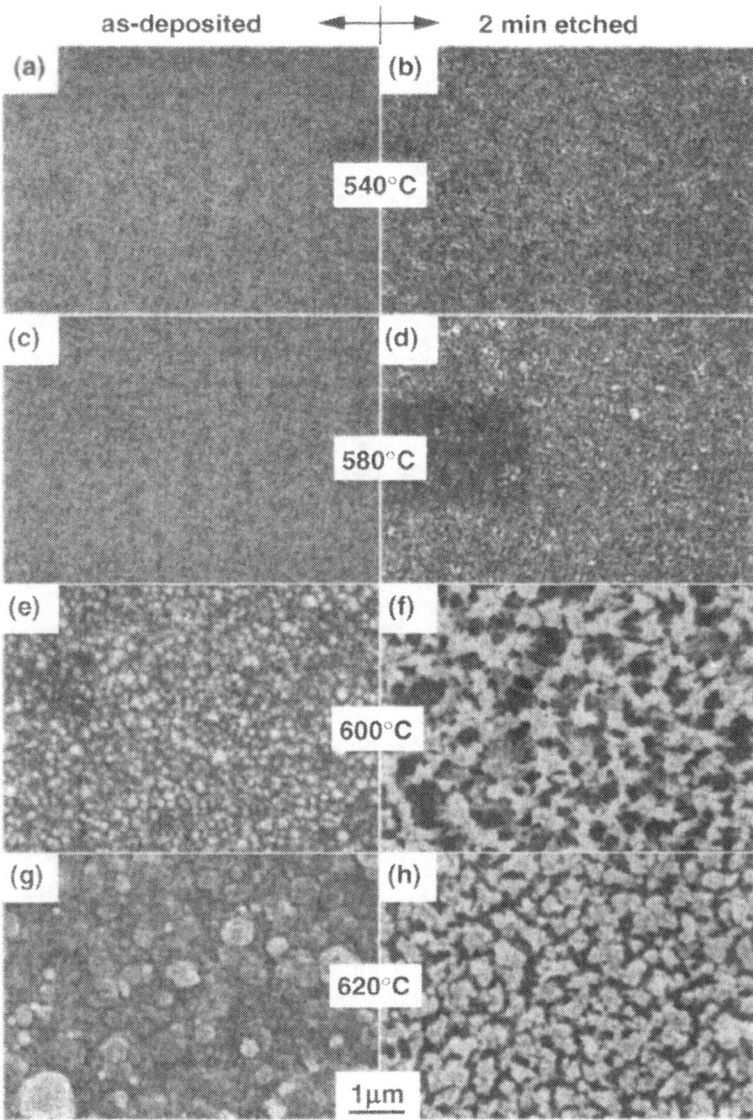

Fig. 2 Surface morphology (SEM) of as-deposited and stain-etched (min) Si films.

appears to change with T_D, as can be observed by comparing Figs. 2(f) and (g). The differences in porosity and morphology of the etched films are related to the differences in etch rate and initial surface morphology of the films, which in turn is related to the crystallinity obtained at various deposition temperatures. It is important to note that, for significantly longer etch times the porosity of films deposited at low temperatures ($T_D \leq 580°C$) increased to the same level as that of that of high T_D films etched for 2 min, as shown in Fig. 3.

As represented in Fig. 4, the films deposited at 580°C (and below) and stain-etched for 2

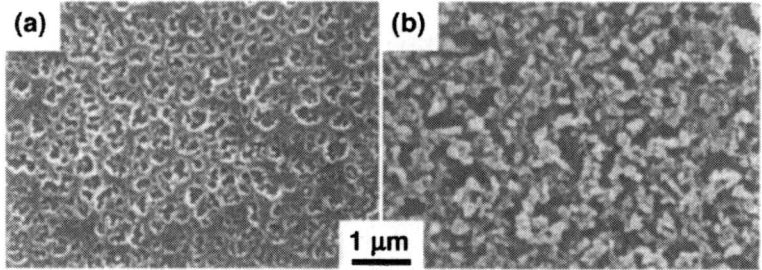

Fig. 3   Surface morphology (SEM) of non-luminescent films
etched for longer periods : (a) 540°C, 31 min; (b) 580°C, 8 min.

min do not exhibit a PL signal. Starting with a deposition temperature of 590°C, the PoSi layers exhibit a broad PL signal with a peak at 660-670 nm. A large PL peak, with a full width at half-maximum (FWHM) of 140 nm, is obtained for the film deposited at 600°C. The background PL level was subtracted for the data shown in Fig. 4.

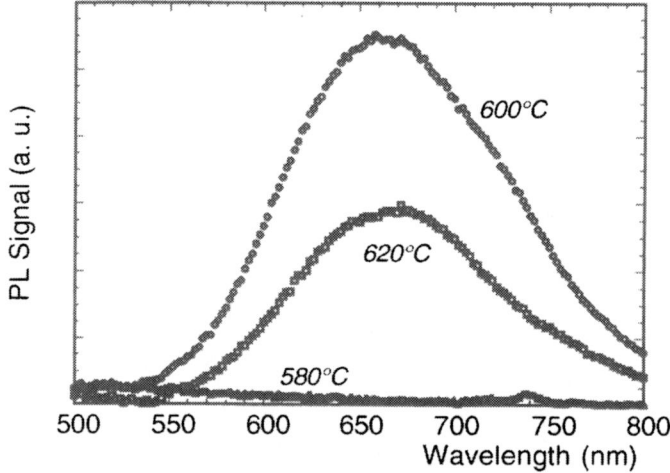

Fig. 4   Room temperature photoluminescence spectra (with background subtracted)
for Si films deposited at 580, 600 and 620°C and stain-etched for 2 min.

The wavelength of peak emission (λ_p) did not vary significantly with deposition temperature, as shown in Fig. 5. The dependence of the PL signal intensity at λ_p (without background subtraction) and of the <220> x-ray signal (which is the strongest XRD component) from the Si films are also given in Fig. 5 as a function of T_D. The stain-etch time for these samples was 2 min. It is important to note that the onset of both PL and XRD signals is observed at a deposition temperature between 580-590°C. Furthermore, the general trend for the two properties is quite similar. A maximum in both PL and XRD signals is obtained for the

films deposited at 600-610°C, with both signals decreasing for films deposited at higher temperatures.
The amorphous films deposited at 540°C and etched for up to 30 min did not luminesce,

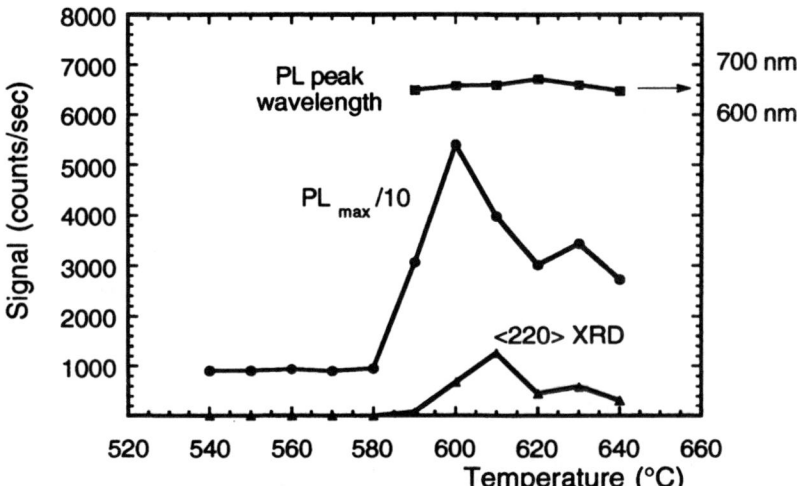

Fig. 5   Effect of Si film deposition temperature on : peak PL wavelength - PL(λ_p);
PL(λ_p) signal (divided by 10) obtained for 2 min etching ; <220> X-ray signal.

even though a significant increase in porosity was produced. This can be seen by comparing Figs. 2b and 3a. For comparison, stain-etched n-type crystalline Si substrates, with morphology and low porosity very similar to the 2 min etched amorphous Si film deposited at 540°C, produced [21] a room temperature PL signal. The films deposited at 580°C and etched for 8 min displayed a barely observable photoemission under UV excitation. These 580°C films had a highly porous structure, comparable to films deposited at 620°C and etched for 2 min (see Figs. 2h and 3b). However, it is important to point out that while the 580°C films had no <220> XRD signal they did exhibit a small <111> signal, and therefore were not completely amorphous.

CONCLUSIONS

In summary, for stain-etched porous Si, we have observed that a minimum level of crystallinity is required for light emission to occur. It appears that porosity by itself is not a sufficient condition for light emission to occur. We conclude that a strong relationship exists between the crystallinity of the starting Si material and the occurrence of photoluminescence upon rendering the material porous.

This work was supported in part by the SDIO/IST and monitored by ARO, under Grant No. DAAL03-92-0290. The authors are pleased to acknowledge the encouragement of L. Lome, R. Trew and J. Zavada, the assistance of M. Phillips and W. Huff with the x-ray diffraction measurements, of W. Bresser and P. Boolchand with the PL measurements, and of G. DeBrabander and J. T. Boyd with the deposition of the Si films.

REFERENCES

1. L. T. Canham, Appl. Phys. Lett. **57**, 1046 (1990).
2. V. Lehmann and U. Gösele, Appl. Phys. Lett. **58**, 856 (1991).
3. A. Halimaoui, C. Oules, G. Bomchil, A. Bsiesy, F. Gaspard, R. Herino, M. Ligeon, and F. Muller, Appl. Phys. Lett. **59**, 304 (1991).
4. N. Koshida and H. Koyama, Jap. J. Appl. Phys. **30**, Pt. 2, L1221 (1991).
5. A. G. Cullis and L. T. Canham, Nature, **353**, 335 (1991).
6. A. Nishida, K. Nakagawa, H. Kakibayashi and T. Shimada, Jap. J. Appl. Phys. **31**, Pt. 2, L1219 (1992).
7. S. M. Prokes, W. E. Carlos and V. M. Bermudez, Appl. Phys. Lett. **61**, 1447 (1992).
8. S. M. Prokes, O. J. Glembocki, V. M. Bermudez and R. Kaplan, Phys. Rev B, **45**, 13788 (1992).
9. P. C. Searson, J. M. Macaulay, and S. M. Prokes, J. Electrochem. Soc. **139**, 3373 (1992).
10. M. S. Brandt, H. D. Fuchs, M. Stutzmann, J. Weber, and M. Cardona, Solid State Commun. **81**, 307 (1992).
11. P. Daék, M. Rosenbauer, M. Stutzmann, J. Weber and M.S.Brandt, Phys. Rev. Lett. **69**, 2531 (1992).
12. E. Bustarret, M. Ligeon, J. C. Bruyere, F. Muller, R. Herino, F. Gaspard, L. Ortega, and M. Stutzmann, Appl. Phys. Lett. **61**, 1552 (1992).
13. K. H. Jung, S. Shih, D. L. Kwong, C. H. Cho, and B. E. Gnade, Appl. Phys. Lett. **61**, 2467 (1992).
14. A. J. Steckl, J. Xu, H. C. Mogul, Appl. Phys. Lett. **62**, 2111 (1993)
15. R. W. Fathauer, T. George, A. Ksendzov, and R. P. Vasquez, Appl. Phys. Lett. **60**, 995 (1992).
16. J. Sarathy, S. Shih, K. H. Jung, C. Tsai, K.-H. Li, D. L. Kwong, J. C. Campbell, S-L. Yau, and A. J. Bard, Appl. Phys. Lett. **60**, 1532 (1992).
17. B. D. Cullity, Elements of X-ray Diffraction, (Addison-Wesley, Reading, MA, 1967).
18. G. Harbeke, L. Kruasbauer, E. F. Steigmeier, A. E. Widmer, H. F. Kappert, and G. Neugebauer, J. Electrochem. Soc. **131**, 675 (1984).
19. T. I. Kamins, M. M. Mandurah, and K. C. Saraswat, J. Electrochem. Soc. **125**, 927 (1978).
20. H. Watanabe, A. Sakai, T. Tatsumi, and T. Niino, Solid State Technology, p.29, (July 1992).
21. A. J. Steckl, J. Xu, H. C. Mogul, and S. Mogren, Appl. Phys. Lett. **62**, 1982 (1993)

NONLINEAR OPTICAL PROPERTIES AND FATIGUE
EFFECT IN POROUS SILICON

Vytautas Grivickas, Jonas Kolenda*, Lino Misoguti,
Pierre Basmaji and Sérgio C. Zilio*

Instituto de Física e Química de São Carlos - USP
Caixa Postal 369, 13560-970 - São Carlos, SP, Brazil

* Vilnius University, Vilnius, Lithuania

ABSTRACT

We report results of photoluminescence (PL) fatigue and nonlinear optical absorption in anodized porous silicon (PS) samples aged in air for a few months. The fatigue strength is found different for PS emitting below 1.65 eV, in 1.65 - 2.05 eV range and above 2.05 eV, revealing the different nature of PL. A comparison with two special glasses is made, and a possible explanation for PL mechanism in PS is discussed.

INTRODUCTION

In recent times, PS layers have attracted considerable interest due to the intense visible luminescence observed in this material. Therefore, chemical composition and PL properties can be varied over a surprisinly wide range and, thus, several different PL mechanisms can be involved [1-3]. In spite of the emission efficiency, rapid PL degradation under short wavelength light illumination has been also observed in this material. In freshly anodized PS layers, PL degradation was shown as a sharp function of illuminating light energy [1-10].

The degradation effect was assigned to optically induced breaking of H-Si bonds influencing passivation of quantum size Si clusters [7-8], restructuring of siloxane layers [1] or as a photochemical reaction on the surface [2]. Nonpermanent degradation (fatigue) effects have been also observed in a-Si:H films [11], Si-SiO$_2$ interfaces [12] and in CdSSe microcrystals embedded in a glass matrix [13]. Further understanding of these effects is important for both device development and for the physical interpretation of the PL mechanism in PS.

EXPERIMENT

PS layers were prepared on p-type crystalline Si wafers of 1-10 Ωcm resistivity by standard anodic etching procedures in HF:H$_2$O solutions. After preparation, samples were passivated in ethanol, dried and aged in ambient air for 1-4 months in order to stabilize the composition of the surface structure [14,15].

PL measurements were performed using a 0.5 m grating spectrometer and a photo multiplier tube. The unfocused Ar laser line $\lambda = 457.9$ nm was used as the excitation source for both PL excitation measurement and for sample irradiation. The absorption coefficient for this wavelength varies between 80-1000 cm^{-1} in aged PS of porosity 85-65% [4]. The PL signal was detected by conventional lock-in technique. Irradiation-induced PL transients have been measured at moderate power range ~ 300 mW/cm^2. PL decay time was measured with a digital oscilloscope. Time response of the system is less than 10 μs.

Transmittance photoinduced changes of PS layers were investigated using a perpendicular pump and probe geometry. The details of this technique are described elsewhere [16]. Samples were excited by second harmonic of Q-switched Nd:YAG laser ($\lambda_{exc} = 532$ nm, pulse duration $\tau_{exc} = 10$ ns, pulse energy 3 mJ) and probed by cw beam of He-Ne laser at wavelength 633, 611 and 594 nm. The probe beam passed the 0.25 nm monochromator and was detected by the photomultiplier. The detected signal was averaged over 2000 pump pulse shots in the digital oscilloscope (LeCroy 9400 A). The time resolution of our measurements was better than 0.1 μs.

The absorptive nonlinearity in ps-pulse regime was investigated using the z-scan technique based on the far-field spatial beam distortion when samples are moved along a tightly focused gaussian beam [17].

Red luminescence at 720 nm is observed in samples with 2-5 μm thickness and yellow-red in the sample with 20 μm thickness. Green emitting layers were obtained by adding Rb or Cs fluorides in the etching solution [14]. These three types of PS will be referred to as, respectively, A-, B- and C-type.

For comparison, we have also investigated two different glasses. Fluoroindate glass (FGL) [18] after treatment in boiling water for 72 h, has the green PL band, similar to that of C-type PS. In addition, IR-transmission spectrum is similar to the PS samples with high porosity [14]. The other, ordinary silica glass (S-GL) was immersed into (CH_3) 3SiCl agent for deposition of a thin < 10 nm siloxane layer on the surface. After silanization, a broad PL band has appeared with a peak at 720 nm, similar to that obtained for siloxane prepared by Kautsky procedure and annealed at 100° C [1].

RESULTS

Fig. 1 shows the temporal luminescence peak and spectral evolutions in PS samples that have been irradiated for the first time. The light is exposed at the elapsed-time. We observed a big change, under several hundred minutes irradiation. Note that the fatigue is sharper for samples luminescing at short wavelength. The fatigue transients may be roughly described by the expression:

$$I_{PL} \sim A exp(-t/t_1) + B(1 + t/t_2)^\beta \qquad (1)$$

where t_1 and t_2, β are fatigue parameters for the fast and the slow decay processes, respectively. The slow fatigue process is also observed in both irradiated glasses. Our investigations indicate that by stopping irradiation for several minutes, the decreased PL intensity does recover. The rapid recovery is observed in B-type PS (Fig. 1) and is slow in A- and C-type samples (not shown). The initial PL intensity value was recovered after keeping the sample for about 12-48h in the dark at room temperature.

The cycling irradiation experiments were performed on several selected samples. The cycle consisted of 90 min exposure at 300 mW/cm^2 followed by keeping the sample in the dark for 24 h. They show that PL enhancement phenomena are no longer observable after one- or two cycles of irradiation. In the samples that do show PL fatigue-effect during first exposure this photo-hardening results in a slight reduction of the β value (the slope of PL drop according to eq. (1), over the next illumination cycles. This result is true regardless of the maximum number 3 to 6 cycles tested. This means that PS samples may be photo-hardened, i.e. the irradiation light following while keeping in air produces more stable luminescing species in the samples. Therefore, we did not find nonreversible PL fatigue effects in glasses.

Fig. 1 - Time evolution of normalized luminescence peak intensity in three PS samples. Ar laser light (λ = 458.9 nm, P = 300 mW/cm^2) is exposed at t = 0 and canceled at the end of the solid curve. Two irradiation pauses (6 min and 18 min) were intentionally produced for the sample with PL peak at 660 nm. Discrete measurements (shown by symbols) have been made with 50-times reduced excitation intensity then the sample is kept in the dark.

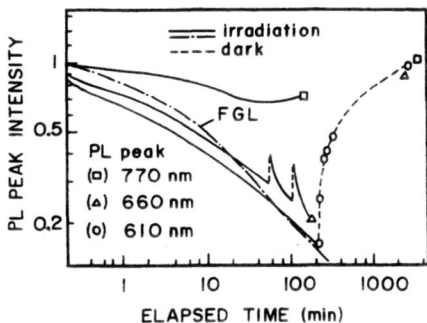

Fig. 2 - Summary of the PL spectra of photo-hardened PS samples (a) and the corresponding fatigue strength under identical experimental conditions (c). The same results of fluoroindate glass (FGL) and the ordinary glass treated with siloxenized-surface-agent (S-GL) are shown in (b,c) for purpose of comparison. Sharp peaks on FGL spectrum in (b) are related to Gd-ion contaminant [18]. All data are corrected to the measurements system response function.

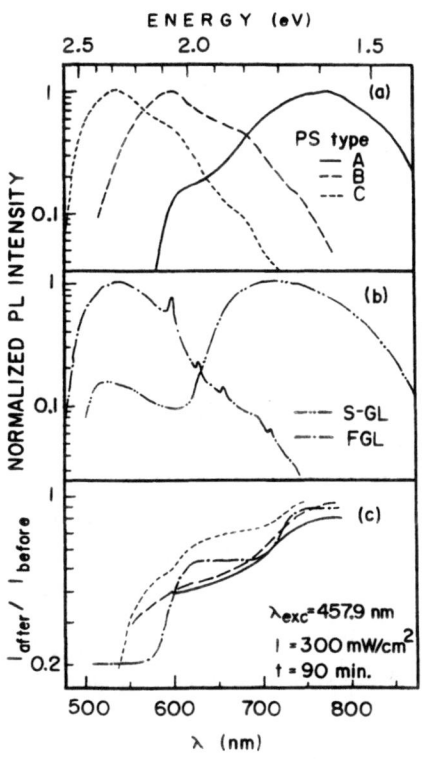

Fig. 2(a) shows the summary of the PL spectra obtained on photo-hardened PS samples. In all cases we observed only the reversible PL fatigue effect. The spectral change caused by illumination (I_{after}/I_{before}) in Fig. 2(c), assigned as the fatigue strength, has shown remarkable difference in three different spectral ranges, i.e. above 720 nm (< 1.65 eV) in the 600-720 nm range (1.65 - 2.05 eV) and below 600 nm (> 2.05 eV).

Fig. 3 shows PL decay time versus emission wavelength. The time constant was obtained at the PL decay tail where an exponential approach may be used. PL decay time increases with λ_{PL} and saturates at 100 μs and 350 μs in B- and A-type PS, respectively. In S-GL decay time has two spectral components of 0.7 ms and 3 ms that are between 1.6 - 1.7 eV (Fig. 2(b)). The PL decay time in FGL is about 2 ms (not shown). This suggests the different recombination mechanisms in glasses and in PS above 1.65 eV. As shown in Fig. 3, the PL decay time decreases up to 20% in PS after the irradiation cycle. This decrease does not accompany the total loss of luminescence by more than a factor of 4. The IR-stretching vibrations of SiHR, groups below 2250 cm^{-1} in A-type PS also do not show any reduction (shown in the inset of Fig. 3). This is a striking difference compared to the results of freshly

Fig. 3 - Luminescence decay time versus emission wavelength in A- and B-type photo-hardened PS samples before and after irradiation cycle. The result of S-GL) sample is also presented. The inset shows $SiH_x R_i$ vibration spectra of B-type P-Si sample.

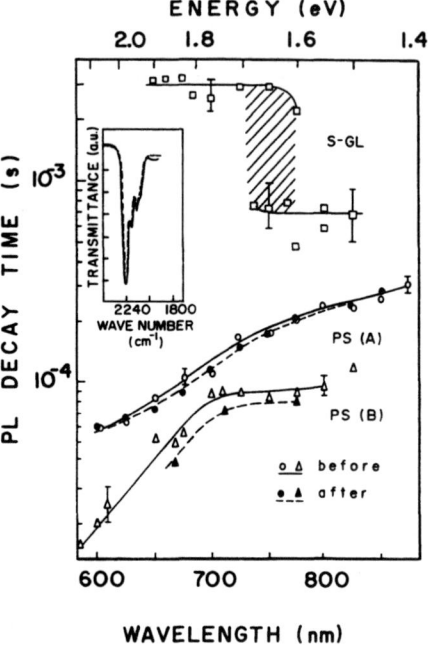

anodized PS where correspondingly, the strong lifetime reduction and the change of Si-H to $SiHR_i$ groups under irradiation have been reported [1,3,7,8,10].

For optical nonlinearity investigations the deliminated PS samples with PL peak situated at 700 nm (porosity in range 70%) and thickness 40 μm were used. Excitation pulse at $\lambda_{exc} = 532$ nm and intensity $I_{exc} = 10$ MW/cm^2 caused strong PL visible by eye. The PL decay is shown in Fig. 4. PL time behaviour shows the influence of fast recombination channels by high photoexcited carrier densities. This corresponds to previous PS photoconductivity investigations [19] showing the shortening photoresponse time by increase of excitation level. The multiexponential PL decay suggests the presence of distributed states as photoluminescence centers [20]. The PL lifetime measured by low excitation level at the same wavelength ($\lambda_p = 594$ nm) is about 40 μs. In the geometry used for the investigations the registered signal is the sum of PL and photoinduced sample absorption change. Figure 5 shows the detected signal U dependence on probe beam intensity. Presence of probe beam lead to decrease of this signal magnitude that one can explane by photoinduced absorption. We suppose that this induced absorption observed in region 633-594 nm probe beam wavelength is caused by excited carrier absorption. The time dependence on the ΔT repeated the shape of PL in time t < 1 μs and then correlated with low excitation level PL lifetime (Fig. 4).

In Fig. 6 the normalized transmittance of open aperture z-scan is shown. We observed the photoinduced bleaching by exciting PS samples above absorption edge and for excitation below the absorption edge PS samples exhibit photoinduced absorption. Absorption edge was determined from Tauc plot $(\alpha h\omega)^{1/2} = f(h\omega)$ for PS samples with porosity 8-% where Eg = 2.1 eV. According to the observed difference of themPL properties in 500-650 nm and 650-850 nm regions [4] we consider that the observed bleaching is caused by saturating of optical transition to the conductivity band of the polymer formed on silicon skeleton.

CONCLUSION

We believe our observations can be understood if we assume that PL is associated with different polymeric layers grown in air on vast surface provided by Si skeleton. The polymeric layers in A-type PS may be attributed to siloxane derivatives that are characterized by stable $SiHR_i$ vibration groups [1] and similar PL decay times

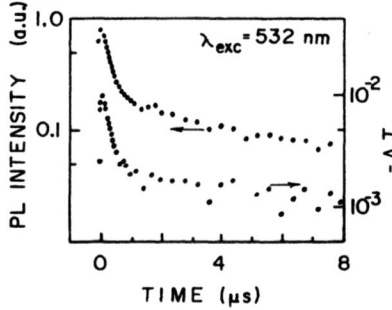

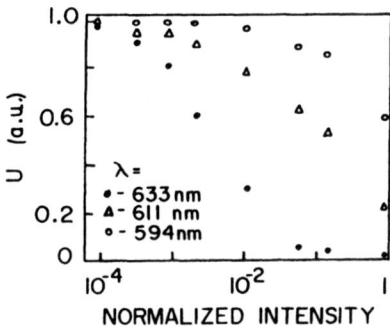

Fig. 4 - PL decay of deliminated PS samples (porosity 70%) and photoinduced transmission change ΔT temporal behavior at wavelength $\lambda_p = 590$ nm. Excitation intensity 10 MW/cm².

Fig. 5 - Detected signal U dependence on the probe beam intensity for three different probe beam wavelengths ($\lambda_{exc} = 532$ nm, $\tau_{exc} = 10$ ns, $I_{exc} = 10$ MW/cm²).

(Fig. 3). The other two compounds are not clear, but some amorphous fluoroindate glass structures or a-Si:H/polysilian derivatives may be compatible. The photoinduced optical density change kinetic of PS samples correlate with PL time behavior and is determined by the silicon skeleton carrier density. However, the mechanism of light-induced restructuring and the origin of charge transfer between surface compounds and Si skeleton remain an open question.

This work is supported by FAPESP, Brazil. J.K. would acknowledge CNPq (RHAE) for research associated ship at IFQSC/USP.

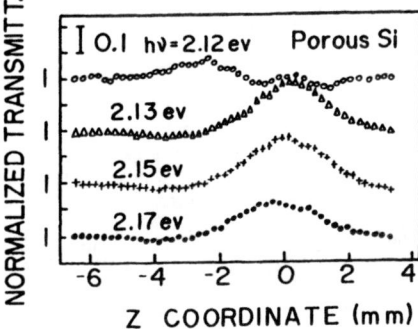

Fig. 6 - The normalized z-scan transmittance of PS samples (porosity 80%). Excitation intensity at point z = 0 was 3.10^5 W/cm².

REFERENCES

1. H.D. Fuchs, M. Stutzmann, M.S. Brandt, M. Rosenbauer, J. Weber and M. Cardona, Phys. Scripta T45, 309, 1992.
2. Z.Y. Xu, M. Gal and M. Gross, Appl. Phys. Lett. 60, 1375, 1992.
3. R.T. Collins, M.A. Tischler and J.H. Stathis, Appl. Phys. Lett. 61, 1649, 1992.
4. P. Basmaji, V. Grivickas, G. Surdutovich, R. Vitlina and V.S. Bagnato, MRS Autumn Meeting Proc., Boston, 1992.
5. A. Bsiesy, J.C. Vial, F. Gaspard, R. Herino, M. Ligeon, F. Muller, R. Romestain, A. Wasiela, A. Halimaoui and G. Bomchil, Surf. Sci. 254, 195, 1991.
6. X.L. Zheng, H.C. Chen and W. Wang, J. Appl. Phys. 72, 3841, 1992.
7. C. Tsai, K.H. Li, J.C. Campbell, B.K. Hance and J.M. White, J. Electr. Mat. 21, 589, 1992.
8. I. Suemune, N. Noguchi and M. Yamanishi, Jpn. J. Appl. Phys. 31, L494, 1992.
9. J. Sarathy, S. Shih, K. Jung, C. Tsai, K.-H. Li, D.-L. Kwong, J.C. Campbell, S.-L. Yau and a.J. Bard, Appl. Phys. Lett. 60, 1532, 1992.
10. M.A. Tishler, R.T. Collins, J.H. Stathis and J.C. Tsang, Appl. Phys. Lett. 60, 639, 1992.
11. M. Yoshida, K. Morigaki and S. Nitta, Sol. St. Commun. 51, 1, 1984.
12. L. Zhong, A. Buczkowski, K. Katayama, F. Shimura, Appl. Phys. Lett. 61, 931, 1992.
13. V. Grivickas, M. Petrauskas, D. Noreika, E. Vanagas and M. Kull, Nucl. Instr. Meth. B65, 397, 1992.
14. V. Grivickas and P. Basmaji (unpublished results).
15. L.T. Canham, M.R. Houlton, W.Y. Leong, C. Pickering and J.M. Keen, J. Appl. Phys. 70, 422, 1991.
16. V.Grivickas, J. Linros, A. Vigelis, J. Seckus, J.A. Tellefsen, Sol. St. Electron. 35, 299, 1992.
17. M. Sheik-Bahae, A.A. Said, T.-H. Wei, D.J. Hagan, E.W. Van Stryland, IEEE J. Quant. Electron. 26, 760, 1990.
18. C.X. Cardoso, Y. Messaddeq, L.A.O. Nunes and M.A. Aegerter, Proc. 8th Int. Symp. on Halide Glasses, France, 1992.
19. V. Grivickas, J. Linros, P. Basmaji, L. Misoguti, V Simpósio Estadual de Lasers e Aplicações, São Paulo, 1992.
20. M.G. Bawendi, W.L. Wilson, L. Rothberg, P.J. Carrol, T.M. Jedju, M.L. Steigerwald, L.E. Brus, Phys. Rev. Lett. 65, 1623, 1990

Si₃N₄/SiO₂ MULTI-LAYER REFLECTING STACKS FOR PHOTONIC SWITCHING APPLICATIONS

D.J. STEPHENS*, S.S. HE*, G. LUCOVSKY*, H. MIKKELSEN**, K. LEO**
AND H. KURZ**
*Department of Physics, North Carolina State University, Raleigh NC 27695-8202, USA,
**Rheinisch-Westfälische Technische Hochschule, 5100-Aachen, Germany

ABSTRACT

We have fabricated stacked-structures comprised of i) fused silica substrates, and ii) near-periodic Si₃N₄/SiO₂ bi-layers by low-temperature, 250°C, remote plasma-enhanced chemical-vapor deposition. Comparing the reflectance of these structures with model calculations, we have been able to identify the effects on the reflectance spectra of departures from i) exact periodicity, ii) not having the constituent dielectric layers each posses an ideal optical path length, OPL, exactly equal to $\lambda_{central}/4$, and iii) the intrinsic dispersion in the dielectric functions of the oxide and nitride materials. We have prepared quasi-periodic structures in which the OPL of the higher index Si₃N₄ layer was > $\lambda_{central}/4$, and in which the OPL of the lower index SiO₂ layer was < $\lambda_{central}/4$. This promotes a second strong reflectance band at an energy that is approximately two times that of the primary band. Calculations have shown that the reflectance values in this band, and near a reflectance minimum on the high energy side of the band, are both very sensitive to changes in the optical properties of the nitride film. We present calculations that demonstrate the effects on the reflectance of this band by a temperature-induced modulation of the optical properties of the oxide and nitride layers.

INTRODUCTION

Multilayer interference reflectance filters known as quarter-wave stacks, QWS, have been fabricated by low-temperature, 250°C, remote plasma-enhanced chemical-vapor deposition [1]. A QWS is a periodic bi-layer structure which has alternating layers of high, H, and low, L, index of refraction materials. Each layer of an ideal QWS structure has an optical path length, OPL, equal to one quarter of the wavelength, λ, of the dominant reflectance band, i.e., the OPL = $\lambda/4$. The OPL is defined as the index of refraction, n, multiplied by the film thickness, d. The OPL changes as a function of wavelength, because of dispersion in the index of refraction in each of the films. Therefore we define, as a reference point for discussion, $\lambda_{central}/4$ to be the wavelength at the center of the first reflectance band. The QWS structures we have studied were comprised of 18 layers of alternating Si₃N₄ and SiO₂ (9 periods of Si₃N₄:SiO₂) with an additional Si₃N₄ layer between the stack and the fused silica substrate, i.e., Si₃N₄:SiO₂/...Si₃N₄:SiO₂/Si₃N₄ or HL/HL/...HL/H, (see Fig. 1.) .

These QWS were investigated analytically with a computer model described in detail in Ref. 2. Using this model we have been able to describe effects in the reflectance spectrum due to i) intrinsic properties of the films and the substrate and ii) extrinsic properties of the structure [3]. Intrinsic properties are the dispersion in the indices of refraction and absorption of the films and the substrate. Extrinsic properties correspond to the layer thicknesses which are i) periodic, but non-ideal layer thicknesses for either (or both) film(s), i.e. OPL ≠ $\lambda/4$, or ii) aperiodic, in the thickness of

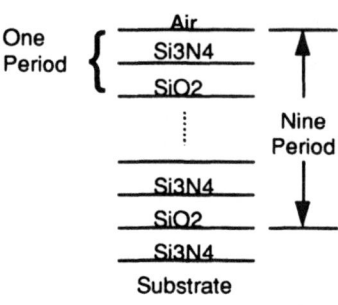

Fig. 1. Schematic of the Quarter Wave Stack multilayer structure.

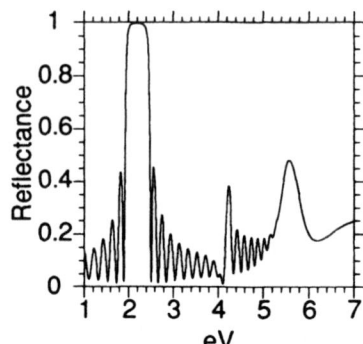

Fig. 2. Reflectance spectrum for an ideal Quarter Wave Stack (2.2 eV peak) with dispersion in indices of refraction

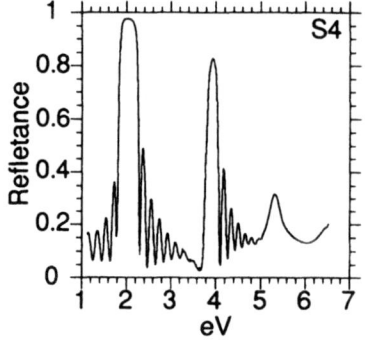

Fig. 3. Reflectance spectrum of an experimental Quarter Wave Stack [3].

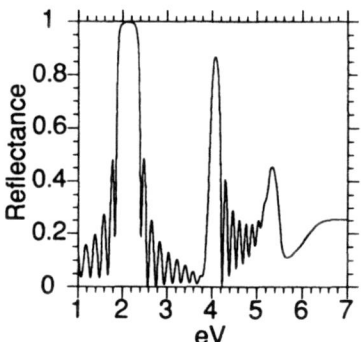

Fig. 4. Calculated reflectance spectrum for non-ideal thickness ratio, and positive vertical dispersion in the nitride [3].

the one (or both) of the layers.

Figure 2 shows the reflectance spectrum for a 19 layer QWS with ideal layer thicknesses on a fused silica substrate that takes into account the intrinsic properties of the films and the substrate. The layer thicknesses were chosen to produce a maximum reflectance at 2.2 eV, E_0, (also defined as $\lambda_{central}/4$) ($n_n = 2.05$, $d_{ni} = 68.6$ nm, $n_o = 1.46$, $d_{oi} = 96.5$ nm: the "n" subscript refers to the nitride layers, the subscript "o", to the oxide layers, and "i" refers to ideal thickness). In the energy range above ~ 4 eV, the Si_3N_4 film becomes absorptive, the higher the energy, the more absorptive the film. A smaller reflectance peak occurs at ~ 4 eV, or ~ $2E_0$, which arises from the fact that there is dispersion in the indices of refraction of Si_3N_4 and SiO_2. If the indices of refraction were constant then this peak would not exist. Even though the peak position is just under $2E_0$ we will still refer to it as the $2E_0$ position for purposes of clarity.

EXPERIMENTAL

Figure 3 shows an experimental QWS where the OPL of the Si_3N_4 film was $> \lambda_{central}/4$, while the OPL of the SiO_2 film was $< \lambda_{central}/4$. This was achieved by simply adjusting the depositions rates of the Si_3N_4 and SiO_2 films so that $d_n > d_{ni}$, and $d_0 < d_{oi}$. An additional extrinsic effect, aperiodicity of the Si_3N_4 layer of the order of 1 nm/layer, is also present due to the deposition process of the Si_3N_4 film. The aperiodicity occurs as a systematic increase in the Si_3N_4 layers with increasing stack thickness. A complete description of the deposition process is described in Ref. 1 and the deposition conditions for this structure in Ref. 3.

ANALYTICAL

With the aid of a computer model we have generated all the features of the experimental QWS's, including both intrinsic and extrinsic effects [3]. Figure 4 shows the reflectance spectrum for a QWS whose maximum reflectance peak is centered at 2.2 eV (E_0). This calculation includes i) dispersion in both the dielectric functions, ii) an ~ 6 % increase in the OPL of the Si_3N_4 layer, $d_n = 72.2$ nm, (compared to an ideal structure $d_{ni} = 68.6$ nm) and a corresponding ~ 6 % decrease in the OPL of the SiO_2 layer, $d_0 = 90.7$ nm, (for ideal structure $d_{oi} = 96.5$ nm) and iii) *spatial dispersion* (aperiodicity) in the Si_3N_4 layer of 1 nm/layer. We define *spatial dispersion* as a continual change in the layer thickness with the layer closest to the substrate being the original thickness. Hence, in this structure $d_n = 72.7$ nm for the layer next to the substrate while $d_n = 81.7$ nm for the top layer. The spatial dispersion or aperiodicity in the Si_3N_4 film is the reason that the minima on both sides of the main reflectance peak do not go to zero. The strength of the $2E_0$ peak comes from the fact that the average OPL of one period is greater than the OPL of an ideal structure, i.e. $n_n d_n + n_0 d_0 > \lambda/2$ (for ideal structure $n_n d_{ni} = n_0 d_{oi} = \lambda/4$). This average increase in the OPL stems from the following reasons i) the values for n_n and n_0 at higher energies are greater than those at lower energy, thereby increasing the OPL at $2E_0$, ii) aperiodicity in the Si_3N_4 layer, i.e., the overall average thickness of the Si_3N_4 film is greater than d_{ni}, and iii) intentionally having a non-ideal thickness for each layer, $d_n > d_{ni}$ and $d_0 < d_{oi}$. This synthesized reflectance spectrum exhibits all the features of the experimental spectrum in Fig. 3.

The computer model was expanded to account for temperature dependence of the thickness and optical properties of the constituent layers. The coefficient of thermal expansion for the Si_3N_4 film is 3.0×10^{-6} °C^{-1} [4] and 5.5×10^{-7} °C^{-1} [5] for the SiO_2 film and substrate. The temperature coefficient of the refractive index dn/dT for SiO_2 is given by:

$$\frac{dn_0}{dT} = 8.16638 \times 10^{-6}\, T + 1.04124 \times 10^{-8}\, T^2 - 5.59781 \times 10^{-12}\, T^3 \qquad (1)$$

where T is the temperature with units of degrees Celsius [6]. dn/dT for Si_3N_4 was not found in the literature. We have assumed that the ratio of the real part of the temperature coefficient of the refractive index for Si_3N_4 to that of SiO_2 equaled the ratio of the coefficients of thermal expansion for Si_3N_4 to SiO_2, so that

$$\frac{dn_n}{dT} = 5.45 \, (8.16638 \times 10^{-6} \, T + 1.04124 \times 10^{-8} \, T^2 - 5.59781 \times 10^{-12} \, T^3) \qquad (2)$$

Using the model calculations, we can calculate the effects of thermally modulated the QWS's. Defining

$$\frac{\Delta \text{Reflectance}}{\text{Reflectance}} = \frac{\text{Reflectance at var. temp. - Reflectance at } 25°C}{\text{Reflectance at } 25°C} \qquad (3)$$

(abbreviated as ΔRefl./Refl.) and plotting ΔRefl./Refl. verses energy one can see where the maximum changes in reflectance are located. All the temperature dependent values increase with temperature (in the temperature range investigated) so that the reflectance spectrum is shifted to lower energy. The higher the temperature the greater the shift to lower energy.

The first calculation in Fig. 5 corresponds to an ideal structure with film thicknesses which are both constant and periodic ($d_{ni} = 68.6$ nm and $d_{oi} = 96.5$ nm), but with dispersion in both dielectric functions. The QWS at each temperature has the same structure as described above, and as shown in Fig. 2. Figure 5 is a plot of ΔRefl./Refl. for this ideal structure at three different temperatures: 50°C, 75°C and 100°C. Figure 7 is a plot of ΔRefl. versus photon energy for each temperature spectra. A constant integer offset was added to the 50°C and the 75°C spectra so as to be able to plot all three ΔRefl./Refl. and ΔRefl. spectra on the same figure, and thereby provide for easy comparisons. The ΔRefl./Refl. and ΔRefl. spectra exhibit the same spectral signatures for each temperature. The greater the temperature modulation, the greater the intensity in this derivative spectrum. The maximum changes in ΔRefl./Refl. are located on both sides of the central peak, ~ 1.9 eV and ~ 2.5 eV. The intensity of these two maximum is mainly due to the division of ΔRefl. by values in the 25°C reflectance spectrum that approach zero (see Fig. 7). As the temperature increases, ΔRefl. increases, therefore the magnitude of ΔRefl./Refl. increases through out the spectrum. A smaller change in ΔRefl./Refl. occurs at ~ 4 eV which corresponds to the low energy side of the second peak at $2E_0$ in Fig. 2. It should be noted that each local minimum in Fig. 2, corresponds to a local maximum in Fig. 5, where the interference fringes that have the highest peak to peak ratio also have the highest ratio of ΔRefl./Refl.

Similarly, ΔRefl./Refl. was calculated for a structure that has i) dispersion in the indices of refraction, ii) a Si_3N_4 OPL $> \lambda_{central}/4$ by ~ 6 %, i.e., $d_n = 72.7$ nm, and a SiO_2 OPL $< \lambda_{central}/4$ by ~ 6 %, $d_o = 90.7$ nm, and iii) a small aperiodicity of 1 nm/layer in the Si_3N_4 film thicknesses, $d_n = 72.7$ nm for the Si_3N_4 layer next to the substrate increasing to $d_n = 81.7$ nm for the top layer. This is the same structure described above, and whose reflectance spectrum at 25°C is shown in Fig. 4 These calculations were done for temperatures of 50°C, 75°C and 100°C, see Figs. 6 and 8. As before an integer offset was added to the 50°C and 75°C calculations for easy comparison. Again, ΔRefl./Refl. exhibits the same signature for the different temperatures, varying in intensity only. The intensity changes as the temperature increases because ΔRefl. increases (see Fig. 8). In this case a pronounced maximum occurs at ~ 4.2 eV, the higher energy side of the $2E_0$ peak. The only other noticeable features are the local maximums corresponding to the interference fringes of Fig. 4.

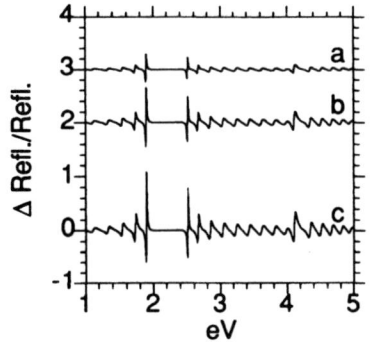

Fig. 5. Δ Reflectance/Reflectance for QWS with ideal layer thicknesses and dispersion in indices of refraction for temperatures: (a) 50°C, (b)75°C, and (c) 100°C.

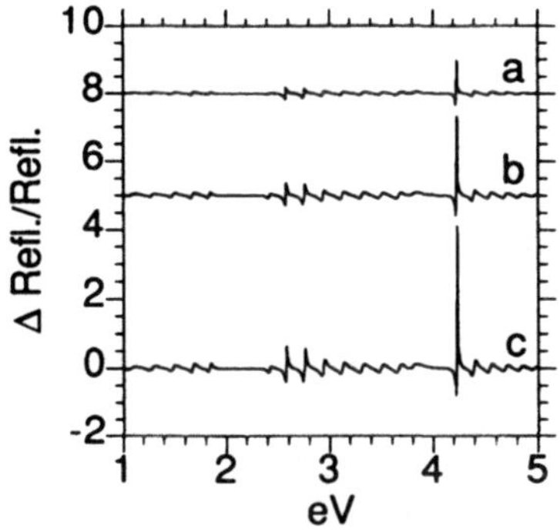

Fig. 6. Δ Reflectance/Reflectance for calculated reflectance spectrum with non-ideal thickness ratio and positive vertical dispersion in the nitride layer for temperatures: (a) 50°C, (b) 75°C, and (c) 100°C.

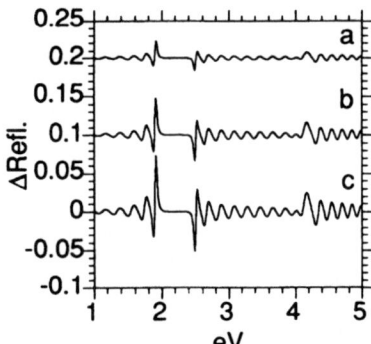

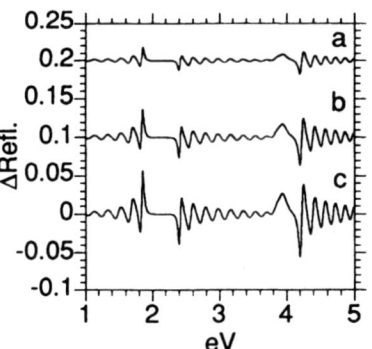

Fig. 7. ΔRefl. for QWS with ideal layer thickness and dispersion in indices of refraction for temperatures: (a) 50°C, (b) 75°C, and (c) 100°C.

Fig. 8. ΔRefl. for calculated reflectance spectrum with non-ideal thickness ratio and positive vertical dispersion in the nitride layer for temperatures: (a) 50°C, (b) 75°C, and (c) 100°C.

The reasons for the second peak to be so pronounced in the ΔRefl./Refl. for a non-idealistic structure are that i) the amplitude of the $2E_0$ peak is increased by over a factor of 2 in Fig. 4 as compared to Fig. 2, hence ΔRefl. is increased, and ii) the minimum on the high energy side of the $2E_0$ peak approaches zero for the non-ideal structure compared to the ideal structure, i.e. Refl. at 25°C goes to zero at ~ 4.2 eV. The combination of these two factors, division of a larger reflectance by a smaller reflectance, accounts for the pronounced spike in the ΔRefl./Refl. spectrum of Fig. 6. Conversely, the division of a smaller number by a larger number accounts for the almost elimination in the ΔRefl./Refl. spectrum at the minimum of both sides of the central peak. For the non-ideal structure the minimum of both sides of the central peak do not go to zero, see Fig. 4. The reflectance spectrum at 25°C for a non-ideal structure has larger values of reflectance than that of an ideal structure at both sides of E_0 by approximately a factor of 20. Therefore a significant reduction in ΔRefl./Refl. is observed.

DISCUSSION

By analytically thermally modulating QWS structures we have demonstrated that an aperiodic structure can have a pronounced spike in the ΔRefl./Refl. spectrum. This spike can be used as a thermal photonic switch due to it's relative intensity as compared with the rest of the derivative spectrum. This sharp feature at just below $2E_0$ is enhanced in a QWS structure in which i) the higher index of refraction layer has an OPL that is $> \lambda_{central}/4$, and whose lower index of refraction layer has an OPL $< \lambda_{central}/4$, ii) the temperature dependent material properties increase with temperature for both materials, with the properties of higher index material increasing faster than that of those of the lower index material and iii) the dispersion in the higher index film also increasing more rapidly with increasing energy than the dispersion in the lower index material. If the last requirement is not met, then the intensity of the $2E_0$ reflectance peak is decreased resulting in a decrease in magnitude of the spike feature in the ΔRefl./Refl. spectrum. Aperiodicity is not a requirement for a these effects, however if it is the right sense, it can enhance the effects noted above. Fortuitously, using the remote PECVD process, and maintaining constant depositions, the thickness of the higher index material, the nitride, increases with increasing overall stack thickness. Therefore, the higher index material's OPLs increase with increasing stack thickness, which in turn increases the strength of the $2E_0$ feature thereby also giving rise to an increase of magnitude of the spike feature in the ΔRefl./Refl. spectrum.

ACKNOWLEDGMENTS

The research presented in this paper has been supported by the National Science Foundation, and the Office of Naval Research in the United States, the DFG in Germany, and through a research collaboration between the states of North Carolina in the USA and North Rhine Westphalia in Germany.

REFERENCES

[1] G. Lucovsky, D.V. Tsu, R.A. Rudder and R.J. Markunas, in "Thin Film Processes II", Ed. by J.L. Vossen and W. Kern, (Academic Press, Boston, 1991), Chapter IV-2.
[2] D.J. Stephens, S.S. He, G. Lucovsky, H. Mikkelsen, K. Leo and H. Kurz, MRS Proceedings 11/1992
[3] D.J. Stephens, S.S. He, G. Lucovsky, H. Mikkelsen, K. Leo and H. Kurz, J. Vac. Sci. Technol. **A11** (1993), in press.
[4] D.R. Messier and W.J. Croft, "Preparation and Properties of Solid State Materials, Volume 7, Growth Mechanisms and Silicon Nitride", Ed. by W. R. Wilcoy, (Marcel Dekker Inc., New York, 1982), p.179 and references there in.
[5] R. C. Waest, "CRC Handbook of Chemistry and Physics, 64th Edition", (CRC Press, Inc., Boca Raton, 1983-1984), p. F-57.
[6] R. M. Waxler and G. W. Cleek, Journal of Research - of the National Bureau of Standards - A. Physics and Chemistry, Vol. 77A, No. 6, Nov. - Dec. 1973, pp. 755-763.

SYNTHESIS OF DISLOCATION FREE $Si_y(Sn_xC_{1-x})_{1-y}$ ALLOYS BY MOLECULAR BEAM DEPOSITION AND SOLID PHASE EPITAXY

GANG HE, MARK D. SAVELLANO and HARRY A. ATWATER
Thomas J. Watson Laboratory of Applied Physics,
California Institute of Technology, Pasadena, CA 91125

ABSTRACT

Synthesis of strain-compensated single-crystal $Si_y(Sn_xC_{1-x})_{1-y}$ alloy films on silicon (100) substrates has been achieved with compositions of tin and carbon greatly exceeding their normal equilibrium solubility in silicon. Amorphous $SiSnC$ alloys were deposited by molecular beam deposition from solid sources followed by thermal annealing. *In situ* monitoring of crystallization was done using time-resolved reflectivity. Good solid phase epitaxy was observed for $Si_{0.98}Sn_{0.01}C_{0.01}$, at a rate about 20 times slower than that of pure silicon. Compositional and structural analysis was done using Rutherford backscattering, electron microprobe, ion channeling, x-ray diffraction, and transmission electron microscopy.

INTRODUCTION

The group IV ternary alloys, including $Si_y(Sn_xC_{1-x})_{1-y}$, hold promise in fabricating silicon-based devices with novel electronic and optoelectronic properties. For example, unstrained or modestly strained group IV single-crystal heterojunction devices on silicon substrate may be possible. It has also been shown that relatively efficient electroluminescence at 1.3 μm can be achieved from a carbon-related isovalent impurity complex, and thus silicon-based optoelectronic devices are among possible applications [1]. Assuming Vegard's law, it is expected that $Si_y(Sn_{0.6}C_{0.4})_{1-y}$ will be lattice-matched to silicon. This lattice-matching not only reduces strain in heterojunction devices but also may enhance the mutual solubility of tin and carbon in silicon, since the solubility of isovalent impurities in silicon is most likely dominated by the strain-induced changes in free energy. In this work, high-quality single-crystal $SiSnC$ alloy films on silicon (100) substrates were synthesized by molecular beam deposition from solid sources, followed by solid phase epitaxy (SPE). When compared with molecular beam epitaxy (MBE) and chemical vapor deposition (CVD), solid phase epitaxy is relatively simple and compatible with silicon VLSI technology. Also alloy synthesis by molecular beam deposition permits greater control over film thickness, composition, and composition profile than does ion implantation.

EXPERIMENTAL

Amorphous $SiSnC$ films were prepared in a custom-designed MBE system with a base pressure of 1×10^{-9} $Torr$. Solid sources were used and the thickness and growth rate were monitored by a quartz crystal monitor. Si (100) wafers were first RCA cleaned ($H_2O : H_2O_2 : NH_3OH$ in $5 : 1 : 1$ at $80°C$) and HF dipped before transferred into the MBE chamber, and a two hour *in situ* prebake at $200°C$ in vacuum was performed. After the substrate was cooled down to about $70°C$, its surface was sputter-cleaned by 500 eV Ar^+ at 0.1 mA/cm^2 for 5 $minutes$ right before the deposition. a $SiSnC$ film 100 nm thick was deposited following an amorphous silicon buffer layer of $200 - 300$ nm thick, all at a growth rate of about 0.2 nm/s. A low temperature anneal at $200°C$ for one hour was carried out before the sample was removed from the growth chamber. The

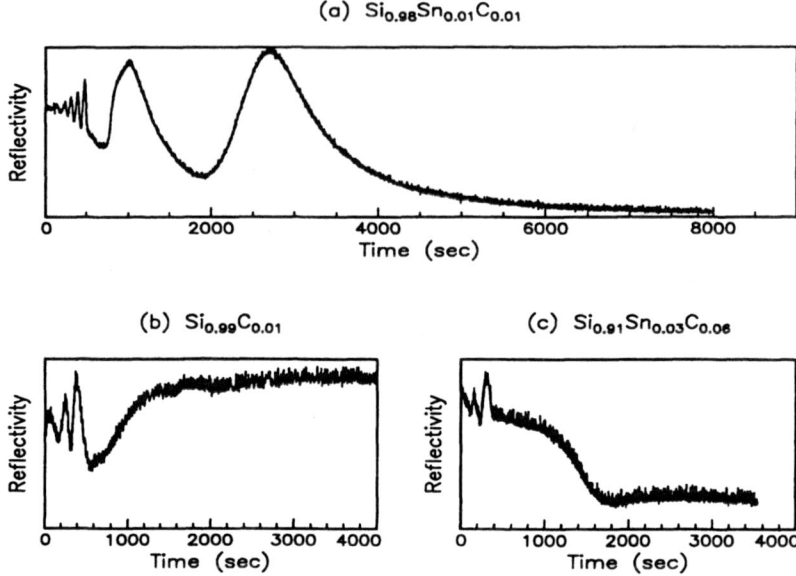

(a) $Si_{0.98}Sn_{0.01}C_{0.01}$

Reflectivity

Time (sec)

(b) $Si_{0.99}C_{0.01}$

(c) $Si_{0.91}Sn_{0.03}C_{0.06}$

Reflectivity

Reflectivity

Time (sec)

Time (sec)

Figure 1: TRR signals of SiSnC samples during annealing at 580°C. The first few relatively fast oscillations corresponded to the planar SPE growth of the silicon buffer layer. (a) $Si_{0.98}Sn_{0.01}C_{0.01}$ sample. The slow oscillation after the buffer layer indicated the planar SPE growth of the 100 nm $Si_{0.98}Sn_{0.01}C_{0.01}$ film, with a SPE rate about 20 times slower than that of the pure silicon buffer layer. (b) $Si_{0.99}C_{0.01}$ sample. The planar SPE stopped when the $Si_{0.99}C_{0.01}$ film was reached. (c) $Si_{0.91}Sn_{0.03}C_{0.06}$ sample. The fast dropping of TRR signal indicated an amorphous to polycrystalline transition in the $Si_{0.91}Sn_{0.03}C_{0.06}$ film.

sample was then annealed at 580°C in air and the solid state epitaxy was monitored by *in situ* time-resolved reflectivity (TRR) using a diode laser working at 670 nm. Rutherford backscattering spectrometry (RBS) and electron microprobe were used for compositional analysis. Ion channeling, x-ray diffraction and transmission electron microscopy (TEM) were used for structural analysis.

RESULTS AND DISCUSSION

Among the various $SiSnC$ film compositions with a range of tin and carbon concentrations (up to 10 $at.\%$), $Si_{0.98}Sn_{0.01}C_{0.01}$ showed very good solid phase epitaxy during annealing, while other tin and carbon concentrations resulted in polycrystalline films or no crystallization, as shown by the TRR signals during the 580°C annealing in Fig. 1. The first few relatively fast oscillations in the TRR signals corresponded to the planar SPE growth of the silicon buffer layer, and the subsequent slower oscillations in Fig. 1(a) indicated the planar SPE growth of the 100 nm $SiSnC$ film, which had a SPE rate about 20 times slower than that of the pure silicon buffer layer. The TRR signals in Fig. 1(b)

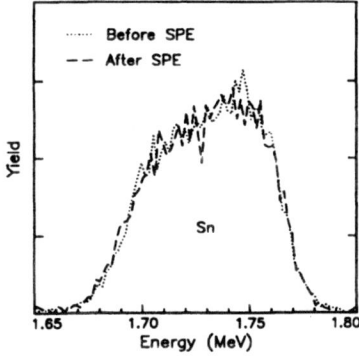

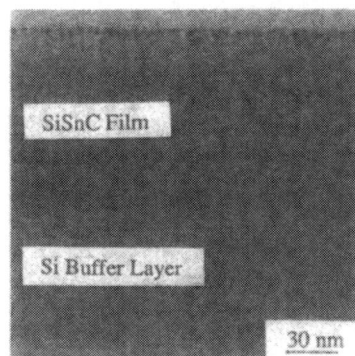

Figure 2: Rutherford backscattering spectra of $Si_{0.98}Sn_{0.01}C_{0.01}$ samples, taken before and after the solid phase epitaxy. No significant Sn redistribution could be seen.

Figure 3: Cross sectional TEM micrograph of $Si_{0.98}Sn_{0.01}C_{0.01}$ sample after annealing at $580°C$ in air. Surface oxide is formed during annealing. No dislocations were present in the micrograph.

and Fig. 1(c) showed no crystallization and polycrystalline films, respectively. Backscattering spectra for the $Si_{0.98}Sn_{0.01}C_{0.01}$ sample were taken both before and after solid phase epitaxy (Fig. 2), and no significant tin redistribution could be seen. The carbon and tin concentrations were also confirmed by the electron microprobe and were in good agreement with the deposition rate controls and the RBS data. It is interesting to note that a sample of $Si_{0.99}C_{0.01}$ did not show solid phase epitaxial regrowth following the epitaxy of the silicon buffer layer, as shown by the TRR signal in Fig. 1c. Compared with the successful epitaxial regrowth of $Si_{0.98}Sn_{0.01}C_{0.01}$, this suggested that the maximum concentration of impurities in silicon without inhibition of solid phase epitaxy might be increased by careful strain compensation.

Cross sectional TEM showed that the regrown $SiSnC$ sample was dislocation free (Fig. 3). Ion channeling spectrum also showed very good crystal quality (Fig. 4). The channeling feature of tin signal in the spectrum indicated good substitutionality of tin in silicon. X-ray rocking curve for (400) diffraction peak (Fig. 5) showed that the peak position of the $SiSnC$ film was slightly shifted relative to that of the substrate, so that the lattice parameter of the film was slightly smaller than that of pure silicon. This suggested that carbon is substitutional in the film, and is consistent with the fact that 1 $at.\%$ carbon is more than enough to compensate the strain induced by 1 $at.\%$ tin. Assuming Vegard's law, it could be calculated from the amount of shift in the x-ray rocking curve that there was about 0.8 $at.\%$ substitutional carbon in the film, which was in good agreement with the carbon concentrations obtained by the other methods.

CONCLUSION

We have achieved good crystal quality $Si_y(Sn_xC_{1-x})_{1-y}$ alloy by molecular beam deposition and solid phase epitaxy with about 1 $at.\%$ of tin and carbon, well above their solid solubility in silicon (about 2×10^{-4} and 6×10^{-6} for tin and carbon, respectively). The film was found to be dislocation free, with good substitutionality of tin and carbon.

232

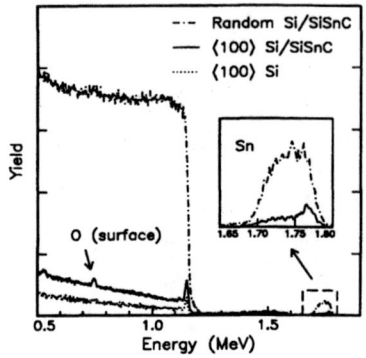

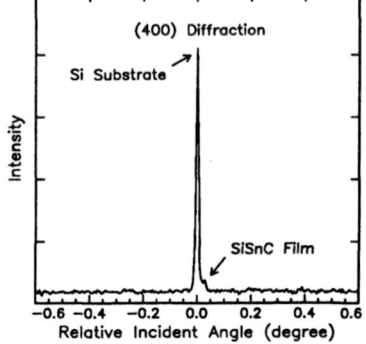

Figure 4: Ion channeling spectra of annealed $Si_{0.98}Sn_{0.01}C_{0.01}$ sample using 2 MeV He^{++} Ion. The channeling feature of Sn signal indicated good substitutionality of Sn in Si.

Figure 5: X-ray rocking curve of annealed $Si_{0.98}Sn_{0.01}C_{0.01}$ sample around (400) diffraction peak. The small shift of the diffraction peak of the $SiSnC$ film relative to that of the substrate indicated that the stain induced by Sn was compensated by, assuming Vegard's law, about 0.8 $at.\%$ substitutional C, whick was in good agreement with the C concentrations obtained by the other methods.

We also suggest that strain compensation may increase the solubility of these impurities in silicon. We note that this is a potentially attractive process for fabrication of Si films with C concentrations exceeding the solid solubility for devices based on electroluminescent emission at 1.3 μm.

ACKNOWLEDGMENTS

We are grateful to John Armstrong and Paul Carpenter for the electron microprobe measurements. This work was supported by NSF under grant DMR 8958070 and the Caltech President's Fund.

REFERENCES

1. L.T. Canham, K.G. Barraclough, and D.J. Robbins, Appl. Phys. Lett. **51**, 1509 (1987).

2. L.T. Canham, M.R. Dyball and K.G. Barraclough, Mat. Sci. and Eng. **B4**, 95 (1989).

3. K. Eberl, S.S. Iyer, S. Zollner, J.C. Tsang, and F.K. LeGoues, Appl. Phys. Lett. **60**, 3033 (1992).

4. M.S. Goorsky, S.S. Iyer, K. Eberl, F.K. Legoues and F. Cardone, Appl. Phys. Lett.**60**, 2758 (1992).

5. S.S. Iyer, K. Eberl, M.S. Goorsky, F.K. LeGoues and J.C. Tsang, Appl. Phys. Lett. **60**, 356 (1992).

6. P. Mei, M.T. Schmidt, E.S. Yang, and B.J. Wilkens, J. of Appl. Phys. **69**, 8417 (1991).

7. G.L. Olsen and J.A. Roth, Mat. Sci. Rep. **3**, 1 (1988).

8. G.L. Patton, J.H. Comfort, B.S. Meyerson, E.F. Crabb , G.J. Scilla, E. deFr sart, J.M.C. Stork, J.Y.C. Sun, D.C. Harame, and J.N. Burghartz, IEEE Elect. Dev. Lett. **11**, 171 (1990).

9. M. Vergnat, M. Piecuch, G. Marchal, and M. Gerl, Phil. Mag. **B51**, 327 (1985).

NANOCRYSTALLINE SiC FORMATION FROM SPUTTER-DEPOSITED NON-EQUILIBRIUM a-Si_xC_{1-x} ALLOYS

HONG WANG, Z. MA, L. H. ALLEN AND J. M. RIGSBEE*
Department of Materials Science and Engineering, University of Illinois, 1304 West Green Street, Urbana, IL 61801
*now at Department of Materials Science and Engineering, University of Alabama at Birmingham, 360 BEC, Birmingham, AL 35294.

ABSTRACT

There exists considerable interest in the synthesis of nanocrystalline SiC particles embedded in an a-Si_xC_{1-x}:H matrix. This study investigated the formation of nanocrystalline SiC by annealing a non-equilibrium a-$Si_{0.35}C_{0.65}$ alloy sputter deposited onto oxidized (100) Si. Evolution of the microstructure was characterized by thin film sheet resistance measurements, glancing angle X-ray diffraction (GAXRD) and transmission electron microscopy (TEM). It was found that films annealed at temperatures below 850°C remained amorphous. Annealing at temperatures above 850°C resulted in the formation of SiC nanocrystallites, as revealed by GAXRD and TEM. The electrical conductivity also showed an abrupt increase around 950°C. Prolonged annealing resulted in a further increase in conductivity, implying that the transformation continued. Annealing at 1100°C for 3 hours increased the film conductivity by two orders of magnitude relative to the as-deposited film.

INTRODUCTION

Wide band-gap hydrogenated amorphous silicon carbide (a-SiC:H) has attracted a great deal of attention due to its potential application in thin film transistors, solar cells and other electronic devices [1-5]. However, the overall performance is severely limited because of the decrease in conductivity as a result of the presence of carbon. Usui et al. [6] and Veprek et al. [7] discovered that better performance can be achieved by introducing microcrystallites into the amorphous network. They reported that the presence of such microcrystals in the amorphous matrix of a-Si:H increased both electrical conductivity and optical band-gap.

Demichelis, Pirri and Tresso [8] recently demonstrated that this idea is also applicable to a-SiC:H. Films contains SiC microcrystallites were grown under suitable conditions from SiH_4 + CH_4 mixtures with a high H_2 dilution and high RF power density using plasma-enhanced chemical vapor deposition (PECVD). In this study, we explored an alternative way of preparing such materials. SiC nanocrystallites were formed by annealing a co-sputtered a-Si_xC_{1-x} film. Our results show that films prepared this way have high conductivities and, in addition, have the thermal stabilities necessary for high temperature applications.

EXPERIMENTAL PROCEDURE

The amorphous SiC films were produced in a diffusion-pumped, RF sputter deposition system using a silicon+graphite composite target. Films about 5600Å thick were sputter deposited onto an oxidized Si (100) wafer, which was mounted on a water cooled substrate holder. Using Auger Electron Spectroscopy, the film compositions were determined to be carbon-rich a-$Si_{0.35}C_{0.65}$ with less than 5% variation across the film. As-deposited films were annealed in high vacuum (10^{-7} torr) over the 750°C to 1100°C temperature range for 1-3 hours. Evolution of the microstructure was studied using GAXRD and TEM. Film resistivity was measured using a four point sheet resistance apparatus.

RESULTS AND DISCUSSION

Microstructural Evolution

       As indicated in a X-ray diffraction spectrum shown in Figure 1, the as-deposited a-$Si_{0.35}C_{0.65}$ film was amorphous. TEM examination of the film in cross-section and planar views shows that the as-deposited film exhibits a columnar-like structure along the growth direction with no evidence of any crystalline phase. For films annealed at 850°C for 1 hour, no formation of SiC crystallites was observed by either GAXRD or TEM. Annealing above 850°C resulted in a small but detectable volume fraction of crystalline SiC particles inside the amorphous matrix. At this early stage, the nanocrystalline β-SiC particles are only resolvable by TEM. Figure 2(a) shows a cross-sectional TEM micrograph for the film annealed at 900°C for 1 hour. It is clearly seen that SiC particles of 5nm in diameter and smaller are uniformly dispersed in the remaining amorphous matrix. Electron diffraction analysis indicates that they are crystalline β-SiC particles. Annealing at higher temperatures causes the formation of more and larger SiC crystallites (see Fig.1). We have not established if SiC crystal nucleation is instantaneous or continuous. But we believe that as a threshold temperature is reached (about 900°C in our case), the β-SiC crystallites nucleate at heterogeneous sites such as voids and amorphous domain boundaries . Extended annealing resulted in the growth of these crystallites, as illustrated by a planar-view TEM micrograph for the film annealed at 1100°C for 2 hours (Fig. 2(b)). It should be indicated that as a result of the formation of SiC, the amorphous matrix becomes increasingly carbon-rich. These results suggest that the nucleation and growth of the SiC nanocrystallites may be controllable by knowing the transformation kinetics of these particles.

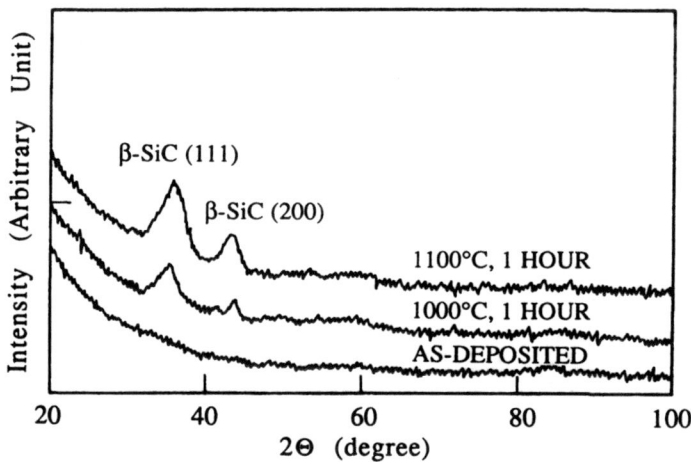

Figure 1. Glancing angle X-ray diffraction of the as-deposited and annealed a-$Si_{0.35}C_{0.65}$ alloy films.

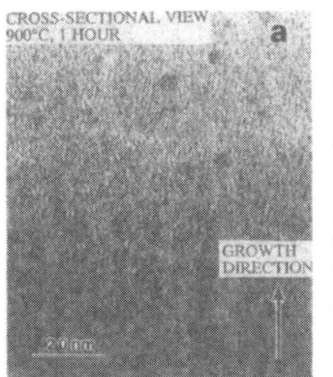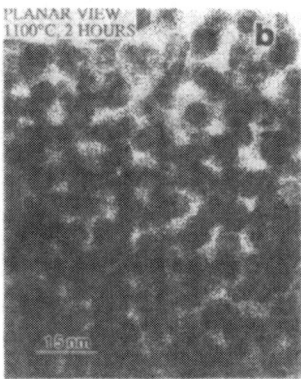

Figure 2. TEM micrographs of the a-Si$_{0.35}$C$_{0.65}$ annealed alloy films.

Thin Film Resistivity Measurement

The sheet resistances of the amorphous Si$_{0.35}$C$_{0.65}$ film after annealing at different temperatures was measured using a standard four point method. The results are summarized in Fig. 3. The sheet resistances for annealed film were converted into the more meaningful conductivity using the definition of sheet resistance:

$$R_s = \rho/t$$

where R_s is the film sheet resistance ($\Omega/\square$), ρ is the film resistivity (Ω-cm), and t is the film thickness (cm). As shown in Fig. 4, the as-deposited film has the lowest electrical conductivity. Annealing below 850°C increases the film conductivity by only a factor of 3-4, confirming that the film remains amorphous. Up to 950°C the conductivity increased slowly. This 850-950°C stage corresponds well to the initial crystalline SiC formation. An abrupt increase in conductivity occurred between 950 and 1000°C, suggesting significant amounts of SiC formed at this temperature. This is consistent with the GAXRD results in Figure 1. At higher temperatures, film conductivity increased rapidly, indicating a rapid increase in nanocrystalline SiC particle formation rate. Prolonged annealing at temperatures above 1000°C resulted in a further increase in film conductivity, suggesting that SiC particle formation is continuing. The critical temperature required for rapid SiC formation appears to be in the 950-1000°C range. Lower temperatures produce slow particle nucleation and growth rates. This high transformation temperature suggests that these materials can be processed to obtain desired microstructures and properties which will be stable at use temperatures below 800°C.

Annealing the a-Si$_{0.35}$C$_{0.65}$ film at 1100°C for 3 hours increased the film conductivity by two order of magnitude. Finally, it is worth noting that the relatively high conductivities obtained in our films may be due to unintentional doping of the film during sputter deposition [8].

238

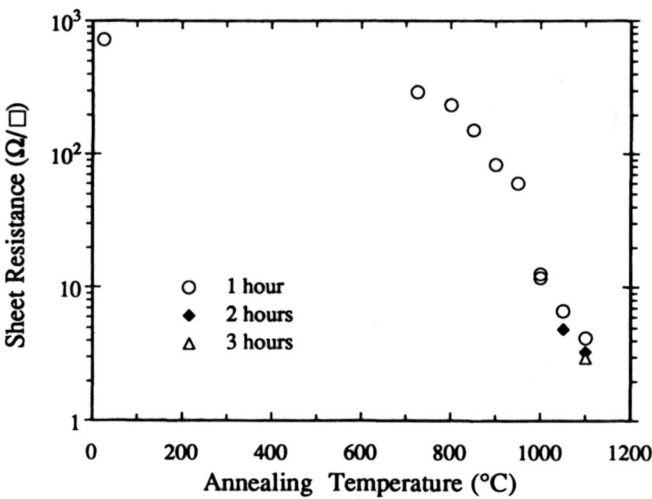

Figure 3. Sheet resistance of the annealed a-$Si_{0.35}C_{0.65}$ alloy films

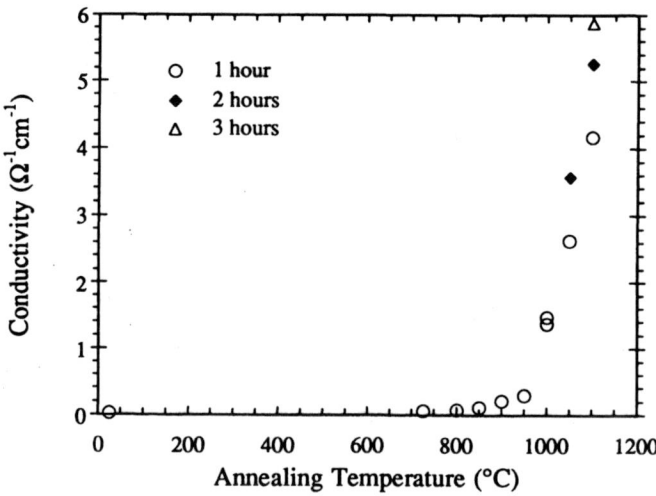

Figure 4. Electrical conductivity of the annealed a-$Si_{0.35}C_{0.65}$ alloy films.

CONCLUSIONS

1. Nanocrystalline β-SiC was formed by annealing a sputter deposited a-Si$_{0.35}$C$_{0.65}$ alloy film. SiC formation starts slowly at approximate 850°C. Above a critical temperature (950-1000°C), the SiC formation increases abruptly. The microstructural evolution was characterized by GAXRD, TEM and four point sheet resistance measurement.
2. The a-Si$_{0.35}$C$_{0.65}$ alloy film conductivity increased with annealing temperature and annealing time. Film conductivity increased by two orders of magnitude after annealing at 1100°C for 3 hours.

ACKNOWLEDGEMENT

We are grateful for the funding support from the Joint Services Electronics Program (JSEP) under contract N00014-90-J-1270 and a grant from the Petroleum Research Fund ACS-PRF#25422-G5 (REF# 91-PRI-A-1729) (Z. Ma and L.H. Allen). Partial support of this work by the U. S. Army Corps of Engineers Construction Engineering Research Laboratory is gratefully acknowledged (H. Wang and J. M. Rigsbee). The materials characterization was carried out in the Center for Microanalysis of Materials at the University of Illinois, Urbana.

REFERENCES

1. Y. Yukimoto, JARECT, edited by Y. Hamakawa (OHMSHA and North Holland, Amsterdam, 1983), Vol. 6, p.136.
2. T.S. Nashashibi, I.G. Austin, T.M. Searle, R.A. Gibson, W.E. Spear, and P.G. LeComber. Phil. Mag. B45, 557 (1982)
3. S. Furakawa, Springer Proceedings in Physics, edited by L.G. Harris and C.Y. Yang, (Springer, Berlin, 1989), Vol. 34, p. 58.
4. Y. Hamakawa and H. Okamoto, Mat. Res. Soc. Symp. Proc. 242, 651-663, (1992).
5. Y. Tawada, M. Kondo, H. Okamoto and Y. Hamakawa, Solar Energy Mat. 6, 299 (1985).
6. S. Usui and M. Kikuchi, J. Non-Crystalline Solids, 34, 1, (1979).
7. K. Veprek, Z. Iqbal, R.O. Kuhne, P. Capezzuto, F.A. Sarott, and J.K. Gimzewski, J. Phys. C, 16, 6241 (1983).
8. F. Demichelis, C. F. Pirri, and E. Tresso, J. Appl. Phys. 72(4), 1327, (1992).

THE STRUCTURE AND THE PHYSICAL PROPERTIES
OF Si/C GROWN BY LASER PYROLYSIS OF $Si(CH_3)_2Cl_2$

HUA CHANG,* LONG JENG LEE*, RONG LI HWANG*, M. S. LIN** AND J. C. LOU**
*Department of Chemistry
**Department of Electric Engineering
National Tsing Hua University, Hsinchu, Taiwan, Republic of China

ABSTRACT

The chemical kinetics of laser pyrolysis of $Si(CH_3)_2Cl_2$ was studied. Besides Si/C deposited on substrate as a needle the main gaseous products were H_2 and HCl. The decomposition rate constants of $Si(CH_3)_2Cl_2$ and the formation rate constants of H_2 and HCl were measured. The chemical and physical properties of the Si/C needles were studied by various methods, such as their outer features, elemental analysis, Auger depth profile spectra, X-ray diffraction. Other useful properties, such as electric resistivity, photovoltaic effect and hardness, were also reported. A model for the pyrolysis and the growth process of this Si/C needles was proposed. It should be emphasized that the samples with desired physical properties could be grown by controlling the growth conditions properly -- the laser power and the pressure of $Si(CH_3)_2Cl_2$.

INTRODUCTION

Laser induced chemical vapor deposition (LCVD) is an important method in electronic industry and also in preparation of special materials. There are two different processes -- photon dissociation (photolytic process) and laser pyrolysis (photothermal process). The characteristics of these two processes have been reviewed by Bauerle.[1] $Ni(CO)_4$,[2] CH_4, C_2H_2 and C_2H_6,[3] SiH_4,[4] and CH_3Br[5] were studied by this laser pyrolysis.

In this laboratory, laser pyrolysis of $Si(CH_3)_xCl_{4-x}$ (x=0-4) had been studied.[6] $Si(CH_3)_2Cl_2$ was chosen to be studied in detail.[7] The growth rates of the Si/C needles were about 0.2 to 1.0 mg/min in weight. Their diameters were in the range of 150 to 500 μm. The decomposition rate of the reactant should be important in controlling the growth rate and the physical properties of the Si/C needle to be produced. The purpose of this study was to study the initial reactions of $Si(CH_3)_2Cl_2$ pyrolyzed by laser radiation. The physical and the chemical properties of these Si/C needles grown in various conditions were reported. Finally, a model for the pyrolysis and the growth process was proposed.

EXPERIMENTAL

The experimental setup is sketched in Fig. 1. A Pyrex tubing was divided into two compartments: reaction compartment and spectrum compartment. After degassed a proper pressure of $Si(CH_3)_2Cl_2$ was introduced into the cell and sealed off. The amount of $Si(CH_3)_2Cl_2$, H_2 and HCl were measured by Raman spectroscopy while the laser beam was focused at the center of the spectrum compartment. When the laser radiation was focused at the surface of the cell, yellowish-red emission appeared. Laser pyrolysis was initiated and Si/C needle began to grow.

The physical and chemical properties of the Si/C needles grown in various conditions were examined by different instruments. Ill-grown Si/C needles were examined too.

RESULTS AND DISCUSSION

The Mechanism of the Pyrolysis.

During the laser pyrolysis the intensities of the Raman bands of $Si(CH_3)_2Cl_2$, HCl, H_2 were measured and calibrated to the pressures. A typical pressure variation of $Si(CH_3)_2Cl_2$, H_2 and HCl measured during the laser pyrolysis is shown in Fig. 2.

The mechanism of the laser pyrolysis of $Si(CH_3)_2Cl_2$ at the hot spot could be as follows:

$$Si(CH_3)_2Cl_2 \xrightarrow{k_1} H_2 + X \quad (1)$$
$$\xrightarrow{k_2} HCl + Y \quad (2)$$
$$HCl + W \xrightarrow{k_4} Z \quad (3)$$

where X, Y, Z and W were the intermediates formed during the pyrolysis. The first two reactions were parallel reactions. The third one was consecutive reaction for HCl.

The decomposition of $Si(CH_3)_2Cl_2$ was a first order reaction. The rate constant k, which should be $k_1 + k_2$, could be obtained directly from the

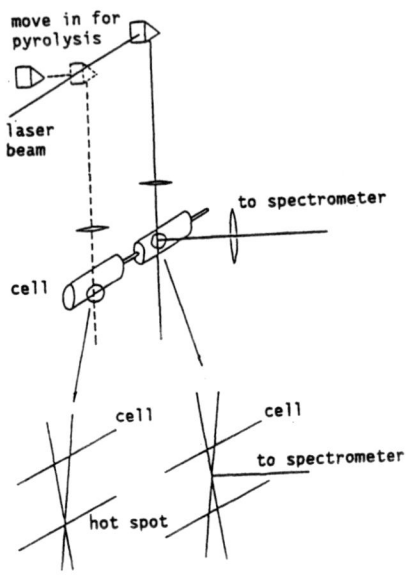

Fig. 1. The experimental setup.

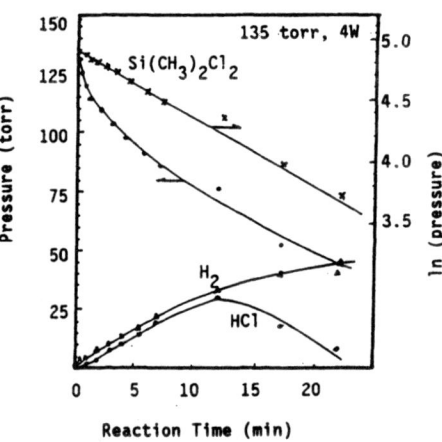

Fig. 2. The variation of pressures of $Si(CH_3)_2Cl_2$, H_2 and HCl during the pyrolysis.

plot of $\ln[P(Si(CH_3)_2Cl_2)]$ vs t. For H_2 and HCl, The formation rate constants k_1 and k_2 could be obtained from the slope at the beginning of the laser pyrolysis. It was also found that HCl involved in Reaction (3). The decomposition to other products was only about one tenth. The rate constants k, k_1 and k_2 for the laser pyrolysis of $Si(CH_3)_2Cl_2$ at the pressures of 135, 107 and 84 torr by 1 to 4 W of 488 nm laser radiation were obtained and only the data at 135 torr were plotted in Fig. 3. The activation energy was 44.3 kJ/mol.

The Chemical and Physical Properties.

The sample --Si/C needles -- were examined by the naked eyes first. It was black. The diameters and surface features were uniform and smooth for the well-grown samples. If the growth conditions varied during the pyrolysis the samples showed swelling or shrinkage and even bend. With a scanning electron

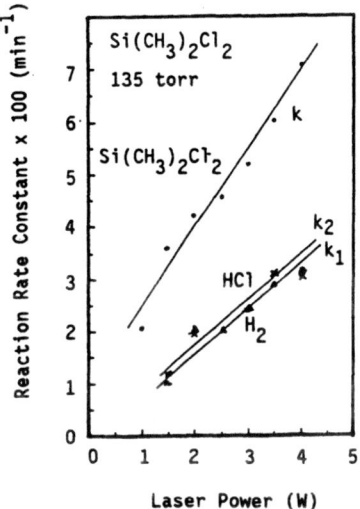

Fig. 3. The reaction rate constants of k, k_1 and k_2.

microscope, a sample with a diameter of 250 μm showed a rather smooth surface as shown in Fig. 4a. The droplets with a size about 20 μm were found to be arranged in rows on the surface as shown in Fig. 4b. Laue diffraction patterns indicated that the sample was polycrystalline. The bending part and the terminal part were the same.

Elemental analysis for Si and C showed that the atomic ratio C/Si were 1.2 and 0.3 for the samples grown under high (3.5 W) and low (1 W) laser radiation. It could be represented by SiC_x (x=0.3 to 1.2).

By scanning Auger microprobe (SAM, model 590 AM from Perkin- Elmer PHI) with a differential ion gun (model PHI-04-303), Auger depth profile spectra of Si, C, O and Cl could be measured in the transverse direction to the axis of the Si/C needle. Fig. 5 was a typical spectrum. Apparently, there are three layers: surface, transition and inner layers, as shown in the figure.

The electric resistivity and the temperature dependence were measured. They showed the character of semiconductor. It showed photovoltaic effect. Above a threshold the electric potential increased linearly with the intensity of the radiation. The hardness of the Si/C needles was in the range of 1800 to 1200 HV between those of Si and WC.

These properties depended on the composition of the samples or the growth conditions.

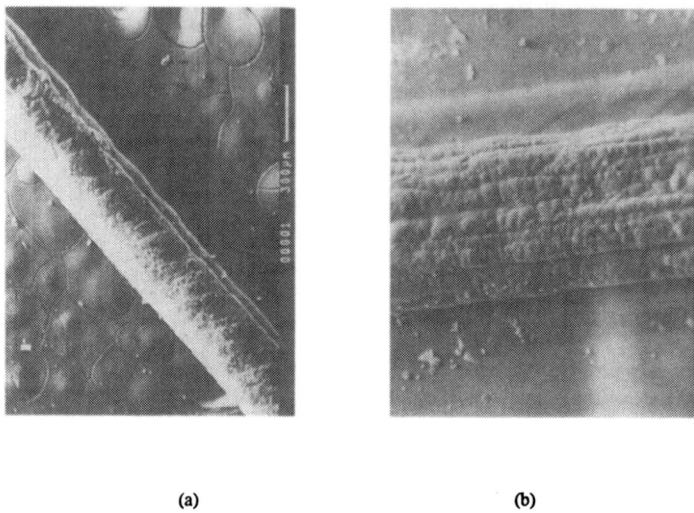

(a)   (b)

Fig. 4. The outer features of two samples under scanning electron microscope.

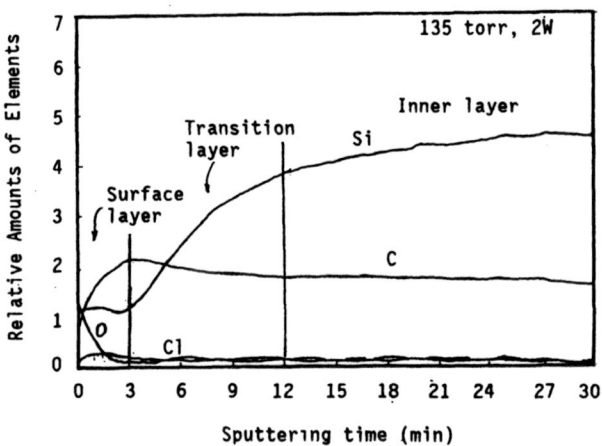

Fig. 5. The Auger depth profile spectrum of sample grown in $Si(CH_3)_2Cl_2$ at 135 torr by 1 W of laser radiation.

MODEL FOR PYROLYSIS AND GROWTH.

Mechanism for Pyrolysis

After focusing, the diameter of the laser beam was reduced to about 19.4 μm. The intensity of the laser radiation at the center of the focal plane was about 27000 W/mm^2 for 4 W of irradiation. The growth could happen over the whole top of the Si/C needle to a diameter of 500 μm. In experiment the growth rate was only 1 mg/min.[7] The efficiency of the growth was only 1/200 if the sticking coefficient was assumed to be 1.

By calculation, the energy irradiated on the sample dissipated mainly by conduction and reflection from the surface. The energy for the chemical reaction was difficult to estimate however it was not much because the production was low.

The initial reactions were the strip of H_2 and HCl. The energy change for the reactions could be calculated from bond energies. They were about 44.5 and 45.8 kJ/mol, respectively, for the reactions between two $Si(CH_3)_2Cl_2$ to give H_2 and HCl. It was the same to our experimental activation energy 44.3 kJ/mol.

Finally, the mechanism for the pyrolysis could be suggested. When the laser beam was focused sharply at the surface of a substrate, a tiny hot spot with a temperature of 1000 to 1800°C was induced. When a gaseous molecule hit the hot spot, it might react with the molecules or radicals on the hot spot and give H_2 or HCl. The decomposition rates of $Si(CH_3)_2Cl_2$ were still first order with respect to the pressure of $Si(CH_3)_2Cl_2$. The reaction continued and finally Si/C formed.

Mechanism for Growth

The properties reported above could also provide a picture for the mechanism of the growth of the Si/C. It can be summarized in Fig. 6. The intensity of the laser beam was Gaussian (as shown in Fig. 6a). According to the intensity distribution there are four regions, A, B, C and D at the tip of the Si/C needle, as sketched in Fig. 6b. The temperature should be proportional to the radiance of the laser intensity. Heat conduction had an effect to average the temperature on the sample. In Region A, the intensity was very high and nearly constant so was the temperature. In Region B, the intensity droped fast. However, temperature should be still high but

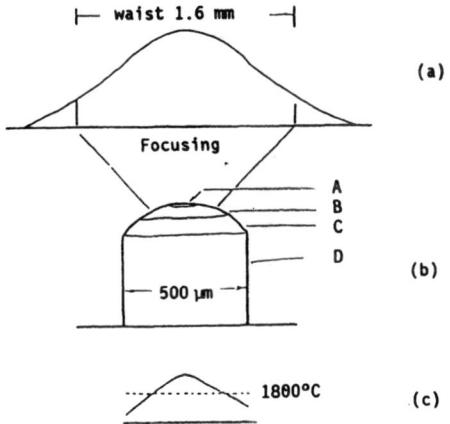

Fig. 6. The mechanism for growth. (a) laser intensity
(b) intensity of radiation on sample and
(c) temperature distribution.

varied drastically due to thermal conduction. Region C was out off the strong laser irradiation, but the heat conducted from Region B kept its temperature high. The distribution of temperature is sketched in Fig. 6c. The measured temperature at the tip of the sample by an optical pyrometer was 1800°C during the pyrolysis by 4 W laser radiation. Region D was the lateral side and far away from the hot tip. The temperature should be low.

In the cell, molecules of the reactant $Si(CH_3)_2Cl_2$ moved randomly. It hit the Si/C needle at various part. When the molecule hit the hot spot, pyrolysis initiated. It should be violent and nearly complete in Region A. The amount of C was low. In region B, the amount of C deposited with Si varied drastically as the temperature was. In Region C, the temperature was lowered further. Fragments could be pyrolyzed for a longer time, therefore the amount of C was decreasing outwards. The above mechanism could explain the Auger depth profile spectra shown in Fig. 5.

The Si and C deposited on the tip should be liquid droplets. As pyrolysis continued on, a layer of droplets formed on the top of the old ones. Therefore, the outer feature of the Si/C needle under the scanning electron microscope showed droplets arranged in rows as shown in Fig. 4.

ACKNOWNLEDGEMENT

This work was supported by the National Science Council of the Republic of China.

REFERENCES

1. D. Bauerle, Chemical Processing with Laser; Springer-Verlag; Berlin, 1986.
2. F. Petzoldt, K. Pigimayer, W. Krauter and D. Bauerle, Appl. Phys., A35, 155 (1984).
3. G.Leyendecker,D. Bauerle, P. Geitter and H. Lydtin, Appl. Phys. Lett., 39, 921 (1981).
4. D.Bauerle,P.Irsigler,G.Leyendecker,H. Noll and D. Wagner, Appl. Phys. Lett., 40, 819 (1982).
5. H. Chang and J. C. Hsieh, Laser Chemistry, in Press.
6. H. Chang and R. L. Hwang, Materials Lett., 12, 517 (1990).
7. R. L. Hwang and H. Chang, Proc. Natl. Sci. Coun. (ROC), 11, 430 (1987).

SIMULATION OF A LOW LOSS OPTICAL MODULATOR FOR FABRICATION IN SIMOX MATERIAL

C K Tang*, G T Reed*, A J Walton*, A G Rickman#
* Department of Electronic & Electrical Engineering, University of Surrey, Guildford, Surrey, UK.
+ Department of Electronic & Electrical Engineering, University of Edinburgh, Edinburgh, UK.
# Departments of Electronic & Electrical Engineering and Physics, University of Surrey, Guildford, Surrey, UK.

Abstract

A novel phase modulator has been designed and analysed, assuming fabrication in a silicon-on-insulator material such as SIMOX. The proposed modulator is based upon a transverse p-i-n structure, utilising the plasma dispersion effect to produce the desired refractive index change in an optical rib waveguide. The device has been studied using the MEDICI two dimensional device simulation package to optimise the injected carrier interaction with the propagating optical mode. Whilst the device is designed to support a single optical mode, it measures several micrometers in cross sectional dimensions, thereby simplifying fabrication and allowing efficient coupling to other single mode devices such as optical fibers. Furthermore the device has an extremely high figure of merit, predicting over 200 degrees of induced phase shift per volt per mm. This implies a short active length together with a low power requirement.

Introduction

For many years silicon has been used as a substrate material for a variety of guided wave structures, due to its stability, good surface finish, and well developed technology, although it was not until relatively recently that silicon was considered as the guiding medium itself. There are a number of other inherent advantages in using a silicon on insulator (SOI) structure for integrated optics applications, such as the resulting strongly confining waveguides, enabling bends and interconnections to be easily produced, the possibility of vertical couplers, and the potential of direct interconnection of optical and electronic circuits. One SOI structure can be formed using the Separation by IMplantation of OXygen (SIMOX) process [1], and it is this waveguiding structure which has the lowest losses of the silicon structures reported to date. Our preliminary work [2, 3] has indicated that SIMOX waveguides can be produced with very low losses (as low as 0.3 dB/cm) if the guiding layer is in excess of approximately $2\mu m$.

The most serious drawback of using silicon as an optical material is the lack of a linear electrooptic (Pockels) effect in the material, although recently a variety of other possible modulation mechanisms have been reported in the literature, such as modulation by carrier injection or carrier depletion, electrorefraction, electroabsorption, or higher order electrooptic effects (e.g. [4,5,6,7]). It should be noted however, that none of these devices have been formed in SIMOX material. In 1987 Soref and Bennett published an important paper in which they studied refractive index changes in silicon [8] due to electrorefraction, the Kerr effect, and charge carrier effects. Their investigation demonstrated that for experimentally reasonable values of applied field (electrorefraction and Kerr effect) and injected carriers (charge carrier) the changes in refractive index were at least two orders of magnitude larger for the latter.

Mat. Res. Soc. Symp. Proc. Vol. 298. ©1993 Materials Research Society

Whilst it is clear that injected charges will produce a change in refractive index, it is also clear that there will be a resultant increase in absorption associated with these carriers. Soref and Bennett quantified the absorption, and demonstrated that submillimeter devices are possible with additional losses of the order of 1dB/cm. In particular they showed that injection levels in excess of 2×10^{17} e-h pairs cm^{-3} are required to achieve useful refractive index changes. Throughout the simulation we have therefore aimed to achieve injection levels in excess of this threshold. Whilst this makes devices practical, it is clearly necessary to use a guiding structure that is initially low loss, such as the SIMOX structure.

Modelling of planar waveguides in SIMOX shows that in order to produce single mode guides, layers of approximately 0.2μm or less are required. These guides are however, too lossy to be of practical interest. Recent experimental evidence reported by Schmidtchen et al [9] and theoretical verification [10] have shown, however, that etched rib waveguides can be fabricated with dimensions of several microns that are single mode. We have also studied such stripe guides at Surrey, and our finding are broadly in agreement with those of Schmidtchen et al, although we have examined a more comprehensive range of waveguide structures. The results will be submitted for publication shortly. These results when considered together with simulation results of the proposed modulator, suggest that a waveguide depth of the order 6.5 μm is useful for a variety of applications, and therefore this work is based upon such a structure (fig1).

Modulator Simulation

The aim of this work was to model a phase modulator in an SOI structure, such as SIMOX. The device was intended to be a practical device such that fabrication is relatively straightforward, based around a standard rib waveguide, but also a device that offers improved performance for a given optimised parameter. The chosen parameter was the efficiency of the refractive index change for a given applied voltage. Clearly such a structure could also be optimised for other changes such as operating speed, but the resultant geometry would be different. By optimising for maximum refractive index, we consequently also optimise for maximum phase change for a given applied voltage. The efficiency of such devices is often characterised by a figure of merit which specifies the induced phase shift per volt per mm. The figure of merit ς may therefore be defined as:

$$\varsigma = \pi/(V_{\pi}L_{\pi}) \quad \text{where} \quad V_{\pi} = \text{the voltage required for a } \pi \text{ radian phase shift}$$
$$\text{and} \quad L_{\pi} = \text{active length (in mm) for a } \pi \text{ radian phase shift.}$$

Typical reported figures for AlGaAs/GaAs devices are approximately 50 degrees per volt per mm [11]. The predicted figure of merit for the proposed device is in excess of 200 degrees per volt per mm.

The structure analysed is shown in Figure 1. The phase modulator design has been based upon the refractive index change due to free carriers. We have used the MEDICI 2-dimensional semiconductor simulation package to determine injected carrier profiles, and converted these concentrations to refractive index changes via the equations produced by Soref and Bennett [7], relating these parameters. Using this data it is clear that in order to achieve a useful refractive index change, an injected carrier concentration of in excess of 2×10^{17} electron hole pairs cm^{-3} is required. We have therefore aimed to achieve this level throughout the region in which the optical mode propagates. Using the MEDICI simulation package, we can obtain cross sectional plots of electron and hole concentration profiles (from which changes in the refractive index profile can be determined), voltage and electric field distributions, and current density distributions. For a given set of geometric parameters and

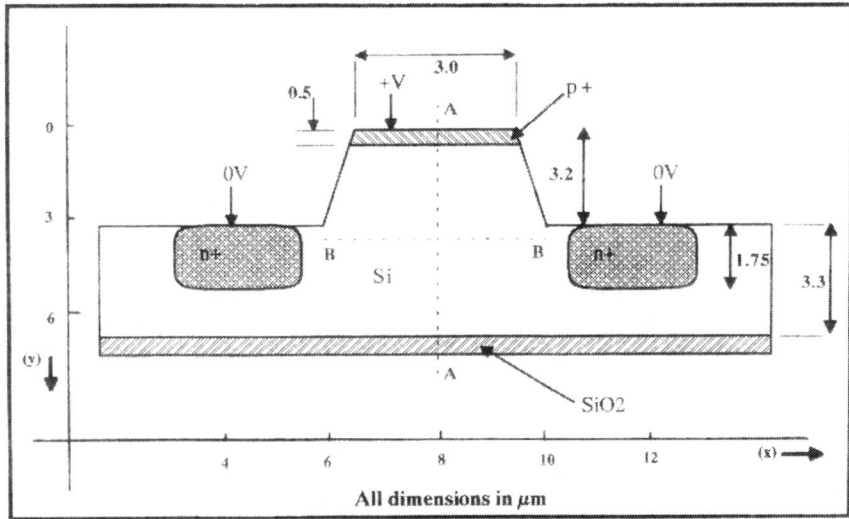

Figure 1. Modulator Geometry

test conditions, the programme produces a grid of mesh points, determining these parameters at each point. By comparing this grid to a theoretical prediction of the propagating optical mode we can ensure that maximum interaction occurs between the refractive index changes and the propagating mode. The approximate optical mode profile has been determined using the method of effective indices (e.g.[12]).

Optimisation of the device has consisted of determining the effects of changes in the geometry of the device, together with electrical and optical properties. This includes for example the effect of different rib widths and heights on the overlap between the induced refractive index change and the electric field of the propagating optical mode. In terms of the geometry of the structure, the angle of the etched rib walls has the greatest effect. Changing from vertical walls, to walls at an angle of approximately 54° gives a vast improvement in overlap. Clearly this is very convenient as it is close to the natural etching angle of silicon. One would intuitively expect an effect of this type, since vertical walls encourage 'clustering' of field lines close to the wall, but the improvement in overlap is much better than had been anticipated. Other variations constitute 'fine tuning' of the device, and will be discussed in the following sections.

Results

Using the MEDICI package we have simulated the device shown in figure 1. The package generates a grid to represent the cross sectional area of the device, and gives information at each grid point. This gives us a vast quantity of data, of which we will present an informative selection. The following results will be given in terms of the co-ordinate system in figure 1, and will be given for a variety co-ordinates to demonstrate the relative uniformity of injected carrier concentrations throughout the central region of the rib structure. Unless otherwise stated for particular results, the following parameters were used for the simulations: applied voltage 0.9V; doped regions 5×10^{18} cm^{-3}; diffusion depth of n$^+$ regions 1.75μm; distance of n$^+$ regions from rib base 0.25μm; rib width 3μm. These parameters correspond to figure 1. The following results compare peak injected carrier concentration for certain parameter changes. Strictly we should compare the entire concentration profile in order to draw conclusions. For the following examples, however, the profile shapes are virtually unaffected by the changes, and it is therefore convenient to compare just the peak concentrations.

Effect of Rib Width

The width of the rib has been analysed since a variety of widths may be useful for different applications, and because there will be a fabrication tolerance. The degree to which the rib width can be varied is of course limited since we wish to maintain single optical mode operation. We have therefore considered increasing the nominal 3μm width shown in figure 1. By way of example consider increases to 4μm or 5μm. The resultant change in injected carrier concentration, for constant applied voltage is very small, although the current density must therefore, by definition, change. It may have been expected that changes in rib width and hence cross sectional area of current flow would cause relatively significant changes in injected carrier concentration. Table (i) shows that this is not the case. Two examples are given, along y = 3.2, and y = 2 (see co-ordinate system in figure 1). This enables us to modify the rib width and remain confident of achieving useful carrier injection levels.

Table (i) : Effect of rib width variation

Rib Width	3μm	4μm	5μm
peak carrier concentration along y=3.2 (cm^{-3})	2.43 x 10^{17}	2.41 x 10^{17}	2.41 x 10^{17}
peak carrier concentration along y=2 (cm^{-3})	2.6 x 10^{17}	2.54 x 10^{17}	2.52 x 10^{17}

The effect of diffusion depth

The regions of n$^+$ and p$^+$ material shown in figure 1 form the p-i-n structure of the device. The position and doping concentration of the n$^+$ regions in particular, will influence the distribution of injected charge carriers. They must also be of sufficient doping density to form ohmic contacts at the surface. The concentration of the profile will clearly vary in a real diffusion, but for the purposes of this simulation we have assumed uniform doping density throughout the region. An enhanced concentration could be used near the surface by various means, to form an ohmic contact if necessary, dependant upon the metal used on the surface. Table (ii) shows, for example, the effect upon injected carriers of varying the depth of the diffusion from 1.2μm to 1.75μm, whilst keeping the concentration constant at 5 x 10^{18} cm^{-3}.

Table (ii) : Effect of diffusion depth

Variation along y=2	Depth of diffusion	1.2μm	1.75μm
	peak injected carrier concentration	2.6 x 10^{17}	2.54 x 10^{17}
Variation along y=3.2	Depth of diffusion	1.2μm	1.75μm
	peak injected carrier concentration	2.44 x 10^{17}	2.41 x 10^{17}

If the diffusion is made even deeper to 2.5 μm, the general trend is for a slight decrease in injected carrier concentration. Table (iii) shows, for example, the variation along y = 5.5:

Table (iii) : Effect of deeper diffusion

Variation along y=5.5	Depth of diffusion	1.2μm	2.5μm
	peak injected carrier concentration	2.29 x 10^{17}	2.25 x 10^{17}

It is therefore advantageous to maintain a shallow diffusion, which also simplifies fabrication.

The effect of lateral translation of the n⁺ regions.

In figure 1 the n^+ regions are shown at a fixed distance from the base of the etched rib walls. A suitable distance is $0.25\mu m$. There will of course, be a tolerance in the fabrication of the device, and we have therefore considered the effect of lateral translation of these regions upon the injected carrier profile. Example results are as follows for a distance increase of $0.6\mu m$ from the rib base:

Table (iv) : Lateral translation of diffused regions

		$0.25\mu m$	$0.6\mu m$
Variation along x=8	Distance from rib base (μm)	$0.25\mu m$	$0.6\mu m$
	peak injected carrier concentration	2.52×10^{17}	2.51×10^{17}
Variation along y=3.2	Depth of diffusion	$0.25\mu m$	$0.6\mu m$
	peak injected carrier concentration	2.41×10^{17}	2.37×10^{17}

As expected the current density reduces slightly for the $0.6\mu m$ spacing, but it is clear from the results of Table (iv) that the distribution of injected carriers is changed little.

Injected Carrier Profiles

In order to achieve a uniform refractive index change across the waveguiding region, a uniform carrier injection profile is clearly required. In order to demonstrate the uniformity, example profiles are given in figures 2 and 3 below. Figure 2 shows the injected electron concentration along $x = 8$ (i.e. at the rib centre), and figure 3 shows the corresponding electron concentration along $y = 3.2$ over the region in which most of the optical power is concentrated. It is evident from these diagrams that the concentration is almost uniform, and this is also true over most of the region containing significant optical power.

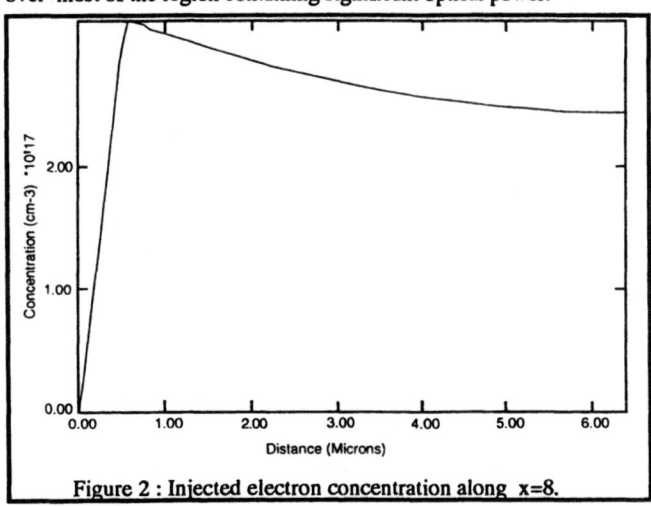

Figure 2 : Injected electron concentration along x=8.

Conclusions

We have modelled an optical phase modulator using the MEDICI semiconductor simulation package, and have optimised the device for maximum phase shift for a given applied voltage. The results demonstrate that the resultant device will have an almost uniform refractive index variation over the region in which the optical mode propagates, and that the device will be extremely efficient. Furthermore the device will be tolerant to fabrication inaccuracies.

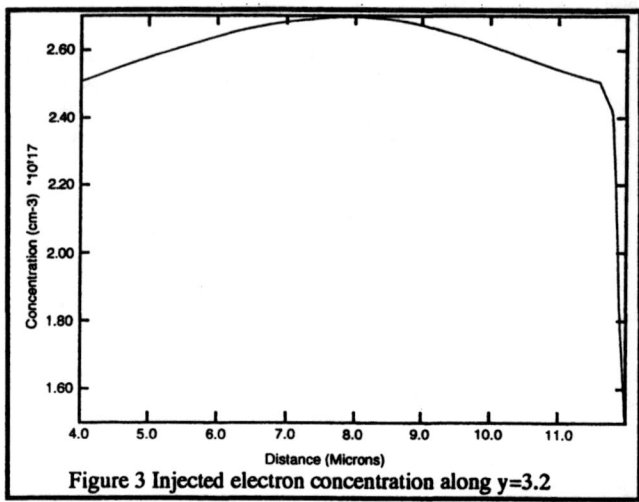

Figure 3 Injected electron concentration along y=3.2

The technique has allowed us to model refractive index changes in the region of the guided optical mode. In this case we have aimed for a uniform refractive index change, but we could also optimise for a particular index profile if required.

Acknowledgements

The authors are grateful to TMA for provision of the MEDICI software. CKT and AGR are grateful to SERC and Bookham Technology Ltd for financial support.

References

1   K J Reeson, K G Stephens, and P L F Hemment, Nucl Instrum and Methods, 290, B37/38, 1989.
2   B L Weiss et al, IEEE Photonics Technology Letters', 3, 19, 1991.
3   G T Reed, A Rickman, B L Weiss, F Namavar and R Soref, Proceedings of MRS Fall meeting, Boston, USA 1991, Symposium J, Optical Waveguide Materials, 1991.
4   R Normandin et al, Can J Phys., p833, 66, 1988.
5   G V Treyz et al, IEEE Electron Device Lett., 12, 276, 1991.
6   R Soref and J P Lorenzo,   IEEE J Quant. Elect. QE-22, 873, 1986.
7   L Friedman et al, J Appl. Phys., 63, 1831,   988.
8   R Soref and B R Bennett,   IEEE J Quant. Elect. QE-23, 123, 1987.
9   J Schmidtchen et al, Elect. Lett., 27, 1486, 1991.
10   R A Soref et al, IEEE J Quant. Elect.
11   A Alping, X S Wu, T R Hausten, L A Coldren,   Appl Phys Lett, 48, 1243, 1986.
12   R G Hunsperger, 'Integrated Optics', Springer-Verlag, 1985, ISBN 0-387-13078-0

ELECTROLUMINESCENCE EVALUATION OF THE
SiO₂-Si STRUCTURES USING AN ELECTROLYTE-SiO₂-Si CELL

A.P.BARABAN*, P.P.KONOROV*, S.A.BOTA** AND J.R.MORANTE**
*Institute of Physics of St. Petersburg University, 198904 St. Petersburg, Petrodvorets, Russia.
**LCMM, Departament de Física Aplicada i Electrònica, Universitat de Barcelona, Diagonal 647, 08028 Barcelona, Spain.

ABSTRACT

The use of an Electrolyte-SiO₂-Si system allows a detailed control of the electron injection from the electrolyte into SiO₂ layer, and makes feasible to reach the electron heating in the conduction band of SiO₂ before to take place the irreversible breakdown.

The injected and heated electrons enhance the probability of the SiO₂ defect excitation as well as the Si-O bond breaking. Both features give raise to the relaxation processes which are responsible of the electroluminescence characteristics of the oxide. So, the measured electroluminescence spectrum presents straightforward information on the defect types, their nature and possible precursor defect behaviour. Results on silicon oxides obtained from different technological processes and treatments, corroborate the above analysis and show the electro-luminescence of SiO₂ as an interesting and powerful method to evaluate the SiO₂ properties.

INTRODUCTION

Nowadays, the scaling down tendency involves working conditions connected with high electric field existence in SiO₂ films. Thus, the electron heating can easily take place depending on the SiO₂ thickness, the values of the applied electric field and the experimental conditions used, leading to a misfunction of integrated circuits operation. As consequence, the understanding of the mechanisms associated with hot electron (HE) processes and their influence on the SiO₂-Si structure properties presents great interest for physics of solid-state electron devices. For these reasons, an interesting feature is the study of the revealing process of the named predefect states and their transformation into positive charged defects [1].

However, the analysis of HE effects is not easy to perform when Metal-SiO₂-Semiconductor (MOS) structures are used. In this case, the small difference between hole and electron barriers causes simultaneous and comparable injection levels of both types of carriers. There is also the difficulty to reach oxide field (E_{ox}) values higher than 8-10 MV/cm due to the oxide breakdown. Besides the metal contact impede to detect the produced photons during the interaction of the hot electrons with defects, and it is necessary a special semitransparent metallization to obtain some information about the electroluminescence (EL) spectrum.

In this context, we present an alternative technique which is based on the use of an Electrolyte-Oxide-Semiconductor system (EOS system) to carry out the controlled injection and heating of electrons in SiO_2 films according to the applied electric field (E_{ox}), and also, to facilitate the EL measurement.

The advantages of the EOS system are related to the specific properties of the Electrolyte-SiO_2 interface. These ones are the high electron barrier, the low density of electronic states at the Electrolyte-SiO_2 interface and the electrolyte transparency in the wide spectral region. Moreover, the EOS system allows us the initial possibility to separate the processes of hole and electron injection; to submit the SiO_2 layer to high electric field regimes, even up to 30 MV/cm depending on thickness, without the appearance of oxide breakdown and, finally, to obtain easily the emitted photon spectrum. Particularly, it is possible to collect the electroluminescence spectrum, produced by the interaction of injected hot electrons with defects or with the atomic matrix of the SiO_2.

A scheme of the EOS system is presented in figure 1. The SiO_2-Si structure under test is placed in contact with an electrolyte in an electrolytic cell -we use as a rule a dissolution of Na_2SO_4 in water, with a pH between 5 and 6-, ohmic contacts to silicon are made by the help of an Indium-Gallium eutectic.

By anodic polarization of the electrolyte, carried out in the electrolytic cell, we can set up a low injection level of electrons from electrolyte into the oxide. Bear in mind that this low injection levels are necessary to prevent the apparition of disruptive oxide breakdown.

The EL spectrum emerging from the SiO_2 bulk is measured by help of a photomultiplier with Sb-Na-K-Cs cathode working in the photon account regime.

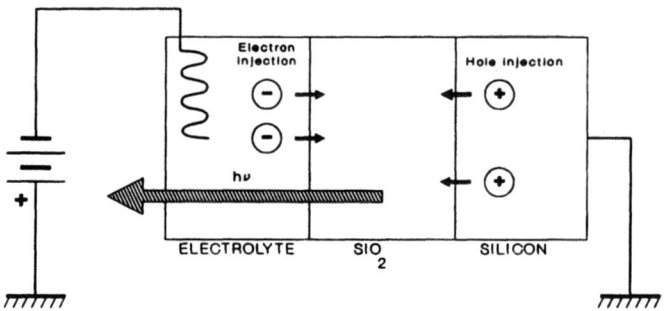

Figure 1. The Electrolyte-SiO_2-Si (EOS) system.

RESULTS

In the figure 2 we present different EL spectra measured in the EOS system, corresponding to several oxides elaborated with different oxidation technologies. Curve 1 of figure 2 corresponds to anodic oxidation, curve 2 corresponds to a dry oxidation process, curves 3,4 and 5 correspond to SiO_2/Si structures grown by wet oxidation at different temperatures (1050 °C, 950 °C and 900 °C respectively), and finally, curve 6 corresponds to dry oxidation with chlorine. Is important to remark that the initial EL spectrum essentially depends on SiO_2-Si fabrication technology and reflects the specific peculiarities of each SiO_2 film structure and defect characteristics.

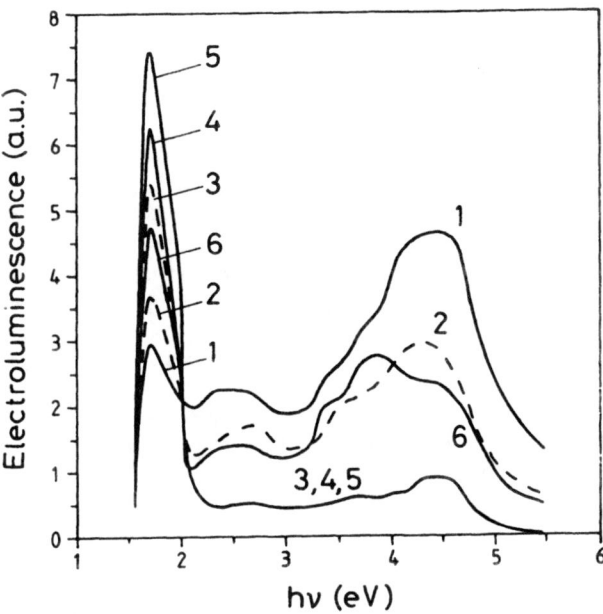

Figure 2. Electroluminescence spectral distributions of SiO_2 layers grown by help of different methods (d_{ox}=100nm).

1.   Anodic oxidation.
2.   Thermal oxidation in dry oxygen, T_{ox}=1100 °C.
3.   Thermal oxidation in H_2O, T_{ox}=1050 °C.
4.   Thermal oxidation in H_2O, T_{ox}=950 °C.
5.   Thermal oxidation in H_2O, T_{ox}=900 °C.
6.   Thermal oxidation in dry oxygen with Cl, T_{ox}=1100 °C.

These spectra consist of a series of characteristic bands. From the point of view of the formation mechanism all the bands are separated into two main groups which can be arbitrarily referred to long-wave and short-wave spectral regions.

DISCUSSION

Long-wave region

In the long-wave region (eV < 2 eV), appears a band located at an energy of 1.9 eV related with the injection process of conditional thermalization of electrons into SiO_2 layer and their subsequent radiation capture in Si-OH traps [2].

$$O \equiv Si\text{-}OH \ + \ e^- \ \rightarrow \ O \equiv Si\text{-}O^- \ + \ H \uparrow$$
$$O \equiv Si\text{-}O^- \ \rightleftarrows \ O \equiv Si\text{-}O + e^- \tag{1}$$

In the transformation process of Si-OH centres into Si-O⁻ traps, hydrogen liberation occurs [3]. The transformation is manifested in the decrease of 1.9 eV band intensity as well as in free hydrogen evolution. Here Si-O⁻ groups seem to be the basis of the defects. The Si-O⁻ centre participates actively in hole conductivity [4]. The latter is manifested by the decrease of SiO_2 layer resistance and by leakage current increase as well by the growth of the number of traps capable of influencing the charge state of SiO_2-Si structures.

The effect indicated by relation (1) is clearly demonstrated in electron injection experiments performed on a-SiO_2 films previously subjected to H_2O diffusion [4]. The most direct way to characterize this defects is via electron spin resonance methods [5] although its limited sensitivity for thin films ($d_{ox} < 500$ nm) supposes a severe handicap [6]. In the figure 2 a correlation between H_2O introduction in the oxidation process and 1.9 eV band intensity is clearly evident.

Short-wave region

The short-wave spectral region, corresponds to EL bands with energies greater than 2 eV. This region contain information related with other defects and precursor states allocated in the SiO_2 film and in the SiO_2/Si interface, a more detailed behaviour is given in figure 3. The bands presented in this figure (located at 2.3, 2.7, 3.3, 3.8 and 4.6 eV) are manifested in the fields E_{ox} greater than 6 MV/cm, resulting in electron gas heating in SiO_2 layers. General peculiarity of these bands is the existence of a threshold values of E_{ox} field which depends on the SiO_2 layer thickness.

These bands seem to be provided by intracenter transitions in atomic defects of SiO_2 matrix excited by hot electrons and in the defects of silicon interface. The 2.3 eV band is related with silicon excess in the structure, that is to say with the trivalent silicon centre. The 2.7 eV band appears in the field corresponding to the onset of SiO_2 matrix collisional ionization, leading to an intense defect formation both in the bulk of SiO_2 layer and in the SiO_2-Si interface region. The growth of the 2.7 eV band intensity is related with the decrease of glow intensity of EL bands at 2.3 and 4.6 eV (see curves 1 to 2 in figure 3).

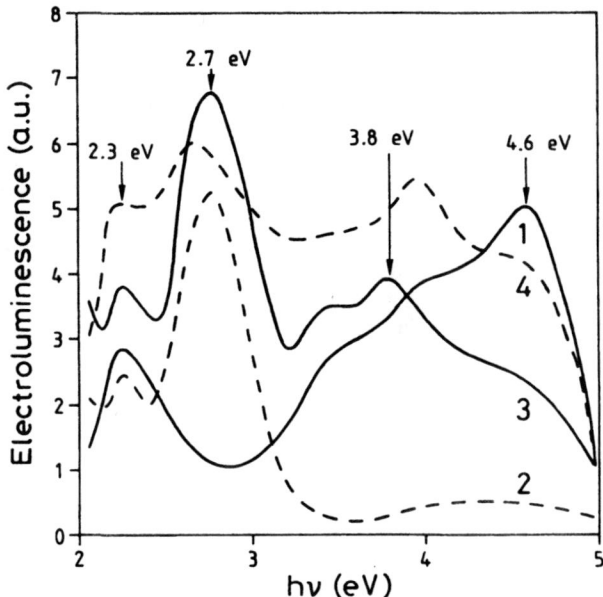

Figure 3. EL Spectral distributions of SiO$_2$-Si structures created by dry oxidation (T$_{ox}$=1100 °C, d$_{ox}$=110 nm). The curves from 1 up to 4 corresponds to the application of increasing E$_{ox}$ fields.

This is the manifestation of new channel of hot electron energy dissipation, connected with the collisional excitation of undistorted SiO$_2$ layer atomic matrix. As a possible glow mechanism in this region one can consider the radiation relaxation process following the restoration of Si-O bonds, peculiar to undistorted SiO$_2$ matrix bonds, previously broken by electrons.

CONCLUSION

The study of EL spectra in the EOS system provides wide possibilities for the study of both the electronic processes proceeding in SiO_2 layers during electron injection, and the defect character and defect formation and transformation under the action of different factors. The long-wave spectral region (characterized by an EL band at 1.9 eV) enables to judge upon the existence of the defects of silanol-group type in SiO_2 and on their transformation under electron injection, resulting in SiO_2-Si structure degradation.

The study of EL spectrum in the short-wave region allows SiO_2-Si structure testing related to the processes taking place under the action of electric field or radiation exposure. The spectrum can be used to reveal the existence of different SiO_2 centres. The presence of 2.3 eV band can be connected with an excess of silicon atoms being the source of fixed positive charge in SiO_2. The study of the EL band at 2.7 eV, permit to judge on the electron heating process effects in SiO_2 layer resulting in the transformation of the existing defects and their charge characteristic variation as well as in collisional ionization of SiO_2 matrix, leading to electron-hole pair generation and new defect formation.

REFERENCES

1. D.J.DiMaria and J.W.Stasiak, J.Appl.Phys, **65**, 2342 (1989).
2. R.J.Singh and R.S.Srivastava. J.Appl.Phys, **54**, 1162 (1983).
3. R.Gale, F.J.Feigl, C.W.Magee, and D.R.Young. J.Appl.Phys, **54**, 6938 (1983).
4. F.J.Feigl, D.R.Young, D.J.DiMaria, S.Lai and J.Calise. J.Appl.Phys, **52** 5665 (1981).
5. D.L.Griscom. Phys.Rev. **B22**, 4192 (1980).
6. R.A.B. Devine. Nucl.Instr & Method. in Phys.Res. **B46**, 244 (1990).

THE EFFECTS OF HALOGEN EXPOSURE ON THE PHOTOLUMINESCENCE OF POROUS SILICON

JEFFREY M. LAUERHAAS AND MICHAEL J. SAILOR

Department of Chemistry, The University of California at San Diego, La Jolla, CA 92093-0506

ABSTRACT

Interaction of I_2 vapor with luminescent porous silicon (PS) results in a mixed hydrogen/iodine terminated surface. The photoluminescence (PL) of PS is completely quenched upon I_2 exposure. Excess iodine can be removed by pumping on the PS for several hours, resulting in partial recovery of the original PL spectrum. Exposing the hydrogen/iodine terminated PS to air results in the accelerated growth of an oxide layer and increased PL. The adsorption of iodine is postulated to induce surface traps that are responsible for the luminescence quenching observed.

INTRODUCTION

It has recently been discovered that electrochemical and chemical etching of silicon for extended periods of time results in a material that photoluminesces at room temperature in the visible region of the spectrum.[1] The etching process produces microporous Si, and the visible PL has been attributed to quantum size effects arising from isolated nanometer-size Si features. Bulk Si, with a bandgap at 1.1 eV, is very weakly luminescent in the near infrared region of the optical spectrum.[1-4] Highly luminescent PS has generated a great deal of interest from a fundamental level and also because of the unique role that Si plays in modern technology. In this work, we report that the luminescence observed from porous Si is completely and irreversibly quenched on iodine adsorption, and present preliminary characterization of the chemistry involved in the process. This chemistry provides control over the luminescence and surface properties of porous silicon, and could be useful in the design of chemical sensors based on luminescent porous silicon.

EXPERIMENTAL

Luminescent PS samples were prepared by galvanostatically (constant current) etching n-Si (0.832 Ω-cm resistivity, (100) orientation, P-doped) in a 50-50 ethanol/HF solution a current density of 5 mA/cm^2 for 30 min., followed by a 20 sec. etch at 50 mA/cm^2. The wafers were illuminated with a 300 watt W lamp during the etch. After preparation of the PS layer, the samples were rinsed with ethanol, dried under a stream of N_2, and placed in a vacuum chamber. The chamber was evacuated to 50 mTorr and backfilled with N_2 three times. With the sample under vacuum, the wafer was exposed for one minute to gaseous iodine, at room temperature and under room lights (less than one Torr of iodine). After one minute, the iodine exposure was stopped and the chamber returned to a 50 mTorr vacuum. The wafers to be used for air exposure studies were evacuated to 50 mTorr for 15 minutes immediately after halogen exposure, and then the sample was air-oxidized at room temperature for 18 hours. The PL was monitored using a 442 nm He/Cd laser (defocused, $5mW/cm^2$) and a $1/4$-m monochrometer/CCD detector setup.

DISCUSSION

A freshly etched PS wafer is hydrogen terminated with very little to no surface oxide present (figure 1, PS(a)).

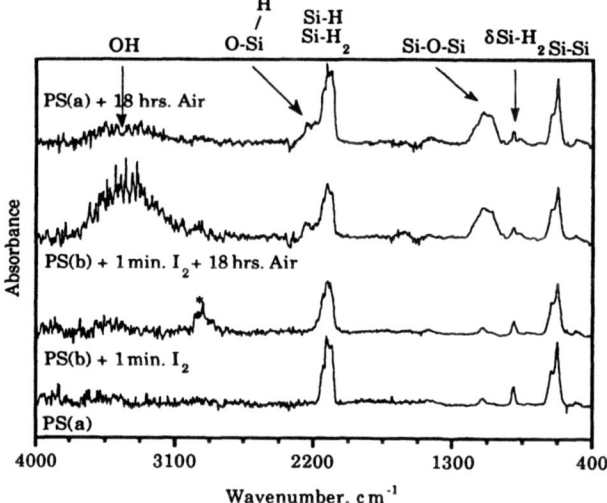

Figure 1. Transmission FTIR of freshly etched porous Si (PS(a)), porous Si exposed to one minute of iodine (PS(b) + 1 min. I_2), porous Si exposed to iodine followed by eighteen hours in air (PS(b) + 1 min. I_2 + 18 hrs. air), freshly etched porous Si exposed to eighteen hours air (PS(a) + 18 hrs. air). The vibrational modes are assigned are as follows: $v(OH)$ 3400 cm^{-1}, $v(O-Si-H)$ 2215 cm^{-1}, $v(Si-H$ and $Si-H_2)$ 2100 cm^{-1}, $v(Si-O-Si)$ 1100 cm^{-1}, $\delta(Si-H_2)$ 910 cm^{-1}, $v(Si-Si)$ 620 cm^{-1}, and (*)hydrocarbon impurity.

As shown in previous experiments, the PL of this hydrogen terminated PS is very sensitive to surface adsorbates.[5, 6] Therefore, in this study we wanted to examine the PL response of a chemically altered PS surface. To do this, a freshly etched wafer was exposed to gaseous iodine. Upon exposure, the light gray surface of the freshly etched PS wafer changes to a dark brown color characteristic of iodine. Pumping on the exposed wafer returns the color of the wafer to light gray. PL monitoring of this process reveals an immediate decrease in the PL upon exposure to iodine (similar behavior is seen for chlorine and bromine). Within seconds of exposure, the PL is reduced to baseline intensity (figure 2).

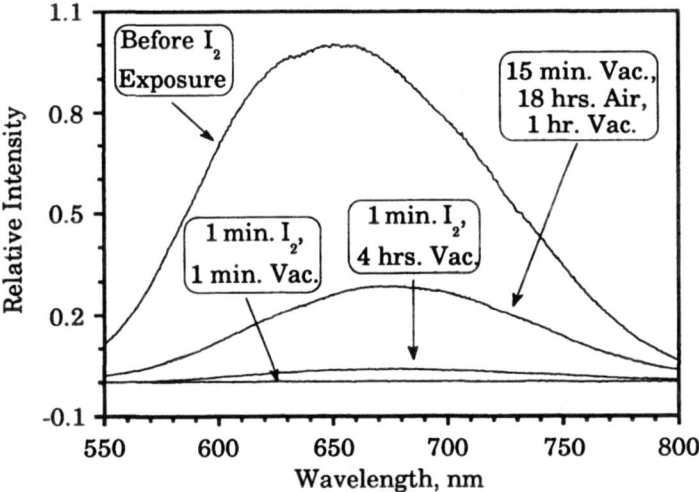

Figure 2. PL spectra of iodine exposed porous Si. Emission spectra before iodine exposure (Before I2 Exposure), immediately after iodine exposure (1 min. I2, 1 min. Vac.), after iodine and vacuum pumping for four hours (1 min. I2, 4 hrs. Vac.), and after air exposure (15 min. Vac., 18 hrs. Air, 1 hr. Vac.).

No change in the PL is observed until the iodine exposure is stopped and the chamber is evacuated. Pumping on the sample results in a slow increase in the PL, but even after long periods of evacuation the original PL intensity is not recovered. A further increase in the PL is obtained upon exposing the wafer to air. Presumably, the iodine introduces non-radiative surface traps into PS. The two step recovery of the PL can be explained by assuming that both chemisorbed and physisorbed iodine can quench the PL of PS. Pumping on the iodine exposed PS removes the physisorbed iodine, resulting in the first increase in the PL from baseline intensity. Electron spectroscopy for chemical analysis (ESCA) and electron microprobe analysis indicates the presence of chemisorbed iodine on the surface (even after long periods of evacuation). The second increase in PL occurs upon air exposure, where the chemisorbed iodide is displaced by oxygen. Oxide formation upon air exposure further increases the PL (see discussion below). Presumably the oxide passivates the surface more effectively than iodide, reducing the number of surface traps available for non-radiative recombination.

To determine the chemistry occurring on the surface of iodine exposed PS, several surface analysis techniques have been employed. ESCA shows the chemisorbed iodine is present on the surface as iodide (I (LiI; ZnI_2) $3d_{5/2}$ @ 619.9 $\pm$ 0.3 eV; Si (SiI_4) 2p @ 102.2 $\pm$ 0.3 eV). FTIR analysis shows that the Si-H stretch at 2100 cm^{-1} remains unchanged throughout the analysis, except for some broadening of the peak (figure 1, PS(b) + I_2). This indicates the iodine does not react with the hydrogen-terminated surface but attacks defect sites or Si-Si bonds on the surface of PS. The Si-O stretch at 1100 cm^{-1} increases only after the iodine exposed wafer has been exposed to air (figure one, PS(b) + I_2 + 18 hrs. air). The ESCA data reveal a surface oxide is present as well on these air-exposed wafers (O (SiO_2) 1s @ 533.0 $\pm$ 0.3 eV; Si (SiO_2) $2p_3$ broad peak @ 100.4 $\pm$ 0.3 eV).

The proposed reaction of iodine with PS is shown below (scheme 1).

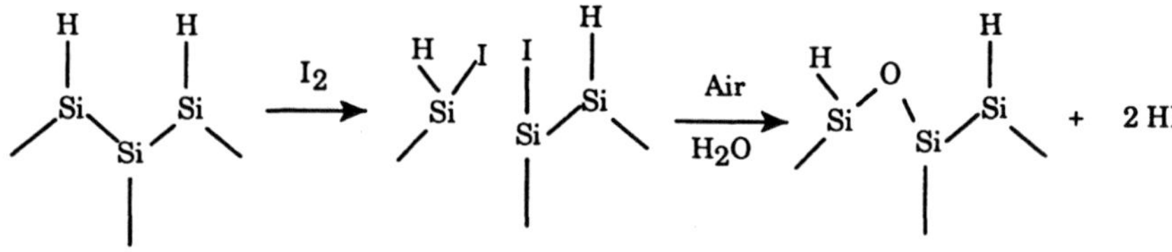

Scheme 1. Reaction of iodine with a Si-Si bond followed by hydrolysis.

This scheme is consistent with the FTIR data, which indicates that the Si-H and Si-H_2 species stay intact upon iodine exposure. The chemistry of iodine with disilanes is well known, with iodine breaking the Si-Si bond to form two Si-I bonds.[7-10] In addition, the Si-I bond in molecular compounds is very prone to hydrolysis, an observation consistent with our data.[7, 11]

This investigation has shown the extreme sensitivity of luminescent PS to the chemical nature of its surface. In the case of iodine, both chemisorbed and physisorbed iodine can quench the PL of PS. The I⁻ can then be displaced on exposure to air to form an oxide layer, with recovery of about 30% of the original photoluminescence intensity.

REFERENCES

1. L.T. Canham, Appl. Phys. Lett. **57**, 1046-1048 (1990).
2. V. Lehmann and U. Gosele, Appl. Phys. Lett. **58**, 856-858 (1991).
3. A. Bsiesy, J.C. Vial, F. Gaspard, R. Herino, M. Ligeon, F. Muller, R. Romestain, A. Wasiela, A. Halimaoui and G. Bomchil, Surf. Sci. **254**, 195-200 (1991).
4. L.T. Canham, M.R. Houlton, W.Y. Leong, C. Pickering and J.M. Keen, J. Appl. Phys. **70**, 422-30 (1991).
5. J.M. Lauerhaas, G.M. Credo, J.L. Heinrich and M.J. Sailor, J. Am. Chem. Soc. **114** 1911-1912 (1992).
6. J.M. Lauerhaas, G.M. Credo, J.L. Heinrich and M.J. Sailor, Mat. Res. Soc. Symp. Proc. **256**, 137-141 (1992).
7. D. Seitz and L. Ferreira, Synthetic Communications **9**, 931-939 (1979).
8. S.J. Band and I.M.T. Davidson, Trans. Faraday Soc. **66**, 406-409 (1970).
9. C. Chatgilialoglu, D. Griller and M. Lesage, J. Org. Chem. **54**, 2492-2494 (1989).
10. H. Sakurai, K. Sasaki and A. Hosomi, Tetrahedron Letters **21**, 2329-2332 (1980).
11. M.E. Jung and M.A. Lyster, J. Org. Chem. **42**, 3761-3764 (1977).

OPTICAL PROPERTIES OF FREE-STANDING POROUS Si FILMS

Y. KANEMITSU[*], H. UTO[*], Y. MASUMOTO[*], T. MATSUMOTO[**], T. FUTAGI[**]
and H. MIMURA[**]

[*]Institute of Physics, University of Tsukuba, Tsukuba, Ibaraki 305, Japan
[**]Electronics Research Laboratories, Nippon Steel Corporation, Sagamihara 229, Japan

ABSTRACT

We have studied optical properties of free-standing porous Si thin films fabricated by electrochemical anodization. The average diameter of Si crystallite spheres is evaluated by Raman spectroscopy and transmission electron microscopy. The blueshift of optical absorption spectrum is observed with a decrease in the average diameter of Si crystallites. However, there is no clear size dependence of the peak energy of broad photoluminescence spectrum. Spectroscopic analysis strongly suggests that the photogeneration of carriers occurs in the c–Si core whose band gap is modified by the quantum confinement effect, whereas the radiative recombination of carriers occurs in the near–surface region of small crystallites.

INTRODUCTION

Recently, a great deal of research effort is focused on nanometer–size crystallites made from Si [1,2] or Ge [3]. Especially, the strong photoluminescence (PL) from Si nanostructures fabricated by electrochemical anodization, often called porous Si, has attracted much attention from the fundamental physics viewpoint and from the interest of the potential application to optical devices. However, the origin and mechanism of visible PL in porous Si are still controversial and several models are presented. It is pointed out that the quantum confinement effect in Si nanocrystallites enhances the oscillator strength of the direct optical transitions and gives efficient radiation from porous Si [4-7]. On the other hand, with a large surface–to–volume ratio in the highly porous structure, surface localized states in Si nanocrystallites are considered to be responsible for the origin of luminescence of porous Si [8-10]. Moreover, silicon based compounds such as siloxene ($Si_6O_3H_6$) derivates [11] or the rearranged Si–Si bonds [12] formed near the surface are also proposed as an origin of the strong PL of porous Si. Although the blueshift of absorption spectra due to the quantum confinement effect is observed in a variety of semiconductor nanocrystallites [13], we have no information about the size dependence of the optical absorption spectrum in porous Si.

In this work, we have studied optical absorption and PL spectra of free–standing porous Si films. We observed for the first time that the blueshift of the optical absorption spectrum of porous Si occurs with a decrease in the average diameter of Si crystallites in porous Si. The peak energy of the PL spectrum is not sensitive to the size of Si crystallite as compared with the absorption spectrum. These results strongly suggest that the photogeneration of carriers occurs in the crystalline Si (c–Si) core with the band gap modified by the quantum confinement effect, while the visible room–temperature PL comes from the near–surface region of crystallites.

Mat. Res. Soc. Symp. Proc. Vol. 298. ©1993 Materials Research Society

EXPERIMENT

The free–standing porous silicon layers were prepared as follows. The substrates were (100)–oriented p–type silicon wafers with resistivities of 0.2, 3.5, 10, 50, and 230 Ωcm. Thin Al films were evaporated on the back of the wafers to form a good ohmic contact. The anodization was carried out in HF–ethanol solution (HF:H_2O:C_2H_5OH=1:1:2) at a constant current density of 20 mA/cm^2 for 20–60 min. After the anodization, we increased the current density abruptly and removed a porous Si layer from the Si substrate. The porous silicon films were finally rinsed in C_2H_5OH solution. The thickness of porous Si films was 20–40 μm.

Microstructures of these porous Si films were studied using transmission electron microscopy (TEM) analysis and Raman spectroscopy. TEM microscope was performed using a JEOL 2010 system operated at 200 keV. Raman scattering measurements were performed at room temperature in a backscattering configuration. Raman spectra were obtained using 514.5 nm laser light and a Raman microprobe measurement system consisting of a 25–cm filter monochromator and a 1–m monochromator. The PL spectra of porous Si films were measured in a vacuum by using 325 nm excitation light from a He–Cd laser. The calibration of the spectral sensitivity of the whole measuring system was performed by using a tungsten standard lamp. The optical absorption spectra were measured using a spectrometer with an integrating sphere in order to reduce the effects of light scattering from porous Si. Spectroscopic data were measured at room temperature.

RESULTS AND DISCUSSION

Our TEM observations of free–standing porous Si films show that porous silicon is a mixture of Si crystallite spheres and an amorphous phase. Si crystallite spheres with various diameters are observed and the typical diameter of Si crystallite spheres is several nanometers. In this work, the average diameter of Si crystallites is determined from Raman spectrum. Figure 1 (a) shows Raman spectra of free–standing porous Si films. Broad and downshift signals are observed in porous Si, as compared with those of crystalline Si (c–Si) bulk. Here, we shall evaluate the average diameter of Si crystallite spheres, L, using a spatial correlation model [14,15]. Finite–size effects relax the q–vector selection rule and this relaxation of the momentum conservation leads to a downshift and broadening of the Raman spectrum [14]. Since the strong phonon confinement expression (with a phonon amplitude of ~0 at the boundary) is usually used for the estimate of the average diameter of various nanocrystals [16], we use this model to determine the crystallite size in porous Si films. In the strong confinement model, the first–order Raman spectrum I(ω) is given by [15],

$$I(\omega) = \int_0^1 \frac{\exp(-q^2 L^2/4a^2)}{[\omega - \omega(q)]^2 + (\Gamma_0/2)^2} d^3q, \qquad (1)$$

where q is expressed in units of $2\pi/a$, a is the lattice constant, and Γ_0 is the linewidth of the Si LO phonon in c–Si bulk (~3.6 cm^{-1} including instrument contributions). We consider the dispersion ω(q) of the LO phonon in crystalline Si to be [17]:

$$\omega^2(q) = A + B\cos(\frac{\pi q}{2}), \qquad (2)$$

where $A = 1.714 \times 10^5$ cm^{-2} and $B = 1.000 \times 10^5$ cm^{-2}. By using Eqs(1) and (2), we can calculate the Raman spectrum of nanocrystallites and estimate the average diameter of Si crystallites. Solid lines in Fig.1 (a) are calculated spectra, which agree well with the experimental results. Figure 1 (b) shows the relationship between the Raman width and the Raman shift with respect to c–Si. The average diameter of crystallite spheres increases with an increase in the substrate resistivity and ranges from ~2 nm to ~10 nm. These average diameters are also supported by TEM observations.

Typical optical absorption and PL spectra of porous Si films with different crystallite sizes [(a) $L \sim 2$ nm, (b) $L \sim 3.5$ nm, and (c) $L \sim 9$ nm] are shown in Fig.2. It is found for the first time that the blueshift of optical absorption spectrum occurs with a decrease in the average diameter of Si crystallites. If we consider the porosity and the optical density of the samples, it is reliable to use a value of the photon energy at $\alpha > \sim 4 \times 10^2$ cm^{-1} as the optical absorption edge [18]. Thus, in this work, we define the optical band gap E_α as the photon energy at $\alpha = 6 \times 10^2$ cm^{-1}. The size dependence of E_α is shown in Fig.3. The solid and broken lines show the exciton energy in Si crystallite spheres calculated using an effective mass approximation [7] and the lowest energy gap of H–terminated Si spheres calculated using tight–binding approximation [19], respectively. The size dependence of E_α is roughly consistent with the theoretical calculations. The blueshift of the absorption spectrum indicates that the quantum confinement effect plays a key role in the absorption process.

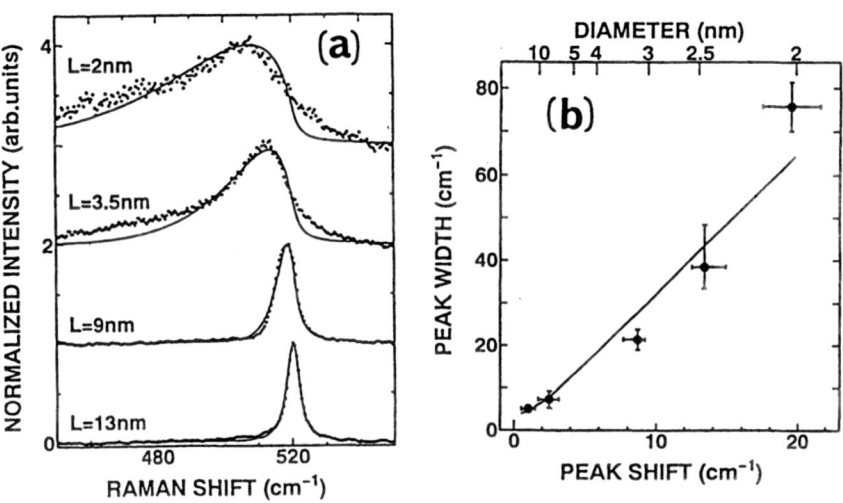

Figure 1   (a) Raman spectra of free–standing porous Si films compared with the spectra calculated for a sphere with diameter L (solid lines). The calculated spectra agree well with experimental ones.

(b) The calculated relationship between the Raman width, the decrease of the Raman peak shift with respect to c–Si. L is the average diameter of Si crystallites.

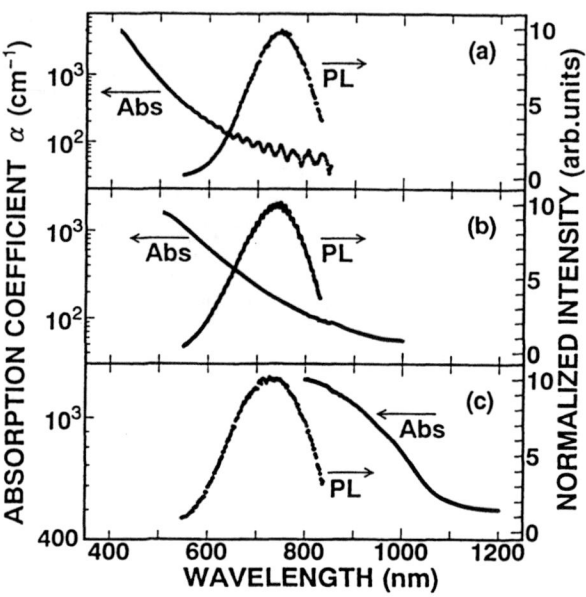

Figure 2   Typical optical absorption and photoluminescence spectra of porous Si films: (a) $L\sim2$ nm, (b) $L\sim3.5$ nm, and (c) $L\sim9$ nm.

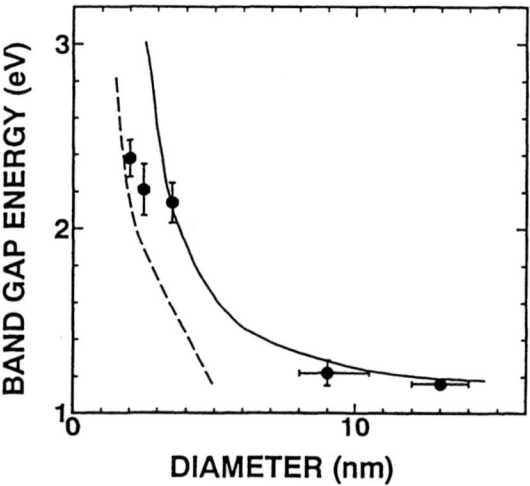

Figure 3   Band gap energy as a function of the diameter of Si crystal spheres. The solid and broken lines are theoretical calculations by effective mass and tight–binding approximations, respectively.

Figure 2 also shows the PL spectra of these films with different crystallite sizes. No significant size dependence of the PL peak energy was observed. The PL peak energy in samples with small average diameters is much smaller than the band gaps both determined from the experiment E_α and the theoretical calculations. The blueshift of the absorption spectrum and the size–independent PL peak energy imply that the site for the photogeneration of carriers are different from that for radiative recombination of carriers. Moreover, in porous Si films having large Si crystallites, there appears the unusual relationship between the absorption and PL spectra: The PL peak energy is higher than the absorption edge energy as shown in Fig.2 (c). In porous Si samples with large crystallite diameter, the PL excitation spectrum entirely differs from the optical absorption spectrum. TEM studies show that there are Si crystallites of various sizes in the sample. Thus, the unusual spectroscopic relation in Fig.2(c) strongly suggests that only small crystallites contribute to the efficient visible PL.

Considering that the surface of Si nanocrystallites in porous Si is terminated by H atoms or OH groups [8,9,11,20], we believe that the electronic properties of the near–surface region are different from those of the c–Si core. To explain the size–dependent absorption and the size–independent PL spectra, we propose a simple model that the photogeneration of carriers occurs in the c–Si core whose band gap is modified by the quantum confinement effect, whereas the radiative recombination occurs in the near–surface region. In our model, the PL efficiency in the near–surface region increases with a decrease in the crystallite size, because both the surface–to–volume ratio and the carrier transfer rate from the core to the surface increase with a decrease in the size of crystallites.

The temperature dependence of the PL intensity also support the above model. In all sample films, the PL intensity increases with raising temperature up to 100 K and then decreases. Along our model, this behavior is explained as follows: At low temperature, the carriers are generated in the core and a part of carriers in the c–Si core transfer to the near–surface region by a thermally–activated diffusion process. At high temperature, the nonradiative recombination in the near–surface region increases and then the PL intensity decreases. The PL efficiency is determined by both the thermal carrier–diffusion process from the c–Si core to the near–surface region and the radiative recombination rate in the near–surface region. The temperature dependence of the PL intensity supports that the near–surface region plays the most essential role in the size–independent visible PL process.

The experimental results obtained in this work can be explained by a simple picture that the photogeneration of carriers occurs in the c–Si core with the size–dependent band gap and the radiative recombination in the near–surface region causes the strong visible PL at room temperature. The strong PL comes from the near–surface region of crystallites with sizes less than a certain diameter. Further theoretical and experimental studies are needed to clarify the electronic structures of the luminescent near–surface region.

CONCLUSION

In conclusion, we studied the size dependence of the optical properties of free-standing porous Si films. The blueshift of the optical absorption spectrum occurs with a decrease in the crystallite size. The electronic structure of porous Si is modified by the quantum confinement effect. The photogeneration of carriers occurs in the c–Si core whose band gap is modified by the quantum confinement effect, whereas the radiative recombination occurs in the near–surface region.

ACKNOWLEDGMENTS

The authors would like to thank Drs. A. Yamamoto, T. Ogawa, K. Shiraishi, and K. Takeda for discussions. Part of this work was done at the Cryogenic Center, University of Tsukuba.

REFERENCES

[1]   L. T. Canham, Appl. Phys. Lett. **57**, 1046 (1990).

[2]   V. Lehmann and U. Gosele, Appl. Phys. Lett. **58**, 856 (1991).

[3]   Y. Kanemitsu, H. Uto, Y. Masumoto, and Y. Maeda, Appl. Phys. Lett. **61**, 2178 (1992).

[4]   S. Gardelis, J. S. Rimmer, P. Dawson, B. Hamilton, R. A. Kubinak, T. E. Whall, and E. H. C. Parker, Appl. Phys. Lett. **59**, 2118 (1991).

[5]   R. Tsu, H. Shen, and M. Dutta, Appl. Phys. Lett. **60**, 112 (1992).

[6]   T. Ohno, K. Shiraishi, and T. Ogawa, Phys. Rev. Lett. **69**, 2400 (1992).

[7]   T. Takagahara and K. Takeda, Phys. Rev. B **46**, 15578 (1992).

[8]   M. A. Tischler, R. T. Collins, J. H. Stathis, and J. C. Tsang, Appl. Phys. Lett. **60**, 639 (1992); J. C. Tsang, M. A. Tischler, and R. T. Collins, Appl. Phys. Lett. **60**, 2279 (1992).

[9]   C. Tai, K. H. Li, D. S. Kinosky, R. Z. Qian, T. C. Hsu, J. T. Irby, S. K. Banerjee, A. F. Tasch, J. C. Campbell, B. K. Hance, and J. M. White, Appl. Phys. Lett. **60**, 1700 (1992).

[10]   T. George, M. S. Anderson, W. T. Pike, T. L. Lin R. W. Fathauer, K. H. Jung, and D. L. Kwong, Appl. Phys. Lett. **60**, 2359 (1992).

[11]   M. S. Brandt, H. D. Fuchs, M. Stutzmann, J. Weber, and M. Cardona, Solid State Commun. **81**, 307 (1992).

[12]   Y. Kanemitsu, K. Suzuki, H. Uto, Y. Masumoto, T. Matsumoto, S.Kyushin, K.Higuchi, and H. Matsumoto, Appl. Phys. Lett. **61**, 2446 (1992).

[13]   See, for example, L.Brus, Appl. Phys. A **53**, 465 (1991).

[14]   H. Richter, Z. P. Wang, and L. Ley, Solid State Commun., **39**, 625 (1981).

[15]   I. H. Campbell and P. M. Fauchet, Solid State Commun., **58**, 739 (1984).

[16]   See, for example, A. Tanaka, S. Onari, and T. Arai, Phys. Rev. B **45**, 6587 (1992).

[17]   R. Tubino, L. Piseri, and G. Zerbi, J. Chem. Phys. **56**, 1022 (1972).

[18]   If we assume that the number of photons absorbed in the sample film depends on the porosity p and is proportional to $(1-p)$, the corrected optical absorption coefficient β is approximately given by $\beta = \alpha - [\ln(1-p)]/d$, where α is the absorption coefficient of the sample film determined from the transmission spectrum without the correction of porosity and d is the sample thickness. Using values of p~0.6–0.8 and d~40 μm, the correction term of $-[\ln(1-p)/d]$ estimated to be ~2–4x10^2 cm^{-1}.

[19]   S. Y. Ren and J. D. Dow, Phys. Rev. B **45**, 6492 (1992).

[20]   T. Masumoto, T. Futagi, H. Mimura, and Y. Kanemitsu, Phys. Rev. B **47** (1993) in press.

CAN OXIDATION AND OTHER TREATMENTS HELP US UNDERSTAND THE NATURE OF LIGHT-EMITTING POROUS SILICON?

P.M. FAUCHET,* E. ETTEDGUI,** A. RAISANEN,*** L.J. BRILLSON,*** F. SEIFERTH,**** S.K. KURINEC,**** Y. GAO,** C. PENG,* AND L. TSYBESKOV*

* Department of Electrical Engineering, University of Rochester, Rochester NY 14627
** Department of Physics and Astronomy
*** Xerox Webster Research Center, Webster, NY 14580
**** Department of Microelectronic Engineering, Rochester Institute of Technology, Rochester NY 14623

ABSTRACT

Using a careful analysis of the properties of light-emitting porous silicon (LEpSi), we conclude that a version of the "smart" quantum confinement model which was first proposed by F. Koch et al [Mat. Res. Soc. Symp. Proc. 283, 197 (1993)] and allows for the existence of surface states and dangling bonds, is compatible with experimental results. Among the new results we present in support of this model, the most striking ones concern the strong infrared photoluminescence that dominates the room temperature cw spectrum after vacuum annealing above 600 K.

INTRODUCTION

Since the discovery of strong light emission from porous silicon, there has been a controversy regarding the physical mechanisms responsible for the high luminescence efficiency in the visible part of the spectrum. The initial proposal[1] was that light was emitted from the columnar crystallites having a diameter below 10 nm which are produced during the electrochemical process.[2] In this very simple model, the bandgap increases well above that of crystalline silicon by quantum confinement and, for very small sizes, the indirect bandgap of crystalline silicon could become a direct bandgap due to zone folding. Thus, the facts that the luminescence is in the visible and is very strong are explained in a natural way. Shortly after this model was proposed, it was challenged by several experimental findings, some of which will be further discussed below. Many alternative models were presented in which the light emission did not require nanometer-size particles. These models attributed the light emission to the presence of an hydrogen-rich amorphous phase[3] or to various chemical complexes located on the internal surfaces of the pores, such as Si-H[4] or siloxene ($Si_6O_3H_6$).[5] At the same time, the quantum confinement model was improved thanks to several rigorous theoretical treatments of the band structure of nanometer-size columns or clusters.

After three years of intense research, we believe that the only models that have survived are those based on quantum confinement. The evidence against the alternative models is strong. All these models require a large concentration of hydrogen, yet porous silicon can be thermally oxidized at high temperature[6] or directly prepared in the electrochemical cell[7] in such a way that FTIR spectra do not detect the presence of hydrogen, despite the fact that the photoluminescence (PL) is strong. In addition, porous silicon samples freshly prepared in the dark show no evidence for the presence of oxygen,[8] and yet their PL is also very strong. Thus, all models requiring the presence of large quantities of hydrogen or oxygen are not consistent with experimental data.

In this paper, we first discuss the experimental and theoretical evidence for the quantum confinement model. When a quantitative interpretation of the experimental facts is attempted, discrepancies emerge, some of which are mentioned below. We then discuss how a "smart" quantum confinement model, first proposed by F. Koch[9] and which includes surface states and dangling bonds, can eliminate most of these discrepancies. We offer new experimental results, including on the infrared PL line, which support the model.

EVIDENCE FOR QUANTUM CONFINEMENT IN POROUS SILICON

There is strong experimental evidence for the presence of nanometer-size crystalline silicon structures in light-emitting porous silicon (LEpSi). Transmission electron micrographs (TEM) of LEpSi directly show the presence of crystalline clusters in the sub-5 nm range.[10] An analysis of the optic phonon Raman lineshape of LEpSi is consistent with the presence of ~ 3 nm Si nanocrystals[11,12] and we have recently shown that in samples prepared with light assistance, there is a correlation between the intensity of the PL and the Raman signature of such nanocrystals.[7] The strongest optical evidence for quantization comes from measurements of the transmission of free standing LEpSi films prepared with different porosities. Samples with a higher porosity start absorbing at shorter wavelengths.[13] Since increasing porosity implies smaller feature sizes, this result is consistent with an opening of the bandgap as the size decreases. Very recently, a correlation was found between the PL peak and the onset of absorption in similar measurements.[14] Surface physics tools have also been used to support the quantum confinement model. For example, X-ray photoelectron spectroscopy measurements have measured an opening of the bandgap in LEpSi that is qualitatively consistent with quantum size effects.[15]

There is also theoretical evidence for the quantum confinement model. The most widely accepted model for the electrochemical etching of porous silicon in HF relies on quantum confinement.[2] In this model, holes must reach the surface to participate in the etching process and the thinning down of the silicon "columns" stops when quantization makes the valence band discontinuity between LEpSi and c-Si so large that holes can no longer be injected from the substrate into the columns. Much theoretical work has been devoted to describing the band structure of silicon columns and clusters in the sub-10 nm size range.[16,17] Although the details vary from work to work, as the size decreases, the bandgap opens and the oscillator strength for radiative recombination increases. In order to emit red light near the bandgap, the size of the nanostructures must be < 5 nm, and below ~ 3 nm, the bandgap starts increasing rapidly with decreasing size. There is a large spread in the calculated lifetimes, but the general trend is a sharp reduction of the radiative lifetime when the size drops below 3 nm. It also appears that the bandgap does not become direct and the radiative lifetime does not drop in the nanosecond regime until sizes much below 2 nm. These results are qualitatively consistent with the blue shift of the PL peak observed (see Fig. 1) after thermal oxidation of LEpSi at > 1000 K:[6,18] in this case, oxidation of a few monolayers of silicon at the surface of the objects decreases their size and thus opens the bandgap. Although there may be other explanations, the theoretical predictions are also consistent with the spectral dependence of the PL lifetime: for a given sample, the PL lifetime increases on the red side of the PL spectrum, where larger objects are expected to luminesce. Finally, the theory also appears to be consistent with the properties of LEpSi after high temperature (> 1300 K) oxidation: PL in the blue with a nanosecond lifetime and no measurable crystalline silicon structures > 1 nm.[19]

EVIDENCE AGAINST THE PURE QUANTUM CONFINEMENT MODEL

The experimental and theoretical support for quantum confinement is strong but it is not without weaknesses. To the best of our knowledge, there has been no *direct proof* that light emission originates from the nanometer-size crystalline objects. Such a proof is extremely difficult to obtain in porous silicon, because it is a highly heterogeneous material. After oxidation, it has been shown that *isolated* silicon nanocrystals in the sub-5 nm range emit bright red light.[20] This shows that the small silicon objects present in LEpSi *can* emit light. In addition, the pure confinement model fails to explain *quantitatively* some observations and it is even in contradiction with other experimental facts. In this section, we present some of this experimental evidence against the pure quantum confinement model.

The decay of the red photoluminescence is not exponential but rather can be fitted by either two components[21] or a stretched exponential.[22] In addition, the PL decay time measured at any given wavelength within the broad PL peak is also not exponential and the characteristic times are longer at longer wavelengths.[21] This behavior is reminiscent of what is observed in

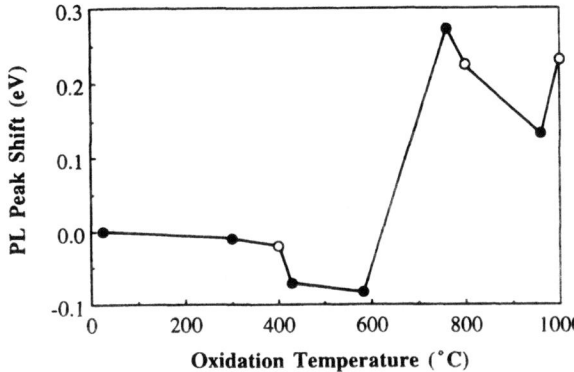

Figure 1

Shift of the red PL peak after oxidation measured in two LEpSi samples. Note the small red shift at low temperatures and the large blue shift above 1000 K.

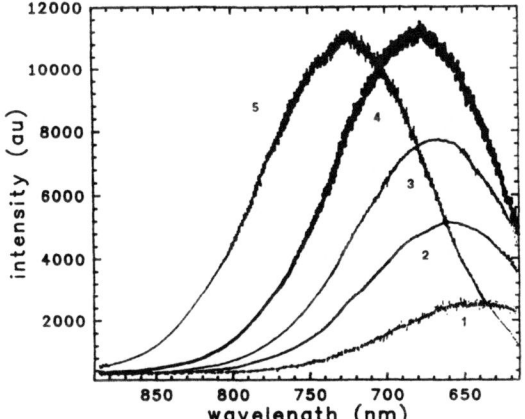

Figure 2

Evolution of the red PL peak measured in-situ when LEpSi is immersed in C_2H_5OH for (2) 1 min, (3) 3 min, (4) 10 min, and (5) 60 min. Curve (1) is for the as-prepared sample.

amorphous silicon, where the PL involves trapped carriers. In contrast, the blue PL is exponential,[19] as would be expected from "free" carriers. The PL peak red shifts during the hydrogen effusion that takes place during heat treatments.[6,18] Within the quantum confinement model, the red shift should be due to emission from larger particles and yet the particle size is unchanged. It is also possible to induce very large, reversible shifts of the PL by selected chemical treatments.[23] Figure 2 shows the evolution of the PL when porous silicon is immersed in alcohol. Such a large red shift is again incompatible with a "pure" quantum confinement model, but rather seems compatible with light emission involving states that can be affected by different species on the surface.

Although the transmission measurements of free standing LEpSi films support qualitatively the quantum confinement model, the shape of the absorption coefficient does not support it quantitatively. Indeed, it has been shown by transmission measurements[14] and by photothermal deflection spectroscopy measurements[9] that the absorption of LEpSi increases exponentially with excess energy above the position of the PL peak. Such an exponential behavior or Urbach tail is again reminiscent of what is observed in amorphous silicon. The X-ray photoelectron spectroscopy measurements that support qualitatively the quantum confinement model also fail to support it quantitatively. Indeed, the bandgap of porous silicon has been measured to be as large as 3 eV, whereas the PL peak stayed in the ~ 2 eV range.[15]

THE IR PL LINE AND THE SMART QUANTUM CONFINEMENT MODEL

A modification of the quantum confinement model was first proposed by Koch et al.[9] This "smart" quantum confinement model has several important similarities with those used to describe the optical properties of II-VI clusters. The blue PL line is associated with recombination involving "free" carriers or "free" excitons, while the red PL line is associated with recombination involving one or two carriers trapped in surface states. This explains immediately the characteristics of the PL lifetime, including the temperature dependence and the non-exponential behavior of the red line, and the temperature independence and exponential behavior of the blue line. It provides an immediate explanation for the large energy difference between bandgap and red PL peak, and for the shape of the absorption coefficient below the bandgap.

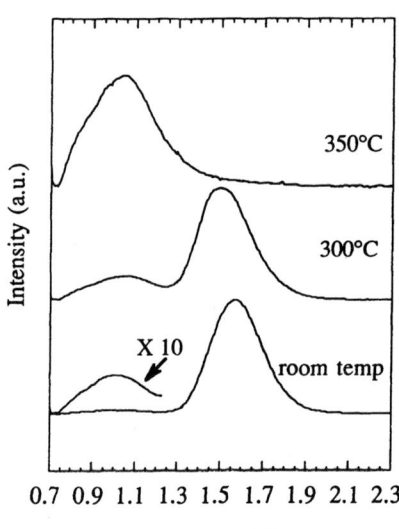

Figure 3

PL spectra measured at room temperature for a typical LEpSi sample held in UHV, before and after vacuum annealing. The IR PL line, which is only a few % of the red PL line in the as-prepared sample, eventually dominates the spectrum. The sample was held at each annealing temperature for 5 min.

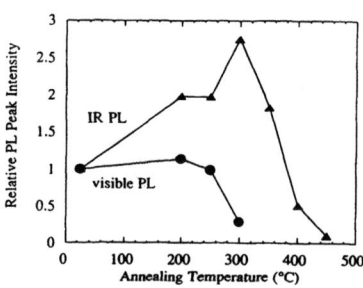
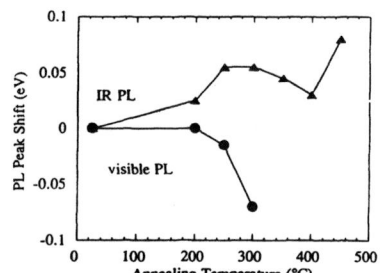

Figure 4

Evolution of the peak intensity and peak position of the IR and red PL lines as a function of the vacuum annealing temperature for the sample of Fig. 3. The IR PL intensity is maximum at the temperature where the red PL disappears and it survives much higher annealing temperatures. The red PL shifts to longer wavelengths and the IR PL shifts to shorter wavelengths with increasing annealing temperature.

A few reports of infrared PL have appeared in the literature.[24,25] These measurements were commonly interpreted as bandgap or near-bandgap PL from crystalline silicon. The "smart" quantum confinement model offers another explanation for this IR PL, in which emission is associated with mid-gap dangling bonds on the surface of the nanocrystallites. We have performed a careful measurement of the IR PL in several LEpSi samples. In our experiments, the samples are placed in a UHV chamber. The 422 nm line of a HeCd laser excites the PL, which is detected by a Ge detector, with a cutoff ~ 0.7 eV. In the results reported below, the PL spectrum is recorded at room temperature, following vacuum annealing at temperatures as high as ~ 800 K for 5 minutes.

Figure 3 shows the PL spectra for one sample before annealing and after annealing at two temperatures. The broad IR PL peak of the as-prepared sample is small but measurable at room temperature. As the annealing temperature is increased, its relative importance increases until it dominates the spectrum. Figure 4 is an analysis of these data, showing the changes in intensity and peak position of the red and IR peaks. Note the small red shift of the red PL peak that is similar to what is observed during low-temperature oxidation (see Fig. 1). These data are representative of most as-prepared LEpSi samples. We were also able to prepare samples for which the IR PL was quite different: its peak energy was ~ 0.7 eV and after annealing around 650 K, its integrated intensity measured at room temperature was comparable to the integrated intensity of the initial red PL peak. These and other results, including the changes in the PL spectrum when the measurement temperature is lowered to cryogenic temperatures, will be discussed in a future publication. From the strength and peak position of the IR PL, we conclude that it is not associated with crystalline silicon bandgap luminescence but rather likely to be produced by mid-gap dangling bonds.

CONCLUSIONS

Experimental evidence strongly suggests that a "smart" quantum confinement model can account for results that cannot be explained within a "pure" quantum confinement model. We have presented results from three experiments, thermal oxidation, chemically-induced PL shift, and vacuum annealing, which can only be explained by the "smart" quantum confinement model. The strong IR PL observed at room temperature suggests that electroluminescent devices made of LEpSi may extend to the important 1.3-1.55 μm region.

ACKNOWLEDGEMENTS

We thank J. Rehm for technical assistance, and F. Koch and V. Petrova-Koch for helpful discussions and preprints of their results. This research was supported in part by grants from Rochester Gas & Electric and the New York State Energy Research & Development Authority (PMF, CP and LT), and the Office of Naval Research (AR and LJB). C. Peng is a Link Fellow.

REFERENCES

1. L.T. Canham, Appl. Phys. Lett. **57**, 1046 (1990)
2. V. Lehmann and U. Gosele, Appl. Phys. Lett. **58**, 856 (1991)
3. D.J. Wolford, J.A. Reimer and B.A. Scott, Appl. Phys. Lett. **42**, 369 (1983)
4. S.M. Prokes et al, Phys. Rev. B **45**, 13788 (1992)
5. M.S. Brandt et al, Solid State Commun. **81**, 307 (1992)
6. V. Petrova-Koch et al, Appl. Phys. Lett. **61**, 943 (1992)
7. L. Tsybeskov et al, *"Comparative study of light-emitting porous silicon anodized with light assistance and in the dark"*, Mat. Res. Soc. Symp. Proc. **298** (this volume)
8. M.A. Tischler and R.T. Collins, Solid State Commun. **84**, 819 (1992)
9. F. Koch et al, Mat. Res. Soc. Symp. Proc. **283**, 197 (1993)
10. A.G. Cullis and L.T. Canham, Nature **353**, 335 (1991)
11. Z. Sui et al, Mat. Res. Soc. Symp. Proc. **256**, 13 (1992)
12. P.M. Fauchet and I.H. Campbell, Crit. Rev. Solid State Mater. Sci. **14**, S79 (1988)
13. P.A. Badoz et al, Mat. Res. Soc. Symp. Proc. **283**, 97 (1993)

14. Y.H. Xie et al, *"Absorption and luminescence studies of free-standing porous silicon films"*, presented at Symposium B of the Spring 1993 MRS meeting and to be published
15. T. van Buuren et al, *"X-ray photoelectron spectroscopy measurements on porous silicon"*, presented at Symposium B of the Spring 1993 MRS meeting and to be published
16. for wires, see for example G.D. Sanders and Y.-C. Chang, Phys. Rev. **B 45**, 9202 (1992) or A.J. Read et al, Phys. Rev. Lett. **69**, 1232 (1992)
17. for clusters or dots, see for example J.P. Proot, C. Delerue and G. Allan, Appl. Phys. Lett. **61**, 1948 (1992) or M. Hirao, T. Uda and Y. Murayama, Mat. Res. Soc. Symp. Proc. **283**, 425 (1993)
18. C. Peng, L. Tsybeskov and P.M. Fauchet, Mat. Res. Soc. Symp. Proc. **283**, 121 (1993)
19. D.I. Kovalev et al, *"The fast and slow luminescence bands of oxidized porous Si"*, to be published
20. K.A. Littau et al, J. Chem. Phys. **97**, 1224 (1993)
21. C. Peng et al, *"Light-emitting porous silicon after standard microelectronic processing"*, Mat. Res. Soc. Symp. Proc. **298** (this volume)
22. X. Chen, B. Henderson and K.P. O'Donnell, Appl. Phys. Lett. **60**, 2672 (1992)
23. K.-H. Li et al, *"The relationship between surface chemistry and photoluminescence of porous silicon"*, Mat. Res. Soc. Symp. Proc. **298** (this volume)
24. C. Pickering et al, J. Phys. C:Solid State Phys. **17**, 6535 (1984)
25. C.H. Perry et al, Mat. Res. Soc. Symp. Proc. **256**, 153 (1992)

Porous Silicon Nanostructure Revealed By Electron Spin Resonance

Y. Xiao*,**, T.J. McMahon*, J.I. Pankove*,** and Y.S. Tsuo*
*National Renewable Energy Laboratory, Golden, CO 80401-3393
**Dept. of EECE, Univ. of Colorado, Boulder, CO 80309-0425

ABSTRACT

Electron spin resonance (ESR) study of annealed porous Si (PS) samples shows the existence of 20% to 30% of oxygen related P_{b1} ($\cdot Si \equiv Si_2 O$) centers, in addition to the P_{b0} ($\cdot Si \equiv Si_3$) centers already reported by other groups. The ^{29}Si hyperfine and superhyperfine lines are well-defined at room temperature. We find that the nanostructure of PS, although twisted as appeared in scanning electron microscopy images, is well aligned with the underlying crystal lattice, and the ESR line broadening is mainly due to the local strain around the dangling bonds and not to the nanoscale column or particle misalignment.

INTRODUCTION

Studies using electron spin resonance (ESR) to identify the defects with unpaired spin in porous silicon (PS) have been reported by several groups recently [1-4]. Through lineshape analysis or g-value mapping, the defects are observed to be similar to the well defined P_{b0} defect at the crystalline Si/SiO_2 interface [5 and references cited therein]. The $g_{//}$ was found between 2.0015 to 2.0044 and $g_\perp$ from 2.0075 to 2.0085 [2-4]. Hyperfine (HF) lines observed from spin dependent photoconductivity [2] also agree with the characteristics of P_b centers. However, for the [110] axis the HF lines lie symmetrically about one main line, while in reality there are two lines. More recently, from room temperature ESR, two sets of nicely resolved lines for the [110] axis about each of the two principle Zeeman lines were observed [6].

In addition to identifying the P_{b0} defect, more information can be extracted by careful analysis of the ESR lines. One important feature is whether the oxygen related defect P_{b1} exists in this system, since it may alter the ESR lineshape significantly. In crystalline Si/SiO_2 interface, the appearance of P_{b1} depends on the oxidation process and is as intense as the P_{b0} lines in certain situations [5]. While in PS, the observation of P_{b1} has not been reported so far.

Another important feature is that, due to the nanoscale nature of PS, the coupling to the crystallographic orientation of bulk Si substrate may not be maintained because of misalignment due to the PS pore or column bending or twisting and the local strain around defect sites caused by bond angle distortion. The misalignment will alter the angle of dangling bonds from their crystallographic direction, therefore forming a distribution of g values [3]; on the other hand, the strain changes the local crystalline field about the dangling bond and also alters the g value. In general, both misalignment and strain can contribute to the ESR line broadening. However, the strain is believed not to contribute to the line broadening when magnetic field is aligned along the dangling bond to the first order approximation [7] and the misalignment is a main contributor to line broadening in this situation. Hence these two facts can be distinguished by studying the anisotropy of line broadening.

In this study we focus on the PS samples that were annealed at 450^0C in vacuum to drive out hydrogen from the surface. Since the annealing creates dangling bonds which quench the photoluminescence (PL), the sample after annealing is an ideal candidate for the study of dangling bonds and their relation to the PL intensity. Furthermore, the annealed samples are highly reproducible and more stable compared to the freshly etched samples [8]. More importantly, it is the initial stage for further remote-plasma processing to recover the PL and after processing the low temperature behavior of PL is also different [9, 10].

EXPERIMENTAL RESULTS AND DISCUSSIONS

All of our PS samples are electrochemically etched from float-zone p-type (100) substrates with the resistivity of 0.2 ohm-cm. Type A samples were etched in electrolyte consisting of equal amounts of ethanol and 48% HF, using a current density of 10 mA/cm^2 for 30 minutes. Type B samples were etched in 2 parts of methanol and one part of 48% HF mixture, using a current density of 1 mA/cm^2 for 90 minutes. The type A samples have lower porosity than that of type B; the PL of type A samples at the freshly etched state is relatively low and peaked in the infrared while that of type B is stronger and in the red region. The scanning electron microscopy (SEM) images shows that the microstructure of the type B samples is much more twisted in appearance than that of the type A, as shown in Figure 1. The ESR was performed at X-band with a resonance frequency near 9.4 GHz. All samples are about 0.3 cm^2 in area with the [011] edge in vertical direction. The D.C. magnetic field is in the plane perpendicular to the [011] axis and its direction is determined by the angle between the [100] direction. The total number of spins of the as-etched samples with very short time in air is undetectable ($<10^{11}$). After annealing at 450 ^{0}C for 15 minutes in vacuum, the total number of spins is about 10^{13} to 10^{14}.

The derivative ESR spectra of sample type A and B when the magnetic filed is parallel to [100] are shown in Figure 2(a). The superhyperfine (SHF) line is seen at the shoulder of the low field side while the high field shoulder is not symmetrical to that at low field and is broader. In an enlarged scale, the HF line is observed with a separation of 112 G and no Si-F related line [2] was found (Fig. 2(b)). Comparing the spectra of sample types A and B, the spectrum of B is noticeably wider with a peak-to-peak width of 5.0 G compared with 3.8 G for type A samples, and the asymmetry becomes more obvious.

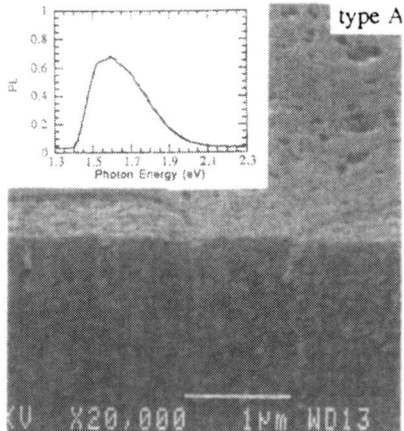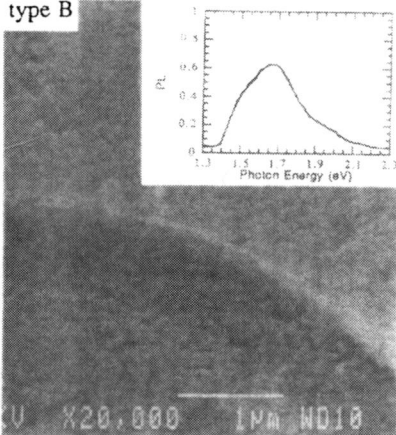

Figure 1   The SEM images of the annealed type A and B samples. The insertions are PL from the as-etched samples.

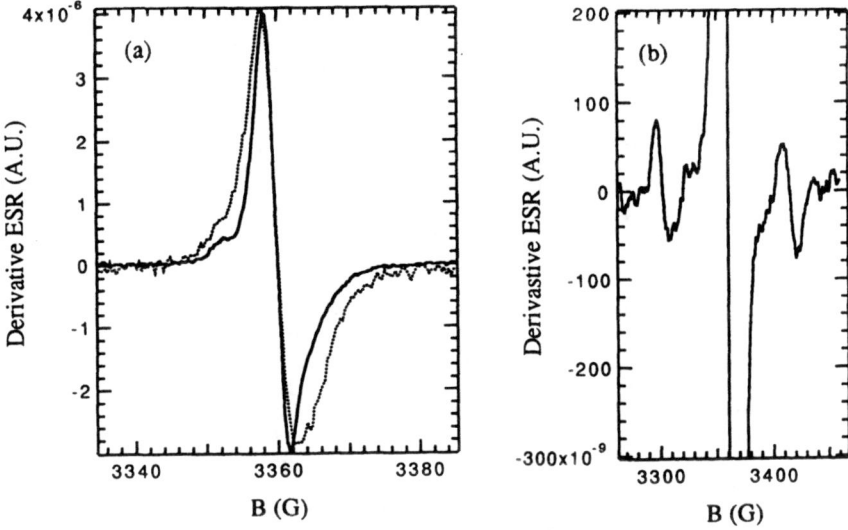

Figure 2 (a) ESR for two types of samples when the magnetic field is parallel to [100], type A: solid line; type B: dashed line, (b) HF line for a type A sample.

To reveal the cause of the asymmetry, curve fittings to the integrated ESR spectra of sample types A and B are performed and are shown in Figure 3. To the first order of approximation, Gaussian lineshape was used for all the Zeeman lines and SHF lines in the fitting. In addition to the P_{b0} Zeeman line and two SHF lines, a fourth line with a peak-to-peak linewidth the same to that of P_{b0} line had to be introduced to fit

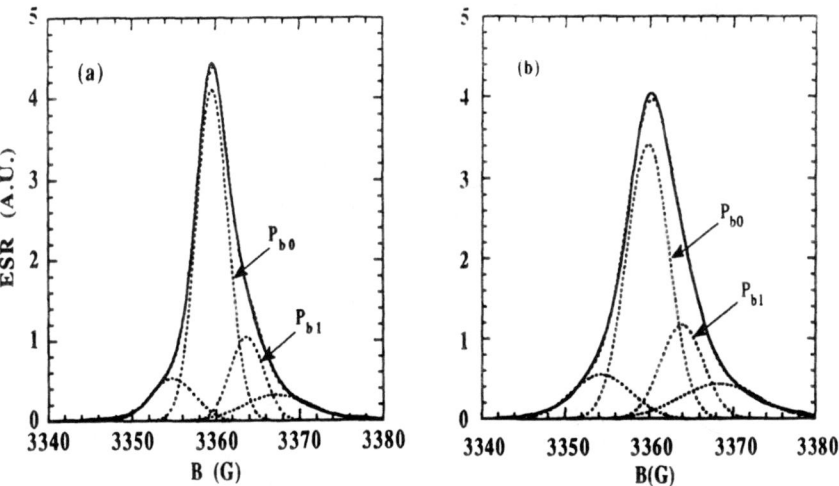

Figure 3 Fitting of ESR to sample (a) type A; (b) type B when the magnetic field is parallel to the [100] direction.

the experimental data. The best fittings were reached when the Δg between the $Pb0$ and the fourth line was about 0.0032, which is at the position of the $Pb1$ center [5]. Therefore we believe that $Pb1$ defect centers exist in our annealed porous Si and the ESR results from other angles, which will be presented later in this paper, also support this assertion. The peak-to-peak widths of the $Pb0$ lines in the fittings are close to that measured from the ESR derivative spectra. The ratios of the number of $Pb1$ to the total number of dangling bonds are about 0.2 and 0.3 for sample types A and B, respectively, which is consistent with that the ESR derivative spectra of sample type B is more asymmetric than that of type A in the high field side. The fractions of SHF in the total spectrum were obtained as 20% and 25% for sample types A and B, respectively, which are slightly larger than one would expect given that there is 4.7% of ^{29}Si in the wafer and three second-nearest neighbors contributing to the SHF line. The fitting was also performed assuming that the $Pb0$ and $Pb1$ Zeeman lines are Lorentzian in shape and it gave very similar results to that fitted with Gaussian lines. The only noticeable discrepancy is that the fraction of SHF in the total spectrum, when fitted with Lorentz line, tended to be smaller than 14.1%. The error might come from the fact that the lineshape is not purely Lorentzian or Gaussian. Instead, it is a convolution of these two components [7].

Figure 4(a) shows the ESR spectra when the magnetic field is perpendicular to the [111] direction. In this case, the $Pb1$ center should locate between the two main Zeeman lines of $Pb0$ from the g-value mapping [5]. The mixing up of two main Zeeman lines of type B samples is consistent with the fact that higher percentage of $Pb1$ exists in them than in type A samples. The $g_\perp$ obtained from the well resolved ESR of type A sample is 2.0080.

In Figure 4(b), the magnetic field is aligned with one of the tetrahedral directions. All four possible tetrahedral directions give the same ESR spectra, indicating that the number of dangling bonds at each tetrahedral direction is the same. The position of the high field parts of the ESR spectra along this direction are determined by $g_{//}$. The width of the $g_{//}$ line for low porosity type A sample is 1.8 G from the experimental curve and less than 1.5 G from the line fitting. The linewidth for

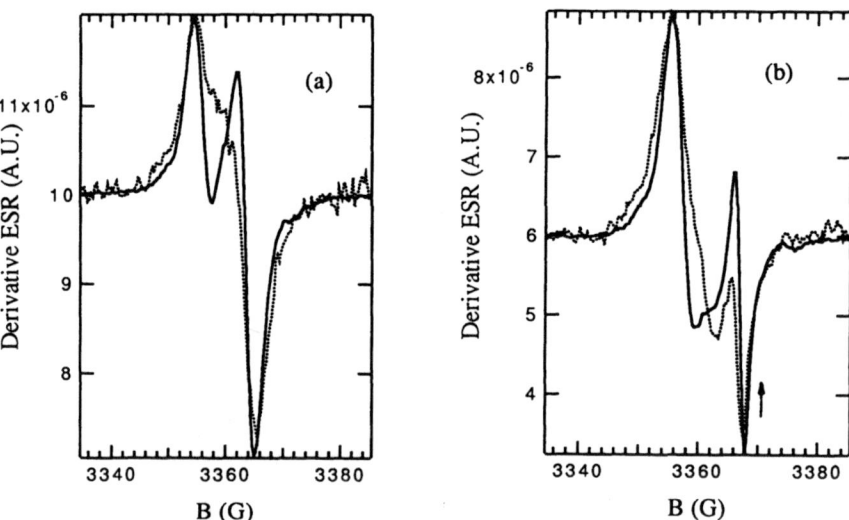

Figure 4   ESR when the magnetic field is parallel to (a) [110]; (b) [111]. Type A samples: solid line; type B samples: dashed line.

high porosity type B samples is only slightly wider than that of type A samples, especially at the side indicated by the arrow, where there is less distortion due to $Pb1$ lines being further away. Considering that the ESR peak-to-peak linewidth in the crystalline Si/SiO$_2$ interface is from 1 to 1.3 G (depending on processing conditions) [7], the linewidth from our experiment suggests that the line broadening due to strain is relatively small when the magnetic field is along the tetrahedral directions. This agrees with the assumption made by Brower [7]. The dangling bonds for both type of samples are aligned with the crystallographic direction of the substrate nearly as well as that in the crystalline Si/SiO$_2$ interface.

The last conclusion seems contradictory to the twisted image of PS shown in Figure 1, especially for type B samples. However, if the PS structure is generated by etching away interstitial materials and subsequent vacuum annealing with remaining materials and defects maintaining the substrate crystallographic direction, and if the twisted appearance is not formed by elastic bending but by forming steps or terraces as shown in Figure 5, then the experimental results from SEM and from ESR can be explained simultaneously. This structure model is extremely simplified. In reality, a more complicated structure with a higher order of crystallographic faces may have to be constructed. However, the physical essence should remain the same; that is, instead of continuous change of structure elastically, the change of structure should be discrete from one defect site to another by keeping the crystallographic orientation unchanged.

Another picture we can get is that, in directions other than [111], the line broadening is also due to the strain, assuming misalignment is isotropic along all directions. Along [100], we obtained the peak-to-peak linewidth of 3.8 G for type A samples and 5.0 G for type B samples, all of which are larger than that of 3.0 G in crystalline Si/SiO$_2$ interface. Hence, the samples with ascending strain around the dangling bonds are crystalline Si/SiO$_2$ interface, low porosity type A samples, and high porosity type B samples.

The structure feature and local strain should have important impacts to many physical phenomena for the nanostructure. The "discrete change" nature provides

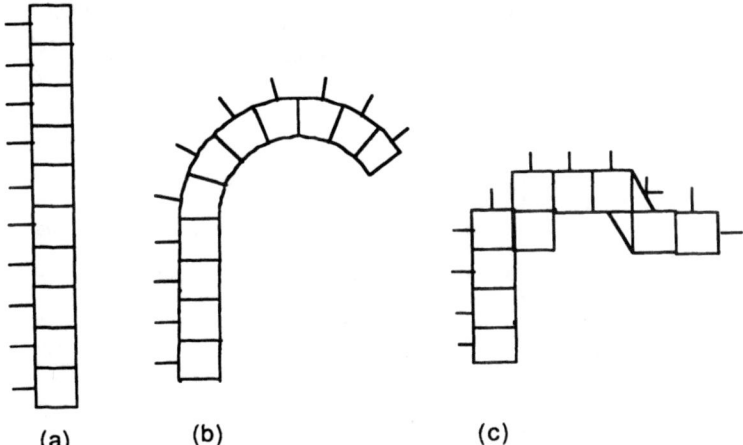

(a)   (b)   (c)

Figure 5 Schematic diagrams showing the (a) straight, (b) bending and (c) steps-terraces nanostructures of porous Si and surface dangling bonds.

more chances for an electron or hole to be confined in the structure, for example, at corners. The scattering mechanism may be different when transport properties are studied; the higher strain of the high porosity samples, which may have smaller nanostructure size, may change the energy band structure, and the oxidation process may be different with the presence of different local strain and deformation of bonding angles.

Finally, the ESR intensity vs. time was studied when both samples were exposed to air. The ESR of the high porosity sample increased rather rapidly when first exposed to air and then decreased, while that of the low porosity samples remained almost unchanged. The exact reason is still unknown, but it is speculated that this might be related to the fact that high porosity samples have more open structure and bear more stress. The continuous oxidation and relaxation process, therefore, might generate or passivate dangling bonds at different time scales. Since all of the treatment conditions are the same, we speculate that the more open structure of the type B samples is easier to be oxidized and the higher local strain or misalignment may be in favor of the formation of $Pb1$.

CONCLUSIONS

The HF and SHF lines of dangling bonds for the annealed PS were observed at room temperature. The neutral dangling bonds are very similar to that at the crystalline Si/SiO_2 interface. The possible oxygen related $Pb1$ centers were identified in our samples and there are more of the $Pb1$ type dangling bonds in the high porosity samples. The line broadening is mainly due to local strain rather than misalignment of dangling bonds from their crystallographic directions. From the line broadening analysis, the dangling bond sites of high porosity samples exhibit more strain than that of the low porosity samples. A simple nanostructure model is proposed to account for the twisted appearance observed by SEM and the highly oriented picture observed by ESR. Its impacts on other physical properties concerning nanostructure are also discussed.

ACKNOWLEDGMENT

We wish to thank R. Matson for the SEM images, and S. Froyen and S.B. Zhang for helpful discussions. This work was supported by the U.S. Department of Energy under Contract No. DE-AC02-83CH10093 through an NREL Director's Development Fund Grant.

REFERENCES

1. S.M. Prokes, W.E. Carlos, and V.M. Bermudez, Appl. Phys. Lett. **61**, 1447 (1992).
2. M.S. Brant and M. Stutzmann, Appl. Phys. Lett. **61**, 1569 (1992).
3. W.Y. Cheung, W.P. Wong, I.H. Wilson, C.F. Kan, and S.K. Hark, Mat. Res. Soc. Symp. Proc. **283**, 155 (1993).
4. J.C. Mao, Y.Q. Jia, J.S. Fu, E. Wu, B.R. Zhang, L.Z. Zhang, and G.G. Qin, Appl. Phys. Lett. **62**, 1408 (1993).
5. E.H. Poindexter and P.J. Caplan, Prog. in Surface Science, **14**, 201 (1983).
6. E.H. Poindexter, F.C. Rong, J.F. Harvey, and G.J. Gerardi, Int'l Workshop on Light Emission and Electronic Properties of Nanoscale Silicon, Charlotte, NC, Feb. 1-3, 1993.
7. K.L. Brower, Phys. Rev. **B33**, 4471 (1986).
8. M.A. Tischler, R.T. Collins, J.H. Stathis, and J.C. Tsang, Appl. Phys. Lett. **60**, 639 (1992).
9. Y. Xiao, M.J. Heben, J.M. McCullough, Y.S. Tsuo, J.I. Pankove, and S.K. Deb, Appl. Phys. Lett. **62**, 1152 (1993).
10. Y. Xiao, M.J. Heben, J.I. Pankove, and Y.S. Tsuo, Mat. Res. Soc. Symp. Proc. **283**, 317 (1993).

EFFECT OF PREPARATION CONDITIONS ON THE SILICON L-EDGE IN ELECTROCHEMICALLY PREPARED POROUS SILICON

T. VAN BUUREN[1], T. TIEDJE[1,2], W. WEYDANZ[3]

[1]Department of Physics, [2]Department of Electrical Engineering, University of British Columbia Vancouver, BC, V6T 1Z1
[3]Department of Physics, Simon Fraser University, Burnaby, BC, V5A 1S6

ABSTRACT

High resolution measurements of the silicon L-edge absorption in electrochemically prepared porous silicon show that the absorption threshold is shifted to higher energy relative to bulk silicon, and that the shift is dependent on how the porous silicon is prepared. When the porous silicon is made from n-type material with light exposure, the blue shift increases logarithmically with the anodizing current. Porous silicon prepared by anodizing p-type silicon exhibits a blue shift in the L-edge which increases with the time spent in the HF solution after the anodizing potential is turned off. The data are consistent with the quantum confinement model for the electronic structure of porous silicon.

INTRODUCTION

Electrochemical anodization of silicon in HF produces a porous material that luminesces efficiently in the visible part of the spectrum. For reasons that are still not completely understood the anodization process preferentially etches the bulk silicon at the bottom of the pores and does not attack the porous silicon (p-Si) film that is left behind on the surface as a product of the reaction. One can conclude that the etching process is self-limiting and stops when the thickness of the residual silicon reaches a certain critical value. The chemical reaction that takes place in the etching process requires holes to be available at the semiconductor electrolyte interface [1]. It has been proposed that the self-limiting behaviour of the etching process is caused by a reduction in the number of holes at the semiconductor-electrolyte interface when the dimensions of the silicon microstructures in the porous material are small enough to produce a quantum shift in the valence band [2].
Recently we have shown that the quantum confinement shift in the conduction band edge of p-Si can be determined directly from the blue shift in the silicon L-edge, measured with synchrotron radiation [3]. To test the hypothesis that the self-limiting behaviour of the etching process is due to the quantum confinement effect and to identify the preparation conditions which produce large quantum shifts, we have explored the effect of various preparation conditions on the silicon L-edge shift in p-Si.

SAMPLE PREPARATION

Porous silicon samples were prepared by anodizing both (100) p-type and (111) n-type 1-30 Ω-cm silicon wafers in a 1:1:2, $HF:H_2O$:ethanol solution. An Al contact was deposited on the back of the wafers by evaporation and annealing. This ensured even current flow in the wafer and a more uniform porous layer. The silicon substrate was pressed against an o-ring which sealed the bottom of a teflon cell filled with the etching solution and a Pt wire counter electrode.
The n-type sample preparation was carried out under illumination, as the reaction was very slow in the dark. During anodization a constant potential difference of 10 V was applied between the Pt wire and the back of the silicon substrate, with the Pt wire negative with respect to the silicon. The samples were initially prepared at a light intensity sufficient to produce a current of 0.8 mA/cm^2, for 10 minutes. This was followed by a 5 minute finishing etch in which the current was raised by increasing the light intensity by variable amounts [4]. For the 10 V bias used in these experiments the current in the electrochemical cell was approximately linear in the light intensity as shown in Fig.1. The data in Fig. 1 was obtained by illuminating the sample in the electrochemical cell by a halogen lamp from a microscope illuminator. The intensity of the light was changed with neutral density filters. Sample preparation for the p-type material is discussed in more detail elsewhere [3]. The samples were removed from the etching solution immediately after the anodization was stopped, blown dry with nitrogen and then immediately loaded into the UHV load-lock for the x-ray absorption experiments. The total air exposure time in the sample transfer

is estimated to be about 5 min. With this procedure no oxygen was detected on the surface of the p-Si either in the silicon L-edge or in the valence band photoemission spectrum [5].

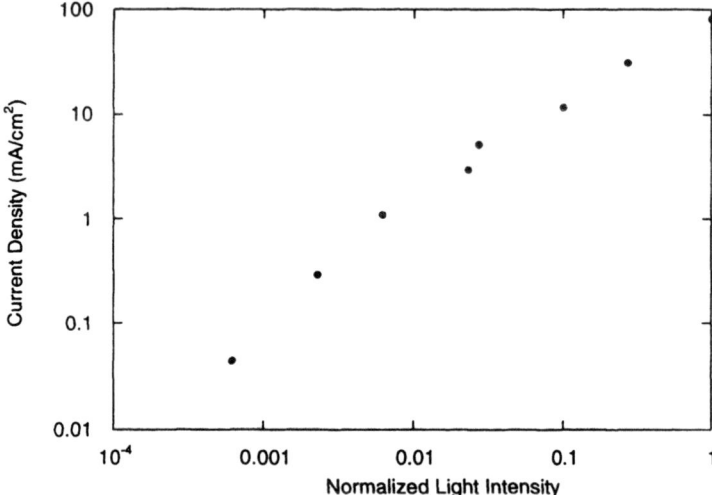

Fig.1 Current density as a function of light intensity during electrochemical etching of n-type 1-30 Ω-cm silicon at 10 V applied bias. A current density of 0.2 mA/cm^2 caused by background light has been subtracted from all current measurements.

X-RAY ABSORPTION MEASUREMENTS

X-ray absorption in the vicinity of the silicon L-edge was measured with synchrotron radiation at the Canadian Synchrotron Radiation Facility at the Synchrotron Radiation Center in Stoughton WI. The absorption spectrum is inferred from the total electron yield measured with a channel plate detector with zero bias potential on the front of the detector. The total electron yield measurements are surface sensitive to a depth of a few hundred angstroms. Typical Si L-edge absorption spectra for porous and bulk Si are shown in Fig. 2. The bulk silicon sample in Fig. 2 was rinsed in HF immediately before loading into the analysis chamber and as a result has an H-terminated surface. Notice in Fig. 2 that the L-edge in the p-Si is shifted to higher photon energy relative to the bulk silicon and that the shift increases with the current density used to prepare the material. We attribute the blue shift to quantum confinement which raises the energy of the bottom of the conduction band relative to the Si 2p core level, which is the initial state in the x-ray absorption transition. The current density indicated in the figure is the current in the final 5 min etch used to finish the sample as discussed above. In addition to a shift to higher energy the double step feature associated with the spin-orbit splitting of the 2p core level, goes away with increasing blue shift. We attribute this effect to broadening due to a distribution of quantum shifts in the inhomogeneous porous material [3].

The energy of the L-edge is obtained by extrapolating the linear part of the absorption edge just above the threshold to its intersection with the baseline formed by a linear extrapolation of the pre-edge part of the spectrum. The quantum confinement shift in the L-edge is then defined as the difference between this extrapolation for the p-Si, and for the bulk Si reference sample. The quantum shift in the L-edge can be measured with an accuracy of about ±0.03 ev in these experiments. In practice the reference spectrum for bulk Si could be obtained simply by moving the sample up or down in the analysis chamber so that x-ray beam was incident on the un-etched perimeter of the silicon wafer. In summary we interpret the shift in the silicon L-edge in p-Si relative to bulk Si as a quantum shift $\Delta\varepsilon_c$, in the bottom of the conduction band. A graph of the quantum shift as a function of the current density in the last 5 min of the anodization is shown in Fig. 3. To first approximation the quantum shift increases logarithmicly with the etching current.

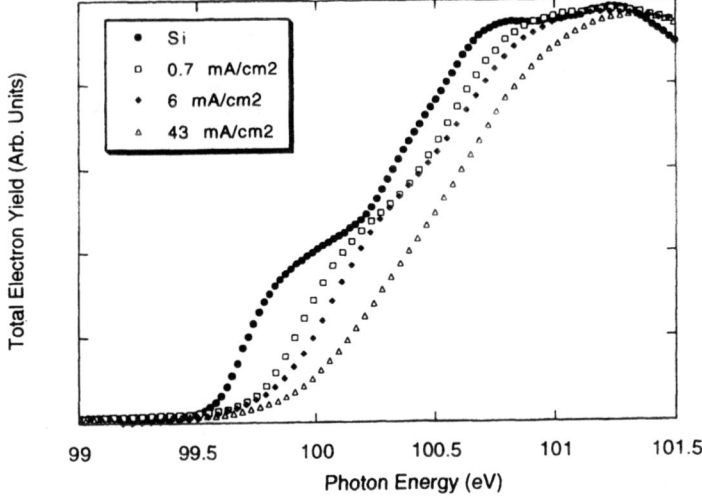

Fig.2 Total electron yield measurements of x-ray absorption at the Si $L_{2,3}$ edge for HF passivated n-type Si (111) (solid dots), and for porous silicon made from n-type Si (111) etched at 0.7 mA/cm^2 (open squares), 6mA/cm^2 (solid diamonds), and 43mA/cm^2 (open triangles).

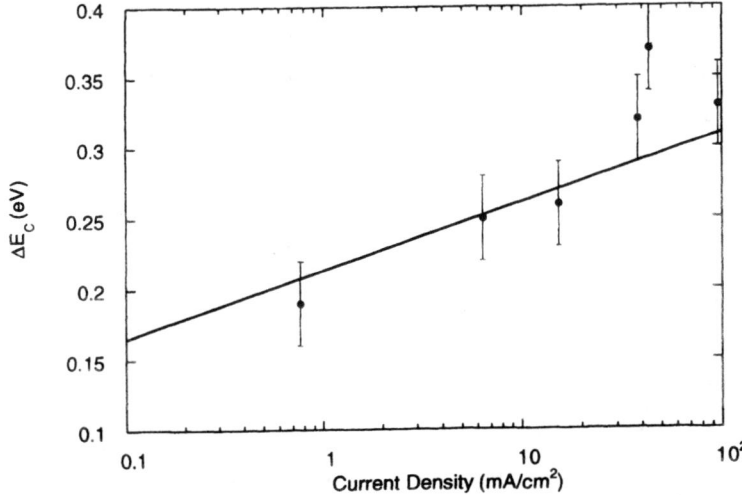

Fig.3 Plot of the conduction band edge shift (L-edge shift) as a function of the current density used to make the p-Si. The solid line is a fit to the data with a slope of 0.6 kT as predicted by the model discussed in the text.

DISCUSSION

In order to explain the observed increase in the quantum size effect with the etching current we propose a simple model, or plausibility argument, that relates the self-limiting dissolution of the p-Si to the light intensity. Although the model involves a number of assumptions that are difficult to justify, it is still instructive as a framework for thinking about the etching mechanism and as guide for further experiments. To begin with we note that there are two types of silicon in the problem: there is the bulk Si at the bottom of the pores which etches rapidly during the anodization, and there is the porous material which we believe has at least one small dimension that has an etch rate that is slow and presumeably self-limiting [1,2]. Based on earlier studies of the etching mechanism we assume that the etch rate is proportional to the density of holes at the semiconductor electrolyte interface [1]. Consistent with this assumption we further assume, as proposed earlier [2], that the quantum size effect is the self-limiting mechanism that is responsible for the survival of the porous material during the etching process. The idea is that the etching proceeds until the Si skeleton becomes so thin that there are no holes left in the valence band, because the band edge is quantum shifted away from the hole Fermi level.

In the experiment the anodization current is found to be proportional to the light intensity, as shown in Fig. 1. Since the semiconductor-electrolyte interface is under strong reverse bias we can neglect recombination in the bulk Si and assume that all of the holes produced by light absorption in the semiconductor are pulled by the applied bias to the electrolyte interface. In this case the density of holes at the semiconductor-electrolyte interface is proportional to the product of the optical flux F and the interfacial charge transfer time τ. In this case the Fermi level for holes at the interface between the bulk Si and the electrolyte (position A in Fig. 4) is given by,

$$F\tau = K_1 \exp[-(E_{fp}-E_v)/kT] \tag{1}$$

where K_1 is a constant independent of the light intensity. As shown in Fig. 4, E_{fp} is the quasi Fermi level for holes and E_v is the energy of the top of the valence band in the bulk Si.

At this point we need a way to estimate the density of holes in the p-Si. The density of holes in the p-Si will be determined by a balance between the amount of light absorbed in the porous material, the recombination rate of electrons and holes in the p-Si and the transfer rate of electron-hole pairs between the bulk Si and the p-Si [6]. Since we do not know how to balance these rates we make the ad-hoc assumption that the quasi-Fermi level for holes is constant in the p-Si and equal to the quasi-Fermi level for holes at the surface of the bulk material as shown in Fig. 4. This assumption is valid, for example, in the limit that; the electron-hole pair transfer between the p-Si and the bulk Si is rapid, the recombination rate in the p-Si is low, and the p-Si is a weak absorber of light. In this case the density of holes, p_{ps} at the surface of the p-Si (position B in Fig. 4) can be related to the hole Fermi level in the bulk silicon as follows,

$$p_{ps} = K_2 \exp[-(E_{fp}-E_{vp})/kT] \tag{2}$$

where K_2 is constant independent of the light intensity and E_{vp} is the energy of the top of the valence band in the p-Si. We have neglected band bending in the p-Si because it is surrounded by electrolyte which will tend to screen out any internal electric field along the length of the p-Si "wires".

If the illumination level is increased during the anodization there will be an increase in the density of holes at the surface of both the porous and the bulk Si, and the rate of the etching reaction will go up. However as the porous material gets smaller E_{vp} in Eq. 2 decreases due to quantum confinement which reduces the surface hole concentration and slows the etch rate. In the experimental situation where the light intensity is allowed to vary and the electrolyte composition and the etching time is held fixed, to first approximation the etching process will terminate at a fixed value of the hole concentration p_{ps} in the p-Si. Setting this critical hole concentration equal to another light intensity independent constant K_3, then we can solve Eqs. 1 and 2 for the light intensity dependence of the confinement shift, $\Delta E_v = E_v - E_{vp}$. The result is,

$$\Delta E_v = kT [\log F + \log (K_2\tau / K_1K_3)] \tag{3}$$

The quantum shift measured in the experiment is actually the conduction band shift ΔE_c and not the valence band shift ΔE_v. However effective mass theory and photoemission measurements of the quantum shift in the valence band show that [5,7],

$$\Delta E_c = 0.6 \, \Delta E_v \qquad (4)$$

Taking this additional factor of 0.6 into account we obtain the solid line shown in Fig. 3, as the predicted current dependence of the quantum shift in the conduction band. The solid line in Fig. 3 has been shifted along the vertical axis to match the data. The slope is in reasonable agreement with the data.

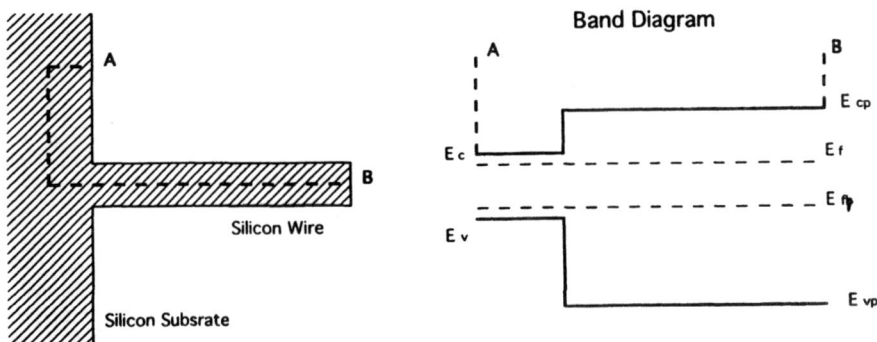

Fig.4 Schematic diagram showing the interfaces between the p-Si and the bulk Si, and between both types of Si and the electrolyte. An energy band diagram along the line connecting A to B is also shown.

The model predicts that there will be a maximum quantum shift in the valence band in p-Si, since the hole Fermi level cannot be made to go below the top of the valence band with reasonable dc light intensities. Thus one would expect the bandgap of Si to be the upper limit on the quantum shift in the valence band for p-Si produced by the electrochemical method. The data in Fig. 3 is consistent with this limit if we convert the conduction band edge shift into the equivalent valence band shift, using Eq. 4.

Larger quantum shifts can be obtained by soaking the p-Si in HF, as shown in Fig. 5. The soaking was carried out in room light. The results in Fig. 5 show that the HF continues to etch the porous silicon very slowly even in the absence of an external potential as has been reported earlier [8]. At very long etch times the apparent quantum shift in Fig. 5 begins to decrease with time. This is likely caused by the complete dissolution of the p-Si and exposure of the bulk Si underneath. If the p-Si consisted of two dimensional Si sheets, rather than wires, the maximum quantum shift in Fig. 5 might correspond to a Si sheet, that is a single atomic layer thick. Using the 0.6 rule from Eq. 4, the data point in Fig. 5 with the largest quantum shift would correspond to a material with a bandgap of 3.1 eV, in reasonable agreement with theoretical estimates of the bandgap of single atomic layers of Si terminated with hydrogen [9].

CONCLUSIONS

In conclusion we find that x-ray absorption measurements at the Si L-edge are a sensitive probe of quantum confinement effects in the conduction band in p-Si, and that these shifts depend on how the p-Si is prepared. Measurements of the Si L-edge are a useful way to probe the effects of different preparation procedures on the electronic structure of p-Si. A model has been proposed to explain the light intensity dependence of the quantum shift in anodized n-type p-Si.

288

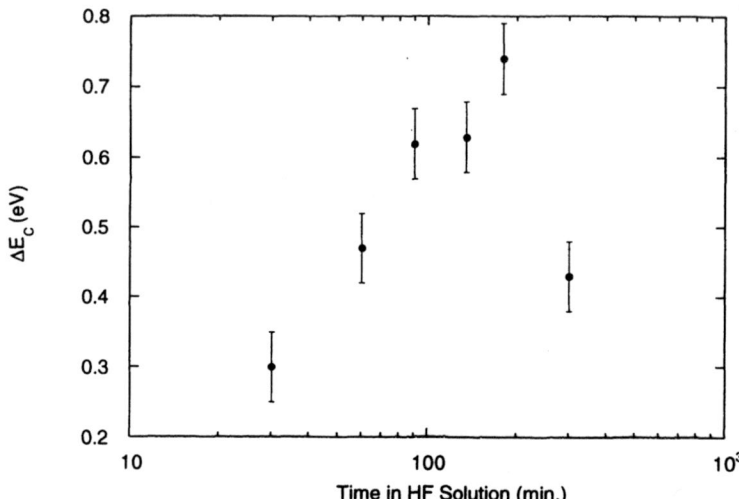

Fig.5 Plot of conduction band shift as a function of the time the porous silicon is soaked in HF solution after anodization. A p-type Si (100) sample was initially anodized in a 50% HF ethanol solution at a constant current density of 20 mA/cm^2 for 20 minutes. After 5 hours in solution the conduction band begins to shift back toward crystalline silicon. This is interpreted as the onset of complete dissolution of the porous silicon and exposure of the bulk silicon underneath.

ACKNOWLEDGEMENTS

We thank J. R. Dahn and B. M. Way for helpful discussions and the Natural Sciences and Engineering Research Council of Canada for financial support.

REFERENCES

1. X. G. Zhang J. Electrochem. Soc. **138**, 3750 (1991); P. C. Searson, J. M. Macauly, F. M. Ross, J. Appl. Phys. **72**, 253 (1992).
2. V. Lehmann, U. Gosele, Appl. Phys. Lett. **58**, 856 (1991).
3. T. van Buuren, Y. Gao, T. Tiedje, J. R. Dahn, B. M. Way, Appl. Phys. Lett. **60**, 3013 (1992).
4. V. V. Doan, M. J. Sailor, Science **256**, 1791 (1992).
5. T. van Buuren, T. Tiedje, J. R. Dahn, B. M. Way (to be published).
6. N. Noguchi, I. Suemune, Appl. Phys. Lett. **62**, 1429 (1993).
7. M. Voos, Ph. Uzan, C. Delalande, G. Bastard, A. Halimaoui, Appl. Phys. Lett. **61**, 1213 (1992).
8. L. T. Canham, Appl. Phys. Lett. **57**, 1046 (1990).
9. C. G. van der Walle, J. E. Northrup Phys. Rev. Lett. **70**, 1116 (1993).

NO CORRELATION BETWEEN POROUS SILICON PHOTOLUMINESCENCE AND SURFACE HYDROGEN SPECIES

S.M. GEORGE, M.B. ROBINSON, AND A.C. DILLON,
Dept. of Chemistry and Biochemistry, Univ. of Colorado, Boulder, CO 80309

Abstract
    The photoluminescence (PL) of porous silicon has been attributed to quantum confinement, amorphous silicon, or surface species such as hydrogen, polysilanes or siloxene. Our research has tested the early claims that surface hydrogen is responsible for PL. Our initial studies examined the effect of thermal annealing on surface hydrogen and PL *in situ* in an ultrahigh vacuum chamber. The results showed that the PL decreased between 450-550 whereas H_2 was desorbed from surface SiH_2 species between 500-575 K. There was no *direct* correlation between the PL and the loss of SiH_2 surface species. Our most recent investigations have monitored PL and surface hydrogen species as a function of HF etching time for electrochemically anodized porous silicon samples that were *not* initially photoluminescent. While the surface hydrogen species continually decreased versus HF etching time, the photoluminescence did not appear until *after* HF etching times of 20-80 minutes depending on initial sample porosity. These results again illustrated that there is no direct correlation between the PL and surface hydrogen species.

Introduction
    Porous silicon is an intriguing material consisting of silicon nanostructures that can be prepared by electrochemically anodizing single crystal silicon in solutions of hydrofluoric acid (HF) (1,2). In 1990, Canham surprised the scientific community by revealing that porous silicon can become photoluminescent after additional HF etching (3). The formation of small silicon structures in the porous layer is hypothesized to give rise to quantum confinement effects and the PL of porous silicon(3). However, the possible mechanisms for this photoluminescence has been extensively debated and include quantum confinement (3-6), the presence of an amorphous silicon layer (7) or surface species such as hydrogen or polysilanes (8-11) and siloxene (12,13).
    Under our preparation conditions, porous silicon has a high surface area enhancement of 400 times that of single crystal silicon (14). Following preparation and storage in vacuum the surface of porous silicon is terminated with *only* hydrogen in the form of SiH, SiH_2 and SiH_3 species (14) The hydrogen desorption from these hydrogen species has been monitored in ultrahigh vacuum (UHV) by transmission Fourier Transform Infrared (FTIR) studies (14). Previous vacuum FTIR investigations have also revealed that hydrogen passivates the porous silicon surface and that the surface is extremely reactive following hydrogen removal (15-17). The hydrogen-passivated porous silicon surface is also susceptible to chemical contamination and instability of the hydrogen-passivated porous silicon surface has led to some confusion regarding the surface species that may influence photoluminescence. For example, the freshly prepared porous silicon surface contains no oxygen; the siloxene mechanism (12,13) should never have been seriously considered.
    Our investigations have sought to explore the relationship between surface hydrogen species and photoluminescence. In one experiment, photoluminescent porous silicon was annealed in vacuum while PL intensity and hydrogen surface species were monitored as a function of annealing temperature. Thermal annealing of photoluminescent porous silicon leads to a degradation of the PL at temperatures that are close to the temperatures for H_2 desorption from the SiH_2 surface species. Earlier thermal annealing results (8-10) reported that PL intensity depended on hydrogen coverage. Our investigations have shown that the photoluminescence is not *directly* dependent on the SiH_2 surface species (18). Recent thermal annealing studies have demonstrated that the PL decreases at temperatures of 450-500 K and does not scale with H_2 desorption from the SiH_2 surface species at 500-550 K (18).

demonstrated that the PL decreases at temperatures of 450-500 K and does not scale with H_2 desorption from the SiH_2 surface species at 500-550 K (18).

In a second experiment, the role of hydrogen surface species was investigated as PL was "turned on" with etching in HF. Initially, non-photoluminescent electrochemically anodized porous silicon was progressively etched in hydrofluoric acid. Infrared absorbance from silicon surface hydrogen species and PL intensity were monitored versus HF etching time (19). These measurements established the relationship between the SiH, SiH_2, and SiH_3 surface species and the photoluminescence. The results again demonstrated that there is no direct correlation between surface hydrogen species and the photoluminescence (19).

Experimental

The experimental apparatus and procedures used to fabricate porous silicon have been previously reported (14). For the HF etching experiments, Si(100) boron-doped crystals with a resistivity of $\rho = .4\ \Omega$ cm were electrochemically anodized in concentrated hydrofluoric acid (48%). This anodization occurred at a current density of 10 mA/cm^2 for 45 seconds. The porous silicon sample were non-photoluminescent as prepared. For thermal annealing studies Si(100) boron doped crystals ($\rho = .2\ \Omega$ cm) were anodized in a 4:1 hydrofluoric acid (48%) : ethanol solution at 100 mA/cm^2 for 15 seconds. After anodization, samples were soaked in concentrated HF for 4-6 hours to yield a bright orange/red PL.

A Nicolet 740 spectrometer and MCT-A and -B liquid nitrogen cooled detectors were employed for the FTIR studies. The 365 nm line of an Oriel 100 W Hg lamp was used to illuminate the samples for the PL studies of thermally annealed porous silicon. HF etched samples were excited with a 325 nm HeCd Omnichrome laser. Emitted light was spectrally analyzed with a SPEX 1/4 m monochromator and a Hamamatsu R928 photomultiplier tube.

The UHV chamber utilized for thermal annealing has been described elsewhere (18,20). Thermal annealing studies were performed in vacuum by increasing the temperature of the sample by increments of 10-20 K with resistive heating. The sample was held at the desired temperature for one minute and then returned to 320 K to record the FTIR and PL spectra. This procedure allowed comparisons between temperature, hydrogen coverage, and PL to be made concurrently *in situ* in high vacuum.

For HF etching experiments, the electrochemically prepared porous silicon samples were not initially photoluminescent. These samples were passively etched in a concentrated HF solution (48%) for periods from 15 minutes to several hours. The etching cycles were performed continuously for periods of 16 to 24 hours to avoid oxidation of the porous silicon surface that occurred with storage in air. The FTIR and PL measurements were taken consecutively after the samples were dried in air for approximately 2 min. following each HF etching period.

To average over the slight spatial inhomogeneities in the PL, the HeCd laser beam was expanded to illuminate an area of approximately .6 cm^2 on the porous silicon sample. The expanded beam was also attenuated to prevent laser-induced degradation of the porous silicon that could affect the PL intensity measurements (21-24). A reproducibility of ± 4% in the PL intensity was obtained for samples repeatedly removed and repositioned in the sample holder. These measurements were used to determine the relationship between PL and SiH, SiH_2, and SiH_3 surface species.

Results

Changes in the infrared and photoluminescent spectra versus annealing temperature are shown in Figs. 1 and 2. Fig. 1 shows the decrease and corresponding red-shift of the SiH_2 scissors mode upon annealing between 320 and 570 K. Fig. 2 shows the concurrent decrease and red-shift of the PL intensity over the same temperature range. The normalized integrated areas of the SiH_2 infrared absorbance and the PL

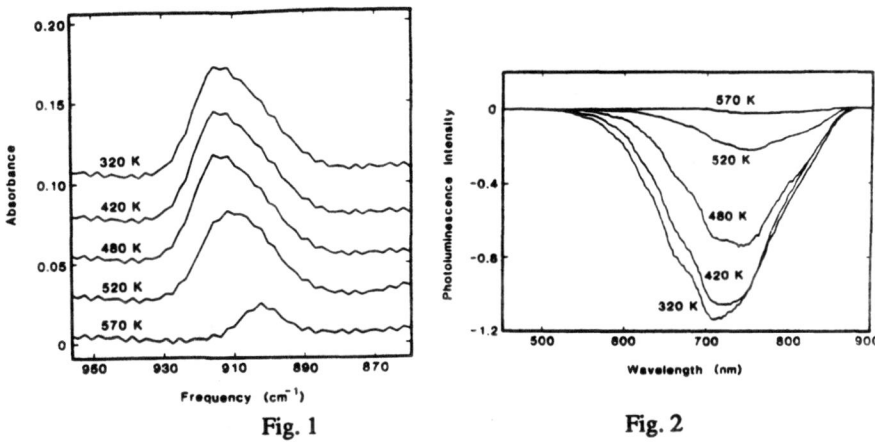

Fig. 1   Fig. 2

intensity are displayed in Fig. 3 The PL intensity begins to decrease by approximately 440 K, whereas H_2 desorption from SiH_2 species does not begin until 470-480 K. Notice that the PL is entirely quenched by 570 K, although 20% of the SiH_2 species still remain on the surface.

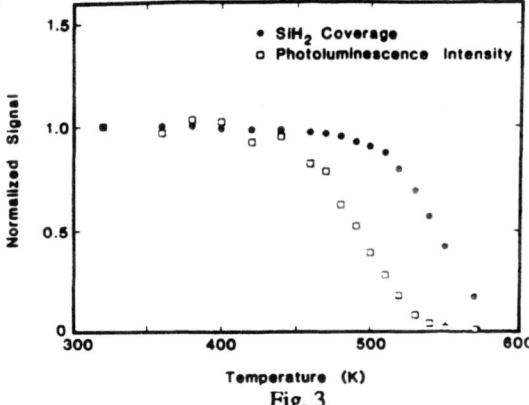

Fig. 3

Fig. 4 shows the PL spectra versus HF etching time of porous silicon samples that are initially non-photoluminescent. After ~100 minutes, the PL intensity appears and is centered at approximately 740 nm. The PL intensity then increases rapidly and blue-shifts versus HF etching time. After reaching a maximum at 150 min., the PL intensity progressively decreases and is eventually extinguished after longer HF etching times.

Fig. 5 shows the absorbance of the SiH_x stretching modes at different HF etching times from 0-610 minutes. These infrared results reveal that the integrated absorbance of the silicon-hydride surface species decreases steadily versus passive HF etching. The integrated infrared absorbance is completely eliminated after long HF etching times that depend on the initial sample porosity.

The absorbances of the SiH, SiH2, and SiH3 species in the SiHx stretching region shown in Fig. 5 were simultaneously fit using three Gaussian lineshapes. The infrared absorbances of SiH, SiH2, and SiH3 stretching modes were obtained by integrating the Gaussian lineshapes and are displayed in Figure 6 versus HF etching time. In addition, the integrated photoluminescence intensity versus HF etching time is given for comparison.

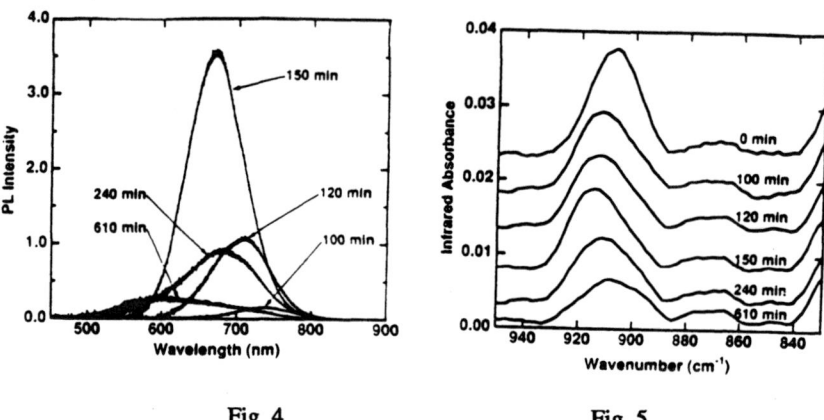

Fig. 4   Fig. 5

Discussion

These experiments raise serious doubts for PL mechanisms directly attributed to SiH2 surface dihydride species or polysilanes (8-11). Thermal annealing studies indicate that the disappearance of PL is not directly correlated with SiH2 surface species coverage. Earlier studies that claimed that the PL was directly related to the SiH2 species were not performed on the *same* sample with *concurrent* FTIR and PL measurements (8-11). The differences between the thermal histories, handling, and environments of the different samples may have precluded an accurate comparison between PL and SiH2 surface species.

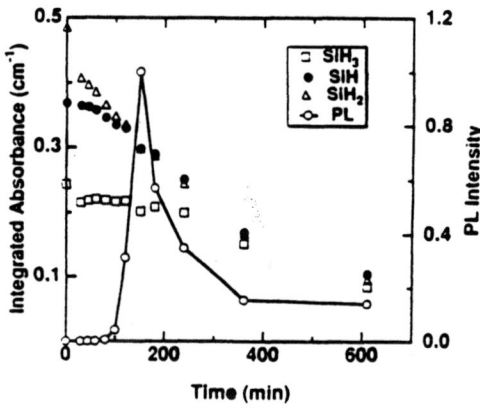

Fig. 6

If the PL loss is not directly related to H_2 desorption from SiH_2 species, what mechanism causes the disappearance of PL intensity at 450-500 K? The PL may be extremely sensitive to non-radiative trapping by surface states that may be induced by thermal annealing (25). Recent electron paramagnetic resonance (EPR) experiments reveal that oxidation creates a measurable dangling bond concentration in PL porous silicon (23). These dangling bonds may serve as surface traps for non-radiative recombination because the EPR investigations demonstrate that the PL intensity and lifetime decrease with oxidation and dangling bond concentration (23).

H_2 desorbs with second-order kinetics from SiH_2 species on porous silicon with E_{des} = 43 kcal/mole and υ_2 = 4.7 x 10^{-2} cm^2/sec (14). Given these kinetics and an estimated initial dihydride coverage of Θ = 2.3 x 10^{14} cm^{-2}, approximately 0.001% of the SiH_2 species will desorb H_2 during a one minute anneal at 475 K (14). This small increase in the dangling bond density upon hydrogen desorption, or the reconstruction of the surface to lower the surface energy (26) could be the mechanism for the degradation of PL intensity with thermal annealing.

HF etching experiments on non-photoluminescent porous silicon samples also reveal that there is no direct correlation between the photoluminescence and hydrogen surface species on porous silicon. At t=0 in Fig. 5, the infrared absorbances of the SiH_x stretching modes are at their maxima and the corresponding integrated photoluminescence intensity is zero. Immediately after the first HF etching periods, the hydrogen species begin to decrease and the PL intensity is still zero. After 100 minutes, the absorbance from the hydrogen surface species continues to decrease progressively as the PL intensity begins to grow. Following the PL intensity maximum after 150 minutes of HF etching, the PL intensity and hydrogen surface species both decrease prior to their disappearance after long HF etching times > 600 -1300 min.

A porosity of ~ 50-60 % is estimated for our initially non-photoluminescent electrochemically anodized porous silicon samples (27). The porosity increases following passive HF etching at a silicon etch rate of ~18Å/hr (28,29). The total surface area will also decrease as the diameters of the silicon structures decrease during HF etching. Assuming an initial distribution of cylindrical silicon columns, the surface area is A = $2\Pi\Sigma r_j \ell_j$, where r and ℓ are the radius and length of the columns. If a fraction of the columns have an initial diameter of ~80 -100 Å, the diameter of these columns would be reduced to ~40 Å after 150 minutes of HF etching time. This size is less than the silicon exciton diameter of 49 Å (30) and in the size regime required for quantum confinement in silicon nanostructures (3).

Initial silicon columns with diameters of 100 Å would be completely etched away and the PL would be extinguished after HF etching times of > 200-300 min. In contrast, Figs. 4 and 6 reveal that the PL is observed for HF etching times between 200-800 min. These PL results suggest that the initially anodized porous silicon contains a distribution of size structures with various diameters. Silicon structures with different initial sizes would be etched to diameters of 40-50 Å for quantum confinement behavior at various HF etching times . Based on the presence of a small PL signal after 1300 minutes of HF etching, the largest silicon nanostructures are probably about 800 Å in diameter assuming HF etch rates of 18Å/min.

The results of the investigation of PL and infrared absorbance of silicon hydrogen surface species versus HF etching and thermal annealing are consistent with a quantum confinement mechanism for PL from porous silicon. These results add to a growing body of evidence that favors quantum confinement instead of a surface species mechanism. Surface coverage is important to passivate the surface and prevent the formation of surface traps for nonradiative recombination. Given the high surface-to-volume ratios of nanostructured materials, the surface coverage may also influence the photoluminescence and electroluminescence properties.

This work was supported by the Office of Naval Research under Contract N00014-92-J-1353.

References

1. A. Uhlir, Bell. Sys. Tech. **35**, 333 (1956).
2. D.R. Turner, J. Electrochem. Soc. **105**, 402 (1958).
3. L.T. Canham, Appl. Phys. Lett. **57**, 1046 (1990).
4. A.G. Cullis and L.T. Canham, Nature **353**, 335 (1991).
5. A. Bsiesy, J.C. Vial, F. Gaspard, R. Herino, M. Ligeon, F. Muller, R. Romestain, A. Wasiela, A. Halimaoui and G. Bomchil, Surf. Sci. **254**, 195 (1991).
6. I. Sagnes, A. Halimaoui, G. Vincent and P.A. Badoz, Appl. Phys. Lett. **pre-print**, (1992).
7. J.M. Perez, J. Villalobos, McNeill, J. Prasad, R. Cheek, J. Kelber, J.P. Estrera, P.D. Stevens and R. Glosser, Appl. Phys. Lett. **61**, 563 (1992).
8. C. Tsai, K.-H. Li, J. Sarathy, S. Shih, J.C. Campbell, B.K. Hance and J.M. White, Appl. Phys. Lett. **59**, 2814 (1991).
9. C. Tsai, K.-H. Li, D.S. Kinosky, R.-Z. Qian, T.-C. Hsu, J.T. Irby, S.K. Banerjee, A.F. Tasch, J.C. Campbell, B.K. Hance and J.M. White, Appl. Phys. Lett. **60**, 1700 (1992).
10. S.M. Prokes, J.A. Freitas and P.C. Searson, Appl. Phys. Lett. **60**, 3295 (1992).
11. S.M. Prokes, W.E. Carlos and V.M. Bermudez, Applied Physics Letters **61**, 1447 (1992).
12. P. McCord, S.L. Yau and A.J. . Bard, Science **257**, 68 (1992).
13. M.S. Brandt, H.D. Fuchs, M. Stutzmann, J. Weber and M. Cardona, Solid State Comm. **81**, 307 (1992).
14. P. Gupta, V.L. Colvin and S.M. George, Phys. Rev. **37**, 8234 (1988).
15. P. Gupta, A.C. Dillon, A.S. Bracker and S.M. George, Surf. Sci. **245**, 360 (1991).
16. A.C. Dillon, P. Gupta, M.B. Robinson, A.S. Bracker and S.M. George, J. Vac. Sci. Technol. A **9**, 2222 (1991).
17. A.C. Dillon, M.B. Robinson, M.Y. Han and S.M. George, J. Electrochem. Soc. **139**, 537 (1992).
18. M.B. Robinson, A.C. Dillon, D.R. Haynes and S.M. George, Appl. Phys. Lett. **61**, 1414 (1992).
19. M.B. Robinson, A.C. Dillon and S.M. George, Appl. Phys. Lett. **62**, 1493 (1993).
20. M.B. Robinson, A.C. Dillon, D.R. Haynes and S.M. George, Mat. Res. Soc. Proc. **256**, 17 (1991).
21. C. Tsai, K.-H. Li, J.C. Campbell, B.K. Hance and J.M. White, J. Elect. Mat. **21**, 589 (1992).
22. I. Suemune, N. Noguchi and M. Yamanishi, Jpn. J. Appl. Phys. **31**, L494 (1992).
23. M.A. Tischler, R.T. Collins, J.H. Stathis and J.C. Tsang, Appl. Phys. Lett. **60**, 639 (1992).
24. R.T. Collins, M.A. Tischler and J.H. Stathis, Appl. Phys. Lett. **61**, 1649 (1992).
25. M.G. Bawendi, P.J. Carroll, W.L. Wilson and L.E. Brus, J. Chem. Phys. **96**, 946 (1992).
26. T. Ito, T. Yasumatsu, H. Watanabe and A. Hiraki, Jpn. J. Appl. Phys. **29**, L201 (1990).
27. M.I.J. Beale, N.G. Chew, M.J. Uren, A.G. Cullis and J.D. Benjamin, Appl. Phys. Lett. **46**, 86 (1985).
28. S.M. Hu and D.R. Kerr, J. Electrochem. Soc. **114**, 414 (1967).
29. W. Kern and C.A. Deckert In *Thin Film Processes*; J. L. Vossen and W. Kern, Ed.; Academic Press: New York, 1978; pp 401.
30. J.P. Wolfe, Phys. Today **March**, 46 (1982).

UNIMPORTANCE OF SILOXENE IN LUMINESCENT POROUS SILICON AS DETERMINED BY NEXAFS & EXAFS

S.L. Friedman,[†] M.A. Marcus, D.L. Adler, Y.-H. Xie, T.D. Harris, and P.H. Citrin
AT&T Bell Laboratories, Murray Hill, NJ 07974

ABSTRACT

Near-edge-- and extended--x-ray absorption fine structure measurements, as well as luminescence excitation and emission spectra, were obtained from samples of porous Si and siloxene. Contrary to a recently proposed explanation for the room temperature luminescence in porous Si, the combined data indicate that siloxene is not principally responsible for the observed effect.

INTRODUCTION

The phenomenon of room temperature luminescence from anodically grown porous silicon (por-Si) holds important promise for potential Si-based light emitting devices. Its origin, however, is still currently under debate. Proposed mechanisms for the luminescence include quantum confined structures,[1,2] surface SiH_x,[3] amorphous Si,[4] and siloxene.[5,6] The latter three materials may be thought of generically as "chemically confined" structures.

This paper reports studies of the local chemical and structural environment of Si atoms in siloxene and porous Si prepared under a variety of conditions. Techniques used include near-edge-- and extended--x-ray absorption fine structure. The data indicate that within the region of porous Si responsible for optical activity, namely the <1000Å penetration depth of the photoluminescence exciting radiation,[7] Si is *not* coordinated to oxygen. With the addition of transmission electron microscopy and emission/excitation measurements, we rule out siloxene luminescence as the mechanism generally responsible for the observed effect in porous Si.

EXPERIMENTAL

The samples were prepared by anodic oxidation of B-doped Si wafers(>50Ω resistivity).[8] Aluminum was deposited on the back of the wafers to facilitate electrical contact. The electrolyte was 30% by volume concentrated HF (49%) in ethanol and the cathode was a Pt stirrer running at about 50 rpm. Two types of porous Si were prepared: one etched at a current density of 25 mA/cm^2 for 12 min and one at 50 mA/cm^2 for 6 min. We refer to the samples by their luminescence colors under UV light, "red" and "orange" respectively. Following etching, rinsing, and drying, the samples were transferred to an Ar-filled container where they were mounted on the holders used in the measurement vacuum chamber. After etching, the sample's total exposure to air was less than 5 min. Cross-section electron microscopy showed that these samples had porous layers of the order microns thick.

The Si K-edge absorption measurements conducted on the AT&T X-15B beamline,[9] at the National Synchrotron Light Source, were obtained using InSb(111) monochromator crystals located downstream from a harmonic-rejecting mirror. Total electron and *KLL*-Auger electron yield detection schemes, with effective sampling depths of <1000Å and ~25Å[10,11] respectively, were used to ensure that the measurements correspond to the optically relevant regions of our porous Si samples. No substantive differences in data were observed between these two detection schemes. Reference measurements using total yield were also obtained from crystalline Si, fused silica, and siloxene prepared according to Wöhler's description.[12] Siloxene data was obtained from both unannealed samples as well as samples annealed[5] at 400°C for 10 minutes in either air, vacuum or nitrogen.

RESULTS

The features in the x-ray data within ~20eV from threshold, referred to as the near-edge x-ray absorption fine structure (NEXAFS), are shown for all the samples in Fig. 1. The prominent peak at ~1845eV in SiO_2 arises primarily from Si $1s \rightarrow O\ 2p^*$ transitions. It is shifted to an energy ~6eV higher than the bulk Si edge due to the more positively charged Si in SiO_2. This peak serves as a sensitive fingerprint for identifying and estimating the amount of SiO_2-like species present in other samples. For example, the data in Fig. 1 from bulk Si show a small peak at ~1845eV due to the native oxide layer present on the surface of that sample. Data from atomically clean Si exhibit a similar feature at this energy;[10] however, due to the absence of the oxide layer the integrated intensity is only about half as large as that in Fig. 1. The thickness of the native oxide layer can be determined by subtracting the intensity of this intrinsic Si feature from the 1845eV peak in the oxide-coated Si data. Using an effective total-yield detection sampling depth of ~500Å,[11, 12] an oxide layer thickness of ~25±10Å is obtained, consistent with typical values.

The NEXAFS from siloxene shows features at energies corresponding to those measured in both the Si and SiO_2 samples. There is also a peak at an intermediate energy of ~1842eV, indicating that the siloxene sample is not simply a mixture of Si and SiO_2. This peak is likely the Si^{+1} species assigned in earlier photoemission work.[13] Note that to avoid clutter in Fig. 1, only the spectrum from unannealed siloxene has been shown. In Fig. 2 we compare data from a siloxene sample before and after annealing. Apart from a barely detectable shoulder at ~1840eV in the unannealed material (undoubtedly due to H-coordinated Si; at 400°C most of the H desorbs), the two spectra are qualitatively similar.

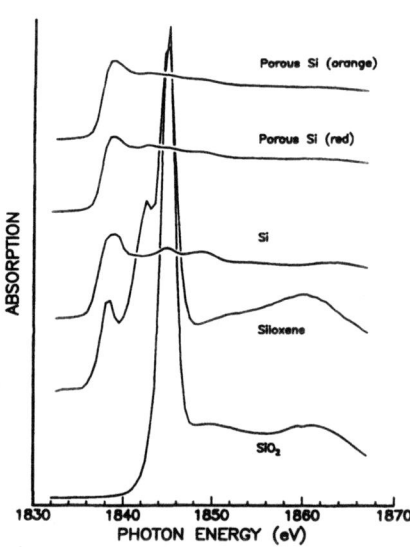 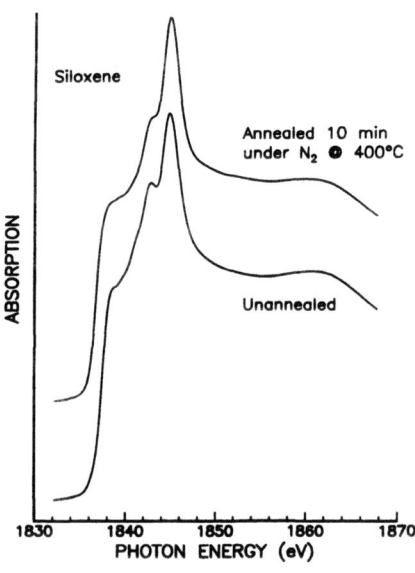

Fig. 1 Si K-edge NEXAFS data from three model systems and two different porous Si samples. The large and shifted $1s \rightarrow O\ 2p^*$ transitions in SiO_2 and siloxene are very small in the porous Si samples.

Fig. 2 Si K-edge NEXAFS data from annealed and unannealed siloxene.

NEXAFS

The lack of prominent peaks in the porous Si data at either ~1845 or ~1842eV indicates that there is very little siloxene in these samples. Before ruling out the importance of siloxene in the luminescence of porous Si, we first quantify the amount of siloxene present, and then evaluate the impact of such an amount. The following analysis of the NEXAFS data places an upper limit on the concentration of siloxene in the porous Si samples. Consider the possibility that our measured siloxene spectrum may contain a component due to some (artifactual) amount of SiO_2-like species in the siloxene sample. Furthermore, assume that if this component were subtracted from the measured data in just the right way, the resulting spectrum of "pure" siloxene would fortuitously mimic the NEXAFS spectrum from the porous Si samples. Fig. 3 illustrates this analysis. The bottom spectrum is the siloxene data reproduced from Fig. 1. The next three spectra represent the results of subtracting

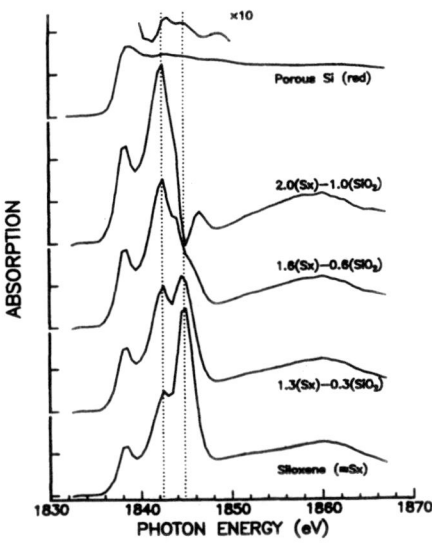

Fig. 3 Analysis using difference spectra to establish upper limit of siloxene (Sx) concentration in porous Si.

increasingly larger fractions (~19%, ~27%, and 50%) of the SiO_2 spectrum from the siloxene spectrum. The overall amount of Si contributing to each spectrum is shown to be the same, i.e., the Si K-edge absorption has been normalized to the same value at an energy >1870eV where the local bonding has little effect on absorption. The red porous Si spectrum and a 10x enlargement of the ~1840-1848eV region are provided at the top of Fig. 3 for comparison (the orange porous Si data are very similar). The difference spectrum most closely resembling the selected region would lie somewhere between 19-27% of subtracted SiO_2. In a manner similar to the procedure for estimating the native oxide layer thickness in bulk Si discussed above, we see that in the region ~1840-1848eV the absorption intensity of the best matched difference spectrum is at least 20 times the absorption in the same region of the porous Si spectrum. We conclude that *no more than 5%* of the Si in the sampled region of the porous Si sample is in the form of siloxene. Note that this is an *upper limit*, for we assumed that *all* structure in the ~1840-1848eV region is due solely to a hypothetical siloxene species and that *none* of the observed structure in the porous Si spectrum is intrinsic to that material.

EXAFS

The data in the region of >50eV above the absorption edge also support the conclusion that there is less than 5%, if any, siloxene present in the porous Si samples. Analysis of the extended x-ray absorption fine structure (EXAFS) establishes the existence or absence of Si-O bonds, which if different from the Si-O bonding found in SiO_2 or siloxene might go undetected in the NEXAFS data. EXAFS measurements,[14] in contrast, are sensitive only to positions of atomic cores, not to details of unfilled valence-derived orbitals. Fig. 4, generated with standard analysis procedures,[14] shows the Fourier-transformed EXAFS data (uncorrected for phase shifts) from spectra of Fig. 1. The first-neighbor Si-O and Si-Si peaks in SiO_2 and bulk Si, appearing at ~1.4 and 2.0Å respectively, are both present in siloxene, as expected.[5] There is, however, no evidence for Si-O bonds in the porous Si data.[15]

Optical Studies

In view of the low concentrations of siloxene found in our porous Si, measurements of the relative quantum efficiency of siloxene and porous Si were obtained. Siloxene, if much more efficient in luminescing than porous Si, could conceivably account for the reported observations[5] even in very small concentrations. To this end, we compared the luminescence emission and excitation spectra of ZnS powder (a high efficiency standard) to our porous Si samples and to siloxene prepared by a variety of methods, including annealing at 400°C.[5] The data were collected at 10° off normal from samples excited at normal incidence in an optically thick geometry using a commercial xenon-lamp fluorometer. All excitation and emission efficiencies were corrected for instrument parameters.

The emission intensities for all the porous Si *and* unannealed siloxene samples were found to be very *comparable*, with small variations observed depending on sample preparation method. The annealed siloxene samples produced significantly lower emission intensities. No sample was greater in yield than ~10% of the ZnS yield (the quantum yield of ZnS was >50% relative to a standard Rhodamine-B quantum converter). From this essential equivalence in yield, along with the 5% limit of siloxene determined above, we might conclude that the emission from porous Si cannot arise from the direct optical excitation and subsequent emission of siloxene. However, this does not preclude the possibility of light absorption first by finely divided Si, followed by energy transfer to siloxene and then subsequent emission by siloxene. To asses this possibility, we compare both the emission and excitation spectra of porous Si and siloxene, shown in Fig. 5. The almost mirror-image shapes of the siloxene spectra and the small Stokes shift between them are reminiscent of molecular luminescence for localized excited states.[16] In contrast, the spectral shapes of porous Si are quite different and they exhibit an extremely large Stokes shift, characteristic of absorption by extended states and emission by trap of localized states. While no evidence exists to prove such localized states in porous Si are specifically not due to siloxene, there are numerous documented cases of trap emission spectra from a variety of other Si containing systems.[17] Therefore, taken separately either the yield data or the mismatched spectral signatures argue against assigning porous Si emission to siloxene; taken together they effectively rule out the assignment of porous Si emission to direct absorption and emission from siloxene.

CONCLUSION

In conclusion, we have shown that the recently proposed explanation of siloxene being responsible for room temperature

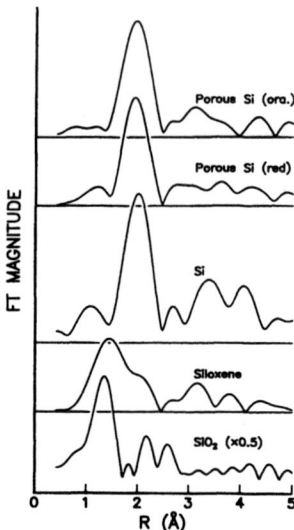

Fig. 4 Fourier-transformed Si K-edge EXAFS data from same systems in Fig. 1. The shorter first-neighbor Si-O bonds in SiO_2 and siloxene, seen as peaks at ~1.4Å, are not observed in the porous Si data, which exhibit only Si-Si first-neighbor bonds, seen as peaks at ~2.0Å.

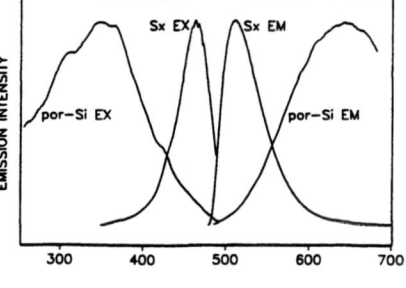

Fig. 5 Luminescence intensity of porous Si and siloxene (Sx) as a function of excitation (EX) and emission (EM) wavelength.

luminescence in porous Si is contradicted by our data. Further work is required to decide which of (or whether) the other proposed luminescence hypotheses is (are) appropriate.

The authors gratefully acknowledge R. Hull and D. Bahnck for the electron microscopy measurements. The x-ray absorption measurements were performed at the National Synchrotron Light Source, Brookhaven National Laboratory, which is supported by the Department of Energy, Division of Materials Science and Division of Chemical Sciences.

REFERENCES

†now at Stanford University, Dept. of Applied Physics, Stanford, CA 94305-4090

1. L.T. Canham, Appl. Phys. Lett. **57**, 1046 (1990).

2. V. Lehmann and U. Gösele, Appl. Phys. Lett. **58**, 856 (1991).

3. S.M. Prokes, O.J. Glebocki, V.M. Bermudez, R. Kaplan, L.E. Friedersdorf, and P.C. Searson, Proc. Mater. Soc. **256**, 107 (1992).

4. R.P. Vasquez, R.W. Fathauer, T. George, A. Ksendzov, and T.L. Lin, Appl. Phys. Lett. **60**, 1004 (1992).

5. M.S. Brandt, H.D. Fuchs, M. Stutzmann, J. Weber, and M. Cardona, Solid State Comm. **81**, 307 (1992).

6. P. Deák, M. Rosenbauer, M. Stutzmann, J. Weber, and M.S. Brandt, Phys. Rev. Lett. **69**, 2531 (1991).

7. D.E. Aspnes and A.A. Studna, Phys. Rev. B **27**, 985 (1983). For 350nm radiation, which is near the peak of excitation efficiency for porous Si, the penetration depth in crystalline Si is <100Å. In lower density porous Si this value is between 2-5 times larger, depending on the degree of sample porosity.

8. Y.-H. Yie, W.L.Wilson, F.M. Ross, J.A. Mucha, E.A. Fitzgerald, J. M. Macaulay, and T.D. Harris, J. Appl. Phys. **71**, 2403 (1992).

9. A.A. MacDowell, T. Hashizume, and P.H. Citrin, Rev. Sci. Instrum. **60**, 1901 (1989).

10. F. Comin, L. Incoccia, P. Lafarde, G. Rossi, and P.H. Citrin, Phys. Rev. Lett. **54**, 122 (1985).

11. T. Guo and M.L. denBoer, Phys. Rev. B **31**, 6233 (1985). Escape depths for total yield EXAFS experiments are actually smaller than those quoted in this work, because elastically scattered electrons do not effectively contribute to the coherent EXAFS signal.

12. F. Wöhler, Lieb. Ann. **127**, 275 (1863).

13. J.A. Wurzbach, in Proceedings of the International Topical Conference on the Physics of MOS Insulators, edited by G. Lucovsky, S.T. Pantelides, F.L. Galeener (Permagon Press, New York, 1980), p. 172.

14. P.A. Lee, P.H. Citrin, P. Eisenberger, B.M. Kincaid, Rev. Mod. Phys. **53**, 769 (1981).

15. The low-R structure seen in Fig. 4 is due to incompletely subtracted background which, unlike the non-artifactual Si-O peaks in the other samples, varies in position with small changes in the analysis. Its intensity, however, remains very small, consistent with the more quantitative limits determined from the NEXAFS data.

16. I.B. Berlman, Handbook of Fluorescence Spectra of Aromatic Molecules, (Academic Press, New York, 1965).

17 For example, see S. Furukawa and T. Miyasato, Jap. J. Appl. Phys. **27**, L2207 (1988); K. Furukawa, M. Fujino, and N. Matsumato, Appl. Phys. Lett. **60**, 2744 (1992).

TWO-DIMENSIONAL EXCITONS IN SILOXENE

M. S. BRANDT*, M. ROSENBAUER** and M. STUTZMANN**
* Xerox Palo Alto Research Center, 3333 Coyote Hill Road, Palo Alto CA 94304
** Max-Planck-Institut für Festkörperforschung, Heisenbergstr. 1, D 7000 Stuttgart 80, Germany

ABSTRACT

The luminescence properties of two different modifications of siloxene are studied with photo-luminescence excitation spectroscopy (PLE) und optically detected magnetic resonance (ODMR). The luminescence of as prepared siloxene, which consists of isolated silicon planes, is resonantly excited at the bandgap, indicating a direct bandstructure. The observation of $\Delta m = \pm 2$ transitions in ODMR shows that triplet excitons contribute to the luminescence process. In contrast, annealed siloxene consisting primarily of six-membered silicon rings shows a PLE typical of a material with an indirect bandgap. The ODMR signal of annealed siloxene and of porous silicon show the same Gaussian line with a typical width of 400 G, which can arise from strong dipolar coupling of an electron and a hole $\approx 5\text{Å}$ apart.

INTRODUCTION

To integrate optoelectronic devices with silicon microelectronics, it would be highly desirable to use silicon-based materials also for light-emitting structures. Due to its bandstructure with an indirect bandgap of about 1.1 eV, crystalline silicon does not show visible light emission at room temperature, so that the bandstructure restrictions have to be relaxed considerably to obtain efficient luminescence. This can either be achieved by structural disorder in wide-bandgap amorphous silicon alloys such as a-SiH$_x$, a-SiC$_x$, a-SiO$_x$ or a-SiN$_x$, or possibly by quantum confinement in crystallites or undulating wires of typically 10-20Å diameter, which has been proposed as the cause for the red photoluminescence from anodically oxidized porous silicon. A third way is to reduce the dimensionality of the silicon backbone from 3 in the case of crystalline or amorphous material to 2 as in planar polysilane (SiH)$_n$ and in siloxene (Si$_6$O$_3$H$_6$)$_n$ or to 1 as in linear polysilanes (SiH$_2$)$_n$. While amorphous silicon and its alloys as well as polysilane have been studied quite extensively, and while porous silicon has received considerable attention in the last two years, the planar silicon modifications have been basically overlooked. Our recent observation of virtually identical luminescence properties of porous silicon and of a specific modification of siloxene, which is characterized mainly by rings made of six silicon atoms which are isolated from adjacent rings by oxygen bridges, has renewed the interest in this kind of material. In the following, we will study some fundamental photoluminescence properties of siloxene using photoluminescence excitation spectroscopy and optically detected magnetic resonance.

PREPARATION OF LUMINECENT TWO-DIMENSIONAL SILICON BACKBONES

To date, the only known way to produce two-dimensional silicon backbones is by starting from CaSi$_2$ [1]. This silicide itself has a two-dimensional structure consisting of the silicon (111) double layers as found in crystalline silicon, which are in this substance seperated from each other by layers of calcium. Inserting CaSi$_2$ in HCl leads to the removal of Ca by the formation of CaCl$_2$, the dangling bonds generated by this procedure are saturated by H- or OH-groups (Fig. 1). The exact percentage of saturation by the respective groups and therefore the stochiometry of the final material depends strongly on the preparation conditions. Numerous attempts have been reported to produce pure planar polysilane (SiH)$_n$ by excluding any oxygen from the wet chemical process sketched above [2,3]. The substances obtained are highly reactive and have therefore not been

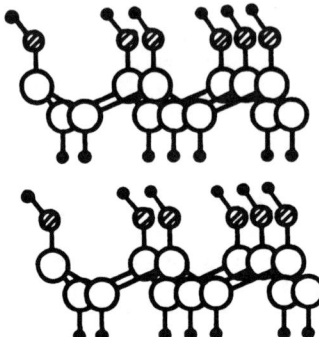

Figure 1: Planar stucture of as prepared siloxene. Open circles indicate silicon, hashed circles oxygen and small full circles hydrogen atoms.

subject to thorough investigation. In contrast, stable materials are usually obtained when water is present during the reaction. One typically finds a stochiometry of $(Si_6O_3H_6)_n$ and the material is called siloxene [4]. The crystallographic structure has been repeatedly established with the help of X-ray diffraction measurements [5,6]. The lattice constant in the Si-layers is decreased by 0.5% with respect to the lattice constant in crystalline silicon, the distance between the layers depends on the preparation conditions but is typically between 5.9 and 7.8Å. The layered structure has recently be confirmed by scanning tunneling microscopy, showing an interlayer spacing of 6Å [7]. The wavelength of the luminescence of siloxene can be shifted by either chemical reactions or brief anneals at 400°C. Infrared spectroscopy shows that the change in the luminescence properties is accompanied by the introduction of oxygen into the silicon planes [8]. If this introduction proceeds in an ordered fashion, rings of six silicon atoms will be formed which are isolated from each other by oxygen bridges. This is the structure of siloxene originally proposed by *Kautsky* mainly based on arguments concerning the substitution chemistry of the material [4].

OPTICAL PROPERTIES

The specific optical properties of the two modifications of siloxene studied here, the as prepared material and the annealed counterpart, become obvious in the compilation of the photoluminescence (PL) and photoluminescence excitation spectra (PLE) in Fig. 2(b). As prepared siloxene shows a luminescence maximum at around 2.4 eV, while the luminescence of the annealed material peaks at about 1.8 eV. The right hand part of the figure shows the corresponding PLE spectra. While the luminescence of the annealed material is best excited at energies above 3 eV, the luminescence of the as prepared material is excited at 2.6 eV, with very little difference between the excitation and the luminescence maxima.

To understand these different behaviours, we have estimated the optical bandgap of the materials from diffuse reflection measurements, assuming that the absorption is approximately given by 1 minus the reflection. From Fig. 2(a) we obtain a gap energy of ≈2.6 eV for as prepared and ≈2 eV for annealed siloxene. When comparing the absorption and the excitation spectra it becomes clear that in the as prepared siloxene containing the silicon planes the luminescence is excited resonantly at the bandgap energy. The obvious explanation is that we observe an excitonic transition as expected at the bandgap energy of a direct semiconductor. This interpretation is corroborated by bandstructure calculations, which predict a direct bandgap of 2.7 eV at the Γ point in the planar siloxene modification [9].

Comparison of the absorption and the excitation spectra for annealed siloxene shows that the luminescence of this material is no longer excited at the band edge, but at considerably higher energies of 3-4 eV. Also, the sharp absorption edge has been replaced by a broad tail, which is exponential over nearly two orders of magnitude with a characteristic slope of 200 meV. X-ray

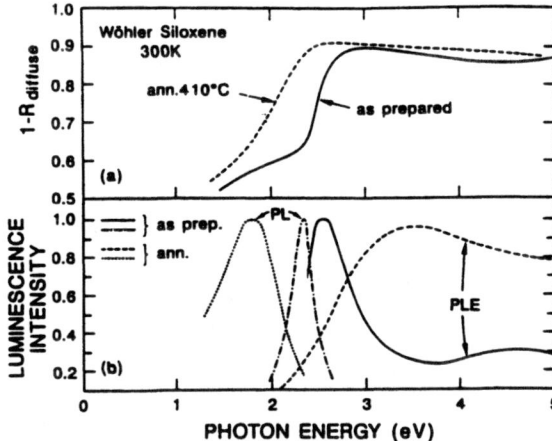

Figure 2: Comparison of the absorption edge as determined from measurements of the diffuse reflectivity (a) and of the photoluminescence (PL) and the photoluminescence excitation (PLE) (b) for as prepared and annealed siloxene.

diffraction shows that the annealed samples consist primarily of an amorphous silicon-oxygen alloy. Calculations for the isolated six-ring structure as the assumed luminescence "center" have indeed shown that the smallest direct transition should be 3-4 eV, with a considerably smaller indirect bandgap. The exact energy of the indirect transition depends on the substitution of the fourth silicon valence, variing from 3 eV for purely hydrogenated rings to 1.9 eV for rings with three hydrogen and three OH-ligands to 1.6 for silicon six-rings completely saturated by OH-groups. The shift of the luminescence to longer wavelengths is therefore accompanied by a significant change in the silicon backbone structure and a change in the type of the lowest transition from direct to indirect.

MICROSCOPIC SIGNATURES OF THE RADIATIVE TRANSITIONS

Further information on the radiative transition can be obtained with the help of optically detected magnetic resonance (ODMR). In this technique, changes in the photoluminescence are detected which occur under electron spin resonance conditions. The basic principle underlying this technique is angular momentum conservation. Only angular momentum conserving transitions are allowed, eg. from certain substates $(m = 1, 0, -1)$ of triplet excitons $(S = 1)$ to the singlet ground state $(S = 0)$. Inducing transitions between the Zeeman-substates will result in a transformation of states not allowed to undergo the radiative transition to states who are allowed to, increasing the overall transition rate of the radiative process and leading to a net enhancement of the luminescence under resonance conditions. The above argument holds for excitons in which both carriers are localized and the orbital angular momentum L is therefore quenched, ie. $L = 0$. In the case of delocalized carriers, the presence of both L and S allows for more complex transitions which would also satisfy angular momentum conservation.

If a paramagnetic defect is involved as a nonradiative recombination center competing with the radiative recombination, one observes due to the same arguments as presented above an increase in the transition rate at the resonance of the defect. Since a process shunting the radiative transition is now made more probable, the net effect is a quenching of the luminescence. Such a resonant quenching at 100 K is shown for as prepared siloxene in Fig. 3. The conventional electron spin resonance spectrum shows a defect resonance at g=2.0036, typical of silicon dangling bonds, where one of the nearest neighbours of the silicon atoms is oxygen [10]. In agreement with the prediction given above, the 2.4 eV luminescence decreases resonantly.

More relevant to our topic is the resonant enhancement of the luminescence observable at

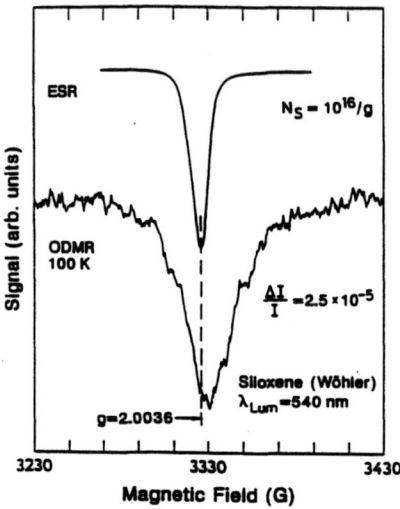

Figure 3: Electron spin resonance (ESR) and optically
detected magnetic resonance
(ODMR) of as prepared siloxene. At 100 K only the
quenching resonance due to
non-radiative recombination
through defects is detectable.

lower temperatures. This is shown in Fig. 4 for the two modifications of siloxene studied here. Note that the magnetic field axis extends from 1000 to 5000 G. Both spectra contain three basic spectroscopic features: The narrow quenching line at about $g \approx 2.005$ due to nonradiative recombination through defects as mentioned above, a broad enhancing line around $g = 2$, which is the microscopic signature of the radiative transition and a feature at half the magnetic field compared to the center of the enhancing resonance, marked as $g = 4$. Temperature dependent measurements show a rapid decrease in the ODMR signal $\Delta I/I$ with increasing temperature for both the broad enhancing lines at $g = 2$ and the half-field resonances, while $\Delta I/I$ remains fairly constant for the defect related quenching signal at $g \approx 2.005$. This fact shows that the two first signals have the same origin, namely a triplet exciton where the $g = 2$ signal corresponds to the spin-allowed $\Delta m = \pm 1$ transitions within the Zeeman-triplet and the $g = 4$ signal to the forbidden $\Delta m = \pm 2$ transitions. The observation of the half-field resonance for as prepared siloxene corroborates beyond doubt the conclusion already drawn from PLE that exitonic states are responsible for the luminescence, since a $\Delta m = \pm 2$ transition cannot occur in weakly coupled electron-hole pairs. The fact that the ODMR of annealed siloxene also exhibits $\Delta m = \pm 2$ transitions shows that triplet excitons are involved in the indirect transitions in this material as well.

ODMR of the luminescence from low-dimensional systems has been studied only for type II-superlattices, both in the GaAs/AlAs and in the Si/SiGe system [11,12]. Generally, due to the spatial separation of the electrons and holes in adjacent layers of the superlattice separate resonances for the two kinds of carriers are observed. In contrast, the structures of siloxene as determined by X-ray diffraction and the bandstructure calculations discussed above suggest that the carriers will be confined to the same silicon plane or six-ring in this material. In addition, it is known from other materials such as polysilane and oxides, that excitons can be self-trapped due to strong electron-phonon-coupling. Indeed, we observe a strong interaction of the electrons and holes which destroys the separate resonances and leads to the broad resonances of Fig. 4. These enhancing lines are typical for luminescence from amorphous hydrogenated silicon and its wide-bandgap alloys. For a-Si:H, an enhancing line with a typical width of 200 G is found in ODMR [13], with variing reports on the exact lineshape. Alloying with either N or C leads to a gradual increase of the bandgap, which is accompanied by an increase of the luminescence peak energy. At the same time, the linewidth of the enhancing ODMR line increases gradually

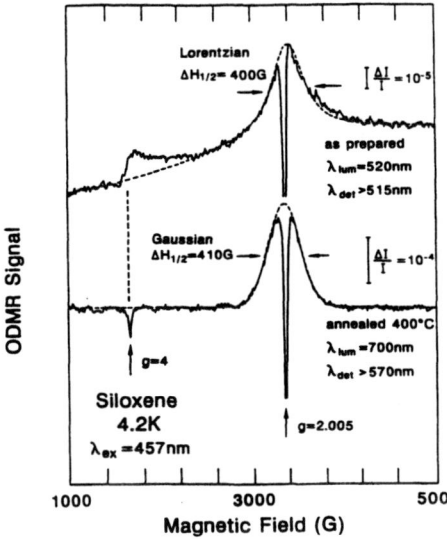

Figure 4: Optically detected magnetic resonance of as prepared and annealed siloxene. λ_{lum}: position of luminescence maximum, λ_{det}: detected wavelengths, λ_{ex}: excitation wavelength. The dashed lines show fits of a Lorentzian and a Gaussian to the allowed transitions at $g = 2$, respectively.

to about 450 G and the line becomes Gaussian [13,14]. The half-field resonances at $g = 4$ are observed as narrow lines as in annealed siloxene (Fig. 4) [13]. The ODMR of annealed siloxene therefore shows a striking similarity to that of wide-bandgap amorphous silicon alloys.

The interpretation of broad featureless resonances is difficult. Some valuable information can be derived from the actual lineshape. The relevant interaction of the electron and the hole forming the exciton can either be the exchange or the dipolar interaction. The exchange interaction, which in the case of s-wavefunctions with Bohr-radius r_0 and distance r is given by $J = J_0 \exp(-2r/r_0)$, leads to a Lorentzian lineshape [15]. Dipolar broadening, in constrast, leads to a Gaussian lineshape. In this case, the typical distance r of the interacting carriers can be estimated from the width ΔB of the observed resonance line, $r = (\mu_0 g \mu_B / 8\pi \Delta B)^{1/3}$. The experimentally determined linewidth of 400 G for the Gaussian line observed in the ODMR of annealed siloxene then yields a typical distance of about 4.2Å. This is in good agreement with the size of an silicon six-ring, which has a diameter of 4.5Å, and thereby corroborates that the luminescence of annealed siloxene originates from states which are localised on the length scale of a six-membered ring. The interpretation of the Lorentzian line and the corresponding step-like half-field resonance will be subject to a forthcoming publication.

As a final note, we would like to state that the ODMR of porous silicon exhibits very similar resonances to that of annealed siloxene, namely an enhancing line at $g = 2$ with a Gaussian lineshape and the half-field resonance [16]. The linewidth of the $g = 2$ resonance varies from sample to sample, the typical value is 500 G.

CONCLUSIONS

We have presented ODMR and PLE studies which allow to address in detail the luminescence properties of the different modifications of siloxene. In as prepared material with a planar silicon backbone, the green luminescence at 2.4 eV can best be excited resonantly at the bandgap of 2.6 eV typical for a material with a direct bandgap. After thermal anneal, the luminescence of 1.8 eV is exited nonresonantly above 3 eV. The observation of $\Delta m = \pm 2$ transitions in optically detected magnetic resonance shows that the luminescence from both modifications is indeed due to strongly bound excitonic states. Porous silicon exhibits the same ODMR resonances as annealed siloxene which adds futher evidence to our previous assignment that the luminescence of porous silicon and of the annealed modification of siloxene have the same origin [17,18].

ACKNOWLEGEMENTS

One of us (MSB) would like to thank the Alexander von Humboldt-Stiftung and N. M. Johnson for support.

REFERENCES

[1] F. Wöhler, Lieb. Ann. **127**, 257 (1863).

[2] G. Schott, Z. Chemie (Leipzig) **2**, 194 (1962).

[3] E. Hengge and G. Scheffler, Monatshefte Chemie **95**, 1450 (1964).

[4] H. Kautsky and G. Herzberg, Z. anorg. Chemie **139**, 135 (1924).

[5] A. Weiss, G. Beil, and H. Meyer, Z. Naturforsch. **35b**, 25 (1979).

[6] J. R. Dahn et al., Bull. Am. Phys. Soc. **38**, 157 (1993).

[7] M. Rosenbauer, P. Molinas-Mata, M. Böhringer, M. Stutzmann, J. Zegenhagen, and M. Cardona, Bull. Am. Phys. Soc. **38**, 157 (1993).

[8] M. Stutzmann, J. Weber, M. S. Brandt, H. D. Fuchs, M. Rosenbauer, P. Deak, A. Höpner, and A. Breitschwerdt, Adv. Sol. State Phys. **32**, 179 (1992).

[9] P. Deak, M. Rosenbauer, M. Stutzmann, J. Weber, and M. S. Brandt, Phys. Rev. Lett. **69**, 2531 (1992).

[10] E. Holzenkämpfer, F. W. Richter, J. Stuke, and U. Voget-Grote, J. Non-Cryst. Solids **32**, 327 (1979).

[11] H. W. van Kesteren, E. C. Cosman, F. J. A. M. Greidanus, P. Dawson, K. J. Moore, and C. T. Foxon, Phys. Rev. Lett. **61**, 129 (1988).

[12] E. Glaser, J. M. Trombetta, T. A. Kennedy, S. M. Prokes, O. J. Glembocki, K. L. Wang, and C. H. Chern, Phys. Rev. Lett. **65**, 1247 (1990).

[13] K. Morigaki and M. Yoshida, J. Non-Cryst. Solids **90**, 139 (1987).

[14] A. J. Lowe and B. C. Cavenett, Phil. Mag. B **57**, 243 (1988).

[15] F. Boulitrop, Phys. Rev. B **28**, 6192 (1983).

[16] M. S. Brandt and M. Stutzmann, Appl. Phys. Lett. **61**, 2569 (1992).

[17] M. S. Brandt, H. D. Fuchs, M. Stutzmann, J. Weber, and M. Cardona, Solid State Commun. **81**, 307 (1992).

[18] M. S. Brandt, H. D. Fuchs, A. Höpner, M. Rosenbauer, M. Stutzmann, J. Weber, M. Cardona, and H. J. Queisser, MRS Symp. Proceed. **262**, 849 (1992).

COMPARATIVE STUDY OF LIGHT-EMITTING POROUS SILICON ANODIZED WITH LIGHT ASSISTANCE AND IN THE DARK

L. TSYBESKOV,[*] C. PENG,[*] S.P. DUTTAGUPTA,[*] E. ETTEDGUI,[**] Y. GAO,[**] P.M. FAUCHET[*] and G.E. CARVER[***]

[*] Department of Electrical Engineering, University of Rochester, Rochester, NY 14627
[**] Department of Physics and Astronomy, University of Rochester, Rochester, NY 14627
[***] AT&T Bell Laboratories, Princeton, NJ 08540

ABSTRACT

In this study, we compare two different types of light emitting porous silicon (LEpSi) samples: LEpSi anodized in the dark (DA) and LEpSi anodized with light assistance (LA). On the basis of photoluminescence (PL), Raman, FTIR, SEM, spatially resolved reflectance (SRR) and spatially resolved photoluminescence (SRPL) studies, we demonstrate that the luminescence in LA porous silicon is strong, easily tunable, very stable and originates from macropore areas. These attractive properties result from passivation by oxygen in the Si-O-Si bridging configuration that takes place during electrochemical anodization. In addition, we have been able to correlate light emission with the presence of crystalline silicon nanograins.

INTRODUCTION

An understanding of the relationship between surface structure, passivation mechanism, local order and PL quantum efficiency from LEpSi is important for optimization of the preparation conditions, PL mechanism studies and possible applications in optoelectronics. The standard procedure for formation of light-emitting porous silicon is to anodize a p-type silicon wafer in an HF-solution using a current density between 1 and 100 mA/cm^2. Unfortunately, PL from as-anodized LEpSi is not stable and a post-formation treatment is required to stabilize the PL. Thermal, chemical and plasma processings of LEpSi have been employed by several groups to alter the chemical composition of the surface of porous silicon [1,2,3]. A few groups have reported the successful production of LEpSi starting with n-type silicon wafers [4,5,6]. This type of anodization is possible with light exposure, because holes play a key role in the electrochemical oxidation. In the case of anodization of p-type Si in the dark, the electrochemical reaction proceeds because holes are injected from the silicon substrate to the porous region across the crystalline Si-LEpSi interface. After some time, the difference in the energy gaps of c-Si (E_g = 1.1eV) and LEpSi (E_g > 2 eV) provides a barrier to hole transport and therefore the reaction stops. For anodization with light assistance, the right concentration of holes can be obtained directly through light absorption and generation of non-equilibrium electrons-holes pairs. LA anodization leads to more efficient and more stable PL and to a PL peak at shorter wavelength [4,5,6]. Recent studies have shown that DA and LA samples have different surface structures and local order [4], but there has been no systematic comparison between LA and DA samples that could explain the origin of this difference.

In this paper, we present the results of PL, Raman, FTIR and SEM measurements for DA and LA samples. We show that the structural and chemical properties of the surface are different in the two types of LEpSi samples. In LA LEpSi samples, we demonstrate a correlation between favorable PL properties, such as stability and intensity, and the presence of Si-O bonds and nanocrystalline silicon grains.

EXPERIMENTAL DESCRIPTION

LEpSi samples were prepared from p-type Si in the dark (DA) or from p- and n-type Si with light assistance (LA), using substrates with resistivity 1-10 ohm-cm and <100> crystal orientation. The anodization was carried out in an HF-solution (HF:H$_2$O: methanol = 1:1:2) for 5-30 min with a constant current density of 1 to 10 mA/ cm^2 in a closed system with continuous solution circulation. The choice of methanol and the relatively small current density minimize the formation of hydrogen bubbles during anodization and lead to good homogeneity over areas larger than 30 cm^2. Both n-type and p-type samples prepared with LA were irradiated with a 300 W white-light halogen lamp. The properties of samples prepared with LA using p-type and n-type substrates having $\rho > 10$ ohm cm were similar.

The PL and Raman measurements were taken with the 458 nm line of an argon ion laser. The laser beam was focused to a 1-2 μm spot after passing through an optical microscope. The microscope also collected the PL and Raman signals. This procedure ensures that a comparison between PL and Raman spectra is meaningful as the signal comes precisely from the same area. Both PL and Raman spectra were recorded using an optical multichannel analyzer (OMA). For the PL measurements, the OMA was connected to single monochromator, while for the Raman measurements, it was connected to a double spectrometer. No light-induced changes were observed during the measurements with an excitation intensity less than 1 W/cm^2. Micro-PL and micro-Raman data were collected from the top of the samples and from the cross-section in cleaved samples.

FTIR measurements were performed with a NICOLET infrared spectrometer. The LEpSi spectra were obtained after subtraction of the crystalline silicon spectra which were independently measured for the same wafer. For SEM work, the samples were cleaved in ambient atmosphere prior to insertion into the chamber, thus allowing surface and cross-sectional studies.

Spatially resolved reflectance (SRR) and spatially resolved photoluminescence (SRPL) maps were recorded under excitation with the 514 nm line of an argon ion laser beam focused to a 1 μm spot. The spot is raster scanned over a 250 μm X 250 μm area. Luminescence is collected at wavelengths longer than 530 nm and detected by either a silicon avalanche photodiode or a germanium photodiode.

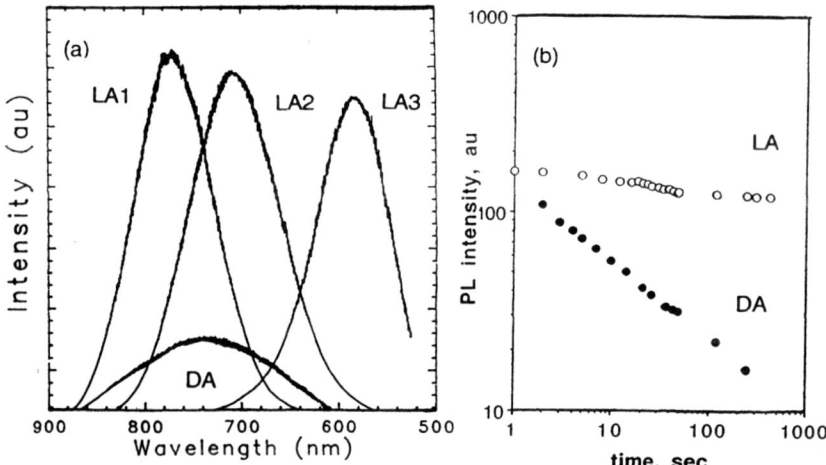

Fig.1   (a) Typical PL spectra for dark (DA) and light-assisted (LA) porous silicon:

sample DA is a p-type, resistivity is $\rho \approx 5$ ohm cm, current density $J \approx 10$ mA/cm^2, anodization time is $t \approx 7$ min;

sample LA 1 is n-type, $\rho \approx 1$ ohm cm, $J \approx 2$ mA/cm^2, $t \approx 5$ min;
sample LA 2 is n-type, same resistivity and current density, $t \approx 25$ min:
sample LA 3 is p-type, $\rho \approx 5$ ohm cm , same current density, $t \approx 40$ min.
   (b) kinetics of photoluminescence degradation for DA and LA (sample 3) porous silicon.
The intensity of the laser was 10 W/cm^2 and the wavelength 458 nm.

EXPERIMENTAL RESULTS AND DISCUSSION

Figure 1a shows the PL spectra from DA and LA samples. The DA LEpSi has a broad peak centered near 740 nm, while for LA LEpSi, the PL spectra are narrower and can be tuned from the infrared to the green by increasing the anodization time. The PL from LA samples is not only more intense but also more stable. Figure 1b shows the kinetics of the PL degradation for two typical samples under a relatively large excitation power P = 10 W/cm². The PL of LA LEpSi is rather stable, whereas the PL intensity from DA LEpSi, I_{pl} (t), degrades with time t as

$$I_{pl}(t) = I_o * t^{-\alpha(I_{ext})}$$

where $\alpha(I_{ext}) = 0.4 - 0.5$ for an excitation intensity I_{ext} from 0.1 W/cm² to 10 W/cm². This fatigue of the PL is reminiscent of the photo-induced degradation in a-Si:H known as the Stabler-Wronski effect [7]. If PL degradation in porous silicon results from the creation of non-radiative recombination centers, such as silicon dangling bonds on the surface, the different PL degradation kinetics between LA and DA LEpSi samples may reflect a different mechanism of surface passivation.

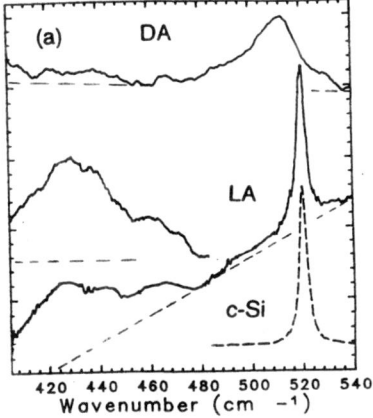 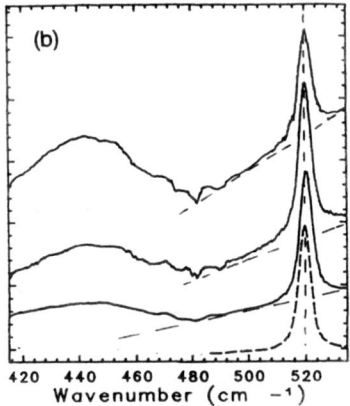

Fig.2 (a) Raman spectra of crystalline Si, DA LEpSi and LA LEpSi (sample 3) . The Raman feature around 420 cm⁻¹ is shown after subtraction of the PL background in LA LEpSi.
(b) Micro-Raman spectra at different locations of sample LA3, showing the correlation between the slope of the PL and the strength of the Si-O Raman feature.

Figure 2a compares unpolarized Raman spectra from LA and DA samples. The broad Raman line of DA LEpSi has a peak near 510 cm⁻¹ and extends to near 480 cm⁻¹. The Raman lineshape can be interpreted as evidence for nanocrystalline silicon objects [8]. In contrast, the LA samples have a narrow peak (6-8 cm⁻¹) centered at 520 cm⁻¹. In addition, the Raman spectrum rides on a strong "tail" of the photoluminescence. There is also a broad feature between 410 and 450 cm⁻¹, as shown after subtraction of the background. In Figure 2b we demonstrate a correlation between the strength of this Raman feature and the intensity of the PL, as measured by the slope of the PL "tail". Galeener [9] has shown that this Raman feature is produced by asymmetric stretching of Si-O-Si bridging bonds.

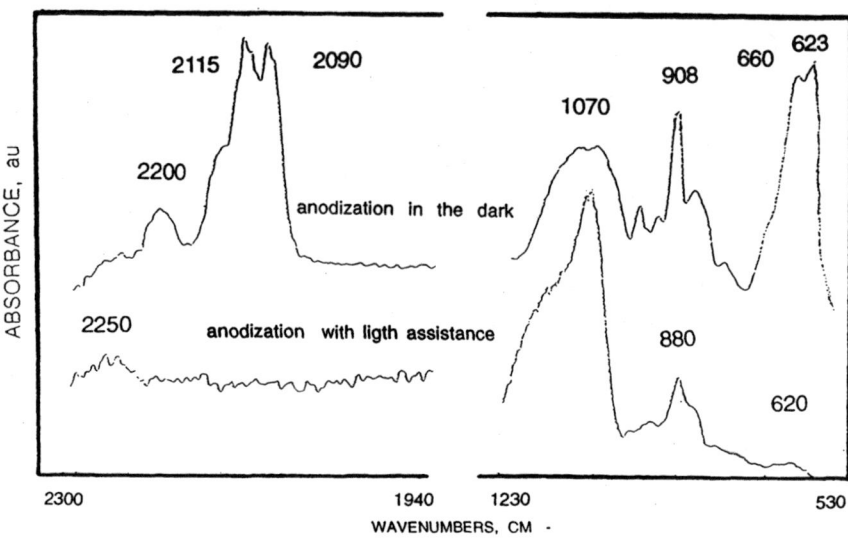

Fig.3 FTIR absorption spectra of DA porous silicon sample and LA (sample 3) porous silicon sample. The silicon-hydrogen, silicon-oxygen and silicon-oxygen-hydrogen bonds are identified in the text.

Additional evidence for such Si-O bonds can be found in the infrared absorption measurements. Figure 3 compares the FTIR spectra of LA and DA LEpSi in regions where the Si-H and Si-O stretching and bending modes appear. The fresh DA samples show the complete set of Si-H bonds in different configurations: complex of stretching bonds near 2000-2200 cm^{-1}, wagging modes at 620-660 cm^{-1}, scissor mode at 980 cm^{-1}, and bending modes in the 850-980 cm^{-1} region [2,9]. The peak at 2200 cm^{-1} has been attributed to the initial oxidation of the stretching bonds and the small feature at 2250 cm^{-1} is evidence for O-Si-H bonds. The Si-O-Si bonds appear as the symmetric doublet at 1070-1110 cm^{-1} [2,10,11]. In contrast, the LA samples have an FTIR spectrum with major contributions from the now-asymmetric doublet around 1080 cm^{-1} and another peak near 880 cm^{-1}, attributed to the O$_3$-Si-H configuration [11]. The FTIR data support the interpretation of the Raman spectra of Fig.2 and demonstrate the strong oxidation of LA samples during anodization. Unlike what is observed in as-anodized DA LEpSi, the FTIR spectra of LA LEpSi do not change measurably with exposure to laboratory atmosphere, for at least several months.

The structure of the surface of LA and DA LEpSi samples is also markedly different. Whereas DA samples have a relatively smooth surface, LA samples usually have a macroporosity with cracks 10 μm in length and oriented along crystalline axes [4] ("crosses" and "stars" of Figure 4a). The micro-Raman spectra obtained from the macropore areas of LA samples (not shown) are very similar to the spectra from DA samples [12], suggesting that nanocrystalline Si objects exist inside the macropores of LA LEpSi. High spatial resolution PL mapping of LA LEpSi reveals the structural features associated with light emission. Figure 4b shows the surface of LA LEpSi as observed by SRPL. The stars observed in SEM and optical microscopy are clearly recognizable. Although the inside of the stars is dark, there is a bright halo in the immediate surroundings which indicates that the PL is stronger near the stars than elsewhere on the sample. The correlation between a stronger PL and the Raman signature of Si nanocrystals suggests that light emission in porous silicon is associated with Si nanocrystalline objects.

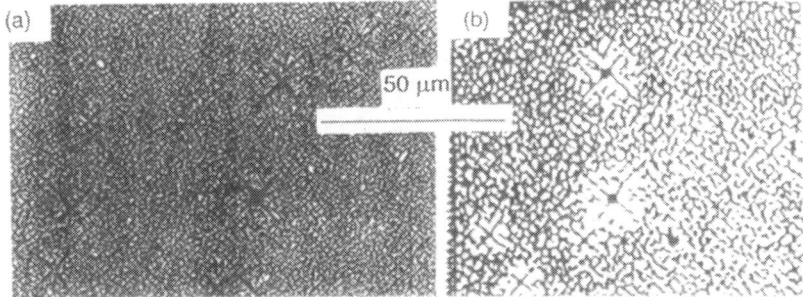

Fig.4   (a) SEM picture of the surface of LA (sample 2) LEpSi, showing cracks and macropores ("stars").
(b) Photoluminescence map of a similar LA LEpSi sample showing the spatial correlation between
bright light emission and the presence of "stars".

CONCLUSION

We have shown that light-emitting porous silicon (LEpSi) prepared by electrochemical
anodization with light assistance (LA) produces samples with attractive properties: strong and
very stable photoluminescence, easily tuned by changing the anodization time. These properties
are due to the passivation of the surface by Si-O bonds that takes place during anodization. The
inhomogeneity of the surface of LA LEpSi has allowed us to correlate light emission with the
presence of macropores or cracks. From micro-Raman spectroscopy, we conclude that silicon
nanocrystals are present in these regions where the PL is strongest. The correlation between
light emission and the presence of silicon nanocrystals supports the concept that light emission
requires nanometer-sized silicon objects.

ACKNOWLEDGEMENTS

This work was supported by grants from Rochester Gas & Electric, the New York
State Energy Research & Development Authority, and Xerox Corporation. We thank K.
Marshall and B. McIntyre for technical assistance. C. Peng is a Link fellow.

REFERENCES

1. J.C. Vial, A. Bsiesy, F. Gaspard, R. Herino, M. Legion, F. Muller, R. Romenstain,
Phys. Rev. B **45**, 14171 (1992).
2. M. Yamada and K. Kondo, Jpn. J. Appl. Phys. **31**, L993 (1992).
3. V. Petrova-Koch, T. Muschik, A. Kux, B.K. Meyer, F. Koch and V. Lehman, Appl.
Phys. Lett. **61**, 943 (1992)
4. E. Ettedgui, C. Peng, L. Tsybeskov, Y. Gao, P.M. Fauchet, G.A. Carver and H.A. Mizes,
Mat. Res. Soc. Symp. Proc. **283**, 173 (1993)
5. P. Steiner, F. Kozlowski, H. Sandmaier and W. Lang, Mat. Res. Soc. Symp. Proc. **283**,
343 (1993)
6. V. Lehmann and U. Gosele, Mat. Res. Soc. Symp. Proc. **283**, 27 (1993)
7. AIP Conf. Proc. **234**, edited by D.L. Stafford (AIP, New York, 1991)
8. P.M. Fauchet and I.H. Campbell, Crit. Rev. Solid State Mat. Sci. **14**, S79 (1988); Mat.
Res. Soc. Symp. Proc. **164**, 259 (1990); I.H. Campbell and P.M. Fauchet, Solid State
Commun. **58**, 739 (1986)
9. F.L. Galeener, Phys. Rev. B **19**, 4292 (1979).
10. C. Tsai, K.-H. Li and J.C. Campbell, Journ. Electron. Mat. **21**, 589 (1992)
11. C. Debauche, C. Licoppe, J. Ficstein, O. Dulac, R.A.B. Devine. Appl. Phys. Lett. **61**,
306, (1992)
12. L. Tsybeskov et al, to be published

Variations in the Photoluminescence Intensity of Chemically and Anodically Etched Silicon Films

P.S. Williams, J.N. Kidder, Jr., H. Yun, D. Crain, T.P. Pearsall, Dept. of Materials Science and Engineering, University of Washington, Seattle, WA; D.T. Schwartz, Dept. of Chemical Engineering, University of Washington, Seattle, WA.

ABSTRACT

We are studying the effects of etch conditions on the surface morphology, chemistry, and luminescent properties of porous silicon (PS) films. Luminescent silicon films are produced by chemical etching using solutions of HNO_3 in HF and by anodic etching using aqueous HF electrolytes. Films produced by both methods are analyzed and compared using photoluminescence (PL), vibrational, and X-ray photoelectron (XPS) spectroscopies. The initial characterization of PS is performed immediately following the etching process, resulting in oxide-free films (as confirmed by XPS). In chemically etched PS films, the luminescent intensity decreases as the vol. % HNO_3 in etch solution increases. Spectral features evolve in the PL spectrum of chemically etched films as the result of aging under ambient conditions and when the films are cooled under illumination. Moreover, we have also found that increased electrolyte convection results in a decrease in photoluminescence intensity of PS films formed anodically. The role of electrolyte flow in modifying the luminescent properties of PS is being evaluated in an etch cell with well-characterized hydrodynamics.

INTRODUCTION

The observation of strong visible luminescence from porous silicon has resulted in a great deal of research activity focused on this new photonic material.[1,2] Porous silicon is formed by either anodic etching in HF based electrolytes or chemical etching in solutions of HF and oxidizing agent. With both these techniques it has been observed that the intensity of the photoluminescence emission is affected by the processing conditions. Recent studies have reported on variations in the PL intensity and spectral distribution with respect to post-etch HF soaking, thermal treatments, substrate microstructure, stain-etch solution oxidation agent concentration, post-etch dry oxidation processing and rapid-thermal oxidation.[3-12] We have previously reported on a dependency between electrolyte HF concentration and the PL intensity peak position in anodic films, where decreasing HF concentration results in a blue shift of the PL peak intensity.[3] This investigation further explores the effects of processing on the photoluminescent properties of porous silicon films by examining the effect of electrolyte flow rate on the morphology and photoluminescent properties of anodically etched films and by studying the effects of variations in solution chemistry on the photoluminescent properties and film chemistry of chemically etched films.

We produce anodic films in an electrochemical cell designed to create complementary electrolyte and current flow. This allows for an added understanding of the effects of electrolyte hydrodynamics in terms of kinetic and diffusional mass transport on the morphology and photoluminescent properties of films produced in this controlled manner. We also study chemically etched films formed by immersion in dilute solutions of HF/HNO_3 where the HNO_3 concentration is varied over three orders of magnitude between 1 and 0.01 volume percent. Our results show that increased electrolyte convection decreases PL intensity and reduces average pore size without affecting film thickness. With chemical etching we have observed that decreasing the etch solution HNO_3 concentration results in films that luminesce with increasing intensity and show increasing absorption associated with a presence of Si-H species.

EXPERIMENTS AND RESULTS

Anodically Etched PS Films:

Single crystal n-type Si (100) substrates with an average resistivity of 10 ohm-cm were anodically etched in the dark under varied electrolyte flow conditions using the etch cell shown in Figure 1 with 17 vol.% HF in H_2O electrolyte at 12 mA/cm² for 1000 seconds then rinsed in DI

Mat. Res. Soc. Symp. Proc. Vol. 298. ©1993 Materials Research Society

water and blown dry with N_2 gas before analysis. The key features of this etch cell are that the electrolyte flow rate is controllable, it has geometrically complementary electrolyte and current flow (shown in the expanded view) and that its PTFE (Teflon®) construction gives added safety in electrolyte handling via HF isolation.

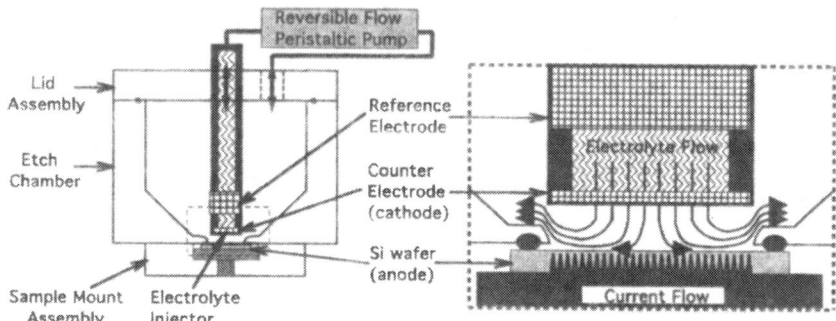

Figure 1: Schematic of electrochemical etch cell with expanded view showing complementary current and electrolyte flow. Key features include variable electrolyte flow rate, geometrically uniform mass and charge transport, and PTFE (Teflon®) construction.

We have discovered that increased electrolyte convection tends to decrease the average pore size of films produced with our etch cell configuration, as illustrated by SEM micrographs 2a and 2b in which we can see that as we increase the electrolyte flow rate from 0.0 mL/sec to 5.6 mL/sec the average pore area decreases from 0.6 μm^2 to 0.2 μm^2. A plot of percent surface pore area vs. flow rate (Figure 3) was generated using computer aided analysis of SEM micrographs, for which the fraction of surface void of material in specimens etched at electrolyte flow rates ranging from 0 and 5.6 mL/sec was measured. The quantity "percent surface pore area" is a quantity derived from SEM micrograph data by simply measuring the fractional area of the surface void of material. This technique is equivalent to multiplying a measured average pore area by the number of pores in the micrograph, which is not a direct measure of porosity, but we believe porosity does increase with an increase in percent pore area of a micrograph. Figure 4 illustrates the connection between the surface view of a large pore and the underlying porosity in which a cross-section of a large pore cavity common in n-type specimens etched at zero flow rate is shown. We have found no dependence of PS film thickness on electrolyte flow rate. Our films had consistent thicknesses of 45 μm.

Figure 2a: SEM micrograph of n-type PS film surface produced at zero flow rate condition showing an average pore area of 0.6μm^2.

Figure 2b: SEM micrograph of n-type PS film surface produced at a flow rate of 5.6 mL/sec showing an average pore area of 0.2 μm^2.

315

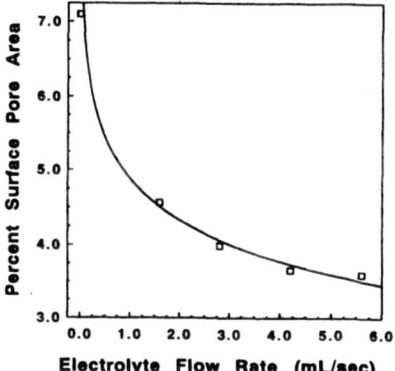

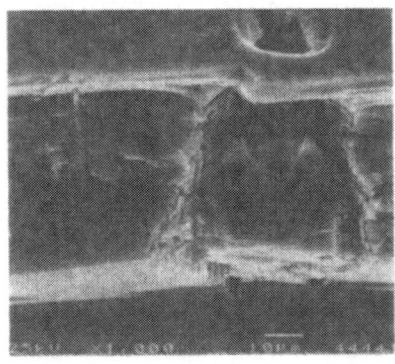

Figure 3: Plot of percent pore area vs. flow rate indicating a decrease in pore area with increased electrolyte convection.

Figure 4: SEM micrograph of a cleaved edge of a PS film produced at a zero flow rate condition showing a large pore cavity common in zero flow specimens.

Photoluminescence spectroscopy (PL) was performed using a 30mW chopped beam of the 476 nm Ar+ laser line and a 3/4 m spectrometer with an S-20 photomultiplier using conventional lock-in techniques. We have found correlation between electrolyte flow rate and PL intensity in that PL intensity decreases with increased flow rate as shown in Figure 5. Integrated PL intensity is plotted vs. electrolyte flow rate in Figure 6 in which we see a sharp decline in PL intensity with relatively moderate flow rates. PL peaks were centered about 6660 Å with little, if any, variation. In comparing our findings of larger average pore size at lower flow rates and higher PL intensity at lower flow rates it is evident that a correlation exists between PL intensity and pore size. This correlation tends to confirm findings by Prokes, et al.[12] in which an increase in PL intensity with pore size is noted.

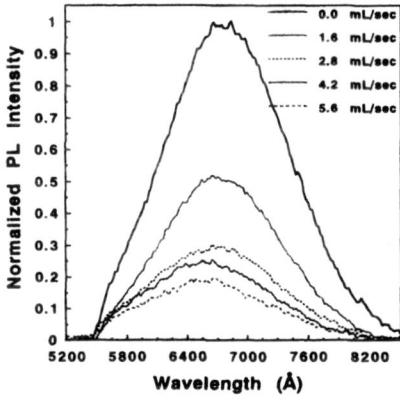

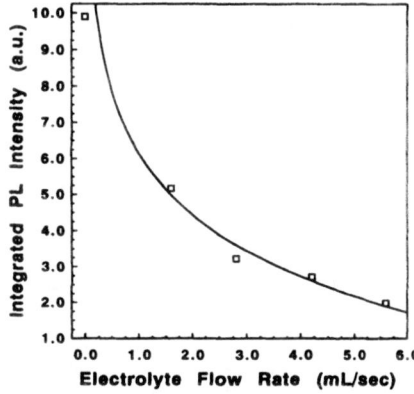

Figure 5: Here we have plotted PL spectra for the range of flow conditions studied. PL intensity decreases with increasing flow rate while the peak intensity position remains unchanged at 6660 Å.

Figure 6: This figure shows the integrated intensities of the spectra shown in figure 5 plotted vs. flow rate where intensity decreases with increasing flow rate.

Our findings of PL peaks consistently centered about 6660 Å regardless of average pore area may raise questions about the simple quantum confinement hypothesis of Canham[1], and thus we have also looked at changes in surface chemistry due to anodic etching as a possible PL mechanism. We have examined spectroscopic data obtained from both XPS and FTIR analysis. XPS shows no significant binding energy shift of the Si_{2p} peak from that of a control specimen when analysis was performed immediately after etching, implying little if any chemical bonding to oxygen. However, the O_{2p} peak is present in our XPS spectra and analysis of control specimens has given evidence which indicates the oxygen presence is due to specimen transport in ambient conditions. This result also raises questions about reports of a siloxene derivative being solely responsible for photoluminescence.[13] However, it is possible that siloxene could contribute to luminescence in some films. XPS analysis was performed immediately after etching with a Perkin-Elmer ESCA-SAM system. Analysis of FTIR spectral features has as yet produced no conclusive evidence linking surface hydride species with PL intensity in anodically etched specimens, as reported by C. Tsai, et al.[5] However, Si-H and Si-O peaks of low intensity have been seen. Samples were measured immediately after etching using a Nicolet FTIR 5DXB spectrometer with gain settings of 1 to 8.

Chemically Etched PS Films:

In our laboratory we also form luminescent silicon films by immersion of single crystal Si substrates in dilute solutions of hydrofluoric and nitric acid. The concentrations range between 1 and 0.01 vol.% nitric acid in hydrofluoric acid. The Si wafers are immersed for 15 minutes, rinsed in DI H_2O and air dried. The solution concentrations are given in terms of volume percent of reagent grade HF (50%) and HNO_3 (70%) solutions. The films were analyzed using PL and FTIR spectroscopy.

We have observed a dependence of photoluminescence intensity of these films on the etch solution concentration. In particular, solutions of decreasing nitric acid concentration produce films that emit PL of greater intensity. Figure 7 shows the PL spectra of four samples formed from an n-type Si wafer of resistivity 10 ohm-cm that were etched in solutions containing different concentrations of nitric acid. The solution nitric concentration is labeled next to the PL spectrum of the corresponding samples. Changes in the etch solution nitric acid concentration are shown to result in changes in the PL spectrum. As the nitric concentration decreases between 0.260 and 0.154 vol.% a shift in the intensity peak position occurs. When the concentration is reduced to 0.101 vol.% a significant increase in the PL intensity occurs. However, decreasing the solution concentration to a value of 0.039 vol.% results in a film that exhibits a much weaker signal. This trend is somewhat expected considering that removal of all the nitric would result in samples that do not luminesce at all.

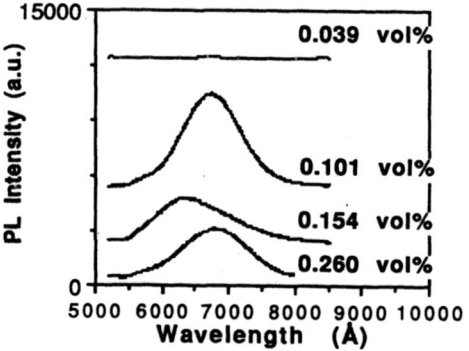

Figure 7 Photoluminescence spectra of four samples formed from an n-type Si wafer of resistivity 10 Ω-cm that were etched in solutions containing different concentrations of nitric acid. The solution nitric concentration is labeled next to the PL spectrum of the corresponding samples.

We have also observed that an increased presence of Si-H species in the films, as detected by IR absorption, is correlated with increases in the PL intensity and changes in the etch solution chemistry. Figure 8 shows a plot of IR absorbance versus wavenumber in a range associated with stretch absorption by Si-H and Si-H_2 bonds. In particular this plot shows peaks at 2139 cm-1, 2115 cm-1, and 2090 cm-1 which have been attributed to Si-H bonds.[14] This plot shows that decreasing the nitric concentration between 0.260 and 0.101 vol.% results in increases in the amount of Si-H species in the films. This trend occurs until the concentration of 0.039 vol.% is reached at which the etched sample shows only a small amount Si-H species absorption. The inset in figure 8 shows a correlation between PL intensity and the presence of

Si-H species in a plot of PL integrated intensity (5200 - 8500 Å) and IR absorbance intensity (2180 - 2014 cm^{-1}) versus HNO$_3$ concentration.

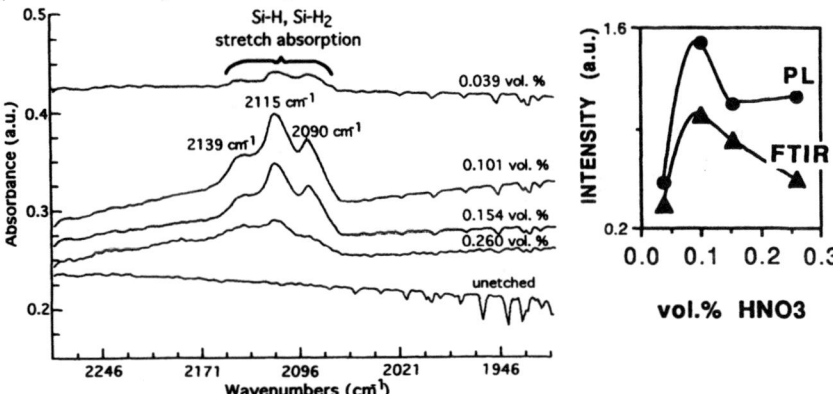

Figure 8 IR absorbance versus wavenumber in a range associated with stretch absorption by Si-H and Si-H$_2$ bonds. The inset shows a correlation between PL intensity and the IR absorbance due to the presence Si-H species.

The evolution of spectral features in the PL spectrum of chemically etched silicon films has been observed. These features appear as two components and are observed as the effect of aging in ambient conditions and also from cooling of the samples under illumination. Figure 9a shows PL measurements done on the same sample etched from a 0.1 ohm-cm Si wafer. The spectra were obtained 15 minutes, 8 days, and 83 days after etching in 0.01 vol.% HNO$_3$ in HF. Immediately following the etch the PL spectrum is a single broad peak with a peak intensity at 6430 Å. After 8 days a shoulder feature has appeared at approx. 7200Å and the peak intensity position has blue-shifted to 6320 Å. After 3 months the shoulder appears at 7000 Å and the peak intensity position has shifted very slightly to 6300Å. The evolution of this spectral feature was further observed upon cooling from 295 K to 52 K under the illumination of 30 mW, 476 nm laser light. The spectra measured during the cooling process are shown in figure 9b. At 295K, under vacuum, the spectrum appears with a shoulder feature at 6900 A and the intensity peak position at 6200 Å. As cooling occurs the shoulder feature becomes more prominent until at 185 K the spectrum appears as a two peaked distribution with intensity peaks at 6120 Å and 6800 Å. Further cooling to 52 K results in the longer wavelength component dominating the spectrum with a intensity peak position of 6750 Å and a shoulder feature at 5900Å.

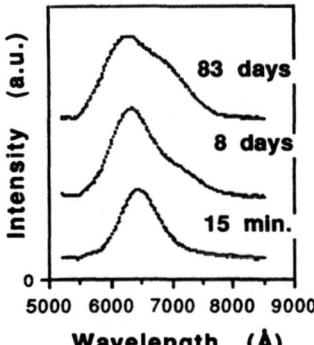

Figure 9a Photoluminescence measurements done on the same sample etched from a 0.1 ohm-cm Si wafer. The spectra were obtained 15 minutes, 8 days, and 83 days after etching in 0.01 vol.% HNO$_3$ in HF.

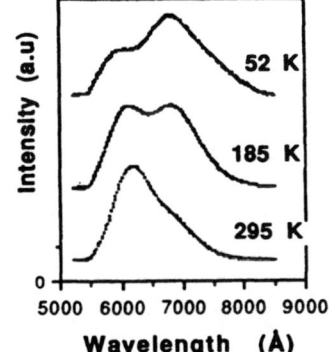

Figure 9b Photoluminescence measurements made while cooling process under illumination at temperatures of 295 K, 185 K and 52 K.

SUMMARY AND CONCLUSIONS

Results from analysis of anodically formed porous silicon films show that PL intensity decreases with increased electrolyte convection while the position of peak intensity remains constant. This decrease in PL intensity also corresponds to a decrease in surface pore area (an indirect measure of porosity), which also decreases with increased electrolyte convection. Film thickness has been determined to be independent of electrolyte convection. Analysis of spectroscopic data has been inconclusive in showing a link between increased surface hydride concentration and increased PL in anodically etched n-type films. XPS spectroscopy performed immediately after etching has shown no silicon-oxide species and a minimal oxygen content attributed to specimen transport in ambient conditions.

Our studies of chemically etched samples have shown that changes in the etch solution composition produces variations in the film chemistry and the photoluminescence intensity. These observations lead to a conclusion that the light emission from these samples is chemistry dependent. The reproducible appearance of spectral features after exposure to conditions that would expectedly alter the chemistry of the films further compounds this conclusion. Further studies are needed to identify the specific chemistry associated with the individual spectral features.

ACKNOWLEDGMENTS

Special thanks to Lee Rumaner for XPS analysis in the surface science laboratory in the Department of Materials Science and Engineering at the University of Washington.

REFERENCES

1. L.T. Canham, Appl. Phys. Lett., **57** (10), 1046 (1990).

2. Light Emission from Silicon, edited by S. S. Iyer, R. T. Collins, and L. T. Canham (Mat. Res. Soc. Symp. Proc. **256**, Pittsburgh, PA, 1992)

3. J. N. Kidder, Jr., P. S. Williams, T. P. Pearsall, D. T. Schwartz and Brett Z. Nosho, Appl. Phys. Lett. **61** (24), (1992).

4. K.H. Jung, et al., J. Electrochem. Soc., **139** (11), 3363 (1992).

5. C. Tsai, K. -H. Li, D. S. Kinosky, R. -Z. Qain, T. -C. Hsu, J. T. Irby, S. K. Banerjee, A. F. Tasch, Joe C. Campbell, B. K. Hance and J. M. White, Appl. Phys. Lett. **60** (24), 14 (1992).

6. K.H. Jung, et al., Appl. Phys. Lett., **61** (20), 2467 (1992).

7. C. Tsai, et al., Appl. Phys. Lett., **59** (22), 2814 (1991).

8. R. W. Fathauer, et al., Appl. Phys. Lett., **60** (8), 995 (1992).

9. V. Petrova-Koch, et al., Appl. Phys. Lett., **61** (8), 943 (1992).

10. Masao Yamada, et al., Jpn. J. Appl. Phys., **31**, 1451 (1992).

11. A. Nakjima et al., Appl. Phys. Lett., **61**, 46 (1992).

12. S. M. Prokes, O. J. Glembocki, V. M. Bermudez, R. Kaplan, L. E. Friedersdorf and P. C. Pearson, An Alternate Mechanism for Porous Si Photoluminescence: Recombination in SiH_x Complexes, edited by S. S. Iyer, R. T. Collins, and L. T. Canham (Mat. Res. Soc. Symp. Proc. **256**, Pittsburgh, PA, 1992) pp. 107-110.

13. H. D. Fuchs, M. S. Brandt, M. Stutzmann and J. Weber, Optical Characterization of The Visible Photoluminescence from Porous Silicon, edited by S. S. Iyer, R. T. Collins, and L. T. Canham (Mat. Res. Soc. Symp. Proc. **256**, Pittsburgh, PA, 1992) pp. 159-162.

14. K.H. Beckmann, Surface Science, **3**, 314 (1965).

MODELS AND MECHANISMS FOR THE LUMINESCENCE OF POROUS Si

F. Koch
Tech. Univ. Munich, Physik-Department E16, D-8046 Garching, Germany

Abstract

Explanations for the efficient, visible luminescence of porous Si fall broadly into three categories. At the extremes are the pure quantum-well point of view and the molecular agents hypothesis. We make here the case for a model in which the dominant absorption characteristics are those of a quantum well, but luminescence occurs via boundary states on the nanocrystalline particles. This so-called "smart quantum-well" mechanism accounts for the multiple emission bands of porous Si in a natural way. In particular, the infrared band is assigned as an electronic transition to a deep level. From this we determine the conduction band shift with particle size.

I. Introduction

Anodically etched Si surface layers luminesce when excited by light. This photoluminescence phenomenon of so-called porous Si has been known for some time /1/. It received renewed and dedicated attention ever after refs. /2,3/ suggested that the effect involves quantum-mechanically confined elementary excitations. Whereas /2/ argues that the luminescence process, the radiative recombination of the photoexcited electron-hole pair, involves quantum-confined states in the volume of the columnar structures of the material, ref./3/ takes a more restrained view. It is proposed there to explain the prominent shift of the absorption edge to higher energy by the quantum mechanics of confinement in nanometer-sized structures. The subsequent emission process is not specified.

There have been other attempts at interpreting and defining the mechanism of light emission. Most notable is the "molecular-agents" approach. Armed with knowledge of the prehistoric literature, some authors /4/ have singled out a specific molecular configuration of Si, 0 and H called siloxene ($Si_6O_3H_6$). Others have decided on polysilane ($SiH_2)_n$ as the luminous agent /5/.

Obviously there is room for arguement between the extremes of pure quantum-well physics and the identification of a specific molecule. There is a broad range of model suggestions that focuses on the internal surface of porous Si. It is more or less loosely suggested that the luminescence originates from electronic states localized on the surface of nanometer-sized crystalline remnants in porous Si. The surface-states idea can be found in various papers /6-11/. Our approach supplements the quantum-well model with a realistic spectrum of surface states. Fauchet has named it the "smart" quantum well.

We have argued in /10,11/ that the luminescing species are locally perturbed Si arrangements on the nanocrystallite surfaces. This approach tries to link porous Si-luminescence with the well-studied light emissions from amorphous, hydrogenated Si (a-Si:H) /12/. a-Si:H photo-luminescence is a highly efficient process at temperatures in the 10 K range. High quality material with the fundamental absorption onset at 1.7 eV, will luminesce at ~1.3 eV. The linewidth is typically ~0.3 eV. The time response is in the range of ms at low temperature and is nonexponential. It is well known that by increasing the H-content or adding some carbon /13/ (a-$Si_{1-x}C_x$:H) both the absorption edge and luminescence peak can be shifted to higher energies, where they would coincide with typical values for porous Si.

Except for the distinctly different temperature dependence of the intensity, most features of the luminescence of porous Si and the various forms of amorphous

Si are the same. Both are sensitively dependent on nonradiative processes that involve the dangling-bond state in Si. Linewidth, quantum efficiency and even the peak position can be made to match up by proper choice of parameters. The temperature-dependent quenching of a-Si:H luminescence is known to originate from the fact that a photoexcited electron-hole pair will drift apart. Except at low temperatures, for which the two will be trapped within a length for which tunneling recombination is possible, a-Si:H will not luminesce. The efficient room temperature process in porous Si can be understood in terms of the particulate, granular nature of the material. This limits the distance over which the excitations can move.

In the next section we shall argue the case for Si-related states localized on the surface of the nanometer-sized crystallites in porous Si. This is followed by the model description of the infrared emission band from which we derive estimates of the band-gap widening /14/. This is the basis for the interpretation of the visible emission-bands in terms of surface state transitions. In a final section we reexamine the role of the fast, blue light emission /15, 16/ in terms of the model.

II. The Energy-Level Spectrum

Structural investigations of porous Si by now draw a fairly consistent picture of this material. Efficiently luminescing Si is formed of a highly random, but interconnected arrangement of nanocrystalline remnants and passivating tissues of either SiH_2 or SiO_2. The fact that the two can be interchanged without substantially changing the emission characteristics /17/ is an important clue for the mechanism. It eliminates at once explanations based on molecular arrangements of Si, H and O. TEM and STM give evidence that the crystalline Si remnants must be of order ~30 Å in order to have efficient luminescence. In the oxidized material known as RTOPS /17,18/ and in luminescing plasma-CVD films /19/ the Si grains are physically, and by-and-large also electrically, isolated. The connectivity of undulating cylinders is therefore not a necessary feature for the light emission process. We choose to discuss the electronic states of an isolated nanometer-sized grain.

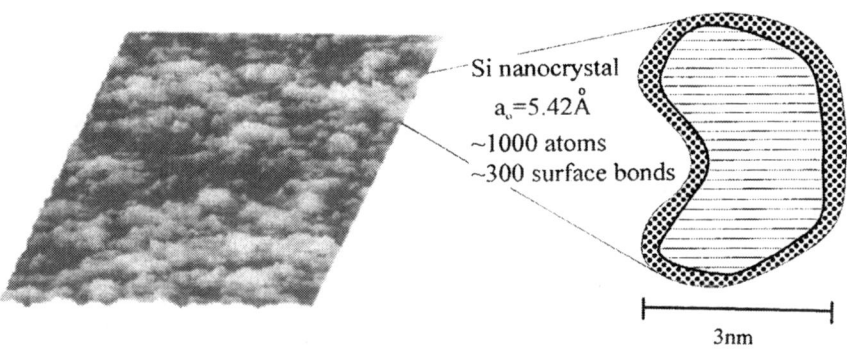

Si nanocrystal
$a_o = 5.42 Å$
~1000 atoms
~300 surface bonds

3nm

Fig.1 Scanning-Tunneling image of the surface of lumenescing porous Si (area 50 nm x 50 nm. Particles have typical dimensions of 30 Å (after Enachescu et al. ref./20/)

Fig.1 is a STM image of efficiently luminescing porous Si. This figure, taken from ref. /20/, shows clearly the fine-grained structure necessary for the emission of light. A typical linear dimension of the particles is ~30 Å. Given the unit cell edge of a Si crystal as 5.42 Å, it ranges ~5 cells across. The nanocrystal is composed of something like 125 cells and about 1000 Si atoms. The essential point to realize is that 300 of these are located on the surface and are bonded to H or O atoms. As in Fig.1, the typical grain has an irregular shape. Its atomic planes, spaced 1.35 Å apart, end abruptly where the bounding atoms terminate in the passivating SiH_2 (or SiO_2) tissue. For the randomly stepped surface with atomic scale ridges and other irregularities, the Si atoms have to accomodate the local chemistry. Bond lengths and angles for the Si atoms will adjust to suit. The reconfigured, relaxed surface region contains Si-Si arrangements that resemble amorphous Si. Given the enlarged gap for nanometer-scale Si it is expected, that some of the arrangements will have energies less than the states extending throughout the core of the crystallite. We argue that the perturbed Si atoms of the boundary layer provide states for which the electronic wavefunction is localized on the nanocrystal surface. These are like the tail states of amorphous Si.

A model calculation of surface states that result from relaxation of the outer Si layers of a planar slab structure can be found in refs./21,22/. It is by now well established that an enlarged gap energy is expected for nanometer-sized Si crystallites /23,24/. As a result the electronic states of specific surface configurations, that for bulk Si are states resonant with the bands /25/, become genuine surface states. This shows the importance of a correct, minimum energy calculation as in /21,22/, as opposed to the unrelaxed cluster considerations in other work /23,24/.

As a qualitative description of confined states in Si the quantum-well model has often been cited. The effective mass type of approach is familiar from its successful application in MBE-grown heterostructures. Such an "ordinary" quantum-well has states $E_n = n^2(2h^2/m^* d^2)$ in one dimension when the confining length is d and the band-edge mass is m^*. The "ordinary" well has a sequence of states spaced as the squares of integer numbers. It is a naive and unrealistic simplification. It should be realized that the basic tenets of effective mass theory do not apply - the potential is not constant on a scale of the particle wavelength and the expected shift of the band edge is not so small as to be described with the constant m^*. For the higher states (n = 2,3 etc.) this is all the more so.

There is a very readable treatment of the problem of the electronic states of a quantum cluster in ref./26/. In building our case for the states of the "smart" quantum-well, we use a picture based on the tight-binding description of the Si valence- and conduction-bands. For this the sp^3 atomic states are first combined symmetrically as bonding orbitals σ and antisymmetrically as antibonding orbitals σ^* (Fig.2). These are the degenerate basis states from which the cluster molecular orbitals of the valence and conduction states are calculated. Given the typical linear dimension of the cluster as five unit cells, there are a finite number of overlapping bonds with various phases to be considered. The resulting number of states is large enough to approximate a continuum. The gap separation exceeds that of bulk Si. For this description the states extend throughout the crystallite with its regularly built-up atomic arrangements. We count these as volume states.

The "smart" quantum-well allows the atoms in the outermost layers to relax and reconfigure according to the local structure and chemistry. Steps, structural irregularities and the local chemistry of bonding to O and H will require specific atomic configurations to minimize the energy. Some of these, in analogy with the calculations in /21,22/, will have energies in the gap of the bulk state spectrum. These are energy levels that confine the electron to the surface, in general a small region of the surface.

A highly localized surface state to be expected for a realistic quantum-well is the dangling bond. Not every valence bond on the arbitrarily shaped particle surface will be passivated. The defect typical for Si interfaces with oxide or hydride will exist where structural irregularities occur. A large planar structure, like the hydrogen-covered (111)Si surface may have a high degree of perfection /27/. The constraints of such regular placement of surface atoms is missing in the small granules of Si. We expect defects and do observe a spin resonance signal that is first-hand evidence for singly occupied bonds /17/. Real quantum-wells of Si formed by the anodic etching process will have dangling-bond surface states. We have included these as a deep level with various charge states $D^{0, \pm}$ in the approximate midgap position of Fig.2.

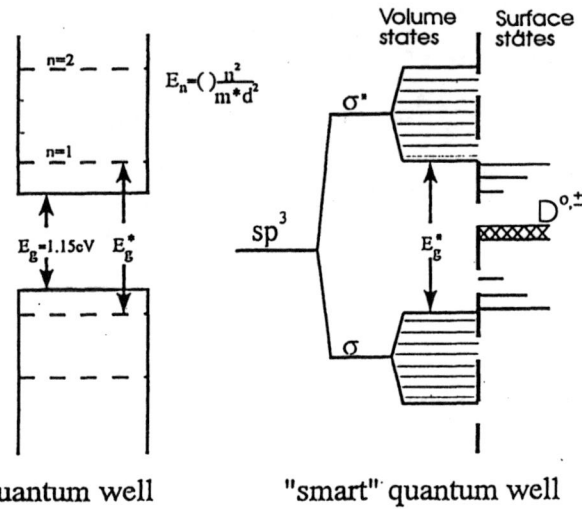

Fig.2 Comparison of the energy levels of a confined system in terms of effective mass and in a molecular orbital picture. In contrast with the "particle in a box" quantum mechanics, the "smart" quantum-well allows dangling bonds D and a spectrum of surface levels in the enlarged gap E_g^*.

There is plenty of evidence that the quantum crystallites which we are discussing here have both shallow and deep level states on their surfaces. We have pointed this out in previous work /10,11/. Evidence for the structural rearrangement and disorder of the surface atoms comes from the width and structure of Si-Si vibration spectrum. The fact that these can be excited by infrared radiation just as in a-Si:H is significant. The problem has also been discussed in /28/. Additional confirmation of disorder in the surface layer comes from the large width of the surface-hydrogen vibration band at 2100 cm^{-1} which is spread over ~80 cm^{-1}. For a perfect crystalline (111)Si surface the vibration linewidth is ~1 cm^{-1} /27/.

Direct evidence for the tail states in the electronic spectrum of the "smart" quantum-well comes from optical absorption measurements. The photothermal-deflection spectroscopy (PDS) measurements that we first introduced in /11/ in order to determine the absorption constant α(E) of porous Si at energies down to 0.8 eV, has since been repeated by other researchers /29,30/. There is no doubt that absorption occurs below the bulk Si band-gap of 1.15 eV. The absorption constant in the spectral region 0.8 to 1.5 eV is comparable to high quality a-Si:H as we have pointed out in /11/. Material preparation, in particular of the a-Si:H, can be expected to greatly influence α in this spectral range. The α values shown in /29/ are for a very low quality a-Si:H and do not allow a meaningful comparison with porous Si. Badoz et al. /30/ give an unrealistically low absorption in the 0.8 to 1.5 eV range for their porous Si. The PDS experiment needs careful calibration. The divergent data may be the result of calibration differences rather than sample quality. We emphasize that the

0.8 eV absorption is an important benchmark relating to the number density of dangling bonds. This can be independently measured by spin resonance. For our porous Si a spin density comparable with but higher than the best a-Si:H (3×10^{15} spins cm^{-3}) applies. This gives confidence to the statement that the subgap absorption of porous Si is comparable with that of good a-Si:H. It follows that the number of electronic states in the relevant energy range for amorphous and porous Si is similar. This is a cornerstone of our arguements for surface states in the light emission process.

III. The Infrared Emission Band

The dangling bond state is an important characteristic of the Si surface. It has a distinct spin signature and g-factor of 2.0055. As a deep level in the energy spectrum of the "smart" quantum-well it will act predominantly as a non-radiative recombination center. Nevertheless, some fraction of the electronic transitions involving the dangling-bond are radiative. The transitions are know to sensitively depend on the spin. In order to identify the deep level contribution to the luminescence it is instructive to reexamine the optical emission spectrum in MOS devices. Fig.3 is obtained as an electroluminescence from an FET device operated under hot electron conditions at 77 K.

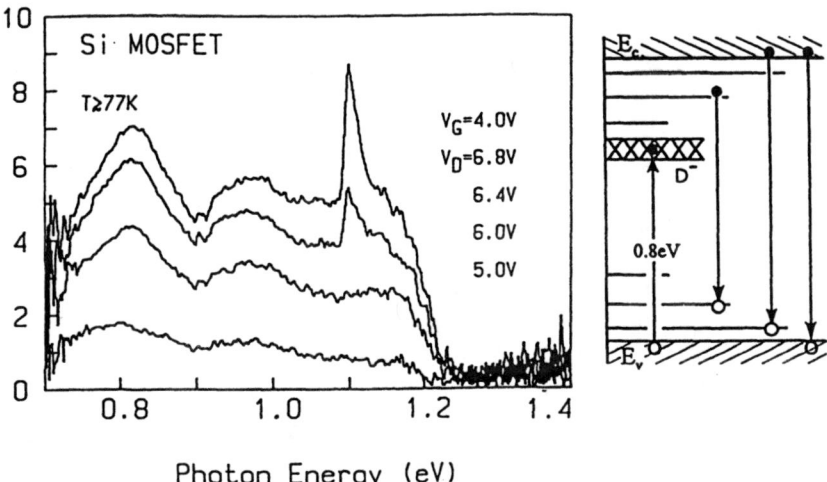

Fig.3 Emission spectrum of a silicon-based field-effect transistor (MOSFET) excited by source-drain current. In ref./31/ the 0.8 eV emission has been arbitrarily assigned as a hole to D⁻-transition. The below band-gap light is interpreted as recombination processes involving the tail states of the energy spectrum.

The sharp peak just above 1.1 eV is the band-gap recombination inside the device where carries are cold. The remainder, in particular the ~0.8 eV peak, comes from radiative processes at the interfacial surface of Si and its oxide SiO_2. The electronic energy spectrum of MOS device interfaces has been studied for many years with sensitive capacitance spectroscopies. The result is found in textbooks. One generally finds a U-shaped distribution of defect states ranging from a low of 10^{10-11} cm^{-2} eV^{-1} in midgap to above 10^{12-13} cm^{-2} eV^{-1} near the band edges. The point is that there are plenty of defect states in the band-gap that can act as recombination centers. We

324

can expect that light in the spectral region 0.7 to 1.15 eV involves surface states which localize the carriers. In ref./31/ it is suggested that the 0.8 eV peak in Fig.3 results when a free hole recombines at a negatively occupied dangling bond, the so-called D⁻-state. Both band-to-bound and bound-to bound transitions can be expected in the transition scheme. The energies above 1.1 eV are likely to be caused by hot carriers in resonant traps above the band edges.

The interpretation of the radiative 0.8 eV transition as hole (h) →D⁻ recombination follows ref./31/. It is generally accepted /32/ that this transition is degenerate with the electron (e) →D⁺ process. This means, that within the resolution of the experiment, one cannot distinguish whether a band-edge hole neutralizes the doubly-occupied dangling-bond state or an electron falls into the unoccupied D⁺-state. This degeneracy is most unfortunate. It leaves an ambiguity in the interpretation that we shall propose below.

The porous Si that most closely resembles the above case is mesoporously etched material. In Fig.4 is shown the photoluminescence of a sample of p⁺ (0.005 Ω cm) Si etched under conditions that are known to produce structures of order

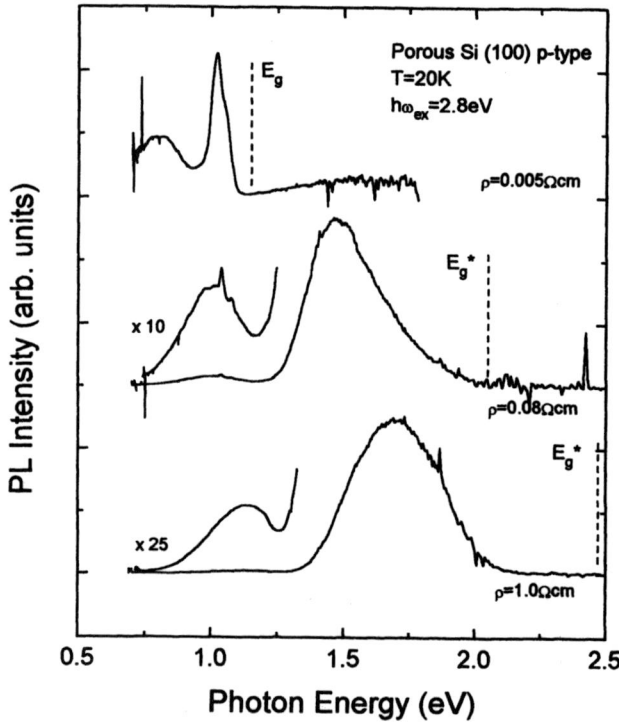

Fig.4 Luminescence spectra of a series of anodically etched p-type Si with different resistivities. The etching conditions (~30 mA/cm²) and layer depths are kept constant. From the position of the infrared band the enlarged band-gap E_g^* is estimated and marked on the figure.

100 Å or more. We note the striking similarity of the deep infrared emissions near 0.8 eV with that in the previous figure. Moreover, the higher energy emission is shifted distinctly below a reasonable value of the gap energy (1.15 eV). Localization at acceptor sites and the phonon replica effects are known to shift the luminescence to energies below the gap. We note that the Stokes shift here is larger than that known for bulk Si and suggest that this is the influence of potential fluctuations on the pore surfaces. We note the wide line that is observed. Being p^+-material the most likely optical transition that explains the 0.8 eV peak is the $e \rightarrow D^+$ process. One cannot totally preclude the alternative $h \rightarrow D^-$ recombination on this basis, but it is unlikely for a material that contains such a high concentration of B atoms.

Starting in Fig.4 with the highly-doped sample makes possible the correlation with macroscopic Si surface luminescence. We follow with p-type samples whose resistivities are 0.08 and 1.0 Ωcm. The sequence of traces suggests that the doublet pair continues to move up in energy as the particle size is decreased. The comparison makes reasonable the assignment of the infrared sattelite in each of the spectra as transitions to the deep level. It becomes increasingly difficult to observe as one procedes from the mesoporous (p = 0.005 cm) to microporous (p = 1.0 cm) etch structures.

The identification as a deep level transition is convincingly confirmed by the recent ODMR experiment. The infrared peak for each of the spectra in Fig.4 is very sensitive to spin excitation. In the experiments /33,34/ it is estimated that saturation of the spin resonance will approximately double the luminescence. This positive and large response means that the infrared light is linked directly with the surface dangling bond. This part of the light emission has its origin in a surface state recombination. The only remaining question is which one of the two possible transitions it is. Is $e \rightarrow D^+$, or the alternative $h \rightarrow D^-$, the radiative channel observed in the experiments?

A deep level on the surface has an energy that is largely unaffected by the neighboring atoms. Its energy, just like that of an internal transition in the inner shells of an atom, is not influenced by the bandstructure effects in the surrounding medium. This is the principle which Langer and Heinrich /35/ applied so successfully in III-V semiconductors to measure band-offsets. We apply it here in order to estimate the band-edge shifts that apply for the data in Fig.4. There is evidence, related to the dependence of the infrared luminescence on background doping, on the spin-density, on temperature and on the variation with porosity and structural size, that uniquely says that $e \rightarrow D^+$ is the radiative transition /36/. We make this assumption in order to derive the conduction band shifts in Fig.5. From the peak positions 0.8, 1.03 and 1.13 eV read off the data (Fig.4), the conduction band shifts are 0, 0.23 and 0.33 eV.

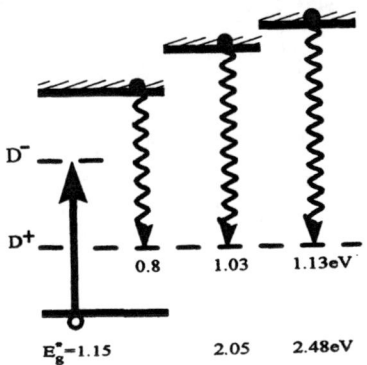

Fig.5
Conduction band shifts and estimated band-gap E_g^* derived from the measured infrared band energies. We identify the $e \rightarrow D^+$ transition as the radiative process.

The conduction band shifts by themselves are of little consequence. But, if we rely on theory to scale to the expected valence band shift then we are in a position to make an estimate for the band-gap widening effect. This is a test for the predictive quality of the infrared data. In spite of the uncertainties connected with tight-binding calculations such as /21-23/, the ratio $\Delta E_v / \Delta E_c$ can be expected to be given with reasonable accuracy. From many observations, including the fact that Urbach tails in a-Si:H, are larger in the valence- than in conduction-band of Si, it must be expected that $\Delta E_v / \Delta E_c$ is substantially greater than unity. From /22/ and /23/ we extract a ratio of three. This results in values of the widened gap E_g^* of 1.15, 2.07 and 2.48 eV. We have entered these in the figure. In each case, they are well on the high energy side of the luminescence. Although the E_g^* estimates are based on assumptions that need yet to be verified, the values are in line with those expected from absorption measurements.

There is a critical test of the $e \rightarrow D^+$ or $h \rightarrow D^-$ hypotheses afforded by the data. Had we assumed the second choice, then the E_g^* values are 1.15, 1.46 and 1.59 eV. For the latter two curves of the data, this would place them squarely on top of the observed luminescences. The surface state hypothesis would have succeeded in disproving itself! The identification of the radiative character of $e \rightarrow D^+$ is an essential part of the present approach.

IV. The Visible Luminescence Band

Having assigned the infrared signal as a conduction-band to deep-level transition, it is a logical next step to identify the visible bands with some sort of localization-perturbed transitions in the level scheme of the "smart" quantum-well. We are guided in this choice by knowledge of the comparably luminescing a-Si:H, the similar relaxation kinetics for this material, and a firm conviction that localization of the primary excitation to a dimension smaller than that of the confining crystallite is necessary. This point can be argued by reference to the oscillator strength calculation in /24/. The thought is simple. If confinement is good for getting light out of Si, than more confinement by localizing in a surface tail-state must be better. In this sense, we have argued in refs./10,11/ previously. The explanation does not preclude the recombination between extended volume states. Only it considers these less likely, when tail-states are available and when the temperature is not too low. The electrons and holes trapped some distance apart on the surface of nanocrystallites recombine via tunneling. At low temperature, in a big particle, this tunneling recombination is exponentially unlikely. The elementary excitations cannot be thermally released to find a more favorable spatial arrangement. The slow rate of recombination at low temperature may lead to a blocking of this mechanism.

We briefly repeat the ideas of the surface-state mechanism with reference to Fig.6. After photoexcitation at a high energy an electron-hole pair extending throughout the crystallite is created. The lowest energy luminescence transitions are those at $\hbar w_2$ for which both partners have relaxed to a surface state. This is also the slowest type of process. Transitions of the type $\hbar w_1$, for which only one of the partners is taken as a localized state, occur at higher energies and are expected to occur faster. Volume transitions of the type $\hbar w_0$ are at the highest end of the energy scale. Because particle sizes created in the anodization are spread over a wide scale, line widths are inhomogeneously broadened. It may well be difficult to extract the individual E_2, E_1 and E_0 contributions from a given visible line spectrum such as that in Fig.4. There is in addition the problem that the processes each have an intrinsic size dependence which will weight their relative contribution to the overall intensity. For example, the 30 Å reference particle may emit mainly via the E_2 type of transition. A smaller particle in the ensemble of the same porous Si specimen may predominantly luminesce via the E_1 process. For this reason it will be difficult to scale peak shifts with particle size.

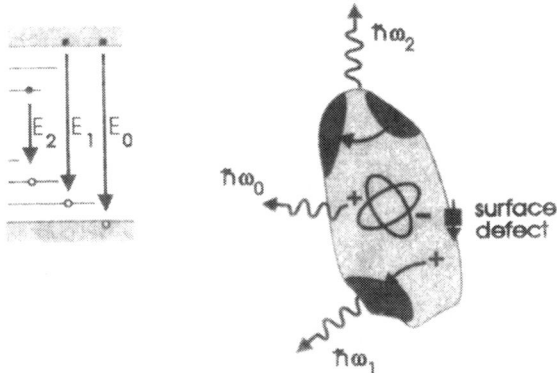

Fig.6 Transition scheme for the surface-state model of the visible luminescence from porous Si. A photoexcited e-h pair in a nanocrystallite came emit in three different modes that are distinguished by their energies and characteristic response times.

It is instructive to derive some numbers for the surface-state model. With reference to the middle curve of Fig.4, for which we had estimated E_g^* as ~2.1 eV, the observed luminescence energy is ~0.3 eV for each of the elementary excitations. By analogy with a-Si:H, we expect a typical binding length to be of order 6-10 Å. A prediction of the model is that posttreatment steps, which have no influence on the particle size can shift the visible luminescence up and down. The up-shift to a maximum of the E_g^* value is possible for the given size. For the middle curve in Fig.4 this is 2.05 eV.

V. The Blue Emission Band. Some Caveats.

The scheme that we have discussed above serves well to explain the infrared and the visible emission in the red to orange-yellow. Even with green at 2.3 eV there is no problem. The blue luminescence that has been the subject of many recent papers /15,16,37-40/ with its peak at 2.7 eV, however, is unintelligible on the basis of the surface state mechanism.

The difficulty lies with the fact that the shifts of the infrared band cover a limited range. The highest energy position of the band that we have observed so far is ~1.2 eV. With the assumptions that we have made so far, this permits us to estimate a maximum value of the gap widening as 1.15 + 4 x 0.4 = 2.75 eV. Emissions in the blue fall at this energy. There is no way to adjust the numbers to make the emission appear below the E_g^* limit. We must conclude that the blue luminescence represents E_0 processes occuring at the band-gap itself. The blue light is not a surface-state emission.

There are many features of the blue light that make it appear as a very different kind of radiative process, one that can not properly be described in terms of Si nanocrystals with surface states. The estimate above shows that it would have to be a direct recombination, not requiring any localization step at all. It appears that the particles have reached an ultimate limiting size and act more like a molecule. The ultrafast response of ~6 ns, the different luminescence excitation spectrum, the fact that the intensity rises strongly when the usual luminescences are quenched, all show the blue light to be an object of a different color /16/. It is interesting to note that it occurs exactly at the limit of the E_g estimates based on an infrared line at 1.2 eV. This value was found for the strongly oxidized samples (T_{ox}~900°C) where the blue luminescence begins to dominate the spectrum.

In the final section of this paper, it is important to put the conclusions into perspective and express some words of caution. The new aspect has been the emphasis on the infrared band, using it to prove the case for the surface-state model. It should be remembered that there is nothing new or radical about the experimental evidence for the infrared emissions. Ref./1/ already has it all. The recently published work in ref./40,41/ adds some specifics. What is new and speculative is our interpretation. This in turn relies heavily on the electroemission spectrum of MOSFETs and on the ODMR experiments in /33,34/. Words of caution are that the dangling bond state is not necessarily so localized that it is unaffected by its neigbors and that the Langer-Heinrich rule may not strictly apply. The statement that $e \rightarrow D^+$ is likely to lead to emission of light, whereas $h \rightarrow D^-$ is mainly nonradiative, is not proven independently. We may be wrong, but when coupled with the $\Delta E_V / \Delta E_C$~3 estimate, there is presently no way to argue surface states without this hypothesis. It is a key requirement for the model to give reasonable numbers.

VI. Acknowledgements

Working as part of a group with a dozen or more colleagues, guest scientists, postdocs and students, ideas circulate freely and their authorship is not easy to trace. What is certain is that the ideas in this paper are not all my own. In particular, Thomas Muschik, Vesselinka Petrova-Koch, Bruno Meyer, Vladimir Gavrilenko, Detlev Hofmann, Andreas Kux, Frank Müller, Moshe Ben-Chorin, Dimitrii Kovalev and others have contributed bits and pieces to this story. They are cited properly in references to the original publications.

The financial support of the Siemens AG and the BMFT is gratefully acknowledged.

References

/1/ C. Pickering, M.I.J. Beale, D.J. Robbins, P.J. Pearson, R. Greef, J. Phys C. 17, 6535 (1984).
/2/ L.T. Canham, Appl. Phys. Lett. 57, 1046 (1990).
/3/ V. Lehmann, U. Gösele, Appl. Phys. Lett. 58, 856 (1991).
/4/ M.S. Brandt, H.D. Fuchs, M. Stutzmann, J. Weber, M. Cardona, Sol. State Commun 81, 307 (1992).
/5/ S.M. Prokes, O.J. Glembocki, V.M. Bermudez, R. Kaplan, L.E. Friedersdorf, P.C. Searson, Phys. Rev. B, 45, 13788 (1992).
/6/ Y.H. Xie, W.L. Wilson, F.M. Ross, J.A. Mucha, E.A. Fitzgerald, J.M. Macauley, T.D. Harris, J. Appl. Phys. 71, 2403 (1992).
/7/ T. Ito, T. Ohta, A. Hiraki, Jpn. J. Appl. Phys. 31, L1 (1992).
/8/ T. Matsumoto, M. Daimon, T. Futagy, H. Mimura, Jpn. J. Appl. Phys. 31, L 619 (1992).
/9/ K. Murayama, S. Miyazaki, M. Hirose, Jpn. J. Appl. Phys. 31, L 1358 (1992).
/10/ F. Koch, V. Petrova-Koch, T. Muschik, A. Kux, F. Müller, V. Gavrilenko, F. Möller, Proc. 21st ICPS (Beijing 1992), World Scientific, in press.
/11/ F. Koch, V. Petrova-Koch, T. Muschik, A. Nikolov, V.Gavrilenko, MRS, Proc. 283, 197 (1992).

/12/ R.A. Street, in Semiconductors and Semimetals 21B, ed. by J. Pankove, (Academic Press, New York 1984), p. 197.

/13/ W. Siebert, R. Carius, W. Fuhs, K. Jahn, phys. stat. sol. (b) 140, 311 (1987).

/14/ F. Koch, T. Muschik, V. Petrova-Koch, E-MRS Strasbourg (1993), J. of Luminescence (to be published).

/15/ V. Petrova-Koch, T. Muschik, D.I. Kovalev, F. Koch, V. Lehmann, MRS Proc. 283, 178 (1993).

/16/ D.I. Kovalev, T. Muschik, V. Petrova-Koch, F. Koch, I.D. Yaroshetzkii, Appl. Phys. Lett., submitted.

/17/ V. Petrova-Koch, T. Muschik, A. Kux, B.K. Meyer, F. Koch, V. Lehmann, Appl. Phys. Lett. 61, 943 (1992).

/18/ A.G. Cullis, L.T. Canham, G.M. Williams, P.W. Smith, O.D. Dosser, MRS Proc. 283, 257 (1993).

/19/ M. Rückschloss, B. Landkammer, O. Ambacher, S. Veprek, MRS Proc. 283, 65 (1993).

/20/ M. Enachescu, E. Hartmann, F. Koch, Appl. Phys. Lett., submitted.

/21/ M. Enachescu, E. Hartmann, F. Koch, E-MRS Strasbourg (1993), J. of Luminescence, to be published.

/22/ V.I. Gavrilenko, P. Vogl, F. Koch, Phys. Rev. B, submitted.

/23/ S.Y. Ren, J.D. Dow, Phys. Rev. B, 45, 6492 (1992).

/24/ B. Delley, E.F. Steigmeier, Phys. Rev. B, 47, 1397 (1993).

/25/ A.I. Shkrebtii, R. Del Sole, Phys. Rev. Lett. 70, 2645 (1993).

/26/ M.G. Bawendi, M.L. Steigerwald, L.E. Brus, Annu. Rev. Phys. Chem. 41, 477 (1990).

/27/ G.S. Higashi, Y.J. Chabal, G.W. Trucks, K. Raghavachari, Appl. Phys. Lett. 56, 656 (1990).

/28/ J.C. Tsang, M.A. Tischler, R.T. Collins, Appl. Phys. Lett. 60, 2279 (1992).

/29/ P. Basmaji, V. Grivickas, G.I. Surdutovich, R. Vitlina, V.S. Bagnato, MRS Proc. 283, 227 (1993).

/30/ G. Vincent, F. Leblanc, I. Sagnes, P.A. Badoz, A. Halimaoui, E-MRS Strasbourg (1993), J. of Luminescence, to be published.

/31/ A. Kux, M. Schels, F. Koch, W. Weber, Proc. INFOS (Liverpool 1991), ed. by W. Eccleston and M. Uren (Hilger, Bristol, 1991) p. 255.

/32/ C. Delerue, G. Allan, M. Lannoo, Phys. Rev. B, submitted.

/33/ B.K. Meyer, D.M. Hofmann, V. Petrova-Koch, F. Koch, P. Omling, P. Emanuelsson, Appl. Phys. Lett., submitted..

/34/ B.K. Meyer, D.M. Hofmann, W. Stadler, V. Petrova-Koch, F. Koch, P. Emanuelsson, P. Omling, E-MRS Strasbourg (1993), J. of Luminescence, to be published.

/35/ J.M. Langer, H. Heinrich, Phys. Rev. B, 55, 1414 (1985).

/36/ F. Koch, T. Muschik, V. Petrova-Koch, E-MRS, Strasbourg (1993), J. of Luminescence, to be published.

/37/ X.Y. Hou, G. Shi, W. Wang, F.L. Zhang, P.H. Hao, D.M. Huang, X. Wang, Appl. Phys. Lett. 62, 1097 (1993).

/38/ P.D.J. Calcott, K.J. Nash, L.T. Canham. M.J. Kane, D. Brumhead, J. Phys. Cond. Matter 5, L91 (1993).

/39/ J.C. Vial, I. Mihalcescu, NATO, Workshop Meylan (1993), ASI Series.

/40/ C.H. Perry, F. Lu, F. Namavar, N.M. Kalkhoran, R.A. Soref, MRS Proc. 256, 153 (1992).

/41/ Y. Mochizuki, M.Mizuta, Y. Ochuai, S. Matsui, N. Ohkubo, Phys. Rev. B, 46, 12353 (1992).

Ion exchange effects in porous silicon

B. Matvienko[a], P. Basmaji[a], V. Grivickas[b] and A.Bernussi[c]

[a]Instituto de Física e Química de São Carlos
Departamento de Física e Ciência dos Materiais
Universidade de São Paulo, Caixa Postal 369
13560-970 São Carlos, SP, Brasil
[b] Vilnus University- Semiconductor Physics 205 Vilnus , Lithuania
[c] CPqD-Telebras - 13085 Campinas, SP , Brazil

Abstract
We report an effect of alkaline metal ion incorporation into porous silicon (PS) structure from the electrolyte during anodic etching. Blue shift of the whole luminescence spectrum was obtained as a consequence of ion incorporation. We suggest the presence of zeolite-type structures in PS, as an explanation to the observed properties.

Introduction

The properties of PS have recently attracted considerable interest since this material exhibits a strong photoluminescence (PL) at room temperature and creates the prospect for optoelectronic applications. PS is typically prepared by electrochemical anodization in HF solution. It consists of an array of randomly spaced pores of different diameters ranging from 500 Å down to "micropores" $< 20 Å$[1]. This property of PS, its structure and chemical composition, although not well understood, can be varied over a surprisingly wide range [2-6]. In order to fully exploit this material, it is important to control its structure as it is generated by electrochemical means.

In this paper, we report some properties of anodized PS with incorporated ions of the alkali metal groups. We show that the ions can be exchanged by rinsing with appropriate chemical soluctions. A noticeable blue shift of the whole luminescence was obtained as a consequence of ion ioncorporation.

Complementing ion exchange we make an attempt to block any possibly present Si-OH groups in an effort to narrow down the nature of the resident charges in PS which take part in ion exchange

Experimental

The used substrates where (100) oriented, 1Ωcm resistivity, boron-doped P-type silicon. The anodization was carried out in different solutions, varying the type of ions (H^+, Na^+, K^+, Rb^+, Cs^+) and using 25% HF solution in water, at a current density of 60 mA/cm^2 for 5 min. After etching, wafers were rinsed in ethanol and dried in ambient air.

The incoporation of the various ions in the PS was checked using energy dispersive X-ray analysis (EDX). In same samples the resident ions were exchanged.

PL experiments were done with PS layers using the 457.9 nm line of an Ar^+ laser as an excitation source. The radiation emitted by PS was dispersed by a 0.5 m single grating monochromator and detected with a cooled S1 type photomultiplier. Typical excitation intensity was in the range between 10 and 200 mWcm^{-2}.

IR spectra were recorded by a Fourier transform Bomem DA8 spectrometer. An unetched Si wafer was taken as the reference for the PS measurements. The surface structure of the PS layers was also studied by scanning electron microscopy (SEM).

Silanizations were performed by exposing samples to tri-methyl-chloro-silane (TMCS) which acts on polymeric silicic acids thus:

$$-(SiO_2)_n - OH + Cl - Si(CH_3)_3 \rightarrow -SiO_2 - O - Si(CH_3)_3 + HCl$$

Some samples were exposed to the vapor of TMCS at ambient temperature during 10 minutes. Other samples were immersed in it for the same duration.

Results

Fig.1(a-d) shows the EDX spectra of freshly anodized PS in the presence of different ions in the anodizing solution. For a further investigation samples etched in the presence of Rb^+, which in a visual inspection showed the most uniform structure, were used in an ion exchange process. This consisted in the immersion of the PS sample into a solution containing NaCl and H_2SO_4, for different time intervals. It is well konwn that H_2SO_4 solutions do not affect Si. Varying the time interval we observed that 5 min was enough to get Rb^+ disappearence and Na^+ ion saturation as seen by the EDX spectra (Fig.1e) Continuing this procedure the same sample was immersed now in a $RbCl + H_2SO_4$ solution for 5 min. Fig. 1(f) shows the re-appearance of Rb^+ ion after the immersion. This result shows that the ions in the PS can be exchanged.

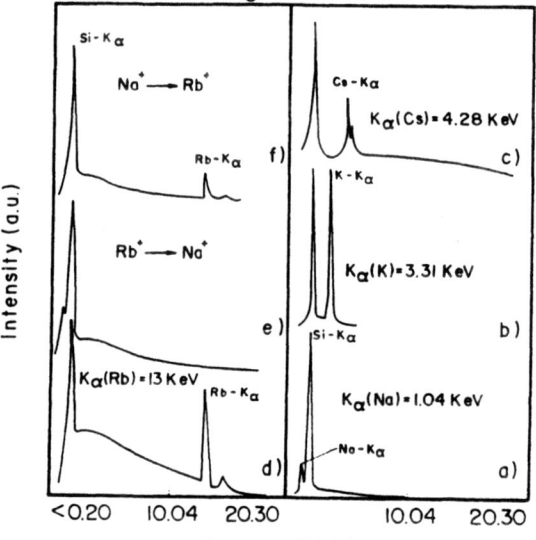

Fig. 1 : Energy (KeV)

EDX spectrum of PS layers a) after etching with sodium Na^+ ions in HF solution. b) etched with potassium K^+ ions. c) etched with cesium Cs^+ ions. d) etched with rubidium Rb^+ ions. As in d after $(Rb^+ \rightarrow Na^+)$ exchange in solutions of H_2SO_4. f) reverse exchange $(Na^+ \rightarrow Rb^+)$ in solution of H_2SO_4. We note, in all spectra the presence of the Si-K_α line.

Fig. 2 shows 300K PL spectra of a) PS as anodized with hydrogen, sample #1, with a peak maximum at 720 nm and a very small shoulder at 600nm. In spectrum b, sample #2 (anodized with Na^+) we observe a shoulder at 675 nm and the peak maximum at 600 nm. For sample #3 (anodized with Rb^+) in spectrum c the shoulder was shifted to high energy in comparison with the shoulder observed in sample # 2 at 675 nm and it became more pronouced, having its maximum at 660 nm in the sample #3 while the peak maximum at 600 nm is present in the both samples.

In spectrum d (anodized with Cs^+) we observed one peak around 700 nm similar to what has been observed in spectrum c, with a small shift toward lower energy. However, the peak located at 600 nm persisted. A new transition appeared at 550 nm with a small shoulder at 510 nm. To demonstrate the ions can be exchanged, the sample #1 was subsequently cut in two halves, sample 1a and sample #1b which was used as a reference. Sample #1a was immersed in a solution containing NaCl + H_2SO_4 at pH = 0.

Fig. 3-a shows the spectrum of as anodized PS, it has a peak maximum at 720 nm. As a consequence of ion exchange, performed in Na^+ + H_2SO_4 solution we observed in Fig 3-b the appearance of a shoulder at 680 nm and a peak maximum at 600 nm due to the Na^+ ions introduced during the exchange. This result is similar to what we observed in the PS sample anodized in the Na^+ + HF solution. This is evidence that ions are exchanged at the surface of PS. Continuing this procedure sample #1-a was immersed now in sulfuric acid solution for 5 min to replace sodium ions by hydrogen ions and thus return to the as anodized state. The resulting PL spectrum shown in Fig 3-c exhibits a peak maximum at 725 nm but the shoulder at 600 nm has almost disappeared. The appearance and disappearance of spectral lines with each exchange is a striking results of the ion exchange.

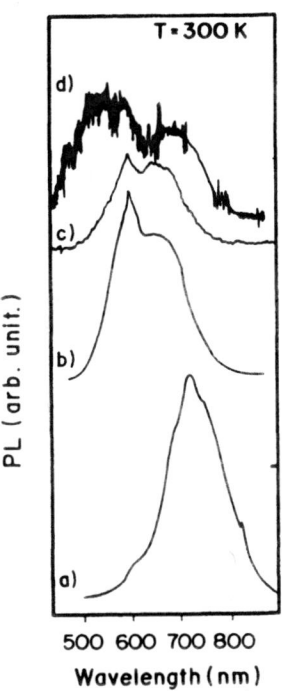

[Fig. 2]: 300K PL spectra of PS sample a) as etched without alkali ions b) in presence of Na^+ions, c) in presence of Rb^+ ions d) in presence of Cs^+ ions.

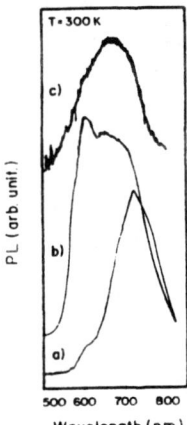

[Fig. 3]: 300K PL spectra of PS sample a) as prepared, b) after immersion in the Na Cl + H_2SO_4 solution, c) after immersion in the sulfuric acid.

The ion exchange experiment is presented through IR spectra in Fig.4. This set of spectra containes vibrational modes of

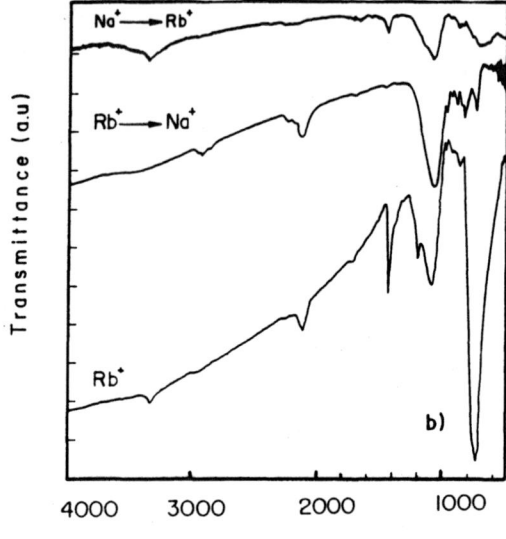

[Fig. 4].

IR absorption spectra of PS layers: The lowest spectrum as prepared, above it as Rb^+ was exchanged for Na^+ ($Rb^+ \rightarrow Na^+$) and at the top where in the reverse exchange Na^+ was replaced by Rb^+ ($Na^+ \rightarrow Rb^+$).

SiH_i (645 cm^{-1}, 870 cm^{-1}, 935cm^{-1} and 2100cm^{-1}), SiH_iR_m modes (2140 cm^{-1}) Si-O-Si modes (1060 cm^{-1}) and H_i modes (1715cm^{-1}, 2850-3050 cm^{-1}). We note that the SiH_i, SiH_iR_m vibrations remain roughly unchanged after the exchange step $Rb^+ \rightarrow Na^+$. The next step, the reverse exchange $Na^+ \rightarrow Rb^+$, restores a set of vibrations which we believe are related to fluorine but their intensity is reduced.

Our attempts to silanize PS samples had negative results: neither the PS spectra nor the IR absorption was affected by it, in particular, no additional O-Si-R related modes appeared in the IR spectra, indicating that no Si-OH groups were originally present.

Discussion

In an attempt to narrow down what type of acid sites could be responsible for holding cations like H, Na, Rb etc and take part in ion exchange, we first list anion species which could be present in our PS samples, observing that to be effective as ion exchangers they ought to be insoluble in acidified solutions.

Of the insoluble kind there are the polymeric silicic acids, as one of them, and a silicon-oxygen-boron network as the other. It is conceivable that besides these two species, which are chemically well defined and can exist in the bulk, an acidic surface compound is present, deriving its negative charges from the crystal grid mismatch between the silicon wafer and the adjoining oxidized layer. In other words, the Si quantum wires could be surrounded by a "crystal defect" in the form of a negatively charged envelope being a consequence of the spacing mismatch of the two adjoining regions. But such a stucture could form at most one monolayer on each quantum wire.

Of the soluble kind, the presence of the following acidic ion can be justified in our PS samples : fluosilicate, fluoride, chloride, borate, sulfate. By themselves they should not be effective as ion exchangers because, due to their solubility they would be washed away, and an attempt to produce a sequence of two ion exchanges in succession would result in the elimination of the potential ion exchanger. But considering that the sample is porous it adsorbs physically the solutes present in the solution. Even so it is unlikely that a significant amount of an adsorbed salt would be left after a sequence of two exchangers and rinsings. For this reason, in our opinion, it is unlikely that ion exchange is due to salts of these soluble acidic compounds.

Considering the two insoluble species capable of ion exchange, either of them could constitute the bulk of the oxidized layer of PS. At first sight the presence of polymerized silicic acid looks plausible: during the formation of PS the silicon wafer is oxidized anodically in a water solution and thus a superficial layer of a substance like silica gel could be expected, i.e. a porous mixture of polymeric silicic acids. However our silanizing experiment shows that there are no Si-OH groups in PS which is probably due to the presence of HF in the anodizing solution. In other words, ion exchange cannot be ascribed to polymerized silicic acids.

Silicon-oxygen-boron networks are compounds which are analogous to the common zeolites. In these there is a regular network of Si-O-Al where the Si and Aluminum atoms are surrounded by oxygen tetrahedra in a way that all O atoms are shared by pairs of neighbors. They can be grown hydrothermally in a process where hydrated alkaline ions are incorporated as charge compensators into an otherwise co-valently bound Si-O-Al network. Depending on the Si to Al ratio and other factors, a large variety of crystal structures is obtained [7-9]. The negative charge on the co-valent network, which requires compensation, is due to the coordination of the forming atoms. Both, Si and Al atoms are surrouded by four oxygen atoms (tetra-coordinated). While this leaves the tetra-valent Si electrically neutral, the tri-valent Al acquires by it a negative charge. The compensating hydrated ion, or sometimes a group of ions, forms large regular cavities, called "cages", inside the zeolites, which remain accessible for ion exchange in the grown crystal. Boron analogues of the common zeolites have been synthesized [10]. They are as stable as the other kind and are as easily formed. One apparent difficulty in postulating their formation in porous silicon is the Si to B ratio in the P-type silicon substrate, which is of the order of 10^5. It sould be born in mind, however, that during PS formation not all the silicon forms oxide and that most of it forms fluosilicate, a species which should not be counted for the Si/B ratio when zeolite formation is concerned. So, an appropriate ratio for zeolite formation could be present. As to the total quantity of boron present, if most of it is used in zeolite formation, several hundreds of molecular layers could be built up around each silicon quantum wire.

The amount of boron liberated during the anodic oxidation, although enough to passivate the pore wall, is insufficient to thicken the layer to the point of pore obstruction, as it happens during the anodization of aluminum [11].

A striking fact about the ion exchange is that besides the appeerence and disappearence of IR absorption lines there is a reduction of intensity with each exchange, as if bringing the structure into a relaxed form that tends to exclude higher mass ions like rubidium.

In summary, we demonstrated the effect of ion incorporation and exchange in PS during etching and subsequent passivation. Our results indicate that a zeolite/silicon interface may be present in the porous silicon structure.

This work was supported by funds form FAPESP and CNPq - Brazil

References

1. L.T.Canham and A.J.Grozek, J.Appl.Phys.*72*,1558 (1992).

2. L.T. Canham, M.R. Houlton, W.Y. Leong, C. Pickering and J.M. Keen, J.Appl.Phys. *70*, 422 (1991).

3. H.D. Fuchs, M. Stutzmann, M.S. Brandt, M. Rosenbauer, J. Weber and C. Cardona, Phys. Scripta T45 309 (1992) and M.S. Brandt, A. Breitschwerdt, H.D. Fuchs, A. Höpner, M. Rosenbauer, M. Stutzmann and J. Weber, Appl.Phys. A *54*, 567 (1992).

4. E.F. Steigmeir, B. Delley and H. Auderset, Phys.Scripta T45, 305 (1992).

5. A. Roy, A. Chainani, D.D. Sarma and A.K. Sood, Appl.Phys.Lett. *61*, 1655 (1992).

6. P. Basmaji, V. Grivickas, A. Bernussi and B.Matvienko (Pittsburgh, PA : Materials Research Society, to be published in 1993).

7. R.M.Barrer, Zeolites and Glay Minerals, Academic Press (1978).

8. D.W.Breck, Zeolite Molecular Sieves,R.E.Krieger company (1984).

9. G.T.Kokotailo et al., Nature. *272*,437 (1978).

10. M.Schenk, Werkstoff Aluminium und seine anodische Oxydation. Published by A.Franke AG. Verlag Bern (1948).P 583.

11. X. Ruren and P. Wengin; The synthesis crystallization and structure of heteroatom containing ZSM-5 type zeolites.In: Zeolites, synthesis, structure, technology and applications. Proceedings, Int. symposium. Portorose. Sept.3-8, 1984. B. Drzaj, et al., Editors. Elsevier, 1985.p.27.

ENERGY LOCALISATION AND SURFACE INTERACTIONS IN THE LUMINESCENCE OF POROUS SILICON

S.GARDELIS, P.DAWSON AND B.HAMILTON
Centre for Electronic Materials & Pure and Applied Physics Department, University of Manchester Institute of Science & Technology, Manchester, M60 1QD, U.K.

ABSTRACT

The fundamental mechanisms controlling the light emission from porous Si remain unresolved. In this paper we report attempts to modify the luminescence using a variety of surface processing steps, such as vacuum annealing with subsequent anneals in nitrogen and oxygen, exposure to hydrofluoric acid (HF) and rapid thermal oxidation. Luminescence, infrared absorption, and electron spin resonance (ESR) have all been used to gain more information on the link between the optical emission and the localisation of the electrons in this material system. We present evidence that the silicon dangling bond is the key component in the non-radiative recombination. This is based on measurements shown that hydrogen coverage of the surface is significant because of saturation of the dangling bonds and a subsequent reduction in the competing non-radiative paths rather than as an active component in the radiative transition. Finally, we focus our attention upon the lower energy band which appears in the luminescence spectrum of porous Si ($\sim 0.9\text{eV}$) by examining its behavior under the surface treatments mentioned above. We found that this luminescence band originates from the surface of the porous layer and its intensity correlates well with increasing oxidation of the porous layer.

INTRODUCTION

Since the discovery of efficient visible light emission from porous Si at room temperature[1], a great deal of work has been carried out in order to understand the origin of the luminescence and ultimately exploit the emission process. Regarding the attributes of the luminescence process, a good agreement now exists. Luminescence spectra from porous Si prepared under identical conditions can be reproduced, time domain and excitation data are in broad agreement. Despite this a real consensus does not exist regarding the nature of the radiative process and evidence for Si quantum wires, Si nanoparticles, amorphous Si, molecules at the porous Si surface such as Si hydrides and possibly siloxene has led to various models being proposed to explain the origin of the luminescence process in porous Si. The well established link between porosity and spectral position for p-type porous Si seems compelling evidence for quantum confinement.[2,3] However, for n-type material, the situation is much less clear, very weakly porous material giving visible emission.[4]

The analysis of the basic physics regarding the emission process in porous Si has proved very difficult due to the complexity of the porous layer. TEM analysis shows in the case of red emitting p-type material[5], for example, well ordered Si single crystal and amorphous Si. Taking into account the enormous surface area of porous Si[6] and the importance of the surface in all electronic processes, one can imagine how important the role of the porous Si surface is in determining the optical properties of such a material. It is this issue we attempt to address in this presentation.

In this paper we examine separately the effect of surface treatments applied to porous Si on the behavior of the low energy band ($\sim 0.9\text{eV}$) and the high energy band of the luminescence spectrum.

Mat. Res. Soc. Symp. Proc. Vol. 298. ©1993 Materials Research Society

HIGH ENERGY PL BAND

Surface modification of porous Si by annealing

We illustrate the importance of surface modification on the luminescence of p^+-type porous Si, by investigating the effect of annealing in vacuum with subsequent anneals in N_2 gas at 300°C. The starting material was anodised to a porosity of approximately 40-45%, giving a luminescence spectrum which peaked at 1.32eV at 10K, as shown in figure 1a. After annealing for 2 minutes in vacuum (10^{-8} Torr) at 400°C, no luminescence was detectable. Two other observations can be associated with the complete disappearance of the luminescence after the vacuum anneal. Firstly, the hydrogen related vibrational modes in the infrared spectrum after the vacuum anneal disappear. Secondly, the average strain in the porous layer measured by double crystal x-ray diffraction is reduced. These two observations are indications that all the hydrogen has been desorbed from the surface after the vacuum anneal at this temperature. It is reasonable to attribute the observed reduction of the luminescence strength to hydrogen desorption, since it is known that hydrogen can passivate surface dangling bonds which can act as non-radiative centres. The difficulty with this observation is that it does not preclude the possibility that hydrogen plays an active role in the emission process, and siloxene based models for the luminescence depend on such a role for hydrogen and other chemical species.

The surface sensitivity of the luminescence, and perhaps a weakening of the role that hydrogen may play emerges when the vacuum annealed sample is subsequently annealed in N_2 gas at temperatures as low as 300°C. After a 5 minute anneal, very weak luminescence was observed. After a 10 minute anneal, a spectrum of nearly identical shape to the starting data, although weaker, was measured as shown in figure 1b. Measurements of infrared absorption after further annealing steps, failed to detect any hydrogen related transitions.

A further crucial observation was that material of this sort, when immersed in HF produced very bright luminescence with the same emission spectrum as observed for the prolonged nitrogen anneal (figure 1c). This demonstrates that the microstructure is preserved and that the final hydrogen coverage simply strengthens the luminescence. We argue below that the strengthening is due to a reduction in the non radiative shunt path.

This data demonstrates that we are getting some recovery of luminescence without hydrogen adsorption. However, it must be stressed that we can only be certain that some gas phase interaction during the prolonged anneal is responsible for the partial recovery. The adsorption of hydrogen at very low levels may be occuring and our optical absorption measurements may lack the sensitivity to provide the information. Clearly, luminescence can be detected under conditions in which no Si hydrides can be measured experimentally. The question then is open as to the exact role of hydrogen, and we have attempted to probe the surface mechanisms further by studying oxygen induced effects on the luminescence.

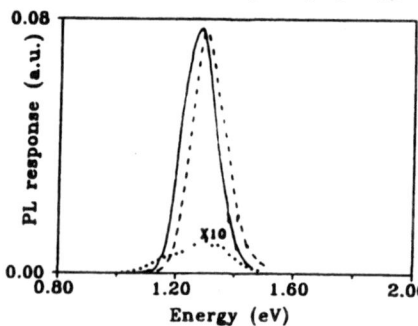

Figure 1 Luminescence spectra associated with the vacuum anneal of porous Si; (a)——— the as anodised material with porosity of approximately 40%, (b)---- after anneal of the vacuum annealed sample in N_2 gas (the luminescence begins to recover), (c)····· recovery of the luminescence after HF treatment

Surface modification by partial oxidation

By another set of experiments, we demonstrate the effect of partial oxidation on the luminescence of a previously vacuum annealed sample from the surface of which hydrogen had been desorbed and the luminescence strength reduced to zero. The starting material was anodised to a porosity of approximately 75-80% giving a luminescence spectrum peaking at 1.64eV at room temperature (figure 2a). After vacuum annealing at 400°C the luminescence was quenched (figure 2b) and IR absorption indicated no hydrogen related vibrational modes. After further annealing in an ambient of 1 part of dry O_2 and 10 parts of N_2 gas at 300°C for 5 minutes weak luminescence began to reappear (figure 3a). After further annealing in an ambient of 3 parts of dry O_2 and 10 parts of N_2 gas at 900°C, for 5 minutes a very clear luminescence band of the same intensity as that of the starting material peaking at 1.52eV could be observed (figure 3b). In all partial oxidation steps an increase in the intensity of the Si-O-Si vibrational mode with increasing oxidation could be observed by infrared absorption. On the other hand no Si hydride vibrational modes could be detected, although a small amount of hydrogen cannot be ruled out. Once more we demonstrate that hydrogen coverage of the porous Si surface is not unique in strengthening the luminescence of porous Si.

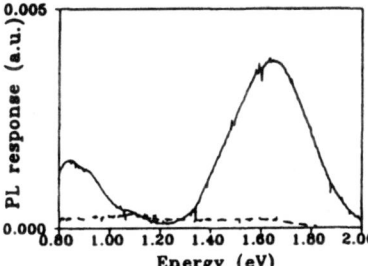 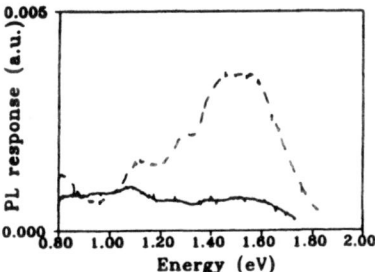

Figure 2 *Luminescence associated with partial oxidation of porous Si; (a)———— the as anodised material with porosity of approximately 75%, (b)---- after vacuum anneal.*

Figure 3 *(a)———— Luminescence spectrum of the vacuum annealed sample after partial oxidation for 5 minutes at 300°C, (b)---- after further oxidation at 900°C.*

Surface modification by rapid oxidation

There are several reports in the literature of the effect of surface treatment with oxygen. These include observations of both luminescence degradation[7] and stabilisation[8,9]. Our aim was to work with p$^+$ Si anodised to a porosity of 40-45%. Such material is not too far above the threshold porosity at which light emission becomes observable, and emits in the near infrared. By choosing low porosities it is then possible to monitor more carefully any spectral shifts to the blue, or other changes that may occur. Material of very high porosity has already a very small microstructure and is in a sense a limiting case .

Rapid thermal oxidation in dry O_2 for 1 minute at 300 and 900°C was carried out. The luminescence spectrum of the rapidly oxidised porous material was found to be rather more complex than that of the anodised material. In general two components are observed in the PL spectrum at low temperature(10K); one close to 1.2eV, and one at higher energies the exact value of the peak energy depending on the oxidation temperature. Note that for the starting porosity used, the new peaks sit either side of the original "as anodised" peak which was typically at 1.32eV.

An example of the oxidised material spectrum as recorded at 10K is given in figure 4. We observed that although the original as anodised porous Si layer did not emit at room temperature, the porous layer oxidised at 900°C emitted light efficiently even at room temperature. We also observed that oxidation at 900°C moved the main peak of the spectrum towards higher energies (~ 1.5eV). The band at 1.2eV may have the same origin as the infra-red band (low energy PL band) described in the next section of this paper.

Cullis et al[10] from TEM measurements on rapidly oxidised porous Si layers at elevated temperatures (> 700°C) observed Si nanocrystals in a sea of Si oxide. This implies that if the luminescence from porous Si is indeed due to quantum confinement of the carriers in the porous Si skeleton then the blue shift with rapid oxidation at 900°C could be a quantum size effect. Perhaps the key aspect of the oxidised sample luminescence data, though, is the strong variation of luminescence efficiency, measured under identical conditions. This is shown in figure 5 where we plot the peak intensity of the high energy band. For the 300°C oxidation, the luminescence intensity is almost zero, and it is significantly recovered for the 900°C oxidation. Infrared absorption of the SiH_x stretching modes provides very sensitive evidence for surface hydrogen coverage. In the starting material, the SiH_x stretching modes were intense. After annealing at 300°C the SiH_x modes are quenched and SiOH vibrational modes appear in the spectrum. After the 900°C anneal (when the luminescence is strengthened) no absorption related to hydrogen can be seen in this region of the spectrum implying that the hydrogen coverage is extremely low. We also measured the Si dangling bond density using ESR during this sequence of oxidation, and the data are also plotted in figure 5. The lowest luminescence corresponds well with the highest dangling bond density, and the recovery of luminescence corresponds with a decrease in the dangling bond population. Thus from these measurements it appears that the oxidation process at the more elevated temperatures results in saturation of the surface dangling bonds, playing a role equivalent to hydrogen in the more usual anodised material. This data is in good agreement with that of Ref.9.

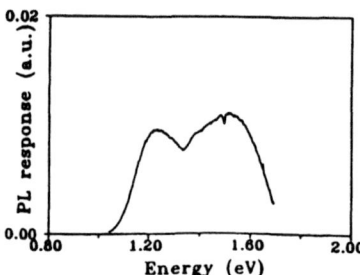

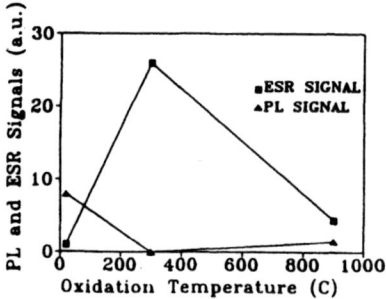

Figure 4 Luminescence spectrum of a porous Si sample oxidised at 900°C for 1 minute.

Figure 5 Luminescence strength and ESR signal of porous Si samples oxidised at various temperatures. The data is in good agreement with that of Ref.9.

LOW ENERGY PL BAND

In all porous Si luminescence spectra we have observed that apart from the main peak associated with the porous layer another band, peaking between 0.8 and 1eV depending on the preparation conditions of the porous layer, is also observed which blue shifts with increasing porosity. An example of the low energy luminescence band is shown in figure 6a for a sample anodised to a porosity of 40-45% and a thickness of 50μm. We have observed that by exciting the porous Si layer using red light (640nm) and green light (560nm) under identical conditions, the intensity of the high energy peak decreases with decreasing excitation wavelength whereas the intensity of the low energy peak remains unchanged. Taking into account the fact that the red light is expected to penetrate deeper into the porous layer than the green light, the above observation implies that the lower energy peak(~ 0.9eV) is associated with the surface of the porous layer.

We have also observed that the low energy peak quenches considerably with desorption of hydrogen from the porous Si surface (figure 6b) and increases again with subsequent partial oxidation of the sample (1 part of N_2 and 2 parts of dry O_2) at a temperature as low as 300°C (figure 6c). When the porous layer is immersed in concentrated HF for 12 hours the low energy band quenches once again (figure 6d).

This is a strong indication that the low energy peak is not associated with the dangling bond created by the hydrogen desorption but with a center which is produced at the oxidised surface of the porous Si layer. A possible candidate is the well known P_{bo} center, a sort of a dangling bond[11] or a floating bond[12] caused by imperfections at the Si/SiO$_2$ interfaces. An indication for the possible existence of such a center in the porous layer is the increasing ESR signal strength due to possibly the P_{bo} when the sample is oxidised (figure 5).

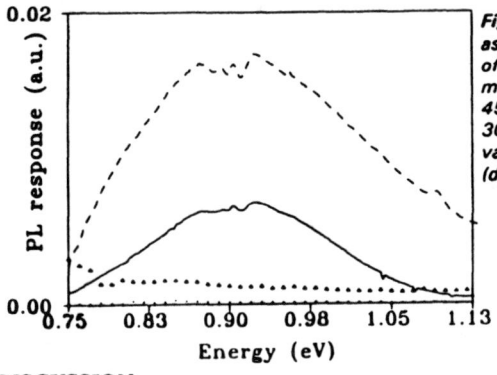

Figure 6 Luminescence spectra associated with the low energy PL band of porous Si; (a)——— the as anodised material with porosity of approximately 45%, (b)····· after anneal in vacuum at 300°C, (c)---- after partial oxidation of the vacuum annealed sample at 300°C, (d)▲ after HF treatment.

DISCUSSION

The difficulty in ascribing precise mechanisms to the luminescence of porous Si stems from the complexity of the modified material. We do not know the precise form of the microstructure, several phases of material may co-exist, the system has enormous surface area per unit mass, also optical absorption and energy migration processes are not fully understood. In this situation it is encouraging that good agreement and reproducibility of data exists. The issue of the role of the surface is a key one, and the debate on whether hydrogen plays a mediating role is of central importance before the exploitation of the light emission phenomenon can be achieved.

The data presented here do not completely resolve the issue, but at least provide evidence that hydrogen is not necessary to ensure light emission from porous Si, whereas the presence of Si dangling bonds can reduce the emission intensity. The recovery and stabilisation of the luminescence after rapid oxidation may present some interesting problems for future device processing. It also poses questions as to the possible involvement of O_2 in the luminescence process.

ACKNOWLEDGEMENTS

This work has been financed by an EEC research grant under the ESPRIT scheme. The authors wish to thank Prof. A.R. Peaker, Director of the Centre for Electronic Materials, UMIST, for his valuable contributions.

REFERENCES

1. L.T. Canham, Appl. Phys. Lett. **57**, 1046 (1991)
2. S. Gardelis, J.S. Rimmer, P. Dawson, B. Hamilton, R.A. Kubiak, T.E. Wall and E.H.C. Parker, Appl. Phys. Lett. **59**, 2118 (1991)
3. S. Gardelis, B. Hamilton, R.A. Kubiak, T.E. Whall and E.C.H. Parker in <u>Light Emission from Silicon</u>, edited by S.S. Iyer, R.T. Collins, and L.T. Canham (Mater. Res. Soc. Proc. **256**, Pittsburgh, PA, 1992)pp.149-152.
4. K. Berwick, MSc dissertation UMIST (1993)
5. S. Gardelis, G. Lorimer and B. Hamilton (unpublished data)
6. V. Vezin, P. Goudeau, A. Naudon, A. Herino, and G. Bomchil, J.Appl.Cryst, **24**,586(1991)
7. M.A. Tischler, R.T. Collins, J.H. Stathis and J.C. Tsang, Appl.Phys.Lett., **60**, 639(1992)
8. S. Shih, C. Tsai, K.H. Li, K.H. Jung, J.C. Campbell and D.L. Kwong, Appl.Phys.Lett., **60**,633 (1992)
9. V. Petrova-Koch, T. Muschik, A. Kux, B.K. Meyer and F. Koch, Appl.Phys.Lett. **61**, 943(1992)
10. A.G. Gullis, data presented at the NATO Advansed Research Workshop, (March, 1st-3rd,1993), Maylan, France
11. E.H. Poindexter, P.J. Caplan, B.E. Deal and R.R. Razouk, J.Appl.Phys.**52**,879(1981)
12. S.T. Pantelides, Phys.Rev.Lett.**57**,2979(1986)

SILICON NANOSTRUCTURES IN Si-BASED LIGHT-EMITTING DEVICES[*]

FEREYDOON NAMAVAR,[*] R.F. Pinizzotto,[**] H. Yang,[**] N. Kalkhoran,[*] P. Maruska[*]
[*]Spire Corporation, One Patriots Park, Bedford, MA
[**]University of North Texas, Denton, TX

ABSTRACT

High resolution cross-sectional electron microscopy and electron diffraction of an np heterojunction porous Si device, capable of emitting light at visible wavelengths, clearly indicates the presence of Si nanostructures within the quantum size regime. These results indicate that the quantum confinement effect is at least partially responsible for photoluminescence at visible wavelengths.

INTRODUCTION

Silicon, through the swift progress of integrated circuit (IC) technology, has become widely recognized as the leading semiconductor material. Silicon's status will likely remain uncontested for many years; unfortunately, silicon's indirect bandgap does prevent its use in optoelectronics. However, true wafer scale integration will become a reality if light emission and light detection can be implemented directly on silicon wafers.

Extensive efforts are in progress to incorporate compound semiconductor light-emitting diode and laser diode technologies into silicon VLSI processes by epitaxial growth. However, the most promising III-V light-emitting compounds, such as GaAs and AlGaAs, are not readily deposited directly onto silicon substrates due to lattice and thermal expansion mismatches. Incompatibility problems of these materials generally degrade the performance of any III-V compound light-emitting device (LED) grown onto silicon. An alternative approach involves mounting GaAs-based devices onto silicon wafers; this appears to be costly, time-consuming, and labor intensive.

Recently, optically excited visible light emission[1] from photochemically-etched, anodized silicon[2] provided us and others[3,4,5] with the possibility of fabricating the first heterojunction visible light-emitting diode based on porous silicon.[6] Our device consists of a heterojunction between electrochemically-etched p-type porous silicon and indium-tin-oxide (ITO), a transparent n-type semiconductor deposited by RF sputtering. On the basis of three types of measurements, current-brightness,[6] current-voltage,[7,8] and current-temperature,[9] our results indicate that observed electroluminescence (EL) originates from minority (electron) carrier injection, the basic mechanism operating in GaAs-based LEDs and other homojunction devices. In addition, these findings show that, once the material quality is enhanced and fabrication processes are optimized, EL should be possible with the same impressive efficiency levels as the photoluminescence (PL) which was observed for porous silicon.

The structure of porous silicon has been extensively studied by cross-sectional transmission electron microscopy (XTEM).[10-19] However, here we report XTEM results for a working porous Si device. XTEM results from a visible light-emitting porous Si device were not only essential to relate electrical and material properties, but also provided a unique opportunity to verify accurately the existence of silicon nanostructures, required for quantum confinement at the porous Si/ITO interface. This was possible because the ITO layer protected silicon nanostructures from oxidation and physical damage during and after TEM sample preparation, thus enabling us to observe the tip of the pores at the interface with ITO.

[*] This work was supported in part by SDIO and monitored by the U.S. Army Research Office.

EXPERIMENTAL PROCEDURE

Porous silicon was produced by anodically etching polished p-type Si(100) and Si(111) wafers, with resistivities ranging from 0.3 to 14Ω-cm, in 1:1 ethanol:HF, at current densities between 2 to 20 mA/cm². The PL of the samples was measured at room temperature using a PL measurement system equipped with an argon laser; the excitation beam was comprised of unfocused blue (488 nm) light with a 5 mW intensity.

The np heterojunction porous Si LED was fabricated by RF sputtering a transparent layer of n-type ITO, 3000Å to 5000Å thick, onto the porous silicon surface. Individual devices were fabricated using two patterning techniques, depositing ITO: 1) through a shadow mask with 3 mm diameter circular windows, and 2) on the entire porous surface. Conventional photolithography and etching steps were then used. Electrical and optoelectronic properties of the devices were studied using standard semiconductor analysis techniques.

XTEM samples were prepared using standard methods. Two pieces of the sample were glued face-to-face. Slices were cut from this block, metallurgically polished, dimpled on one side using a ball grinder, and thinned to electron transparency using ion milling.

EXPERIMENTAL RESULTS

Figure 1 shows a typical porous Si layer with a thickness of about 6.5 μm. The porous layer is fairly uniform, though its thickness varies slightly from point-to-point.

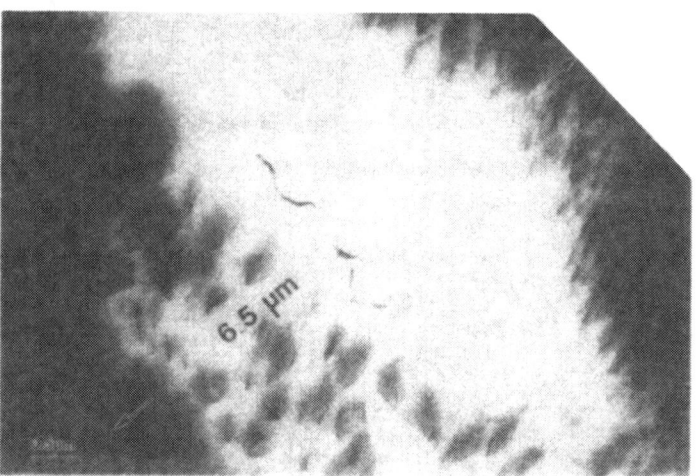

Figure 1   *XTEM of a typical porous Si layer without ITO.*

Figure 2a shows a schematic cross section of an np heterojunction porous Si LED capable of emitting light at visible wavelengths. Figure 2b shows an XTEM of the actual working device. The dark layer on top is the ITO which does not thin as readily as the silicon underneath it. The ITO layer has remained electron-opaque in this particular area of the sample. In other areas of the sample where the ITO is electron-transparent, we are able to distinguish its internal microstructure. Below the ITO layer is the porous silicon layer which consists of two regions. Adjacent to the ITO layer is ~1000Å thick region composed of

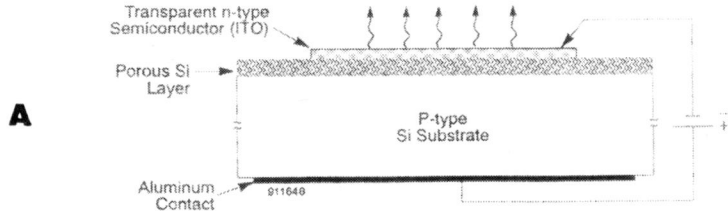

Figure 2   *Cross-sectional view of a porous Si LED a) schematic and b) TEM micrograph.*

primarily vertical pores perpendicular to the original sample surface. Below this region is the remaining porous Si. This region consists of an amorphous material with small Si nanocrystals embedded throughout its volume.

Figure 3 shows a high magnification lattice image of the porous silicon layer. The background microstructure is primarily amorphous; however, embedded within this amorphous material are a number of very obvious "dark spots." These dark spots are small single silicon crystal particles with average diameters of about 20 to 50Å. Their diameters are within the quantum confinement size regime. Note that all of the particles are still aligned with the silicon substrate.

Figure 4 is a room temperature PL spectrum for the ITO/porous Si device whose XTEM microstructure is shown in Figures 2 and 3. The spectrum shows fine structures superimposed on a larger main peak. We have found[20] that the spacing between the smaller maxima (fine structures) is reproducible and independent of the main peak energy and sample history. In addition, the energy steps between the smaller maxima correlate with differences in the exciton energies due to the change in the sizes of the quantum wires by complete monolayers of atoms. Optical interference[21,22] at thin surface-layers cannot be responsible for the small variable peaks, as optical interference would produce equidistant peaks due to the constant layer thickness of a sample. Furthermore, the peak-distance should change from sample to sample because of different layer thicknesses; however, neither this nor optical interference were observed.[20]

Figure 5 is an electron diffraction pattern of the porous silicon region. The spots observed in the electron diffraction pattern clearly indicates the presence of single crystals in the porous region. The orientation shown in this micrograph is [110], as we are observing a cross section of [100] material. However, [200] spots, normally seen in thick materials, are not observed. The [200] spots are caused by double diffraction, which requires multiple scattering of electrons. The 20 to 50Å diameter Si nanocrystals in the porous-Si layer are too small for double diffraction to occur.

346

Figure 3   *XTEM micrograph showing the lattice image of a porous Si layer.*

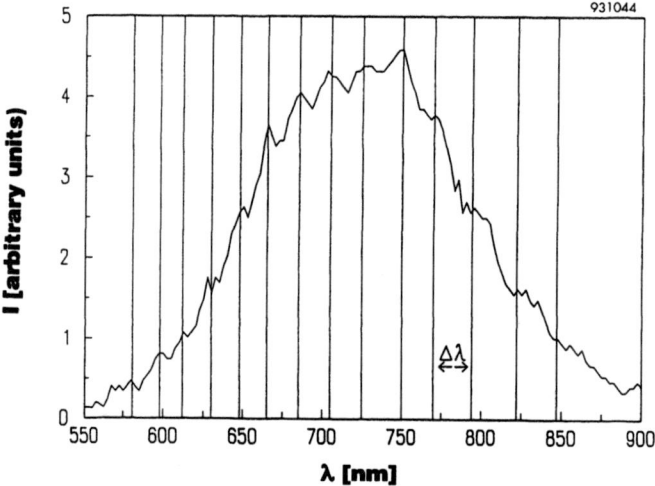

Figure 4   *Room temperature PL spectrum for a working porous Si device.*

Similar effects have been observed for Si nanocrystals fabricated using other procedures. The lack of [200] spots further supports our idea that the particles are very small and that the diffraction observed was only from the particles and not due to a residual Si substrate.

Examination of the single crystal silicon particles in Figure 5 shows the spots slightly spread out into arcs. There are two explanations for this observation:

Small Particle Size - The size of the diffraction spots in reciprocal space is inversely related to the particle size, *i.e.*, the smaller the particles causing the diffraction, the larger the spots observed in the diffraction pattern.

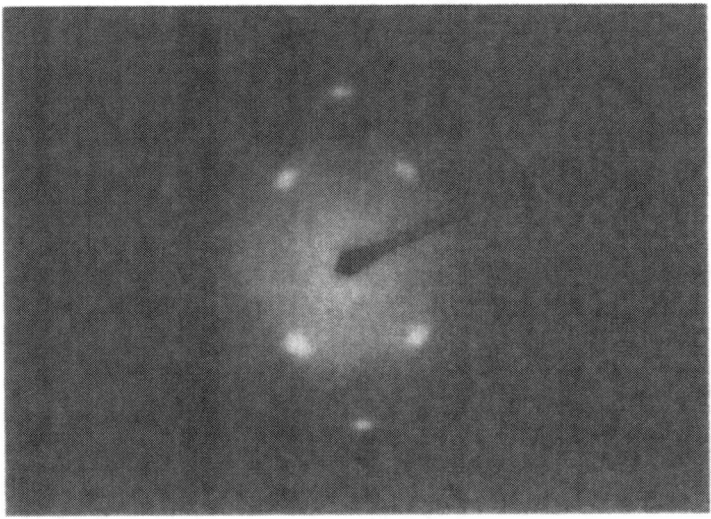

Figure 5   *Electron diffraction pattern of the porous Si layer.*

Shifting Particle Orientation - Slight misorientation of the Si nanoparticles may be caused by the volumetric expansion caused by oxidation of the silicon so that they are no longer exactly oriented along [110]. This would also result in larger diffraction spots as observed.

Figure 6 shows a high magnification view of the interface between the ITO and the porous silicon. As expected, the surface of porous silicon is very rough. The ITO layer covers all the hills and valleys of the surface and replicates it quite well. In addition, there is no evidence of voids or any other layer between the ITO and the porous silicon. Lattice images of the ITO were obtained; however, we were unable to obtain lattice images of the nanoparticles (silicon) in this area. Assuming the dark particles in the figure are actually single crystal silicon nanoparticles, a direct contact appears to have been made between the conducting ITO and quantum-confined particles. This is consistent with our recently developed tunneling model[8,9] based on our electrical measurements.

SUMMARY

XTEM micrographs of an np heterojunction porous Si/ITO light-emitting device indicated that the porous silicon layer consists of two layers. The top 1000Å contains vertical pores perpendicular to the sample surface. The rest of the porous Si layer consists of an amorphous matrix plus embedded Si nanocrystals. High magnification lattice imaging of the porous silicon layer clearly indicated the presence of single crystal particles with dimensions of 20 to 50Å which were aligned with the original silicon substrate orientation. This microstructure is consistent with detailed analyses of electron diffraction patterns of the porous region as well as fine structures observed in room temperature PL spectra. These results clearly indicate that the quantum confinement effect is at least partially responsible for visible luminescence of porous Si.

We would like to acknowledge A. Cremins for her assistance with the preparation of this manuscript.

348

Figure 6   *XTEM micrograph showing the interface of the ITO and porous Si.*

REFERENCES

1. L.T. Canham, *Appl. Phys. Lett.* **57**, 1046 (1990).
2. R.L. Smith and S.D. Collins, *J. Appl. Phys.* **71** 8, R1 (1992).
3. A. Richter, W. Lang, P. Steiner, F. Kozlowski, H. Sandmaier, *MRS* **256**, 209 (1991).
4. N. Koshida and M. Katsuno, *Appl. Phys. Lett.* **60**, 347 (1992).
5. T. Futagi, T. Matsumoto, M. Katsuno, Y. Ohta, H. Mimura, K. Kitamura, *MRS* **283**, 389 (1993).
6. N.M. Kalkhoran, F. Namavar, H.P. Maruska, *MRS* **256**, 89 (1992).
7. F. Namavar, P. Maruska, N. Kalkhoran, *Appl. Phys. Lett.* **60**, 2514 (1992).
8. H.P. Maruska, F. Namavar, N.M. Kalkhoran, *Appl. Phys. Lett.* **61**, 11 (1992).
9. H.P. Maruska, F. Namavar, N.M. Kalkhoran, *MRS* **283**, 383 (1993).
10. H. Takagi, H. Ogawa, Y. Yamazaki, A. Ishizaki, T. Nakagiri, *APL* **56** 24, 2379 (1990).
11. J.L. Heinrich, C.L. Curtis, G.M. Credo, K. Kavanagh, M.J. Sailor, *Science* **253**, (1991).
12. P.C. Searson, *Appl. Phys. Lett.* **59** (7), 832 (1991).
13. A.G. Cullis and L.T. Canham, *Nature* **353**, 335 (1991).
14. M.W. Cole, J.F. Harvey, R.A. Lux, D.W. Eckart, R. Tsu, *APL* **60** (22), 2800 (1992).
15. J.M. Macaulay, F.M. Ross, P.C. Searson, S.K. Sputz, R. People, *MRS* **256**, 47 (1992).
16. S. Shih, K.H. Jung, T.Y. Hsieh, J. Sarathy, C. Tsai, K.-H. Li, J.C. Campbell, D.L. Kwong, *MRS* **256** (1992).
17. K.H. Jung, S. Shih, J.C. Campbell, D.L. Kwong, *MRS* **256** (1992).
18. Y.H. Xie, W.L. Wilson, F.M. Ross, J.A. Mucha, E.A. Fitzgerald, *MRS* **256** (1992).
19. A.G. Cullis, L.T. Canham, O.D. Dossier, *MRS* **256** (1992).
20. R. Behrensmeier, F. Namavar, G.B. Amisola, F.A. Otter, J.M. Galligan, *Appl. Phys. Lett.* **62**, 19 (1993).
21. G.W. t'Hooft Y.A.R.R. Kessener, G.L.J.A. Rikken, A.H.J. Venhuizen *Appl. Phys. Lett.* **61**, 2344 (1992).
22. T.S. Nashashibi, I.G. Austin, T.M. Searle, R.A. Gibson, W.E. Spear, P.G. LeComber, *Phil Mag.* **B45**, 554 (1982).

PHOTOLUMINESCENCE STUDY OF RADIATIVE RECOMBINATION IN POROUS SILICON

CHUN WANG, FRANCO GASPARI AND STEFAN ZUKOTYNSKI
Department of Electrical and Computer Engineering, University of Toronto
Toronto, Ontario, M5S 1A4, Canada

ABSTRACT

Photoluminescence has been studied in porous silicon. Two types of radiative recombination centers have been identified. One gives rise to luminescence at about 820 nm and is believed to be related to Si-H bonds. The second gives rise to luminescence at about 770 nm and is likely associated with S-O bonds. Above about 20 K radiative recombination is assisted by excited states of the recombination centre located about 10 meV above the ground state. The Si-H recombination centre is a single electron center whereas the Si-O center appears to be a multi-electron center.

INTRODUCTION

There have been several reports of strong visible photoluminescence (PL) in chemically and electrochemically etched single crystal silicon (so called porous silicon).[1-3] Photoluminescence with an integrated intensity comparable to that observed in high quality AlGaAs/GaAs multiquantum wells has been reported.[3] Electroluminescence has also been observed.[4-5]

Luminescence in porous Si has been explained either in terms of quantum confinement,[1] or in terms of the formation of a wide band-gap phase such as hydrogenated amorphous silicon[6] or siloxene.[7]

We present results of a study of photoluminescence in porous Si produced by electrochemical anodization of p-type, 10 Ωcm crystalline silicon. Our results suggest that radiative recombination occurs at extrinsic localized centers associated with both SiH and SiO complexes.

EXPERIMENTAL

Samples of porous silicon were prepared by anodic etching of boron doped, p-type, (100) oriented, single crystal Si wafers. Nominally 10 Ωcm material, obtained from several manufacturers was used. The electrolyte consisted of 1 part 49% HF and 1 part deionized water. The current density was varied from 0.01 to 1 mA/cm² and the corresponding etch time from 15 to 3 hours. After etching the samples were rinsed in deionized water, blow dried with nitrogen, and mounted in a closed-loop optical cryostat.

The PL spectra were measured using a 3/4 m monochromator equipped with a 574 lines/mm grating. The monochromator was set to a band pass of 5 nm. Optical pumping was provided by the 476.5 nm (2.6 eV) line of an argon ion laser. The light beam was sine-wave modulated, at 50 Hz for PL intensity measurements or in the range from 10 to 2×10⁵ Hz for frequency resolved spectroscopy, using an acousto-optic modulator. The luminescence signal was detected using a silicon photodiode and a lock-in amplifier. The spectra were normalized to the throughput of the optical system. For most measurements the pump power was kept below 0.5 mW/cm² to avoid rapid degradation of the porous Si sample.[8]

RESULT AND DISCUSSION

Figure 1 shows a typical low temperature PL spectrum for a sample prepared using a high current density and a short etch time (1 mA /cm² and 3 hours). The PL spectrum exhibits a single peak at 820 ± 5 nm. We find that the 820 nm PL peak is present in all "fresh" samples, regardless of the preparation conditions. It has been observed that a freshly anodized Si surface is mainly composed of silicon hydrides.[8-10] We attribute the PL peak at 820 nm to a Si-H complex. We note, that this result is somewhat reminiscent of PL in hydrogenated amorphous silicon (a-Si:H).[11] Possibly our preparation conditions which involve etching at low current density give rise to a material in which PL is dominated by Si-H complexes that resemble those of a-Si:H.

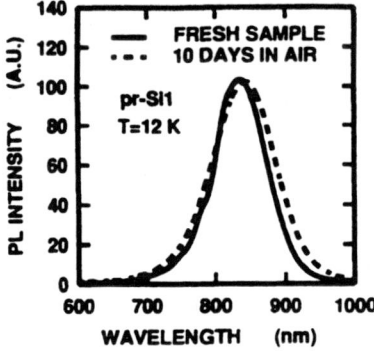

Figure 1. Photoluminescence of sample pr-Si1 shortly after it was prepared and 10 days later (the latter ×2.7).

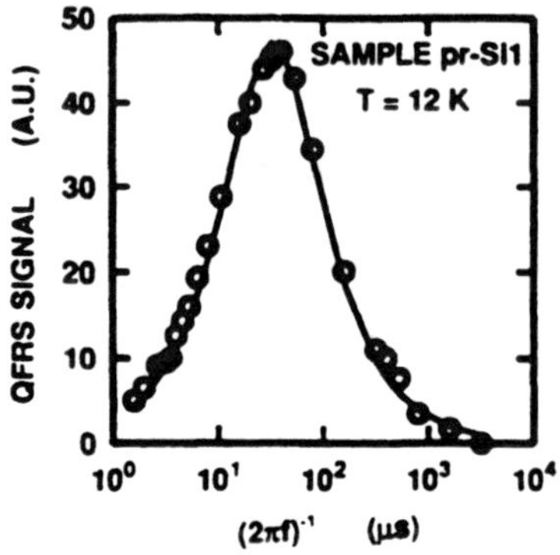

Figure 2. Quadrature frequency resolved spectroscopy for sample pr-Si1 measured at 820 nm.

The PL life-time, τ, was studied using quadrature frequency resolved spectroscopy (QFRS).[13] Figure 2 shows the QFRS signal for the same sample. The solid line is the theoretical fit assuming a single decay time constant of 300 µS at 12 K. The good fit between the experimental data and the theoretical curve indicates that the PL decay is controlled by a single lifetime. We note, that this result is in contrast to a-Si:H which exhibits a much broader QFRS spectrum.[11,12]

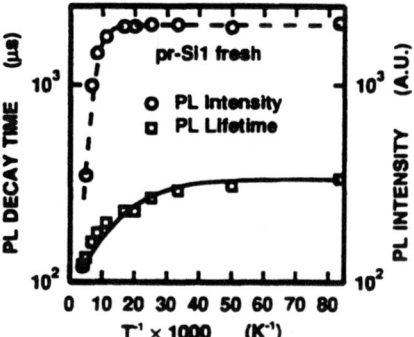

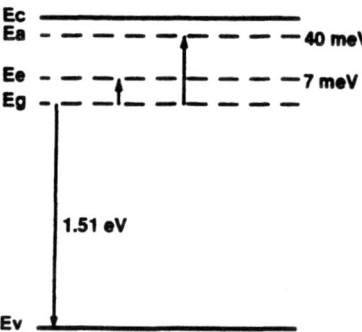

Figure 3. PL lifetime and PL intensity at 820 nm as function of temperature for sample pr-Si1.

Figure 4. Energy level diagram of the recombination centre.

Figure 3 shows the temperature dependence of the PL lifetime and PL intensity again for the same sample. The lifetime increases with decreasing temperature and reaches the value of $300 \pm 10\,\mu S$ at about 20 K and then remains constant. This behaviour can be explained assuming cascade recombination.[14] According to this model an electron captured at a recombination centre may recombine with a hole from the valence band either directly from the ground state of the recombination centre or via an excited state of the recombination centre, as shown in Figure 4. The first process is characterized by the lifetime τ_o. Recombination via an excited state involves emission from the ground state to the excited state, characterized by an emission time constant τ_e. The electron then recombines radiatively with a hole from the valence band. Emission is the limiting process and determines the lifetime associated with this recombination path. Accordingly, the effective lifetime τ is given by:

$$\frac{1}{\tau} = \frac{1}{\tau_o} + \frac{1}{\tau_e} \tag{1}$$

The emission time decreases with temperature according to the relation:

$$\tau_e = t_e \, exp \left(\frac{E_e - E_o}{k_B T} \right) \tag{2}$$

Where t_e^{-1} is the transition rate, E_o is the ground state energy and E_e is the energy of the excited state. Combining Equations 1 and 2 we get:

$$\tau = \frac{\tau_0}{1 + \frac{\tau_0}{t_e} \, exp \left(- \frac{E_e - E_o}{k_B T} \right)} \tag{3}$$

The solid curve for PL decay time in Figure 3 is the theoretical fit to the experimental data using Equation 3. We obtain $E_e - E_o = 7 \pm 2\,meV$, $\tau_0 = 300 \pm 20\,\mu S$ and $t_e = 80 \pm 10\,\mu S$.

Figure 3 shows the temperature dependence of the PL intensity for the sample pr-Si1. Below about 150 K the PL intensity is almost constant. At higher temperatures it drops quickly. The decrease in luminescence intensity is due to an increase in nonradiative recombination. At elevated temperature an increasing number of electrons is excited to higher energy levels of the recombination centre. The wave function of these excited electrons may extend to include nonradiative recombination centres. This will increase the probability for nonradiative recombination. The nonradiative recombination can be characterized by a effective activation energy, $E_a - E_o$, corresponding to electrons being excited from E_o to E_a. The temperature dependence of luminescence intensity is expressed as:

$$I = \frac{I_o}{1 + \frac{g_a}{g_o} \, exp \left(- \frac{E_a - E_o}{k_B T} \right)} \tag{4}$$

Where g_a and g_o are the degeneracy of the excited and ground state, respectively and E_a is the energy of the excited state. By fitting Equation 4 to the experimental data we obtain $E_a - E_o = 40 \pm 5\,meV$ and $g_a/g_o = 75 \pm 5$.

Provided a recombination level is degenerate, we expect a decrease in PL intensity at very low temperature due to Auger recombination.[14] The PL intensity in sample pr-Si1 is constant at low temperature. This indicates that Auger recombination plays no role in the sample and suggests that the recombination centre is a single electron centre.

It has been reported that the PL in porous Si samples is subject to degradation under strong illumination and on exposure to air. Infrared and secondary ion mass spectroscopy have shown that hydrogen in porous Si samples is gradually replaced by oxygen.[8, 9, 15] It has also been shown that on exposure to air the amount of carbon in the porous Si samples increases.[15] ESR studies have shown that the dangling bond density increases with time in samples exposed to the air.[8]

We find that the changes in PL can be divided into two stages. During the first few days of exposure to air of the porous silicon sample the PL intensity decreases and the peak position moves slightly to longer wavelengths. The dashed curve in Figure 1 shows the PL spectrum of sample pr-Si1 after the sample had been exposed to air for 10 days. The PL intensity has dropped to about one third of its original value and the peak has moved 14 ± 4 meV to the red. The temperature dependence of the PL intensity and PL lifetime of the aged sample are shown in Figure 5. The PL activation energy is reduced by 21 ± 5 meV. We note that at the same time the activation energy for PL lifetime at 820 nm is unchanged. The change of activation energy for luminescence intensity is likely related to the increase in defect density. The reduction in the activation energy of 20 ± 5 meV is consistent with the observation that the PL peak energy is reduced by 14 ± 4 meV.

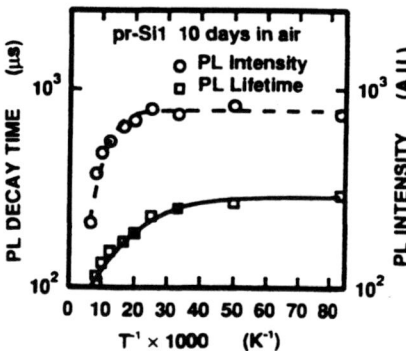

Figure 5. PL lifetime and PL intensity at 820 nm vs. temperature for pr-Si1 10 days after it was prepared.

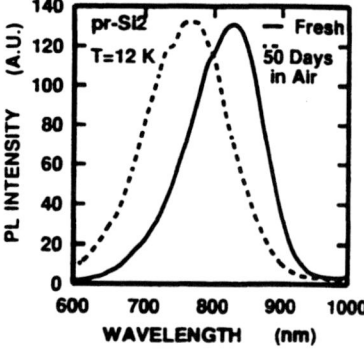

Figure 6. PL intensity of sample pr-Si2 shortly after it was prepared and 50 days later.

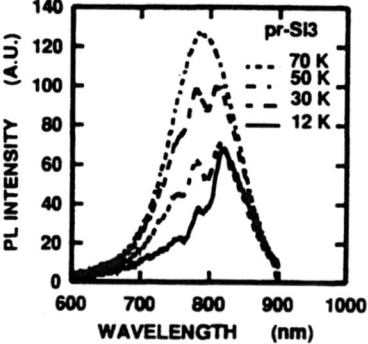

Figure 7. PL intensity of sample pr-Si3.

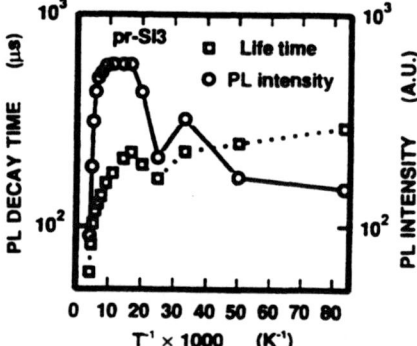

Figure 8. PL lifetime and PL intensity at 780 nm as function of temperature for sample pr-Si3.

In the second phase of ageing the PL peak at 820 nm is replaced by a peak at about 770 nm. This shift is illustrated in Figure 6. We believe that the PL peak at 770 nm is related to a new recombination center, likely the SiO complex. The PL response of samples prepared using long etching time supports this assumption. For very long etching time we expect that oxygen which is present in the etching solution may be incorporated into the porous Si film. Figure 7 shows a set of PL spectra of a porous Si sample prepared under slow etching conditions (current density of 0.01 mA/cm^2 and an etching time of 15 hour). In addition to the peak at 820 nm two other peaks, at 780 nm and 760 nm, can been seen. The PL intensity and PL lifetime of the 780 nm peak show a similar behaviour as that of a sample exposed to the air for a long period of time. Figure 8 shows the temperature dependence of PL intensity and PL lifetime of the slowly etched sample. The PL intensity generally increases as the temperature increases form 12 to 70 K, but with a dip in PL intensity at about 40 K. From 70 to 120 K the PL remains almost constant and above 120 K the PL drops sharply. The PL intensity shows an activation energy of about 100 meV above 150 K. The PL lifetime also shows a dip at about 40 K. Below 70 K the PL lifetime shows an activation energy of 3 meV increasing to about 7 meV in the range from 70 to 120 K. Above 150 K activation energy for the PL lifetime is about 100 meV. The decrease of PL intensity at very low temperature could be due to the Auger recombination. This would suggest that the recombination center that is responsible for the 780 nm PL peak is a multi-electron center.

CONCLUSION

Two types of radiative recombination centers have been identified in porous Si prepared by electrochemical etching of p-type, 10 Ωcm crystalline silicon. One is believed to be related to Si-H bonds and the other to the S-O bonds. Above about 20 K radiative recombination is assisted by emission to excited states some 10 meV above the ground state. The center related to Si-H bonds is a single electron center whereas the center related to Si-O bonding appears to be a multi-electron center. The PL spectrum of freshly prepared samples bears some similarity to the PL of a-Si:H. However, it is shifted to higher frequency and the recombination is characterized by only a single decay time constant.

REFERENCES

1. L.T. Canham, Appl. Phys.Letter. **57**, 1046 (1990).
2. A. Bsiesy, J.C. Vial ,F. Gaspard, R. Herino, M. Ligeon, F. Muller, R. Romestain, A. Wasiela, A. Halimaoui and G. Bomchil, Surface Sci., **254**, 195 (1991).
3. K. H. Jung, S. Shih, T. Y. Hsieh, D. L. Kwong and T. L. Lin, Appl. Phys, Lett. **59**, 3264 (1991).
4. G. Namavar, H. P. Maruska and N. M. Kalkhoran, Appl. Phys. Lett. **60**, 2514 (1992).
5. N. Koshida and H. Koyama, Appl. Phys. Lett. 60,347, (1992).
6. S. M. Prokes, O. J. Glembocki, V. M. Bermudez and R. Kaplan, Phys. Rev. **B45**, 13788 (1992).
7. M.S. Brandt, H.D. Fuchs, M. Stutzmann and M. Cardona, Solid State Comm. **81**, 307 (1992).
8. M.A. Tishler, R.T. Collins, J.H. Stathis and J. T. Tsang, Appl. Phys. Lett. **60**, 639 (1992).
9. Y.H. Xie, W.L. Wilson, F.M. Ross, J.A. Mucha, E.A. Fitzgerald, J.M. Macaulay and T. D. Harris, J. Appl. Phys. **71**, 2403 (1992).
10. C. Tsai, K.-H. Li, D.S. Kinosky, R.-Z. Qian, T.-C. Hsu, J.T. Irby, S.K. Banerjee, A.F. Tasch, J.C. Campbell, B.K. Hance and J. M. White, Appl. Phys. Lett. **60**, 1700 (1992).
11. R.A. Street, Advances in Physics, **30**, 593 (1981).
12. C. Wang, J.M. Perz, F. Gaspari, and S. Zukotynski, Proceedings of the Canadian Conference on Electrical and Computer Engineering, Toronto, September 13-16, 1992, p.

TM9.13.1.

13. S. P. Depinna and D. J. Dunstan, Phil. Mag. **B50**, 579 (1983).

14. P. T. Lansberg, Recombination in Semiconductors, (Cambridge University Press, 1991), p. 144.

15. L. T. Canham, M. R. Houlton, W. Y. Leong, C. Pickering and J. M. Keen, J. Appl. Phys. **70**, 422 (1991).

ACKNOWLEDGEMENT

The authors wish to acknowledge the financial support of the Natural Sciences and Engineering Research Council of Canada, the University Research Incentive Fund of Ontario and Ontario Hydro. One of us (C. W.) would also like to acknowledge the financial support provided by the Connaught Foundation and by the University of Toronto.

VISIBLE PHOTOLUMINESCENCE FROM POROUS SILICON CARBIDE

T. Matsumoto*, T. Tamaki*, T. Futagi*, H. Mimura*, and Y. Kanemitsu**
*Electronics Research Laboratories, Nippon Steel Corporation,
5 − 10 − 1 Fuchinobe, Sagamihara, Kanagawa 229, Japan
**Institute of Physics, University of Tsukuba, Tsukuba, Ibaraki 305, Japan

ABSTRACT

We have fabricated porous silicon carbide using single crystal 6H−SiC that has a wider indirect bandgap than silicon crystal prepared by electrochemical anodization. We have observed intense blue−green luminescence with the peak wavelength of around 500 nm at room temperature. The luminescence intensity is about five hundred times stronger than that of free electron to acceptor recombination in 6H−SiC crystal. This porous SiC offers an intense blue−green luminescent material. The results of structural analysis (secondary electron microscope analysis) and optical measurements (photoluminescence spectrum, Raman spectrum, and picosecond luminescence decay) suggest that the origin of the intense blue−green luminescence is the same as that of the intense red luminescence of porous silicon.

INTRODUCTION

Porous Si fabricated by electrochemical anodization has attracted much interest because it exhibits strong−red luminescence at room temperature [1]. Recently, rapid−thermally−oxidation (RTO) techniques are applied for anodized−porous Si to obtain stable oxygen terminations [2]. Blue−green luminescence is observed from this RTO− porous Si at the oxidized temperatures more than 1000 °C [3]. However, the luminescence intensity is lower than red one and the sample preparation technique is more difficult than anodization.

In this paper, we report that we have fabricated porous silicon carbide using single crystal 6H−SiC that has a large indirect−bandgap of 433 nm (2.86 eV) [4, 5] prepared by electrochemical anodization. We have observed intense luminescence with the peak wavelength of around 500 nm. The luminescence peak shifts to higher−energy with increasing anodization current density. However, it is situated below the bandgap of 6H−SiC crystal. The luminescence intensity is about five hundred times stronger than that of free electron to acceptor recombination in 6H−SiC crystal. This porous SiC can offer an intense blue−green luminescent material.

Figure 1. Photograph of porous SiC obtained with electrochemical anodization at current density of 40 mA/cm².

EXPERIMENTS

The substrates were (0001)−oriented n−type (0.1 Ω cm, nitrogen−doped) 6H−SiC wafers. Thin Ni films were evaporated on the back of the SiC wafers to form good ohmic−contact . Anodization was performed with light irradiation at a current density of 20 mA/cm², 40 mA/cm², and 60 mA/cm² in HF−ethanol solution (HF : H_2O : C_2H_5OH = 1 : 1 : 2).

Figure 1 shows a photograph of porous SiC using the above method. The central dark−brown region is

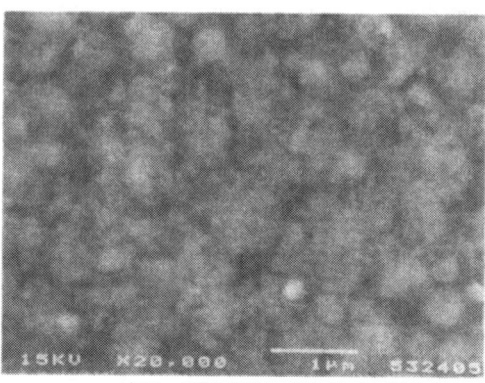

Figure 2. SEM image of porous SiC layer. Anodization current density is 40 mA/cm².

porous SiC layer fabricated by electrochemical anodization. The annual black region is a Ni film evaporated on the back of the wafer. The surrounding green region is the 6H−SiC wafer.

Figure 2 shows a secondary electron microscope (SEM) image of porous SiC layer. There are many pores in this layer and the structure resembles porous Si. The thickness of this layer was about 0.3 μm for 60 minutes anodization duration. The anodized depth for SiC substrates is smaller than that for Si substrates. This may originate from the mobility of carriers : both electron and hole mobilities in SiC crystal are about ten times smaller than those in Si crystal from our measurements.

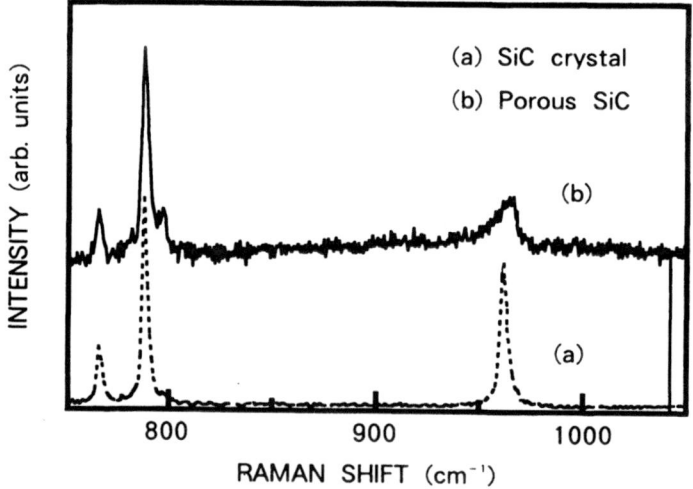

Figure 3. Room temperature Raman spectra from porous SiC (solid line) and crystalline SiC (broken line).

Figure 3 shows the Raman spectra obtained at room temperature using an Ar—ion laser as an excitation with a wavelength at 514.5 nm. The solid line shows the Raman spectrum of porous SiC and the broken line shows that of SiC crystal for reference. There is a broad base spectrum in Raman signal due to the luminescence from porous SiC. There is a broadening in each Raman peak (especially 960 cm⁻¹ mode). However, each peak is almost situated at the same position. These results suggest that a dissolution process not only occurs selectively to Si atoms but also to C atoms and that the stoichiometry of SiC does not change. Raman spectrum calculation for porous SiC is rather more complicated than that for Si because of the phonon dispersion curve [6]. However, we suppose that the broadening originates from the microstructure in porous SiC due to the phonon confinement.

Figure 4 shows the photoluminescence spectra from porous SiC with the anodization current density of (a) 60 mA/cm², (b) 40 mA/cm², and from (c) SiC substrate using a He—Cd laser excitation (325 nm). It is clear that luminescence becomes intense and luminescence peak shifts to higher —energy with increasing anodization current density : 500 nm for 20 mA/cm², 480 nm for 40 mA/cm², and 460 nm for 60 mA/cm². The luminescence from SiC substrate originated from the free electron to acceptor recombination is very weak at room temperature. The blueshift of the luminescence peak with increasing current density is similar to that generally observed in porous Si [7]. However, all the luminescence peaks are situated below the SiC bandgap (2.86 eV at room temperature) [Ref. 4, 5] and this redshift is different from the blueshift of the luminescence peak in porous Si.

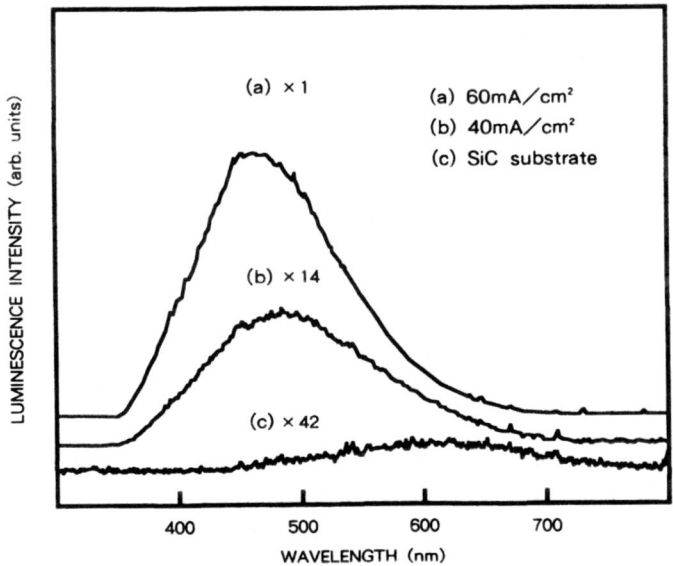

Figure 4. Luminescence spectra from porous SiC for the anodization current density of (a) 60mA/cm², (b) 40mA/cm², and from (c) 6H—SiC crystal.

Figure 5 shows the picosecond luminescence decay curves at 500 nm emission wavelength for anodization current densities of (a) 60 mA/cm² and (b) 40 mA/cm². With luminescence being intense, the picosecond luminescence decay becomes slow. The picosecond luminescence decay behavior of intense luminescent sample (the sample of a current density of 60 mA/cm²) is similar to that of porous Si : the decay shows the nonexponential behavior with the decay components from 100 ps to nanosecond and it becomes faster with increasing emission energy [8]. However, in the case of weak luminescent sample, the decay becomes faster. This may be due to the existence of dangling bonds by imperfect hydrogen termination or due to the remaining of the indirect−bandgap character.

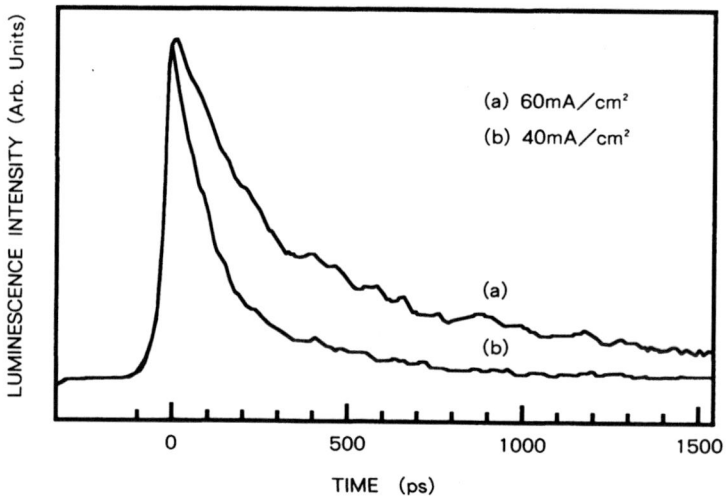

Figure 5. Picosecond luminescence decay curves at 500 nm emission wavelength for current densities of (a) 60 mA/cm² and (b) 40 mA/cm².

DISCUSSION

The luminescence−peak shift in porous SiC as a function of anodization current density is similar to that in porous Si [7]. Picosecond luminescence decay also shows similar behavior with that of porous Si. However, the luminescence peaks (∼2.7 eV) are situated below the bandgap of 6H−SiC crystal (2.86 eV). This is different from the peak−shift in porous Si. Our experimental results suggest that the luminescence both from porous Si and porous SiC can be attributed to the luminescence from hydrogenated amorphous Si or SiC rather than from Si or SiC microcrystals. However, luminescence intensity of the porous structure is higher than that of amorphous semiconductor. It is likely that the origin of both luminescences is the same however, the existence of the microcrystals plays an important role in the absorption and the carrier transfer processes [9−11]. We consider that the substance of the intense−luminescent origin is the molecular−like clusters or reconstructed bonds prevailing near the surface on Si or SiC microcrystal cores [12].

The structure of porous SiC is similar with that of porous Si from SEM observations. The Raman spectra also suggest the existence of microstructures with the same stoichiometry as SiC crystal. However, there is no reason why the anodization process maintains the stoichiometry between Si and C. There are many problems hidden in anodization processes (One is the creation of condensed nanometer order Si microcrystals.) and this offers a new problem for electrochemistry. We need further investigations such as a transmission electron microscope analysis or a X—ray photoemission spectroscopy to confirm the microstructures in porous SiC.

CONCLUSION

We have formed the porous SiC prepared by electrochemical anodization. We have observed the visible luminescence from this with a peak wavelength of around 500 nm at room temperature. The luminescence peak is situated below the bandgap of 6H—SiC crystal. Our experimental results suggest that the origin of the intense blue—green luminescence of porous SiC is the same as that of the intense red luminescence of porous silicon : the near—surface region around microcrystals is responsible for the intense luminescence. The intensity of this blue—green luminescence is about five hundred times stronger than that of free electron to acceptor recombination in 6H—SiC crystal. This porous SiC will open the door to intense blue—green light emitting diodes.

REFERENCES

[1] L. T. Canham, Appl. Phys. Lett. 57, 1046 (1990).
[2] V. Petrova—Koch, T. Muschik, A. Kux, B. K. Meyer, F. Koch and V. Lehmann, Appl. Phys. Lett. 61, 943 (1992).
[3] V. Petrova—Koch, T. Muschik, A. Kux, F. Koch, and V. Lehmann, Extended Abstracts of the 1992 International Conference on Solid State Devices and Materials (Tsukuba, August 26 – 28, 1992) p. 372.
[4] H. R. Philipp and E. A. Taft, "Silicon Carbide" (edited by J. R. O'Corner and J. Smiltens), p. 366, Pergamon Press (1960).
[5] T. Kimoto, H. Nishino, T. Ueda, A. Yamashita, W. S. Yoo, and H. Matsunami, Jpn. J. Appl. Phys. 30, L289 (1991).
[6] S. Nakashima, H. Katahama, Y. Nakakura, and A. Mitsuishi, Phys. Rev. B33, 5721 (1986).
[7] N. Koshida and H. Koyama, Jpn. J. Appl. Phys. 30, L1221 (1991).
[8] T. Matsumoto, M. Daimon, T. Futagi, and H. Mimura, Jpn. J. Appl. Phys. 31, L619 (1992).
[9] T. Matsumoto, T. Futagi, H. Mimura, and Y. Kanemitsu, Extended Abstracts of the 1992 International Conference on Solid State Devices and Materials (Tsukuba, August 26 – 28, 1992) p. 478.
[10] T. Matsumoto, T. Futagi, H. Mimura, and Y. Kanemitsu, Phys. Rev. B47 (1993) in press.
[11] Y. Kanemitsu, H. Uto, T. Matsumoto, T. Futagi, and H. Mimura (unpublished).
[12] Y. Kanemitsu, K. Suzuki, H. Uto, Y. Masumoto, T. Matsumoto, and H. Matsumoto, Appl. Phys. Lett. 61, 2446 (1992)

CHARACTERIZATION OF PHOTOLUMINESCENCE FROM ANODICALLY ETCHED SiC/Si HETEROSTRUCTURES

A. J. Steckl, J. N. Su, J. Xu, J. P. Li, C. Yuan, P. H. Yih and H. C. Mogul
Nanoelectronics Laboratory, Department of Electrical and Computer Engineering
University of Cincinnati, Cincinnati, OH
45221-0030

ABSTRACT

Patterned SiC/Si heterostructures have been treated by anodic etching in HF/ethanol solutions at 25°C. The anodic process used a current density ranging from 2-10 mA/cm^2 and times from 2-5 min. The SiC/Si samples had SiC regions and exposed Si regions of dimensions from 2.5 μm to ~500 μm. For n-Si substrates, short etching times (<3 min) result in selective-area UV-induced visible photoluminescence (PL) being observed at 25°C from only the SiC regions. Longer etching times (≥3 min) render both the SiC-protected and the exposed Si regions photoluminescent, with nearly identical spectral characteristics and with a peak located at 660-670 nm. The selective-area PL is based on the rapid lateral etching of the n-Si surface under the SiC layer due to either high surface stress caused by the lattice mismatch between SiC and Si and/or to a higher SiC conductance. This is confirmed by the time progression of the PL images in the SiC regions. The PL degradation with UV exposure time has been shown to be substantially reduced by a passivation procedure involving HNO$_3$/H$_2$O. For SiC/p-Si structures, the exposed Si regions become photoluminescent upon very short anodization time, while the SiC regions did not luminesce at all for even the longest anodization times used (5 min).

INTRODUCTION

Photoluminescence from porous Si [1-4] (PoSi) has triggered considerable interest in exploring various Si structures which can provide useful light emitting devices. In this paper, we report on light emission from 3C-SiC/Si heterostructures. 3C-SiC (or β−SiC) is a semiconductor with a diamond cubic structure (as is Si) and with a band-gap energy of 2.2 eV, which makes it transparent to visible light. In addition, it can be heavily doped to produce a low resistivity and it is impervious to most, if not all, common chemicals. This combination of properties makes SiC an ideal candidate for a transparent electrode to PoSi. For example, poly-crystalline SiC deposited on Si after it is rendered porous has been utilized [5] in the fabrication of electroluminescent p-n junction diodes. A similar approach for a wide band-gap injector into PoSi has been reported [6] by Campbell et al. using GaP.

The SiC/Si structure is initially formed by the rapid thermal chemical vapor deposition (RTCVD) of a thin β-SiC layer on a Si substrate. RTCVD of SiC by the carbonization of the Si surface with various hydrocarbons has been shown [7] to produce crystalline thin films. Next, the SiC film is patterned by reactive ion etching (RIE) using CHF$_3$/O$_2$ under conditions designed [8] to provide a residue-free etch field. As shown in Fig. 1, this step defines exposed Si regions and complementary Si regions protected by the SiC film. Then anodic etching of the patterned structure is performed at 25°C in an HF:ethanol:H$_2$O solution with a ratio of 1:2:1. In the electrochemical process the SiC/Si sample is the anode, with the cathode being a platinum wire. During anodization the SiC-covered surface is exposed to the electrolyte (see Fig. 1c). The anodization current density was generally varied from 2 to 10 mA/cm^2.

RESULTS AND DISCUSSION

The pattern of the SiC/Si heterostructure is shown in the SEM photograph of Fig. 2a. This sample had a 5-7 Ω-cm (100) n-Si substrate and had been anodized for 2.5 min. Both large (>100µm) and small (few µm's) SiC-masked regions and exposed Si regions are evident. Using a UV fluorescence microscope, with a Hg lamp emitting at 360-370 nm, the anodized SiC/n-Si structure exhibits visible wavelength photoemission at 25 °C only in those regions where SiC is present. This selective-area photoemission is shown in Fig. 2b for the sample whose SEM photograph is contained in Fig. 2a. A comparison between Fig. 2a and Fig. 2b clearly indicates that only the SiC-masked regions are emitting. Uniform emission is obtained in these regions, with even the narrowest line of 2.5 µm exhibiting a well-resolved emission pattern. It is interesting to compare these results obtained with heterostructures on n-Si with those obtained on p-Si substrates. The SiC/p-Si heterostructures exhibit a reversal in the photoemission pattern. In this case, the exposed Si regions readily emit after even short anodization times, while the SiC-masked region do not emit even after 5 min anodization, as seen in Fig. 2c.

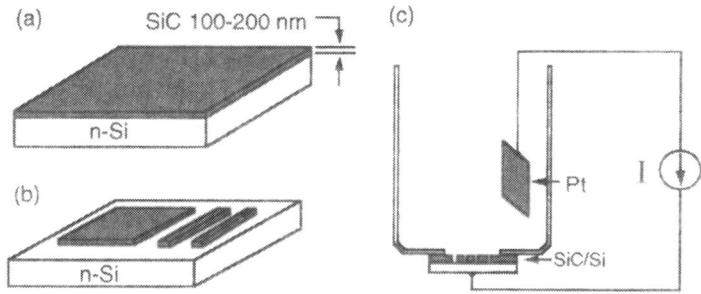

Fig. 1 SiC/Si sample fabrication and anodic etching.

To understand this selective-area emission process, the emission patterns from SiC/Si samples anodized for different times were obtained. Fig. 3 contains images obtained with conventional microscopy (reflected white light) and UV fluorescence microscopy from the patterned SiC surface of various samples. Figs. 3a and b show the reflected white light and UV-induced images of the same SiC/n-Si sample anodized for 2 min. In the UV-induced image it appears that, while the smaller (20 µm) features emit fairly uniformly, the larger (80 µm) features have a central region which does not emit. For a longer anodization time of 2.5 min, Fig. 3c, the dark regions in the center of the large SiC-masked regions disappear, resulting in uniform emission. If the anodization time is increased further to 3 min, the entire surface emits under UV excitation, as shown in Fig. 3d. The reflected white light image of a SiC/p-Si sample with same pattern is shown in Fig. 3e. The corresponding UV-induced image is shown in Fig. 3f. Even though this sample had been anodized for 5 min, no emission is observed from the SiC-masked regions.

Photoluminescence (PL) spectroscopy was performed on various SiC/n-Si samples using the same Hg UV source. As shown in Fig. 4a, under conditions (2.5 min anodization) of selective-area emission the SiC/Si region emits in a broad spectrum peaked at a 650 nm. This type of PL spectrum is very similar to that obtained with PoSi [3, 4]. The Si region of the sample only produces a "background" spectrum. Increasing the anodization from 2.5 min to 3 min produces the PL spectra contained in Fig. 4b. Under this anodization condition the emission was shown in

Fig. 3d to be no longer area selective. Indeed, we observe in Fig. 4b that the PL spectra from the two regions are now nearly identical.

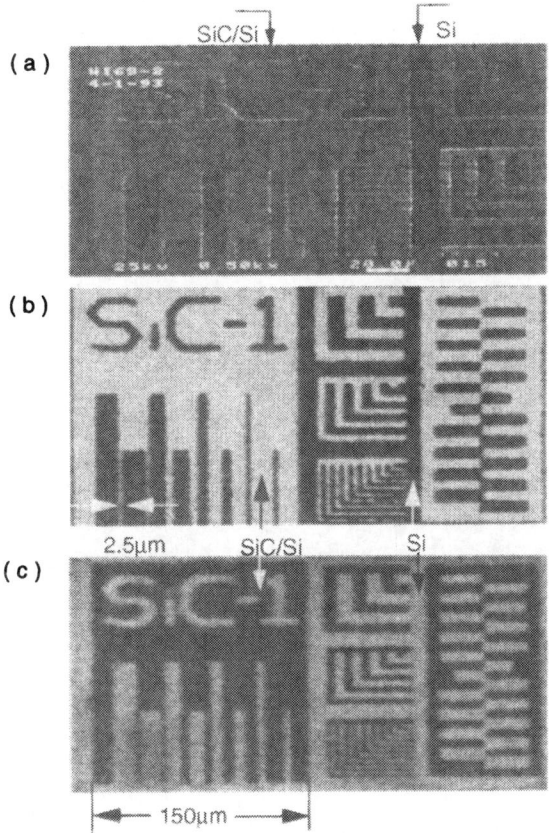

Fig. 2   Images of patterned and anodically-etched SiC/Si samples: (a) SEM microphotograph, n-Si, 2.5 min etch time; (b) UV-induced emission, n-Si, 2.5 min; (c) UV-induced emission, p-Si, 5 min.

Based on the results contained in Figs. 2-4, we can begin to explain some of the observed phenomena. For the SiC/n-Si samples it appears that anodization is first ocurring preferentially at the SiC/Si interface. The process aparrently starts at the exposed edge of the SiC/Si mesa and proceeds laterally under the SiC-masked region, as seen from the UV-induced images on Fig. 3b and 3c. For a sufficiently long anodization time, the entire surface area (including both exposed and SiC-protected Si regions) becomes photoemissive, as shown in Fig. 3d. The source of the photoluminescence is PoSi formed underneath the SiC layer. This is confirmed by the fact that when both regions emit they exhibit an essentially identical spectrum, see Fig. 4b. Possible reasons for the initial preferential anodization of the SiC-covered regions include: (a) stress-induced charge carrier generation at the interface; (b) a SiC layer more conducting than the n-Si substrate. Turning to the SiC/p-Si structure, the presence of abundant holes at the top surface of the exposed Si regions apparently prevents the lateral anodization underneath the SiC-masked regions.

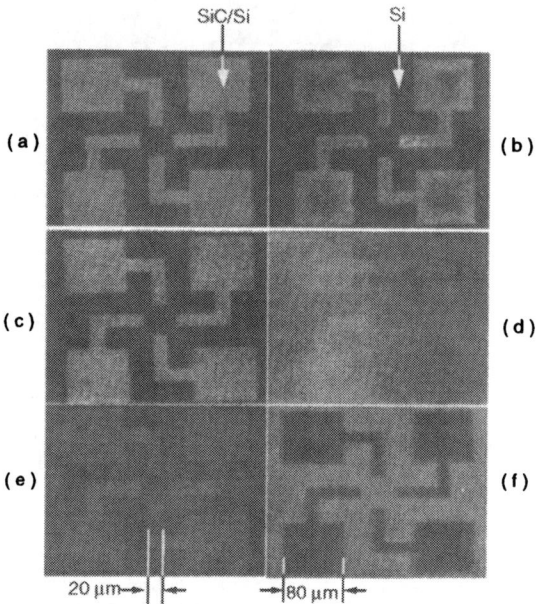

Fig. 3 Optical images of anodically-etched SiC/Si samples as a function of anodization time : (a) reflected white light, n-Si, 2 min; (b) UV-induced emission, n-Si, 2 min; (c) UV-induced emission, n-Si, 2.5 min; (d) UV-induced emission, n-Si, 3 min; (e) reflected white light, p-Si, 5 min; (f) UV-induced emission, p-Si, 5 min.

One of the problems associated with PoSi currently receiving considerable attention is the degradation in PL intensity with UV exposure time [9, 10]. This is believed to be due to the removal of hydrogen and exposure of dangling bonds on the PoSi surface, which can act as surface traps. One approach reported [11, 12] to stabilize the PL intensity is a slight oxidation of the PoSi by either chemical or thermal means. This is believed to passivate the PoSi by oxidizing the surface and removing the surface traps [12]. We have applied the chemical passivation method to the SiC/n-Si structure using a 25°C solution of $HNO_3:H_2O$ with a 1:4 ratio. The PL intensity was measured as a function of UV exposure time for various levels of passivation time. As seen from Fig. 5, there are two important aspects of the PL decay : (a) an immediate fast decay component; (b) a much slower decay component, leading to a stable level after several minutes of UV exposure. The chemical oxidation treatment appears to ameliorate both aspects of the PL degradation. Increasing the passivation time to 30 sec or more results in both a slower initial decay time constant and a higher PL equilibrium value.

CONCLUSIONS & ACKNOWLEDGEMENTS

In summary, we have reported selective-area visible photoemission under UV excitation from anodically etched SiC/n-Si structures. We conclude the emission is associated with PoSi preferentially formed in the Si substrate under the SiC-masked regions. The localized visible light is readily transmitted through the SiC layer, which has twice the band-gap energy of Si. The SiC/Si heterostructure appears to be a promising candidate for the fabrication of electroluminescent PoSi devices.

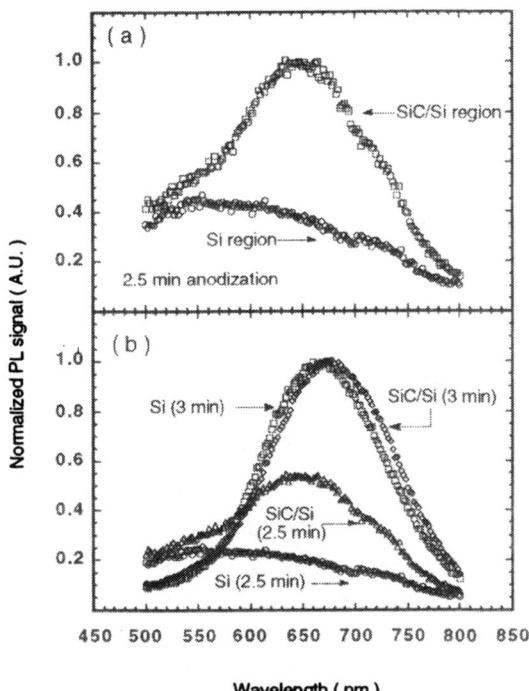

Fig. 4 Photoluminescence spectra from Si and SiC/Si regions anodically-etched for (a) 2.5 min ; (b) 2.5 and 3 min

This work was supported in part by SDIO/IST and monitored by ARO and by EMTEC. The authors are pleased to acknowledge the encouragement of L. Lome, R. Trew and J. Zavada, and the assistance of W. Bresser and P. Boolchand with the PL measurements.

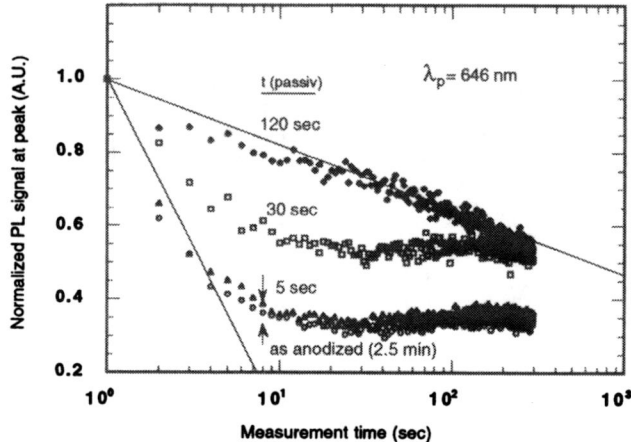

Fig. 5 Normalized PL intensity versus measurement time for various values of passivation time.

REFERENCES

[1]   L. T. Canham, Appl. Phys. Lett. **57**, 1046 (1990).
[2]   V. Lehmann and U. Gösele, Appl. Phys. Lett. **58**, 856 (1991).
[3]   A. Halimaoui, C. Oules, G. Bomchil, A. Bsiesy, F. Gaspard, R. Herino,
       M. Ligeon, and F. Muller, Appl. Phys. Lett. **59**, 304 (1991).
[4]   N. Koshida and H. Koyama, Jap. J. Appl. Phys. **30**, Pt. 2, L1221 (1991).
[5]   T. Futagi, T. Matsumoto, M. Katsuno, Y. Ohta, H. Mimura, and K. Kitamura, Jap.
       J. Appl. Phys. **31**, L616 (May 1992).
[6]   J. C. Campbell, C. Tsai, K-H. Li, J. Sarathi, P. R. Sharps, M. L. Timmons, R.
       Venkatasubramanian, and J. A. Hutchby, Appl. Phys. Lett. **60**, 889 (Feb. 1992).
[7]   A. J. Steckl and J. P. Li, IEEE Trans. Electr. Dev. **39**, 64 (Jan. 1992)
[8]   A. J. Steckl and P. H. Yih, Appl. Phys. Lett. **60**, 1966 (Apr. 1992).
[9]   M. A. Tischler, R. T. Collins, J. H. Stathis and J. C. Tsang, Appl. Phys. Lett. **60**,
       639 (Feb. 1992).
[10]   Z. Y. Xu, M. Gal and M. Gross, Appl. Phys. Lett. **60**, 1375 (March 1992).
[11]   G. Mauckner, T. Walter, T. Baier, K. Thonke and R. Sauer, MRS Symp. Proc.
       283, 109 (Dec. 1992).
[12]   C. Peng, L. Tsybeskov and P. M. Fauchet, MRS Symp. Proc. **283**, 121 (Dec. 1992).

THE ROLE OF SILICON IN OPTOELECTRONICS

DENNIS G. HALL
The Institute of Optics, University of Rochester, Rochester, New York 14627

ABSTRACT

This invited paper reviews some of the ways in which silicon technology has been applied in integrated optics and optoelectronics.

INTRODUCTION

Silicon has been called the ideal semiconductor material for microelectronics, but there is universal agreement that it is much less ideal for applications in optoelectronics. This is the case for two principal reasons. First, the silicon crystal structure (diamond lattice) is centrosymmetric, therefore silicon does not exhibit a linear electro-optic effect (Pockel's effect). Second, silicon is an indirect bandgap semiconductor, which means that direct, radiative, band-to-band transitions are forbidden by the k-conservation selection rule. The former makes materials such as $LiNbO_3$ (lithium niobate) or $LiTaO_3$ (lithium tantalate) more attractive choices than silicon for applications that require light modulation. The latter makes direct-bandgap materials like GaAs or InP more attractive choices than silicon for the development of light-source technologies. In spite of these serious limitations, there continues to be a great deal of interest in finding ways to make use of silicon in optoelectronics. The idea that one might take advantage of a mature silicon technology to combine optical and electronic components on a single substrate is a compelling one.

A survey of the published literature shows that efforts to date to find a place for silicon in optoelectronics can be placed in two categories. In the first category, the objective has been to define passive components and devices that simply avoid the two difficulties mentioned in the previous paragraph. For example, optical waveguides made of silicon or any of a number of other materials can be formed on a silicon substrate. Once a waveguide technology is in hand, one can, at least in principle, make use of waveguide diffraction gratings, waveguide lenses, and other passive components to develop such systems for applications such as the optical interconnection of electronic devices, or wavelength demultiplexing in communications. Passive systems are comparatively easy to conceive of because doing so involves working within the bounds of what is now known. In the second category, the objective has been to try to find ways to overcome the limitations imposed by the indirect bandgap and the lack of a linear electro-optic effect in silicon. One can consider, for example, circumventing the k-conservation selection rule either by altering the band-structure of silicon or by using a mechanism other than band-to-band recombination to generate light. One might achieve the former, for example, by using superlattices made up of layers of silicon, germanium, and/or their alloy, or by using quantum confinement in silicon nanostructures, as might be one mechanism at work in porous silicon. One might achieve the latter, for example, by introducing radiative impurity complexes such as erbium into conventional electronic-grade silicon, or into silicon-germanium alloys grown on silicon. Likewise, in the absence of a Pockel's effect, one might consider using the free-carrier effect in silicon as a mechanism for modulating light. For the most part, developing alternative approaches to active devices, sources and modulators, requires more than incremental steps beyond what is already known. Genuine breakthroughs are needed if a viable active-device technology based in silicon is to emerge.

This paper reviews and discusses several examples of both passive and active silicon-based optical or optoelectronic devices and systems. It would take a very long article to describe every such device or system that has been demonstrated or proposed to date. This article instead examines a few representative examples that illustrate common approaches, fundamental principles, or particular device structures.

WAVEGUIDES

It is reasonable to expect that any integration of optical and electronic components is likely to involve an optical waveguide. Figure 1 illustrates five different classes of silicon-based optical waveguides. In the first, Figure 1(a), a thin layer of glass or a similar material is deposited onto an oxidized silicon substrate. The high refractive index of Si (n ~ 3.5) limits the choices for an overlayer that can function as an optical waveguide when deposited directly onto Si. A waveguiding layer must ordinarily have a higher refractive index than the adjacent media. An SiO_2 (refractive index n ~ 1.5) layer with a thickness of approximately one micron makes an excellent buffer layer. This approach has the advantage of applicability for a wide range of waveguide materials, but the disadvantage that the optical wave confined in the waveguide is physically separated from the silicon, which functions merely as a convenient substrate. A similar approach would be to eliminate the glass layer and instead change the composition of the upper portion of the oxide layer to create a higher-index region atop the lower-index oxide.

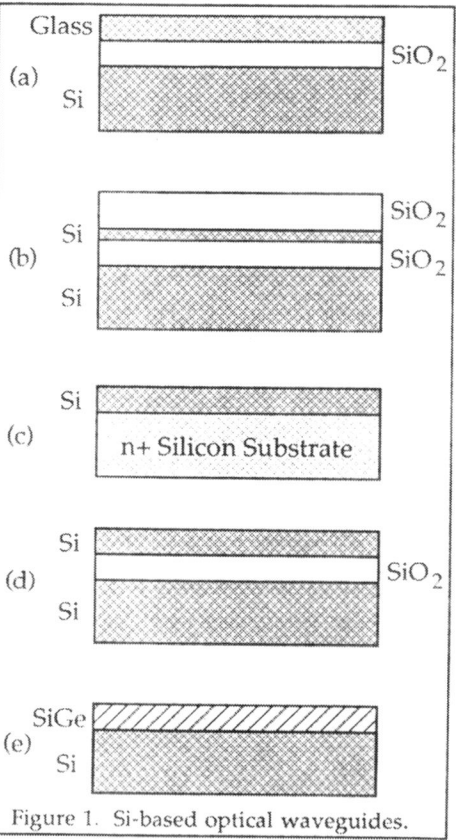

Figure 1. Si-based optical waveguides.

Figure 1(b) illustrates the so-called ARROW (anti-resonant, reflecting optical waveguide). In this structure, light is confined in the upper oxide layer by the combined action of total internal reflection and what is essentially a

Fabry-Perot effect. [1] It is this latter effect that enables the oxide layer to confine the optical wave even though it rests on a thin layer of higher-refractive-index polycrystalline silicon. Optical losses of 1 dB/cm or less, the usual benchmark value, have been demonstrated for the structures in both Figs. 1(a) and 1(b).

Figures 1(c). (d), and (e) illustrate waveguide concepts in which the optical wave is confined *within* the semiconductor. In each case, of course, the wavelength must be selected to fall in the transparent region of the spectrum for the material in question (photon energy < bangap energy). The scheme in Figure 1(c) was first demonstrated by Soref and Lorenzo [2]. The heavy doping of the Si substrate reduces, by the free-carrier effect, the refractive index below that for undoped or more lightly doped silicon. Waveguiding is thus permitted in a less heavily doped layer of epitaxial silicon grown on the n+ substrate. Unfortunately, the exponential tail of the waveguide mode penetrates significantly into the substrate, resulting in an unacceptably high loss due to the mode's interaction with the carriers in the substrate. Losses in the vicinity of 10 dB/cm or more have been observed for operation at wavelength $\lambda = 1.3 \ \mu m$.

Figure 1(d) illustrates a silicon-on-insulator (SOI) structure that consists of a thin layer of single-crystal silicon separated from the silicon substrate by a layer of silicon dioxide. The use of this structure as an optical waveguide was proposed by Kurdi and Hall [3] in 1988. Their analysis showed that optical losses of 1 dB/cm should be achievable for the appropriate range of thickness parameters. There are at least two ways in which SOI waveguides have been fabricated: the SIMOX (separation by the implantation of oxygen) process and the bond-and-etchback process. [4,5] Measured losses as low as 2 dB/cm in planar SOI waveguides [6] and 0.5 dB/cm in rib SOI waveguides [7] have been reported at wavelengths $\lambda = 1.3 \ \mu m$ and 1.55 μm.

Figure 1(e) illustrates another possibility, the use of a silicon-germanium (SiGe) alloy layer deposited onto a silicon substrate. Ge has a significantly higher refractive index (n ~ 4.3 for λ~ 1.3 μm) than Si, as does a SiGe alloy. A variation of this basic scheme makes use of a higher-index diffusion layer formed by diffusing Ge into a silicon substrate. [8]. There is a 4% difference between the lattice constants of Si and Ge, however, so the alloy layer indicated in Figure 1(e) is typically a strained layer. Measured optical losses of approximately 2 dB/cm in multimode rib SiGe/Si waveguides grown by chemical-vapor deposition [9] and 3 – 5 dB/cm in strained SiGe/Si ridge waveguides grown by molecular-beam epitaxy [10] have been reported, both for wavelength λ~ 1.3 μm.

The sketches in Figure 1 show only the basic layer structure, i.e. planar waveguides. Two-dimensional rib (or ridge) waveguides can be, and have been, made by the usual masking and etching techniques, the details of which will not be discussed here.

PASSIVE DEVICES AND SYSTEMS

The waveguide geometry in Figure 1(a) was the first to attract serious attention for the integration of several optical/optoelectronic components on a common silicon substrate. Because the waveguide layer is physically isolated from the silicon substrate, the arrangement in Figure 1(a) makes it possible to use the waveguide in a spectral region in which Si is highly absorbing. By making the oxide buffer layer thickness at least 1.5 μm thick, wavelengths near

0.8 μm, the wavelengths emitted by the AlGaAs/GaAs family of semiconductor lasers, can propagate with relatively low loss.

Figure 2 shows a sketch of a four-channel wavelength demultiplexer (DMUX) fabricated using the planar glass/oxide/Si system of materials. [11] The DMUX system is essentially a multi-channel fiber-optic receiver formed on a two-inch diameter silicon substrate. Its function is as follows. An optical fiber

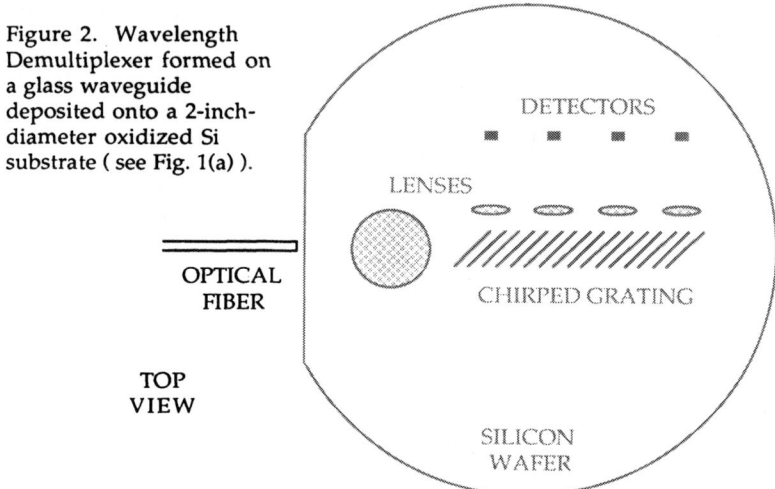

Figure 2. Wavelength Demultiplexer formed on a glass waveguide deposited onto a 2-inch-diameter oxidized Si substrate (see Fig. 1(a)).

is mounted in a silicon V-groove to align its core region with the edge of the glass waveguide. The fiber carries several distinct information channels, each one associated with a source operating at a different optical wavelength. This multi-wavelength signal enters the glass waveguide where it is collimated by a waveguide lens. The collimated beam then enters a region of the waveguide that has been prepared with a surface corrugation, a grating. The grating lines are oriented at a 45-degree angle with respect to the the original direction of propagation. Further, the grating period (separation between adjacent lines) is "chirped," which means the period varies (almost linearly) along its length, from short period to long period. Each wavelength component of the incident light travels along the original direction until it comes to that section of the grating where the local period satisfies the Bragg condition for that wavelength, at which point it is diffracted through a 90-degree angle, passes through a focusing lens, and is routed to a p-i-n detector formed in the Si substrate. Each wavelength component, and hence each information channel, is diffracted from a different portion of the grating than the others. [12] The system therefore passively separates the N information channels, routing each to its own Si p-i-n detector, producing N electrical signals.

Figure 3 shows a photo of a completed eight-channel demultiplexer. [11] In a complete package, wire bonds would lead from the individual detectors to connectors mounted on the housing. There is clearly plenty of room left on the Si substrate for the fabrication of preamplifiers or signal-processing electronics.

There are many other examples of systems based on the waveguide configurations similar to that of Figure 1. Henry et al [13], in 1990, reported the fabrication and operation of a wavelength demultiplexer similar in spirit to

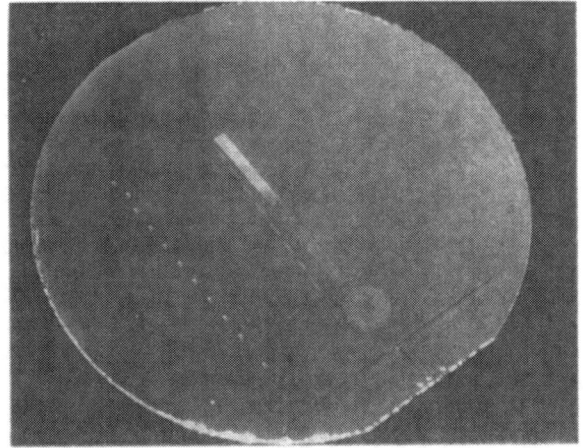

Figure 3. Photo of the completed Si-based wavelength demultiplexer. The Si wafer is two inches in diameter.

that in Figs. 2 and 3, but different in detail. Instead of the planar glass waveguide used in the earlier device, the more recent version used channel waveguides made of phosphorus-doped SiO_2 (8% P) surrounded by SiO_2. The device was based on a dispersing element that consisted of an elliptical Bragg reflector, which is illustrated in Fig. 4.

Channel waveguides A and B are placed at the focal positions F of the confocal ellipses that make up the elliptical Bragg reflector (EBR). An optical wave leaving waveguide A is focused by the grating onto the leading edge of waveguide B if the wavelength satisfies the first-order Bragg condition. The fact that the Bragg condition is highly wavelength dependent means that several different EBRs can be arranged sequentially. Those designed for one wavelength will be essentially transparent to a different wavelength, just as one section of the chirped grating in Figure 2 is transparent to longer wavelengths than the one for which it is designed to be Bragg reflecting. The system described in Ref. 13 was designed for four-wavelength-channels of operation.

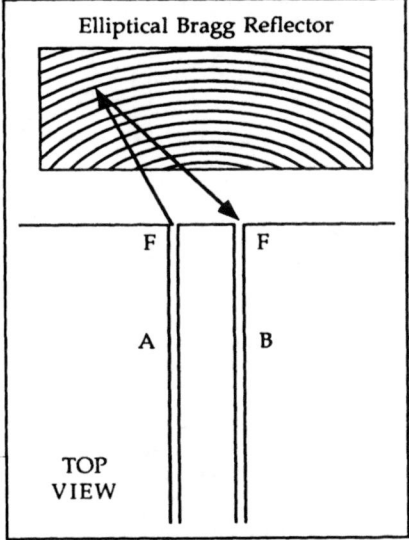

Figure 4. Elliptical Bragg reflecting element.

One key difference between the two approaches to wavelength demultiplexing in Figs. 2 and 4 is that the former used detectors integrated into the Si substrate whereas the latter did not. The system reported in Ref. 13 was designed for operation at wavelengths near 1.5 μm, for which Si is transparent.

Even more than in Figure 2, then, the EBR system of Ref. 13 consigns Si to the role of a convenient substrate.

ACTIVE DEVICES AND SYSTEMS

As mentioned in the Introduction, silicon lends itself less well to active optical devices than to passive ones. Hybrid approaches, in which one attaches to a silicon substrate active components made from other materials, have attracted interest from several groups. Various attempts have been made, for example, to grow III-V semiconductor lasers on Si substrates [14], something that has met with only limited success. The lattice mismatch produces lasers with significantly reduced lifetimes. Another technique called epitaxial lift-off (ELO) [15] makes use of highly selective etching to remove III-V epitaxial structures from their substrates and affix them to a variety of substrates (glass, $LiNbO_3$, silicon) [16,17]. To date, the ELO method has been used primarily to attach photodetectors to various substrates.

An example of a relatively straightforward, hybrid, Si-based approach is the chip-to-chip interconnection system shown in Figure 5. This system was first demonstrated and described by researchers at NTT in 1987. [18,19] Each Si

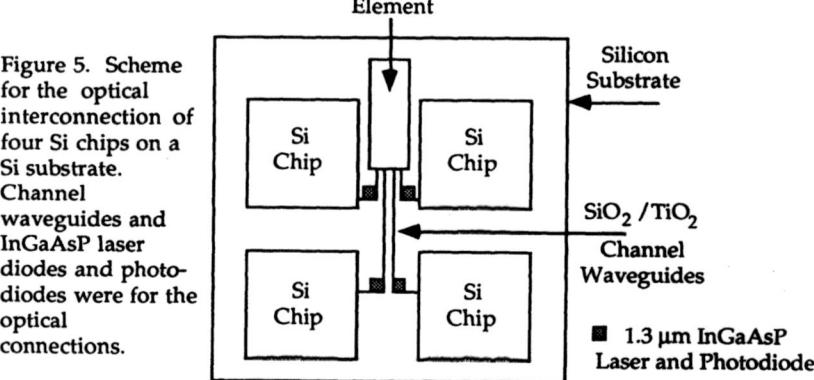

Mixing
Element

Figure 5. Scheme for the optical interconnection of four Si chips on a Si substrate. Channel waveguides and InGaAsP laser diodes and photo-diodes were for the optical connections.

Silicon Substrate

SiO_2 / TiO_2
Channel Waveguides

■ 1.3 μm InGaAsP Laser and Photodiode

chip is electricallyconnected to its own InGaAsP laser diode and photodetector. These are, in turn, optically connected to each of the three other such modules by means of SiO_2 / TiO_2 channel waveguides and a reflective mixing element. The system operates in a broadcast mode, which means that the signal from each Si chip is shared with all the others. To fabricate the waveguides, the Si substrate is first coated with a 20 μm thick layer of SiO_2 followed by an 8 μm thick layer of SiO_2 / TiO_2. The SiO_2 / TiO_2 layer is then patterned and etched (reactive-ion etching) to produce channel (or ridge) waveguides with widths ranging from 14 μm to 200 μm depending on the circuit element. Finally, a 8 μm thick cladding layer of SiO_2 was deposited over the entire surface by chemical vapor deposition. [19] The InGaAsP circuit elements were attached to metal electrodes deposited onto the oxide by first aligning them in the appropriate positions and then melting the electrode to essentially weld the component in place. The authors report operating the system at a 340 Mbps data rate and a 250 MHz clock frequency. [19]

Modulation Schemes

Several investigators have examined the use of the free-carrier effect in silicon as a mechanism for modulating light. In the free-carrier effect, the real and imaginary parts of the refractive index are functions of the electron and hole densities. Carrier injection or removal can therefore be used to change the amplitude or phase (or both) of an optical wave propagating in, or reflected from, the semiconductor. The size of the (real) refractive index changes associated with the free-carrier effect in Si and and GaAs are comparable; both are relatively small, typically less than 0.001. GaAs with its direct energy gap, however, exhibits much smaller carrier lifetimes than Si, making faster devices possible in GaAs.

The prototype devices demonstrated to date have required high injection current densities to produce relatively modest degrees of light modulation, as indicated in Table 1, below. In the Table, AM and PM refer to amplitude and

Table 1. Modulator Characteristics				
Author, Year	Type	λ (μm)	Current Density (A/cm^2)	Reference
Grodnenskiy et al, 1982	AM	1.15	3,600	20
Lorenzo and Soref, 1987	2×2	1.3	1,260	21
Hemenway et al, 1989	PM, REF	1.3	16,600	22
Xiao et al, 1990	REF	1.3	2,000	23
Treyz et al, 1991	AM	1.3	3,200	24
Treyz et al, 1991	AM & PM	1.3	1,600	25
Treyz, 1991	PM	1.3	NA (Thermal)	26

phase modulation, respectively, and REF designates a reflection modulator. The current densities all exceed 1,000 A/cm^2, a number that is large even in comparison to semiconductor lasers, for which threshold current densities on the range 100 – 200 A/cm^2 are relatively routine. The last entry in the Table made use of a thermal effect to change the refractive index in a silicon-on-insulator waveguide. The thin-film heaters deposited on the waveguide surface consumed 30 mW of power to achieve a 40% modulation depth.

Figure 6 illustrates as an example the Mach-Zehnder interferometric modulator reported by Treyz et al [25], the second to last entry in Table 1. The modulator makes use of the plasma dispersion effect and is fabricated using rib waveguides. As mentioned in connection with Fig. 1(c), the heavy doping reduces the refractive index of the Si substrate relative to the more lightly doped (p⁻, in this case) epitaxial layer. The refractive index of the lightly

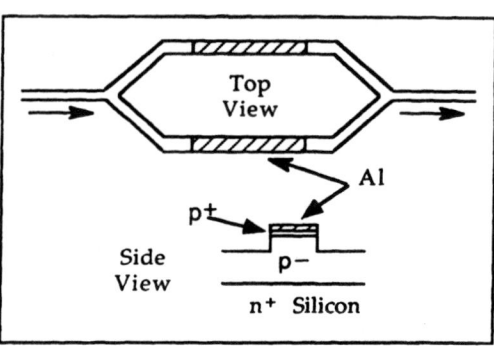

Figure 6. Mach Zehnder Si-based light modulator for operation at wavelength 1.3 microns. See Ref. 25 for details.

doped p⁻ region (the waveguiding region) is modulated by carrier injection by forward biasing the p-i-n device. [25] Light entering the left side of the device is divided at the junction into two optical waves, which proceed through the upper and lower waveguide sections. Carrier injection under the aluminum (Al) electrodes changes the refractive index in each section, allowing the relative phase of the two waves to be varied. The two waves come together (or interfere) at the second waveguide junction. If they interfere constructively, a transmitted signal appears at the device output on the right. If they interfere destructively, there is no transmitted signal. Modulation is achieved by varying the control voltage on the electrodes. The authors report modulation depths of -4.9 dB (on/off ratio of ~ 1/3) with response times < 50 ns for a current density of 1,600 A/cm².

This performance can be compared with that of a similar commercial product. A 1986 Mach Zehnder interferometric modulator of the same basic design, but fabricated in $LiNbO_3$ instead of silicon, required only four volts to achieve a modulation bandwidth of 3 GHz, and an on/off ration of 1/100. $LiNbO_3$, of course, has a strong electro-optic effect (Pockel's effect), something that is absent in silicon, and is a material with a very fast response time.

Light Sources

Dumke pointed out in his early analysis that direct bandgap semiconductors like GaAs were more likely candidates than indirect bandgap semiconductors like Si or Ge for the development of semiconductor lasers. [27] His predictions have certainly proved true. III-V semiconductor lasers have reached an extraordinary state of development. It is impressive that the telephone companies' stringent standards for an acceptable semiconductor laser for fiber-to-the-home (or -curb) applications are routinely met. A few of the requirements are: a cost of less than $50 per fully packaged (including fiber pigtail) laser, operation over the full temperature range of $-40°C \leq T \leq +85°C$, a mean-time to failure of one million hours, 5 – 10 mW of laser facet power at wavelength 1.3 µm, maximum laser threshold less than 70 mA over the full temperature range and lifetime of the laser, and spectral width less than 10 nm. [28] III-V semiconductor lasers set the standard for general-purpose semiconductor light sources. Genuine breakthroughs are needed if a Si-based light-emission technology is to stand up to even friendly comparisons with III-V sources.

Several approaches present themselves. Porous silicon [29], band-to-band recombination in SiGe/Si superlattices [30,31], and rare-earth-doped silicon [32] have been discussed in many papers in this Symposium (B). One mechanism for optical emission that has received less attention in this Symposium is the radiative decay of excitons bound to isoelectronic (or isovalent) impurity complexes in Si, SiGe alloys, and SiGe/Si superlattices. This mechanism is known to produce strong luminescence at room temperature in the indirect bandgap semiconductor gallium phosphide (GaP). Nitrogen, zinc-oxygen, and cadmium-oxygen complexes are well known examples in GaP. [33,34,35]

Isoelectronic-bound-exciton (IBE) emission is also known to occur in silicon. The best known example is that of beryllium pairs in silicon. [36] The Be-complex in Si produces intense (external efficiency ~ 1%) luminescence at wavelength 1.15 μm (see Figure 7), but for relatively low temperatures (T < 80K). It is, nevertheless, the prototype impurity complex for IBE emission. The Si:Be emission can be easily excited optically or electrically, and can be introduced into wave-guides by either diffusion or implantation [37]. The spectrum in Fig. 7 was actually obtained for the case of Be introduced by ion implantation into an epitaxial silicon optical waveguide of the type shown in Figure 1(c). The second, smaller peak in the spectrum, near 1.3 μm is a phonon replica. The Be complex can also be formed in SiGe alloys [38] and SiGe/Si [39] super-lattices [40] grown by molecular beam epitaxy (MBE).

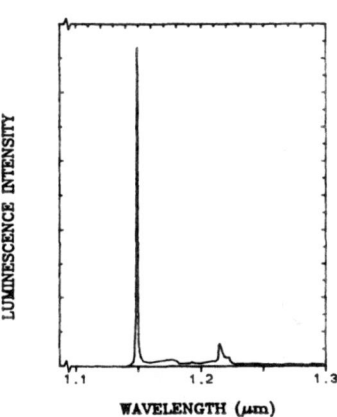

Figure 7. Si:Be Spectrum.

Chalcogen impurities in silicon appear to produce isoelectronic impurity complexes in silicon. Both sulfur- and selenium-related complexes have been reported to produce luminescence in the 1.3 μm wavelength range. [40,41] The sulfur-related emission exhibits an external efficiency of several percent and can be excited either optically or electrically. In fact, a Si:S light-emitting diode (LED) fabricated in a silicon epitaxial waveguide was demonstrated and reported in 1989. [42] Figure 8 shows both the photoluminescence (PL) and the electroluminescence (EL) spectra for Si:S at T = 60K. The LED exhibited an external efficiency in the 0.2% - 0.5% range at temperature T ~ 77K. The temperature dependence of the luminescence from the Si:S and Si:Se complexes is better than that for the Si:Be complex discussed above. The sulfur-related emission persists to nearly T = 200K, with maximum emission occurring near T = 77K. These results are promising, but there remains the need to achieve room-temperature operation if the IBE mechanism is to prove useful as the basis for a general-purpose silicon device technology.

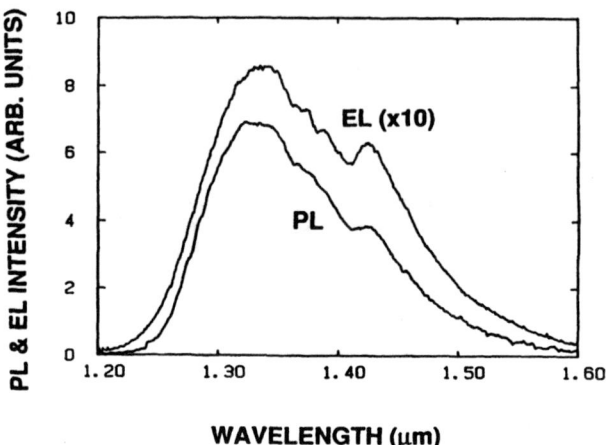

Figure 8. EL and PL spectra from Si:S LED at T = 60K.

Concluding Remarks

Passive optical or optoelectronic systems with some measure of device integration have been demonstrated. The design and fabrication of such systems requires only incremental advances beyond the current state of the art. Much the same can be said for hybrid systems. Active components and devices remain a challenge. The indirect bandgap and the lack of a Pockel's effect in silicon are formidable obstacles to the development of a silicon-based optical or optoelectronic technology. Another issue not even discussed in the preceding paragraphs is the availability of detectors. Applications in the communications arena create a special interest in devices that operate at wavelengths in the 1.3 μm and 1.55 μm spectral regions. Si is transparent at these wavelengths. SiGe alloys and SiGe/Si superlattices constitute one approach to the detector problem, and such detectors have already been demonstrated for wavelength 1.3 μm. [43] Genuine fundamental breakthroughs are needed for the eventual development of active Si-based optoelectronic devices and systems.

The insterested reader can consult several review articles for further discussion of the subject of Si-based integrated optics/optoelectronics. [44,45,46]

REFERENCES

1. M. A. Duguay, Y. Kokubun, T. L. Koch, and L. Pfeiffer, Appl. Phys. Lett. **49**, 13 (1986).
2. R. A. Soref and J. P. Lorenzo, IEEE J. Quantum Electron. **QE-22**, 873 (1986).
3. B. N. Kurdi and D. G. Hall, Opt. Lett. **13**, 175 (1988).
4. K. Izumi, M. Doken, and H. Ariyoshi, Electron. Lett. **14**, 593 (1978).
5. W. P. Maszara, G. Goetz, A. Caviglia, and J. B. McKitterick, J. Appl. Phys. **64**, 4943 (1988).
6. A. F. Evans, D. G. Hall, and W. P. Maszara, Appl. Phys. Lett. **59**, 1667 (1991).
7. J. Schmidtchen, A. Splett, B. Schuppert, and K. Petermann, Electron. Lett. **27**, 1486 (1991).
8. B. Schuppert, J. Schmidtchen, and K. Petermann, Electron. Lett. **25**, 1500 (1989).
9. R. A. Soref, F. Namavar, and J. P. Lorenzo, Opt. Lett. **15**, 270 (1990).
10. A. Splett, J. Schmidtchen, B. Schuppert, K. Petermann, E. Kaspar, and H. Kibbel, Electron. Lett., **26**, 1036 (1990).
11. J. D. Spear-Zino, R.R. Rice, J.K. Powers, D.A. Bryan, D.G. Hall, E.A. Dalke and W.R. Reed, Proc. SPIE **239**, 293 (1980).
12. A. C. Livanos, A. Katzir, A. Yariv, and C. S. Hong, Appl. Phys. Lett. **30**, 519 (1977).
13. C. H. Henry, R. F. Kazarinov, Y. Shani, R. C. Kistler, V. Pol, and K. Orlowsky, J. Lightwave Technology. **8**, 748 (1990).
14. H. Z. Chen, A. Ghaffari, H. Wang, H. Morkoc, and A. Yariv, Appl. Phys. Lett. **51**, 1320 (1987).
15. E. Yablinovitch, T. Gmitter, J. P. Harbison, and R. Bhat, Appl. Phys. Lett. **51**, 2222 (1987).
16. W. K. Chan, A. Yi-Yan, T. J. Gmitter, L. T. Florez, J. L. Jackel, D. M. Hwang, E. Yablinovitch, R. Bhat, and J. P. Harbison, IEEE Photonics Lett. **2**, 194 (1990).
17. A. Yi-Yan, W. K. Chan, C. K. Nguyen, T. J. Gmitter, R. Bhat, and J. L. Jackel, Electron. Lett. **27**, 89 (1991).
18. M. Kobayashi, M. Yamada, Y. Yamada, A. Himeno, and H. Terui, Electron. Lett. **23**, 143 (1987).
19. Y. Yamada, M. Yamada, H. Terui, and M. Kobayashi, Opt. Eng. **28**, 1281 (1989).
20. I. M. Grodnenskiy, I. N. Dyuzhikov, I. O. Ogloblin, Yu. F. Sokolov, and K. V. Starostin, Radio Eng. Electron. Phys. **26**, 147 (1991).
21. J. P. Lorenzo and R. A. Soref, Appl. Phys. Lett. **51**, 6 (1987).
22. B. Hemenway, H. K. Heinrich, J. H. Goll, Z. Xu, and D. M. Bloom, IEEE Electron. Device Lett. **EDL-8**, 344 (1987).
23. X. Xiao, J. C. Sturm, P. V. Schwartz, and K. K. Goell, SOS/SOI Technol. Conf., pp. 171-172 (1990).
24. G. V. Treyz, P. G. May, and J-M. Halbout, IEEE Electron. Device Lett. **12**, 276 (1991).
25. G. V. Treyz, P. G. May, and J-M. Halbout, Appl. Phys. Lett. **59**, 771 (1991)
26. G. V. Treyz, Electron. Lett. **27**, 118 (1991).
27. W. P. Dumke, Phys. Rev. **127**, 1559 (1962).
28. Private Communication, Dr. L. D. Hutcheson, Raynet Corp., Menlo Park, CA.
29. L. T. Canham, Appl. Phys. Lett. **57**, 1046 (1990).
30. U. Gnutzmann and K. Clausecker, Appl. Phys. **3**, 9 (1974).

31. S. Satpathy, R. M. Martin, and C. G. Van de Walle, Phys. Rev. B38, 13237 (1988).
32. H. Ennen, J. Schneider, G. Pomrenke, and A. Axmann, Appl. Phys. Lett. 43, 943 (1983).
33. D. G. Thomas, J. J. Hopfield, and and C. J. Frosch, Phys. Rev. Lett. 15, 857 (1965).
34. T. N. Morgan, B. Welber, and R. N. Bhargave, Phys. Rev. 166, 751 (1968).
35. C. H. Henry, P. J. Dean, and J. D. Cuthbert, Phys. Rev. 166, 754 (1968).
36. R. K. Crouch, J. B. Robertson, and T. E. Gilmer, Jr., Phys. Rev. B5, 3111 (1972).
37. T. G. Brown, P. L. Bradfield, D. G. Hall, and R. A. Soref, Opt. Lett. 12, 753 (1987).
38. R. A. Modavis, D. G. Hall, J. Bevk, B. S. Freer, L. C. Feldman, and B. E. Weir, Appl. Phys. Lett. 57, 954 (1990).
39. R. A. Modavis, D. G. Hall, J. Bevk, and B. S. Freer, Appl. Phys. Lett. 59, 1230 (1991).
40. T. G. Brown and D. G. Hall, Appl. Phys. Lett. 49, 245 (1986).
41. P. L. Bradfield, T. G. Brown, and D. G. Hall, Phys. Rev. B38, 3533 (1988).
42. P. L. Bradfield, T. G. Brown, and D. G. Hall, Appl. Phys. Lett. 55, 100 (1989).
43. H. Temkin, T. P. Pearsall, J. C. Bean, R. A. Logan, and S Luryi, Appl. Phys. Lett. 48, 963 (1986).
44. J. T. Boyd, R. W. Wu, D. E. Zelmon, A. Naumaan, H. A. Timlin, Opt. Eng. 24, 230 (1985).
45. D. G. Hall, IEEE Computer 20, 25-32 (1987).
46. R. A. Soref, IEEE J. Quantum Electron QE-22, 873 (1986).

OPTICAL STUDIES OF POROUS SILICON

T. LIN, M. E. SIXTA, J. N. COX, and M. E. DELANEY
Intel Corporation, Components Research, Santa Clara, CA

ABSTRACT

The optical properties of both electrochemically anodized and chemically stain-etched porous silicon are presented. Fourier transform infrared (FTIR) spectroscopy showed that absorbance in stain-etched samples was 3x and 1.7x greater than in anodized samples for the SiH/SiH_2 stretch and scissors-bending modes, respectively. Also, oxygen is detected in stain-etched samples immediately after formation, unlike anodized samples. Photoluminescence measurements showed different steady state characteristics. Electrochemical-etched silicon samples stored in air increased in photoluminescent intensity over time, unlike the stain-etched samples. A photoluminescent device made by anodization on epitaxial p-type material (0.4 Ω-cm) on n-type substrate (0.1 Ω-cm) did not exhibit electroluminescence.

INTRODUCTION

Porous silicon has recently been the subject of intensive study both theoretically, to understand the light emission mechanisms, and experimentally, to construct a light-emitting device. The phenomenon is important because it offers the possibility of silicon-based optoelectronics.

The first model to describe this photoluminescence attributed the light emission to quantum confinement [1]. In this model, anodization in hydrofluoric acid (HF) creates a network of H-passivated quantum-sized wires. These wires confine the carriers and the passivating layer prevents non-radiative surface recombination. Supposedly, additional post-anodization etching in hydrofluoric acid further decreases quantum-wire sizes resulting in increased radiative recombination efficiency and band gap widening. Microstructural investigations revealing large amorphous regions have led other researchers to suggest that hydrogenated amorphous silicon is responsible for the band gap widening and light emission [2-4]. Researchers have also suggested that the light emission is not a bulk effect at all, but rather is due to chemical species such as SiH_2 or siloxenes [5,6]. While a mechanism invoking surface chemical species is still a possibility, recent work has provided evidence supporting the quantum confinement mechanism. In particular high voltage sparks can sputter away silicon leaving quantum-sized silicon wires that emit light [7]. This photoluminescent material is free of hydride or siloxene molecules. Researchers at Philips have used lifetime measurements to strengthen the quantum confinement argument [8].

Stain-etching has been used primarily as a method of coloring defects on wafers. A stain-etch is a simpler way of forming photoluminescent porous silicon without an external current. Supposedly, the etching process occurs through the creation of localized anodes and cathodes on the surface of the wafer [5].

In this paper, the authors compare characteristics of both electrochemically and chemically (stain-etch) anodized samples using FTIR and steady state photoluminescent measurements. Additionally, initial experiments to measure electroluminescence were conducted on light emitting devices fabricated from epitaxial p-type material on n-type substrate using the electrochemical technique.

EXPERIMENTAL

Electrochemical Anodization

For this study, Czochralski grown boron-doped, p-type silicon (100) with a resistivity $\rho = 0.4$ - 0.6 Ω-cm and eptiaxial boron-doped, 2μm p-type silicon ($\rho = 0.30$ - 0.35 Ω-cm) grown on a phosphorus-doped, n-type substrate ($\rho = 0.01$ Ω-cm), were the starting materials. The 10 mm x 10 mm samples were cleaved from whole wafers. The samples were anodized locally by limiting the exposed area utilizing a 7 mm o-ring in the anodization cell. Anodization was conducted at current densities ranging from 10 - 200 mA/cm^2 for 15-200 seconds in an electrolyte mixture consisting of 1:1 [HF] (49%) : [ethanol] (99%). After anodization, the samples were removed and rinsed in DI H_2O to minimize open circuit etching, then dried with a N_2 gun. Unlike other studies, the samples were not soaked in concentrated acid following anodization [9].

Chemical Stain-etch

Bare wafers were dipped in a 1:5:10 mixture [HF]:[HNO$_3$]:[H$_2$O] for times of 3-15 minutes. To shorten the initiation time, a priming wafer was placed in the acid mixture for 15 seconds prior to adding water. The concentrated acid was then added to the water and the priming wafer was left in the dilute acid for another 30 seconds. Initially these films appeared uniform, however, gas evolution on the surface of the wafer caused non-uniform film formation. This resulted in a spotted appearance on the surface of the wafer. Stirring or using an ultrasonic oscillator did not prevent bubbles from remaining on the surface of the wafer.

Characterization

Transmission FTIR was performed on both anodized and stain-etched samples. Prominent absorbance peaks showed a significant amount of Si-H bonding for the etched samples. Etched samples showed prominent Si-H bonding absorbance peaks. Table I lists the relevant peak assignments [10-17].

Photoluminescence data was collected using a PTI LS1 photoluminescence spectrophotometer. A xenon lamp excited the sample with ultraviolet light in a narrow band-width centered at 350 nm; a PMT detector measured the intensity of the light emission from the sample in the visible spectrum (450-880 nm).

An SEM cross-section was used on cleaved wafers to determine the porous silicon film thickness. FTIR integrated-peak areas, expressed in absorbance units, were normalized for variations in the film thickness by invoking Beer's law. Relative changes in the various bond-contents were quantitatively compared. Beer's law states that $A = \varepsilon\, t\, c$. A is the integrated absorbance; ε, the absorptivity of the material; t, the thickness; and, c, the concentration of absorbing species.

Table I: FTIR features in porous silicon

MODE	POSITION
SiH$_x$	616 cm-1
SiH$_2$ bend-scissors	910 cm-1
Si-O-Si stretch	1107 cm-1
SiH/SiH$_2$ stretch	2104-2087 cm-1

RESULTS

FTIR

The infrared spectra of the anodized and stain-etched samples, collected immediately following etching and 5 days later, are shown in Fig.1. The normalized spectra for the anodized sample showed strong absorbance of the SiH/SiH_2 2104-2087 stretch mode, SiH_2 bend scissors 910 mode, and SiH_x 616 stretch mode. Compared to the bare wafer, no additional Si-O-Si bonding at 1107 cm^{-1} was observed from the anodization process, indicating a uniform passivating layer consisting of Si-H species. However, after 5 days, growth of an oxide was seen at 1107 cm^{-1}. Additionally, the SiH/SiH_2 and SiH_x absorbance decreased by 66% and 43% respectively. Fig. 2 indicated a linear growth for the oxide and linear decrease in the Si-H species, while the SiH_2 bend scissors mode remained constant with time.

The behavior of the stain-etched samples deviated from the anodized as illustrated in Fig. 2. Significant oxide was present immediately following the etch process, showing that the surface was not fully passivated by SiH_2. The normalized SiH/SiH_2 stretch mode at 2100 cm^{-1} was ~ 3 times stronger than the absorbance from the anodized sample, but the features were not as sharp, indicating greater chemical inhomogeneity of the stain-etched porous silicon film. The 910 cm^{-1} SiH_2 bend scissors mode was ~ 1.7 times stronger than that seen in the anodized samples. The IR spectra taken after 5 days showed virtually no absorbance in the SiH/SiH_2 stretch mode. At the same time, the SiH_x peak at 616 cm^{-1} remained constant with time. The SiH_2 bend scissors mode cannot be assessed after 5 days since the features were masked by the growth of a peak at ~ 864 cm^{-1} believed to be an Si-OH mode. Accompanying the loss of Si-H species was the appearance of peaks at ~2250 and 2200 cm^{-1}. It has been suggested that these peaks are a silicon hydride stretching mode where the silicon atom is bonded to at least one oxygen [18].

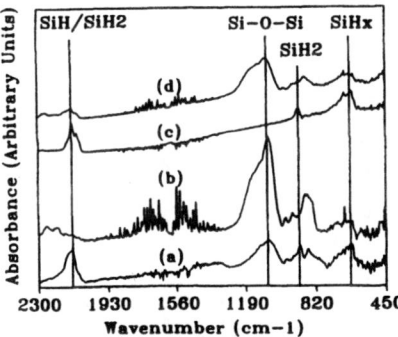

Fig. 1 IR spectra normalized by thickness and bare wafer subtracted. (a) stain-etch t=0 (b) stain-etch t=5 days (c) anodized t=0 (d) anodized t=5 days.

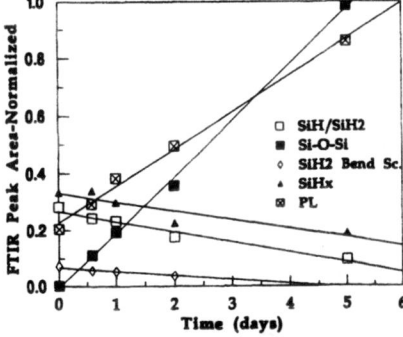

Fig. 2 Time dependence of FTIR (four chemical bonds) and steady state photoluminescence (PL) spectra of anodized samples over 5 days (stored in air).

As shown in Fig. 3, the time dependence of the FTIR spectra for the several chemical species vastly differs between the anodized and stain-etched samples. Whereas a linear relationship was evident in all of the species for the anodized samples, the stain-etched porous silicon demonstrated non-linear behavior.

Photoluminescence Measurements

The PL spectra produced by the anodized and stain-etched photoluminescence samples are compared in Fig. 4. Immediately following the etching process, the features of the anodized samples displayed a peak wavelength at ~775 nm and a full width half maximum (FWHM) of 185 nm. Stain-etched spectra showed a FWHM of 125 nm and a peak wavelength of 660 nm. The stain-etched porous silicon films generated PL spectra which were blue shifted relative to the anodized samples, and the peak was much sharper. Initially, the intensity of the stain-etched samples was approximately 3 times greater than the anodized sample.

Fig. 4 shows typical PL spectra immediately after preparation and how they have changed after five days. Figs. 2 and 3 show in more detail the time dependence of the integrated PL peak areas over 5 days (labelled "PL"). A linear growth rate over the 5 day period was observed for the anodized porous silicon. In contrast, the integrated PL peak areas of the stain-etched samples displayed a non-linear drop of 40-60% before leveling off.

Characteristics of Porous Si Formed on Epitaxial Wafers

Epitaxial silicon was investigated for use in electroluminescence experiments. The n-type substrate was planned to be a source of electrons. No illumination was applied during anodization other than normal fluorescent room light. CZ silicon anodization times had little effect on the photoluminescent behavior. Times up to 500 seconds were used to generate porous layers on CZ material. Anodization time appeared to have a large impact on the photoluminescent properties of porous silicon layers formed on epitaxial wafers. The PL intensity peaked at ~ 10 seconds and decreased as the anodization time approached 15 seconds. Initial attempts to etch down to the p/n junction were unsuccessful. Apparently, the formation characteristics for making porous silicon in comparable p-type layers depended strongly on whether the layer was part of an epi-stack or part of a bulk-CZ wafer.

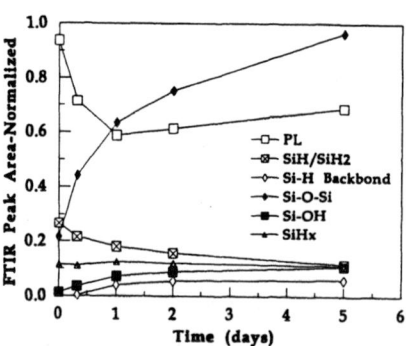

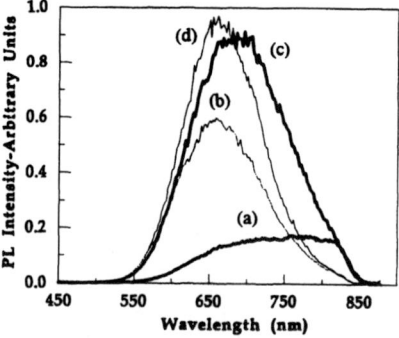

Fig. 3 Time dependence of FTIR (five chemical bonds) and steady state photoluminescence (PL) spectra of stain-etched samples over 5 days (stored in air).

Fig. 4 Measured room temperature PL spectra for (a) anodized t=0 (b) stain-etch t=5 days (c) anodized t=5 days (d) stain-etch t=0 days.

Electroluminescence

Preliminary efforts for creating current-induced light emission from porous silicon involved constructing an electroluminescent device using the CZ and epi-silicon [19-22]. In both cases, electrons were injected as minority carriers into the p-type porous region where recombination occurs and photons are emitted. The electron injection in the CZ porous silicon device occurred via the gold electrode. In the epitaxial porous silicon, electron injection occurred either from the n-substrate or from the metal electrode at the surface, depending on the bias. A schematic for both devices is shown in Fig. 5. Despite the strong photoluminescence behavior of the samples, we did not observe light emission by eye for positive or negative biases. A current flow was measured through the samples and is shown in Fig. 6, but any light emission was not intense enough to be seen with the naked eye.

DISCUSSION

Characterization of electrochemically anodized and chemical stain-etched porous silicon layers revealed significantly different properties. Features of the FTIR spectra suggested an Si-H passivated surface for anodized silicon, but an incomplete hydrogen passivated layer for stain-etched porous silicon. Additionally, the growth of Si-OH and hydrogen back bonding was present in the stain-etched samples after 5 days. Neither of these properties were present on anodized wafers.

Behavior of the PL intensity varied with time for both preparation techniques. Whereas the photoluminescent intensity increased linearly with time over the 5 day test period for anodized wafers, the intensity dropped on the average 50 % after 5 days for porous silicon layers formed by a stain-etch. Additionally, the stain-etched PL spectra were blue shifted and significantly sharper than anodized samples. In an attempt to restore PL intensity, stain-etched porous silicon was dipped in a 1:20 mixture of [HF]:[H$_2$O] for 10 seconds to remove the native oxide. However, post-dip FTIR spectra revealed that this dip had removed the existing porous layer, and the PL intensity dropped to virtually zero.

Over 5 days, the SiH/SiH$_2$ stretch mode dramatically decreased for both the anodized and stain-etched samples while the PL increased for the anodized and decreased for stain-etched porous silicon . Additionally, the characteristics of the FTIR and PL spectra were significantly different. The results suggest that there is a different mechanism responsible for photoluminesce.

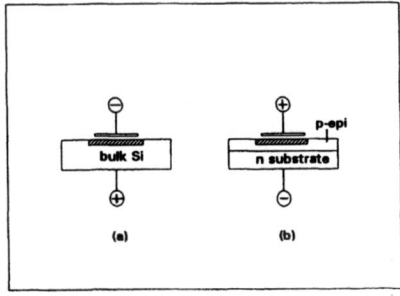

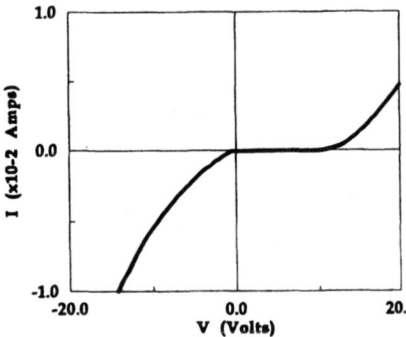

Fig. 5 Schematic structure of EL devices (a) CZ-silicon (b) epi-silicon.

Fig. 6 IV characteristics of CZ porous silicon device.

photoluminescence in each of the two cases. Possibly, chemical species are responsible for photoluminescence in the stain-etched porous silicon layers, while quantum confinement may be the PL mechanism in anodized samples.

CONCLUSION

Two methods for generating porous silicon layers are anodization in a HF:ethanol electrolyte at a current density of 10-200 mA/cm^2 and a chemical stain etch in a mixture of HF:HNO$_3$:H$_2$O. The characteristics and stability of samples produced by these alternative processes differed significantly. Anodized samples appeared to be stable, even increasing in PL intensity over time, and the surface appeared uniform. However, the PL dropped approximately 50% over 5 days in stain-etched porous silicon layers, while specular appearance of the surface caused by gas evolution during the etching process suggested a non-uniform porous layer. The varying characteristics of the PL and FTIR spectra for anodized and stain-etched porous silicon suggested that mechanisms responsible for photoluminescence were different for samples prepared by the two techniques. The anodization of epi-silicon did not parallel that of CZ silicon, and the photoluminescence was quenched for long anodization times. Devices to test for electroluminescence were constructed. Although photoluminescence was easily demonstrated from porous silicon, electroluminescence was not readily observed.

ACKNOWLEDGEMENTS

The authors are grateful to Brad Anders for assistance in EL experiments, Kim Gupta for useful discussions on anodization, Gabi Neubauer and R'Sue Caron for AFM and SEM work, respectively.

REFERENCES

1. L.T. Canham, Appl. Phys. Lett., 57, 1046 (1990).
2. T. George, M.S. Anderson, W. T. Pike, T. L. Lin, R. W. Fathauer, K.H. Jung, and D.L. Kwong, Appl. Phys. Lett. 60, 2359 (1992).
3. D.J. Wolford, B.A. Scott, J.A. Reimer, and J.A. Bradley, Physica B 117/118, 920 (1983).
4. R.A. Street in *Semiconductors and Semimetals*, edited by J.I. Pankove (Academic, Orlando, FL,1984), Vol. 21, Part B, P. 197.
5. C. Tsai, K.-H. Li, D.S. Kinosky, R.-Z. Qian, T.-C. Hsu, J.T. Irby, S.K. Banerjee, A.F.Tasch, Joe C. Campbell, B.K. Hance, and J.M. White, Appl. Phys. Lett. 60, 1700 (1992).
6. H.D. Fuchs, M.S. Brandt, M. Stutzmann, and J. Weber, Mat. Res Soc. Symp. Proc., Vol. 256, (1992).
7. R.E. Hummel and Sung-Sik Chang, Appl. Phys. Lett. 61, 1965 (1992).
8. G.W.'t Hooft, Y.A.R.R. Kessener, G.L.J.A. Rikken, and A.H.I.J. Venhuizen, Appl. Phys.Lett. 61, 2344 (1992).
9. M.B. Robinson, A.C. Dillon, D.R. Haynes, and S.M. George, Mat. Res Soc. Symp. Proc., Vol 256, (1992).
10. Y.J. Chabal, G.S. Higashi, K. Raghavachari, and V.A. Burrows, J. Vac. Sci. Technol. A, 7, 2104 (1989)
11. H. Wagner, R. Butz, U. Backes, and D. Bruchmann, Solid State Commun. 38, 1155 (1981).
12. F. Stucki, J. A. Schaefer, J. R. Anderson, G.J. Lapeyre, W. Gopel, Solid State Commun. 47, 795 (1983).
13. W. Kaiser, P.H. Keck, and C.F. Lange, Phys. Rev. 101, 1264 (1956).
14. H.J. Hrotowski and R.H. Kaiser, Phys. Rev. 107, 966 (1957).
15. P. Gupta, V.L. Colvin, and S.M. George, Phys. Rev. B. 37, 8234 (1988).
16. Y.J. Chabal, K. Raghavachari, Phys. Rev. Lett. 53, 282 (1984).
17. Y.J. Chabal, K. Raghavachari, Phys. Rev. Lett. 54, 1055 (1985).
18. C. Tsai, K.-H. LI, J.C. Campbell, B.K. Hance, and J.M. White, J. Electronic Mat., 21, No. 6, 589 (1992).
19. N.M. Kalkhoran, F. Namavar, and H.P. Maruska, Mat. Res Soc. Symp. Proc., Vol 256, (1992).
20. A. Richter, W. Lang, P. Steiner, F. Kozlowski, H. Sandmaier, Mat. Res Soc. Symp. Proc., Vol 256,(1992).
21. N. Koshida and H. Koyama, Mat. Res Soc. Symp. Proc., Vol. 256, (1992).
22. H.P. Maruska, F. Namavar, and N.M. Kalkhoran, Appl. Phys. Lett. 61, 1338 (1992)

ELECTROLUMINESCENCE FROM μc-SiC / POROUS Si PN JUNCTIONS

H.Mimura*, T.Futagi*, T.Matsumoto*, Y.Ohta*, K.Kitamura*, and Y. Kanemitsu**
*Electronics Research Laboratories, Nippon Steel Corporation, 1618 Ida, Nakahara-ku, Kawasaki 211, Japan
**Institute of Physics, University of Tsukuba, Tsukuba, Ibaraki 305, Japan

ABSTRACT

We have measured electrical properties and electroluminescence (EL) characteristics of the light-emitting diode (LED) based on a pn junction of n-type microcrystalline silicon carbide (μc-SiC) and porous silicon (PS). The μc-SiC/PS pn junctions showed rectification behavior, and a uniform red EL was observed at a forward voltage larger than 15V. From the relationship between the EL intensity and the forward current, the EL mechanism is interpreted as the recombination of electron-hole pairs doubly injected into the PS layer. No degradation was observed in the EL intensity during the measurements over 8 hours. These results means that μc-SiC serves well as a electron injector to the PS.

INTRODUCTION

Si has historically been the dominant material for the semiconductor industry, but light-emitting devices have not been achieved in Si because of the poor radiative efficiency of its indirect band structure. However, as the Si technology is well established and comparatively cheap, the idea of integrating optoelectronic and conventional electronic devices on Si chips (OEICs) has always been highly attractive due to the projected benefits in cost, reliability, and functionality. The recent emergence of visible light-emitting porous Si (PS) opens up prospects of such devices [1].

For the first step of the realization of Si-OEICs, it is necessary to fabricate the device in which minority carriers are effectively injected into the PS. One of such devices is a light-emitting diode (LED) using a pn junction. However, for the pn junction using the PS, there are important requirements for junction materials: A low temperature preparation process (below 400°C) is required because of the thermal quenching of the light-emission of the PS [2]. Since the band gap of the PS is estimated to be larger than about 2eV [3], junction materials with a band gap wider than 2 eV as well as a conductivity higher than about 10^{-3} S/cm are preferable for effective injection of minority carriers.

Therefore, as a junction material, we developed n-type microcrystalline silicon carbide (μc-SiC) films prepared by electron cyclotron resonance plasma chemical vapor deposition (ECR plasma CVD) [4]. The n-type μc-SiC has an optical band gap from 2.1 to 2.4 eV and a dark conductivity from 10^{-3} to 1 S/cm. Its deposition temperature is less than 300°C.

In this paper we will show the structure, electrical properties and electroluminescence (EL) characteristics of the LED based on a pn junction of the n-type μc-SiC and the PS.

μc-SiC PREPARED BY ECR PLASMA CVD

The ECR plasma CVD apparatus is a conventional one and its details are described elsewhere [5]. Fig.1 shows relationship between dark conductivities and optical band gaps of the n-type μc-SiC. For the comparison, that of n-type amorphous silicon carbide (a-SiC) prepared by conventional rf plasma CVD is also shown in Fig.1. The typical deposition conditions of the n-type μc-SiC were as follows; the microwave power was 300W, the deposition temperature was 300°C, the gas pressure was about 5mTorr, the gas ratio was SiH4:CH4:PH3:H2=1:2:0.01:190. As shown in Fig.1, dark conductivities of the μc-SiC films still remains higher than 10^{-3} S/cm even for the optical band gaps of about 2.4eV. Raman measurements reveal that the μc-SiC films have a structure consisting of Si microcrystallites embedded in a-SiC. From photoluminescence (PL) measurements, we confirmed that the μc-SiC films showed no PL at room temperature when it was excited with an Ar^+ laser (488nm) which had a larger energy than the optical band gap energy of the μc-SiC. This is because the hydrogen content of the μc-SiC films is less than several per cent.

STRUCTURE OF μc-SiC/PS PN JUNCTIONS

Figure 2 shows the schematic structure of μc-SiC/PS pn junctions. Al was used as an ohmic contact material. The fabrication procedure of the pn junctions was as follows: The substrates were (100) oriented, 0.2~0.4 Ωcm resistivity, p-type crystalline silicon (c-Si) wafers. The PS layers were formed for 3~5 minutes by electrochemical anodization in HF-C_2H_5OH solution (HF:H_2O:C_2H_5OH=1:1:2) using a constant anodic current of 20 mA/cm^2. Immediately after anodization, the samples were transferred into an ECR plasma CVD apparatus, and then n-type a-SiC (thickness, about 20Å) and n-type μc-SiC (thickness, about 500Å) were deposited onto the PS layers in sequence. a-SiC serves as a buffer layer for the hydrogen-plasma damage [6].

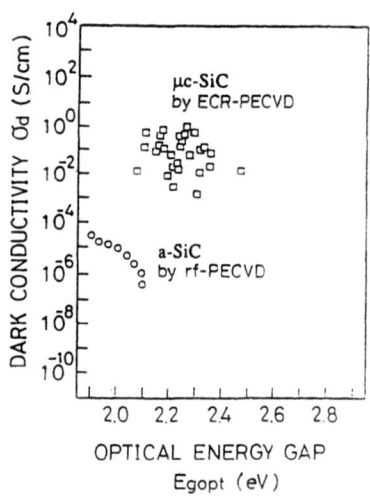

Fig.1 Relationship between the dark conductivity and the optical band gap of n-type μc-SiC. For the comparison, that of n-type a-SiC prepared by conventional rf plasma CVD is also shown in Fig.1.

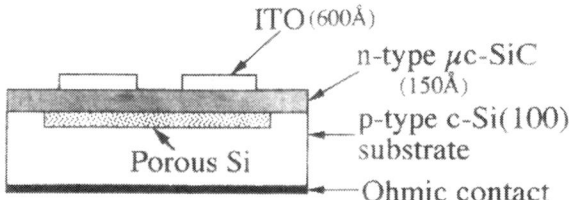

Fig.2 Schematic structure of μc-SiC/PS pn junctions.

After the deposition of the μc-SiC, ITO (thickness, 600Å) was evaporated as a transparent electrode. The thickness of the PS layer anodized for 5min was about 2μm.

Figure 3 shows the TEM micrograph of the interface between μc-SiC and PS for a sample without ITO layer. It is found that the surface of PS is uniformly covered with the μc-SiC films.

ELECTRICAL PROPERTIES OF μc-SiC/PS PN JUNCTIONS

Figure 4 shows a typical current-voltage (J-V) characteristic of the μc-SiC/PS pn junction. Open squares and solid squares refer to the positive bias applied to a p-type c-Si substrate (forward direction) and negative bias (reverse direction), respectively. The rectification behavior was clearly observed for all samples, and the rectification ratios were from 10 to 10^3 at ±5V. As for rectification ratios, larger ratios were obtained when we used c-Si substrates with higher resistivity [7].

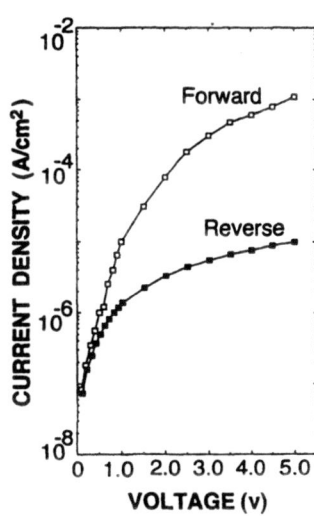

Fig.4 Typical J-V characteristic of a μc-SiC/PS pn junction.

——————— 100nm

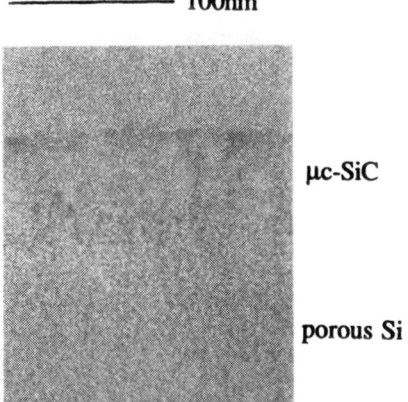

μc-SiC

porous Si

Fig.3 TEM micrograph of the interface between μc-SiC and PS for a sample without ITO layer.

Fig.5 Photograph of the typical uniform red light emission. The applied voltage and the current density were 29V and 20mA/cm^2, respectively. The light emitting area was 1cm^2.

EL CHARACTERISTICS OF µc-SiC/PS PN JUNCTIONS

A uniform red light emission from a whole electrode area of 1cm^2 was typically seen with the naked eyes at a forward voltage larger than 15V. However, a threshold voltage and current of the light emission showed a variety of values from 15 to 30V and from 5 to 50mA/cm^2, respectively even for the samples which had a similar fabrication condition. This is probably ascribed to the extent of natural oxidation of the PS.

Figure.5 shows a photograph of the typical uniform red light emission. The brightness was lower than 1Cd/m^2.

Figure 6 shows the spectrum of the uniform red EL. For the comparison, PL spectrum of the PS which had a similar c-Si substrate and anodization condition is also shown in Fig.6. The EL exhibited a very broad spectrum from 580 to 820 nm with a peak of 700 nm, which is roughly similar to that of the PL of the PS.

Figure 7 shows the relationship between the EL intensity and the forward current. As shown in Fig 7, the EL intensity increased as the square of the forward current. This results is similar to that obtained in ITO/PS junctions [7], suggesting that the EL is not dominated by the monomolecular recombination, but is dominated by the recombination of electron-hole pairs doubly injected into the PS layer . We also confirmed no light emission in the reverse direction even at the breakdown voltage of about 60V, which supports that the uniform red light emission is not due to an intrinsic type EL, but due to an injection type EL.

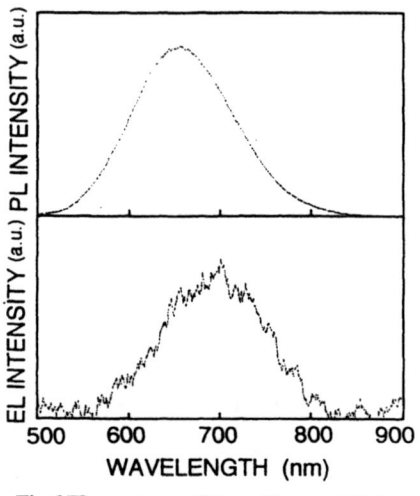

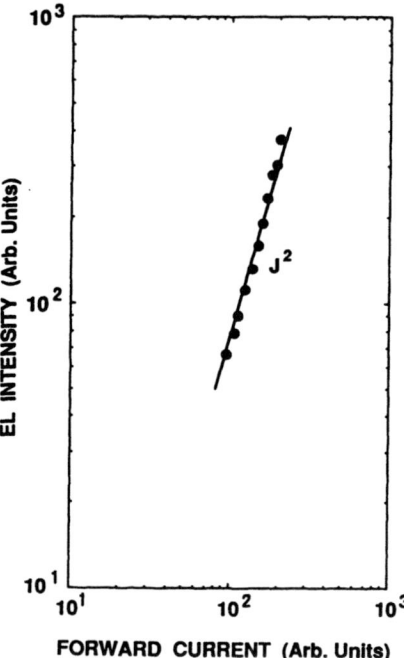

Fig.6 EL spectrum of the uniform red light emission from a µc-SiC/PS pn junction. The applied voltage and the current density were 29V and 20mA/cm^2, respectively. For the comparison, PL spectrum of the PS which had a similar c-Si substrate and anodization condition. The excitation light source was an Ar$^+$ laser (488nm).

Fig.7 Relationship between the EL intensity and the forward current.

Figure 8 shows the EL intensity as a function of time. No degradation was observed during the measurement over 8 hours. These results shown in Figs. 3-8 imply that the n-type µc-SiC serves well as an electron injector to the PS.

In our experiments, besides the uniform red EL described above, a weak white EL and a strong red EL were also sometimes observed for the samples which had a similar fabrication condition. Figure 9 shows the spectrum of the weak white EL. From comparing its spectrum with that of the uniform red EL, it is found that the peak wavelength of the EL spectrum is short and the width is broader. Fig 10 shows the the spectrum of the strong red EL. It is almost similar to that of the uniform red EL. This strong red EL was clearly noticed under a conventional fluorescent lamp. However, the light emitting area was restricted to a spot smaller than 1mm^2. For a weak EL and a strong red EL, we are now performing further experiments to understand the mechanism of these ELs.

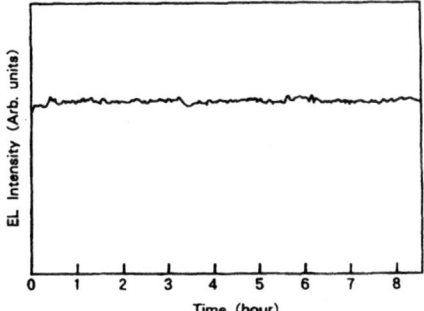

Fig.8 EL intensity as a function of time.

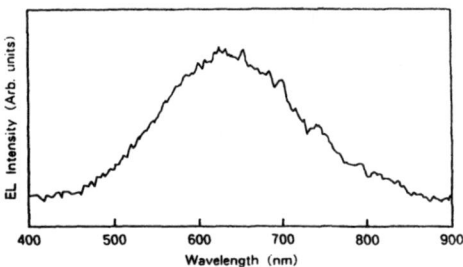

Fig.9 Spectrum of the weak white EL.

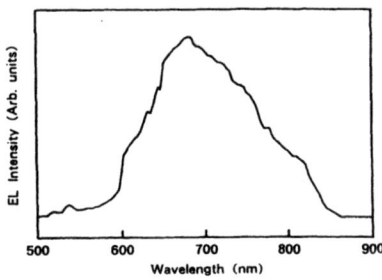

Fig.10 Spectrum of the strong red EL.

CONCLUSION

We described the fabrication procedure, structure, electrical properties and EL characteristics of the LED based on a pn junction of n-type μc-SiC and PS. From the TEM micrograph of the interface between μc-SiC and PS, it was found that the surface of PS was uniformly covered with the μc-SiC films. The μc-SiC/PS pn junctions showed rectification behaviors, and a uniform red EL was observed at a forward voltage larger than 15V. The EL intensity increased as the square of the forward current, which suggests that the EL is not due to the monomolecular recombination, but due to the free carrier bimolecular recombination. The EL was stable and no degradation was recognized in the EL intensity during the measurements over 8 hours. These results means that the μc-SiC is useful to the junction materials for the PS.

ACKNOWLEDGEMENTS

The authors would like to thank Professor N.Koshida of Tokyo University of Agriculture and Technology for his useful discussion.

References
[1] L.T.Canham, Appl. Phys. Lett. 57, 1046 (1990).
[2] C.Tsai, K-H.Li, J.Sarathy, S.Shih, J.C.Campbell, B.K.Hance, and J.M.White, Appl. Phys. Lett. 59, 2814 (1991).
[3] H.Koyama, M.Araki, Y.Yamamoto and N.Koshida, Jpn.J.Appl.Phys. 30, 3606 (1992).
[4] T.Futagi, M.Katsuno, N.Ohtani, Y.Ohta, H.Mimura, and K.Kawamura, Appl. Phys. Lett. 58, 2948 (1991).
[5] H.Mimura, M.Katsuno, T.Futagi, Y.Ohta, K.Kitamura, in 10th Symp. Photoelectronic Image Devices, London 1991 (IOP Pub., Bristol, 1992) p.371-376.
[6] T.Futagi, T.Matsumoto, M.Katsuno, Y.Ohta, H.Mimura and K.Kitamura, MRS Symp. on Microcrystalline Semiconductors: Materials Science and Devices, F21.4 Boston, Dec.4, 1992.
[7] H.P.Maruska, F.Namavar and N.M.Kalkhoran, Appl. Phys. Lett. 61, 1338 (1992).
[8] T.Futagi, T.Matsumoto, M.Katsuno, Y.Ohta, H.Mimura, and K.Kitamura, Jpn. J. Appl. Phys. 31, L616 (1992).

PROSPECTS FOR INFRARED ELECTROLUMINESCENCE FROM POROUS SILICON

F. Koch and A. Kux
Physik-Department E16, Tech. Univ. München, 8046 Garching, Germany

ABSTRACT

Efficient visible luminescence from porous Si requires the 3-dimensional confinement of charges in structures with typical ~3nm size. Such microporously etched Si acts as an intrinsic wide-gap material and is highly resistive. The material does not have the good transport properties consistent with an efficient electrical excitation. We instead suggest to employ mesoporously etched, p^+-type Si with its better conductivity in electroluminescence application. The material luminesces in two spectral bands centered about 0.8eV and 1.0eV in the infrared. Both emissions originate from surface-bound states. We report on the temperature dependence of luminescence, on transport and first attempts to generate infrared light by the injection of electrical current.

INTRODUCTION

The emission of visible light from porous Si when passing a current through the material is by now well established [1, 2, 3, 4]. Extensive work has been done on the electroluminescence phenomenon in a search for possible device applications. The energy conversion efficiency is poor. One reason is the high degree of quantum confinement in isolated ~3nm crystalline grains. Emission in the visible spectrum requires strong confinement, which is basically inconsistent with good transport. The resistance is highly nonlinear, with values of order 10^5-10^6 Ω-cm in electroluminescence application [5]. Confinement in cylindrical columns, such as the two seminal publications on porous Si [6, 7] postulated, has never been demonstrated. Moreover, carrier injection into the wide-gap porous Si is required for electroluminescence. For these reasons good transport for the efficient electroluminescence of porous Si in the visible will be difficult to achieve.

In this paper we intend to examine the prospects for infrared electroluminescence from the mesoporously-etched p^+-type (~0.005 Ω-cm) Si. Such material is presumably anisotropic with conduction in structures with linear dimensions of ~10nm. We estimate from the measurements that the resistivity of mesoporous Si is in the 10^4 Ω-cm range at room temperature. The material will support a higher current density than the visibly luminescing microporous Si.

Efficiency of electrical transport in mesoporous Si and the possibility of making good contacts in which minority carrier electrons can be injected into the mesoporous layer, are a distinct advantage for electroluminescence. The better transport properties, however, must be traded off against the less efficient infrared emission characteristics. In the next section we discuss the intensity, spectral distribution and temperature dependence of the infrared luminescence. This is followed by observations related to the conductivity. In the final section, we consider the electroluminescence.

INFRARED LUMINESCENCE

We have etched anodically a (100) Si wafer with resistivity 0.005 Ω-cm to a depth of ~1μm. Current density was in the 30mA/cm² range. The electrolyte was the conventional mixture of 50% HF and C_2H_5OH (1:1 by volume). The coarse grained texture of the mesoporous layer, with typical linear dimensions of 10nm or greater, leads one to expect that the emission characteristics resemble that of an extended Si-SiO$_2$ interface. The planar, passivated Si-SiO$_2$ heterostructure, as it exists in a MOSFET can be excited by current passing through the device [8]. A particularly distinct emission spectrum is observed in the gated-diode configuration under reverse bias condition [9]. For this geometry the gate-voltage is set to cause avalanche conditions with holes attracted to the Si-SiO$_2$ interface.

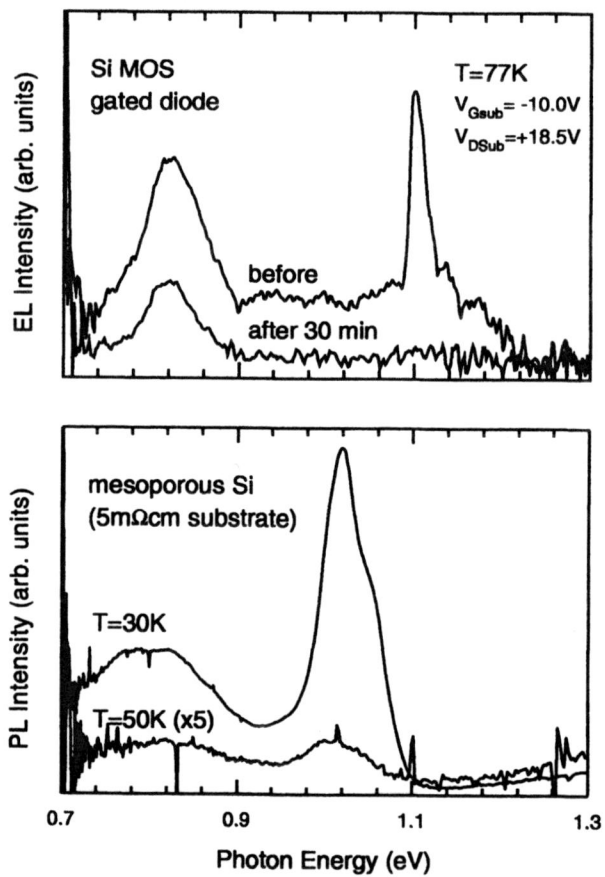

Figure 1: IR electroluminescence spectrum from a MOSFET operated in a gated diode configuration (top) and PL spectrum of mesoporous Si (bottom)

The 0.82eV peak in Fig.1 is related to the creation of dangling bond defects. There is a broad emission about the 1.1eV gap energy of bulk Si. The entire light, as it is recorded at 77K, comes from an area of order $1\mu m^2$ only. The electric fields and current density are quite high. The luminescence decreases with time in Fig.1 because the oxide accumulates positive charge reducing the avalanching field (before, after 30 min).

The photoexcited luminescence of porous Si in Fig.1 closely resembles the electroluminescence in the FET device. The illuminated area is 1mm² of mesoporous Si. The 0.8eV peak is a radiative emission process involving the dangling bond. This has been convincingly demonstrated in optically detected magnetic resonance (ODMR) experiments [10]. The emission in the 1.0eV range involves localization, probably via a surface process. It falls well below the band-gap energy of 1.12eV presumably because of the binding to surface potential fluctuations. A reduction of order 0.1eV is also observed in ref.[11]. It reflects a typical value of the binding energy. Both of the emissions are best observed at low temperatures. There is a significant decay of intensity above 70K. The emissions can hardly be observed at room temperature.

CURRENT-VOLTAGE CHARACTERISTICS

We have contacted the mesoporously etched layer with Al-contacts with area 2mm². The curves in Fig.2 are observed to be asymmetric with a forward direction for negative voltage at the metal. The temperature dependence indicates that there is a series resistance. This can only be the porous layer. The curves in Fig.2 differ in voltage from those reported in the microporously etched material by a large factor [5]. An exact comparison with that reference shows a factor of 30 lower voltage in the present case (300K). The characteristic of Fig.2 is that of a diode with a series resistance and forward direction for negative voltage. Such a behaviour is expected for a barrier of order 0.6eV between the depleted mesoporous layer and the p⁺-substrate. It is reasonable to argue that it is the effect of depletion which stops the continued chemical dissolution of the layer when a certain internal area has been generated. The remaining material is depleted by compensation of the acceptors with surface defects.

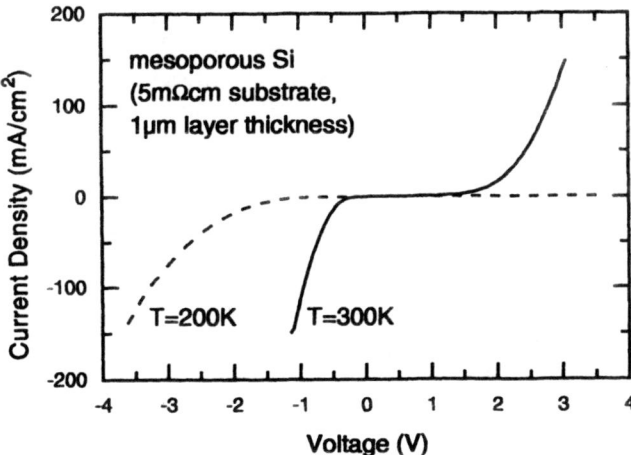

Figure 2: Current-voltage characteristics of mesoporous Si measured at different temperatures.

We sketch in Fig.3a the situation that we believe explains the I-V characteristics. The Al-contact lies close to the midgap position so that no strong band-bending occurs at the metallic contact side to the depleted material. Between the mesoporous layer and the p$^+$-substrate the bands must rise by ~0.6eV. This accounts for the observed diode characteristic.

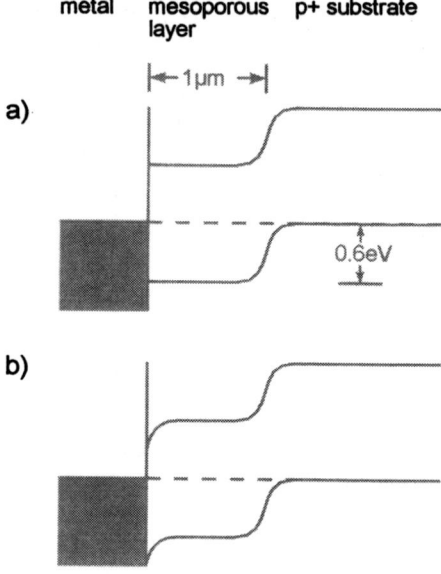

Figure 3:
a) Suggested band alignment diagram between a metal (Al, NiCr) whose Fermi energy falls at the midgap position, mesoporous Si and the p$^+$-substrate.
b) Band diagram if the metal acts to inject electrons and block holes.

ELECTROLUMINESCENCE ?

For mesoporous Si the transport of charges occurs along the columnar structures with their ~10nm diameters. When both the elementary excitations, electrons and holes together, are constrained to move along the narrow tubes one may expect that they will efficiently recombine. Processes involving the surface defects and surface localization will dominate. This is expected from the observations in Fig.1.

Working with Al and NiCr contacts, we have searched for an electroluminescence with the spectral shape to match the photoexcited case. Using a sensitive Ge-detector (North Coast EO817) we have worked diligently but without success so far. Only a broad and weak background signal was found at low temperatures.

With the knowledge provided by the transport properties, it now is obvious that using Al or NiCr contacts does not represent a favourable case for electroluminescence. The band scheme suggested in Fig.3a represents a majority carrier device. With forward bias holes stream from the substrate into the etched layer and are neutralized by electrons of the metal. The alignment of band edges in Fig.3a allows holes to freely enter the metallic contact.

What is needed for electroluminescence is to replace the metals used by an electron-injecting contact. We need a barrier to block the holes. Fig.3b suggests such a case. Taking a metal whose Fermi energy falls above midgap one increases exponentially the chances of an electron to enter the depleted layer. Simultaneously the holes would be blocked. By proper choice of this contact an equal number of electrons and holes could be made to transverse the device.

The microscopic process by which the electro-injected electrons and holes recombine, will involve the internal surface of the etched layer. For the dangling-bond defect emission band at $\sim$0.8eV this is obvious because such states exist only at the surface. Only a fraction of such recombination processes are likely to be radiative. Low temperatures may be essential for reasonably efficient luminescence from these centers.

There is reason to believe that near band-gap emission also involves an initial trapping and localization step. The large internal surface of the pores and expected potential fluctuations will provide the necessary localization. Random imperfections and the structural irregularities in the columns would help to localize the carriers along their paths through the mesoporous layer.

CONCLUSIONS

The infrared emissions discussed here are an alternative to the visible electroluminescence from porous Si devices. They are in a way related to the emissions from some of the Si-Ge devices also discussed in this volume. It is clear that more work needs to be done in order to realize an infrared electroluminescence diode and to explore its potential. The temperature dependent quenching of the luminescence appears to be a fundamental problem. It will be necessary to introduce just the right kind of traps along the conductivity paths to achieve room temperature operation. The device that we have discussed here is simply a Si diode with a huge amount of internal surface. The approximately $1\mu m^2$ surface in the MOS device (Fig.1) serves to give a measurable amount of infrared light. The $1\mu m$ thick mesoporous layer with $2mm^2$ cross sectional area has 10^8 times the amount of surface area. We expect to see the infrared light some day in the near future.

ACKNOWLEDGEMENTS

We have received support for this work by the Siemens AG and the Bundesministerium für Forschung und Technologie. Stimulating discussions with M. Ben Chorin are gratefully acknowledged.

REFERENCES

1. A. Richter, P. Steiner, F. Kozlowski, and W. Lang, IEEE Electron Device Letters 12, 691 (1991)

2. E. Bassous, M. Freeman, J.-M. Halbout, S.S. Iyer, V.P. Kesan, P. Munguia, S.F. Pesarcik, and B.L. Williams in Light Emission from Silicon, edited by S.S. Iyer, R.T. Collins, and L.T. Canham (Mater. Res. Soc. Proc. 256, Pittsburgh PA, 1992) pp. 23-26

3. N. Koshida and H. Koyama, Appl. Phys. Lett. 60, 347 (1992)

4. F. Namavar, H.P. Maruska, and N.M. Kalkhoran, Appl. Phys. Lett. 60, 2514 (1992)

5. P. Steiner, F. Kozlowski, H. Sandmaier and W. Lang in <u>Microcrystalline Semiconductors: Materials Science and Devices</u>, edited by P.M. Fauchet, C.C. Tsai, L.T. Canham, I. Shimizu, and Y. Aoyagi (Mater. Res. Soc. Proc. **283**, Pittsburgh PA, 1993) pp. 343-351

6. L.T. Canham, Appl. Phys. Lett. **57**, 1046 (1990)

7. V. Lehmann and U. Gösele, Appl. Phys. Lett. **58**, 856 (1991)

8. M. Herzog and F. Koch, Appl. Phys. Lett. **53**, 2620 (1988)

9. A. Kux, M. Schels, F. Koch and W. Weber in <u>Insulating Films on Semiconductors 1991</u>, edited by W. Eccleston and M. Uren (Adam Hilger, Bristol, 1991), pp. 255-258

10. B.K. Meyer, D.M. Hofmann, W. Stadler, V. Petrova-Koch, F. Koch, P. Omling, and P. Emanuelsson, Appl. Phys. Lett. (submitted)

11. C. Pickering, M.I.J. Beale, D.J. Robbins, P.J. Pearson, and R. Greef, J. Phys. C **17**, 6535 (1984)

A phenomenological investigation on

porous silicon electroluminescence

P.Steiner, W.Lang, F.Kozlowski, H.Sandmaier
Fraunhofer Institute for Solid State Technology
Munich, Germany

Abstract

The processing of light emitting diodes in porous silicon with green/blue electroluminescence spectrum is described. The spectral behaviour and the degradation are investigated. A phenomenological theory for the luminescence is given.

1. Introduction

Electroluminescence in porous silicon has been known since 1991. Meanwhile, a number of devices have been presented. Recently, the fabrication process for green and blue LEDs in porous silicon has been reported (1). We have made measurements on the luminescence behaviour of these structures. On the basis of these experiments, an experimental description for the green/blue electroluminescence is given.

2. Preparation of the samples

The technology and the performance of porous silicon electroluminescent (EL) structures has already been described in previous works (1,2,3). The details of the performance measurement are given by (4). The material used is n-doped silicon (P, 1 - 2 Ωcm). To get an ohmic contact on the back of the wafer, ion implantation is used (P, 100 keV, 5 * 10^{15} /cm^2). Etching is done in the double cell described in (1,2), 25% HF is used. A current of 10 mA/cm^2 is applied for 10 - 60 minutes. During etching the wafer is illuminated from the front side. A nanoporous layer with a thickness of about 1 μm is formed. For contact, pads of 17 nm gold are deposited by evaporation.

It is known that the illumination during the etching process is very important for the luminescent properties. The task of this investigation is to look at the influence of different illumination techniques. Two sets of wafers are processed. The wafers of the first set are illuminated with a halogen lamp which emits mainly visible light. This type of samples is called VIS-type porous silicon. To understand the influence of the wavelength of the light applied during the etching process, another set of wafers is etched which is illuminated with ultraviolet light (UV-type porous silicon). A Hg glow discharge lamp was used for this purpose.

3. Measurements

3.1. Electroluminescence behaviour and spectra

When a current is applied, we find a rectifying behaviour of the current/voltage characteristics, showing that the device is a light emitting diode. The minimum voltage for measureable light emission of the UV-type samples is 1,35 V. The quantum efficiency of these devices is about 10^{-5}.

Spectroscopy shows that the VIS-type and the UV-type samples show different colours of EL. Fig. 1 shows three measured EL spectra. The samples of VIS-type show the normal red/orange emission which has been reported many times for porous silicon. The UV-type spectra show a shorter wavelength in luminescence. The EL light ranges from the green to the blue. The two spectra given are from two different wafers processed with the UV-process. The spectra in fig. 1 clearly indicate a strong dependence of the EL on the light applied during anodization.

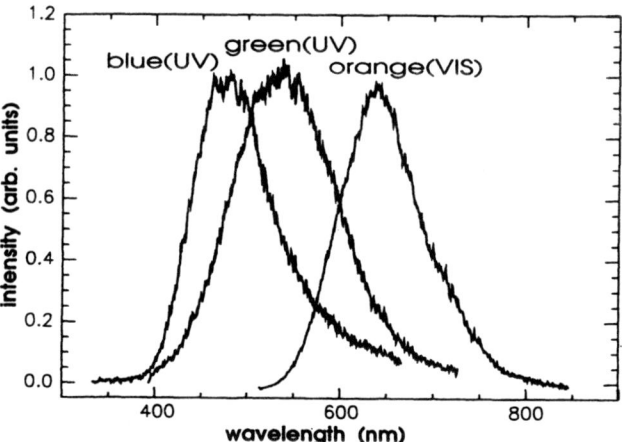

Figure 1: EL spectra with different spectral distributions. Orange: VIS-type.   Green/blue: UV-type.

3.2. Photoluminescence

The photoluminescence (PL) spectra of the VIS-type is well known. It shows red/orange light. The PL of the UV-type samples is given in Fig. 2, upper curve. The light is in the red/orange region, too. In this case, the PL and the EL spectra do not coincide. Another important observation is made, when the PL is measured after an electric current flow. A pad is contacted and a current is applied, EL light emission occurs. After 15 minutes the current is cut and this pad is used for a PL measurement. The spectrum measured is given in the lower curve of fig. 2. The PL is almost quenched. The little remaining light is still red/orange. For the UV-type samples, the PL is quenched by the electric current, while the EL is not.

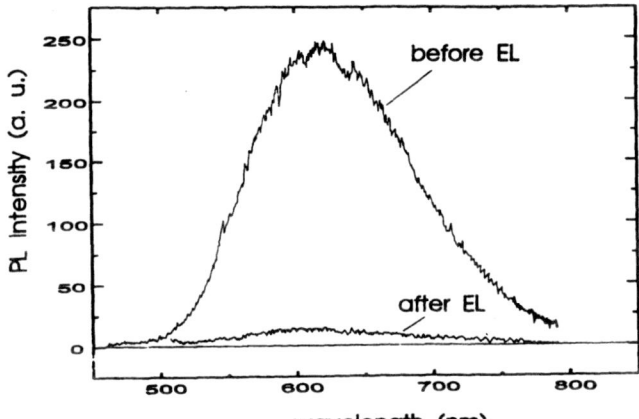

Figure 2: PL of a UV-type sample as prepared (upper curve) and after current flow (lower curve).

3.3. Current induced change in electroluminescence

We have observed that the UV-type spectra show a development of EL with time. A freshly contacted pad will first show red light for a moment, then the blue light will come up and stay. To measure that, time resolved spectroscopy is used. Fig. 3 shows the development of the spectrum with time. Time resolution is 1 second. At first, red light is emitted. This is seen at around 700 nm for about 10 seconds. Then the red emission is reduced

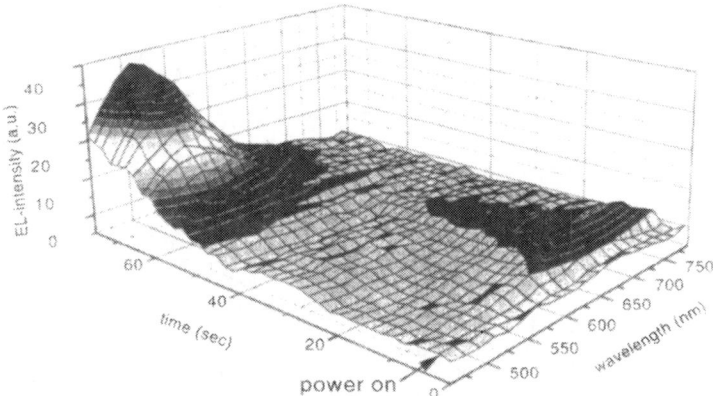

Figure 3: Time development of electroluminescence in a UV-type sample. Spectral intensity versus wavelength and time.

and the green/blue emission gets strong and forms the large peak at 520 nm that comes up after 40 seconds. Within the first seconds, the red light decreases and a strong green/blue luminescence rises. The spectra show that there is not a shift from red to green, but a decrease of one mode and an increase of another. When the pad is contacted for a second time, the green/blue emission is seen from the beginning.

While the green line gets stronger, the current through the film increases as well (when a constant voltage is applied). This increase of the conductivity is similar to the breakdown of an oxide. At an oxide between two nanoparticles the local field strength is very high. There, damage of the oxide occurs and a current path is established. Like the "waer out" of oxides the effect increases with the electric charge applied.

4. Interpretation

The observations allow us a number of conclusions on the film formation and the electroluminescence mechanism.

4.1. Morphology of the UV-type layers

The model described by (5) explains the formation of porous silicon by a self-adjusting electrochemical etching process for p-type silicon. To dissolve silicon in HF, a hole current has to flow through the sample. If the silicon structures are small enough, however, quantum size effects cause an increase of the band gap in the small silicon particles. Thus a barrier for the carriers is build up and the etching process stops locally.

In order to produce nanoporous n-silicon, we have to create the holes necessary for the etching process by illuminating the sample during the preparation. Short wavelength light creates electron/hole pairs even in smaller silicon structures. The use of UV-light during the anodization process leads to smaller structure sizes than using visible light. Smaller structures cause a shift to blue of the luminescence. We have not yet been able to perform an independent measurement of the particle size, but all the same the blueshift of the EL is an experimental evidence in favour of the quantum confinement model of the etching process given by Lehmann and Gösele (5).

4.2. Electroluminescence mechanism

The experiments do not show a continuous shift, but two different modes of luminescence. The determination of the excited mode is influenced by the etching process and by electric degradation, too. When we look at the observations described above, we may set up an phenomenological model for the EL mechanism. Models for the crystallites, the internal surfaces and the influence of hydrogen are given in (6,7,8,9). A general model taking into account the internal surface as well as the core of the crystals is given by F.Koch (10).

Fig. 4 shows a simplified band structure of four different kinds of silicon:

a) A crystal surface of bulk silicon. The band gap (thick line) is at 1.1 eV in the infrared, luminescence is very small due to the indirect gap. The surface is covered with oxygen and hydrogen. Since the surface is small and the surface states are above the band gap, there is almost no luminescence from them.

b) A nanocrystal in porous silicon, which is either n-silicon (VIS-type) or from p-silicon. The crystals are in the 3 to 5 nm region, the internal surface is covered with oxygen and hydrogen. There are hydrogen passivated surface states (dashed line at 1.8 eV). The band gap is increased due to quantum confinement. The surface states and the band gap are at roughly the same energy. A luminescent transition is possible and has an energy of 1.8 eV, which is measured by PL and EL experiments.

c) A crystallite of UV-type porous silicon, as prepared. The size is smaller than in b), the internal surface is covered by hydrogen and oxygen as before. The band gap is further increased; now the band gap energy with about 2.5 eV is considerably higher than the hydrogen surface states. The band gap would cause a green/blue light, but the states which would do that are empty. The excited carriers are attached to the hydrogen states of the surface, which show a lower energy. Therefore, red light is still seen in PL as well as in EL.

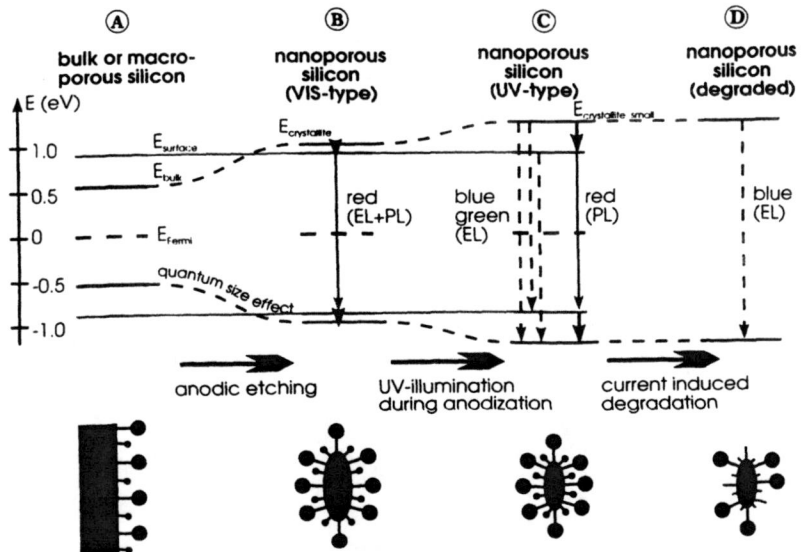

Figure 4: Simplified band structure of four kinds of silicon:
A) Bulk silicon crystal surface.
B) Nanoporous silicon VIS-type.
C) Nanoporous silicon UV-type.
D) Nanoporous silicon UV-type after electric degradation.

d) A crystallite of UV-type after electrical degradation. Obviously the surface is changed by the current. A voltage of some V is enough to break up the hydrogen bonds to the surface. The hydrogen leaves the crystallite, a dangling bond remains. This has an impact on luminescence: The red luminescence is gone. The remaining luminescence is green/blue due to the high band gap.

Another possible mechanism can be proposed: The growing of the blue light can be due to localised centers, which might be in the oxide. Due to the change of the surface and the wear out of the oxide those states would become dominant for the electroluminescence. The basic argument of a current induced change of the internal surface would hold then, too.

Why do we find a green/blue EL, but no green/blue PL? This can be understood looking at the mechanism of excitation, which is different for PL and EL. In the film shown in fig. 4d, the absorption of the light used for PL experiments is small. First, the structures are small and, therefore, the geometrical interaction is decreased. Then, the absorption edge is also shifted and the absorption of small structures is reduced, they become more and more transparent (5). EL excitation finds another situation. Let us imagine the path of an electron through a nanoporous film. The film is imagined as a system of crystallites with undulations and oxide barriers (11). On its way the electron passes crystallites, constrictions between them and oxide barriers from one crystallite to another. Even when the band gap is increased in a constriction or in an oxide barrier, the electron is forced to cross it by the external field. This way, unfavorable places of the current path are crossed by the electron, too, but they are not excited by the photons of the PL measurement, which just cross them without interaction. Therefore the PL is quenched by the removal of hydrogen, while the EL is only quenched in the red component, but not in the blue one.

Acknowledgement
We thank Prof. I. Ruge for his support and K. Marusczyk for technological assistance. Financial support has been given by the BMFT.

Literature

1. P.Steiner, F.Kozlowski, H.Sandmaier, W.Lang: MRS Meeting, Boston, Dec 1992
2. A.Richter, P.Steiner, F.Kozlowski und W.Lang: IEEE Electron Dev. Lett., 12, No.12, pp. 691, (1991).
3. A.Richter, W.Lang, P.Steiner, F.Kozlowski, H.Sandmaier: Proc. MRS Vol 256, 1991
4. F.Kozlowski, M.Sauter, P.Steiner, A.Richter, H.Sandmaier, W.Lang: EMRS Strasbourg 1992, Thin solid films, 222 (1992)
5. V. Lehmann und U. Gösele: Appl. Phys. Lett., 58, (1991), pp. 856.
6. D.J.DiMaria, J.R.Kirtley, E.J.Pakulis, D.W.Dong, T.S.Kuan, F.L.Pesavento, T.N.Theis and J.A.Cutro: J.Appl.Phys. 56(2), 15. July (1984)
7. C.Tsai, K.-H.Li, D.S.Kinoski, R.-Z.Qian, T.-C.Hsu, J.T.Irby, S.K.Banerjee, B.K.Hance, J.M.White: Appl. Phys. Lett. 60 (14), 6. April 1992
8. Y.Takeda, S.Hyodo, N.Suzuki, T.Motohiro, T.Hioki, S.Noda: J.Appl.Phys. 73 (4) 15. Feb. 1993
9. M.S.Brandt, H.D.Fuchs, M.Stutzmann, J.Weber, M.Cardona: Solid State Commun. 81, 307, (1992)
10. F.Koch, V.Petrova-Koch, T.Muschik, A.Nikolov, V.Gavrilenko: MRS Meeting, Boston, Dec 1992
11. F.Kozlowski, W.Lang: J.Applied Phys., 72 (11), 1. Dec 1992

MONOCRYSTAL DISLOCATIONLESS Si:Ge,
GROWN FROM THE MELT WITH Gd IMPURITY.

Borschensky V.V., Brinkevich D.I., Petrov V.V., Prosolovch V.S.
Byelorussian State University, Department of Physic, Scorina
Avenue, 4, 220080 Minsk, Republic Belarus.

ABSTRACT

The properties of Si grown from the melt having impurities
of germanium and gadolinium have been studied by IR-absorption
and Hall effect methods.It was stated that Ge and Gd are effective getters for technological impurities of oxygen and
carbon in silicon melt. It has been shown that the combined
doping by rare earth and isovalent impurities allows to increase the thermostability of dislocationless monocrystals of
silicon.

1.INTRODUCTION

It is known that doping of silicon by rare earth elements (REE) increases its thermal and radiation stability [1].
It has been stated earlier that isovalent impurity (IVI) Ge in
silicon suppresses generation of thermodonors (TD) introduced
at 450 ^{o}C [2], but practically does not effect the formation of
high temperature thermal donors (HTTD) [3]. Thus the study of
the properties of silicon doped simultaneously by the REE and
IVI impurities is of great interest.

2.EXPERIMENTAL ASPECTS

In this paper dislocationless n-Si monocrystals grown
from the melt involving Ge and Gd impurities by Czochralsky method have been studied. Their resistivity was ~ 10 Ohm.cm. Concentrations of IVI (N_{Ge}) and REE (N_{Gd}) were measured by the method of neutron-activated analyses.Note that N_{Gd} in all samples
was ≤ 10^{13} cm^{-3},i.e. beyond the limits of detection. Concentration of the interstitial oxygen and carbon in the state of re-

placement was found by the IR bands of absorption at 1106 and 607 cm^{-1} using empirical coefficient of $3,3.10^{17}$ and $2,2.10^{17}$ cm^{-2} correspondingly. Measurements of Hall effect and conductivity were made in the temperature range of 77-300 K. Thermal treatment (TT) have been made in air at 450 and 650 ^{o}C for period to 100 hours.

3. RESULTS AND DISCUSSIONS.
A.Gettering

The results of measurements (Table 1) show that Ge and Gd are getters for oxygen and carbon in the melt and this effect is the most prominent for the latter technological impurity. The noticeable decreases of oxygen concentration in the doped REE and IVI samples were observed only for the lower parts of the ingots. It should be noted that the simultaneous introducing of Ge and Gd into the melt efficiently increased the rate of purifying of monocrystals from C. Gettering effect

Table 1.

Original parameters of silicon ingots.

Ingots	Number of the wafer	$N_{Ge} \cdot 10^{20}$, cm^{-3}	$N_0 \cdot 10^{17}$, cm^{-3}	$N_c \cdot 10^{16}$, cm^{-3}	$N_{TD}^{G} \cdot 10^{14}$, cm^{-3}
Test material	1U	–	8,6	6,7	1,53
	1P	–	6,5	16,0	–
	1L	–	8,3	42,2	–
Si:Ge	2U	1,06	8,7	6,0	0,6
	2P	1,14	6,3	13,5	–
	2L	1,27	7,8	35,3	–
Si:Gd	3U	–	8,6	5,3	0,2
	3P	–	6,4	7,4	–
	3L	–	7,6	13,2	–
Si:Ge,Gd	4U	0,80	8,5	3,3	0,2
	4P	0,87	6,5	5,0	–
	4L	0,94	7,1	7,5	–

(* U,P,L– upper part, middle part, lower part of the ingot, N_{TD}^{G}– concentration of "grown" thermodonors.)

at addition of REE into the melt occurred for background technological impurities of Au,Cu,Na,W [1] as well.

Measurements of the profile of oxygen distribution showed that in all crystals N_0 decreased when approaching the edge of the wafer due to evaporation.Introducing of Ge into the melt did not change the radial homogeneity of oxygen distribution: $\Delta N_0/N_0 \approx 0,06$ into Si:Ge and test material. Addition of Gd caused its decrease in Si:Gd ($\Delta N_0/N_0 \approx 0,12$) and Si:(Ge,Gd) ($\Delta N_0/N_0 = 0,30$).Radial distribution of carbon impurity was highly homogeneous ($\Delta N_0/N_0 \approx 0,02$).

The effect of IVI and REE described above may be caused by some reasons:

1.Change of coefficients of segregation of oxygen and carbon.

2.Binding of the given impurities in the melt of germanium or REE followed by the formation of the complexes of $Z_x O_x$ and $Z_n O_m$ (Z =Ge, Gd) types, expelled into the melt and precipitated at the bottom of the crucible as a slug.

3.Additional solution of silicon crucible resulted in the increase of the silicon melt by oxygen.

According to [1, 4] and the data given above, mechanisms of gettering in Si:Ge and Si:Gd should differ.For Si:Ge the decrease of the magnitude of segregation of C and O is dominant because germanium did nit produce noticeable effect on the solution of the crucible;the formation and precipitation of slugs also did not occur.Introducing of REE into the melt enables the increase of N_0, increasing the solubility of silicon crucible [4]. Lanthanoids, binding oxygen and carbon in the melt prevent their introduction into the growing crystal that enables the gettering effect in Si:Gd.Its increase and decrease of the concentration of germanium in Si:(Ge,Gd), shows the interaction of Ge and Gd impurities in the melt.

B. Thermal Donors

From the analyses of the experimental data (Table 2) one can conclude that Si:Ge, Gd hasthe increased thermostability as compared to the test material.This effect reveals itself

in all tupes of TT in the samples cut from the lower parts of
ingots as well as that from the upper parts.Kinetical curves of
generation of TD and HTTD in this paper are not presented as
they are analysed in details in Ref [5,6]. Duration of the used
TT was taken sufficient for the formation of TD in maximum concentration bet us note the fact stated before [1] that impurity
of Gd at the concentrations used in the present paper does not
show the noticeable effect on the duration of TD at $450^{\circ}C$ and
$650^{\circ}C$.

One can explain the specific characteristics of TD accumulation at $450\ ^{\circ}C$ by the fact that the impurities of carbon
at $N_0 \geq 5.10^{16}\ cm^{-3}$ [7] and germanium at $N_{Ge} \geq 10^{19}\ cm^{-3}$ [2] essentially suppress the introducing of these defects.Taking all
the facts into account one can understand why the concentration
of TD was lower in the lower part of the ingot than in the upper part, and the initial rate of introducing (IRI) and the maximal achieved concentration (MAC) of thermodonors had minimal
values in Si:Ge, and the maximal - in Si:Gd (Table 2).

Table 2

Initial rates of introducing and maximal achieved concentrations of thermal donors

Num-ber of pea-te	TD				HTTD		
	−		SA		−	HT $450^{\circ}C$ 85h	SA+HT $450^{\circ}C$ 85h
	IRI $\times 10^{-13}$, cm^{-3}/h	MAC $\times 10^{-14}$, cm^{-3}/h	IRI $\times 10^{-13}$, cm^{-3}/h	MAC $\times 10^{-14}$, cm^{-3}/h	MAC$\times 10^{-14}$, cm^{-3}		
1U	11,0	16,57	0,32	6,34	1,62	4,14	5,58
1L	~0,1	2,6	−	~1,0	6,2	8,08	10,80
2U	0,6	2,73	−	<1,0	2,15	4,51	5,06
2L	−	<0,8	−	<1,0	5,1	6,96	7,6
3U	12,0	12,33	0,65	6,24	0,54	1,61	2,34
3L	~0,1	2,4	−	~1,0	2,8	6,04	8,62
4U	3,8	6,28	0,28	3,57	0,56	1,54	1,92
4L	−	<0,8	−	<1,0	0,3	1,10	1,40

The impurities studied at generation of HTTD have another behavior. It is known that carbon stimulates it [6] and Ge even at $N_{Ge} \sim 10^{20}$ om^{-3} does not show noticeable effect [3]. Thus, gettering effect of REE in the melt (Table 1) causes the lower of accumulation of HTTD in Si:Ge,Gd and Si:Gd.

Stabilizing annealing (SA) at 650 ^{O}C during 30 min made after the growth of an ingot, suppressed the further generation of TD introduced at 450°C. Its strongest effect revealed itself in the test material (Table 2). Taking into account that the given effect was caused by the annealing of the centers of generation of TD (8),one can come to the conclusion that in Si:Ge their concentration is lower. On the other hand SA caused the increase of the concentration of HTTD at the two-stage TT (450..650°C). The given effect revealed itself most of all in Si:Gd (the increase of MAC was approximately 45%) and in the test material (approximately 35%). SA did not practically show any effect on the samples of Si:Ge and Si:Ge, Gd (Table 2). It is most probable that stabilizing annealing enables the formation of the centers of HTTD generation, and Ge suppresses their generation.These centers are not low temperature TD,because SA, as it has mentioned before, decreases its concentration.

4.CONCLUSION

On the bases of the experimental data presented in this paper one can come to the following conclusions:

-impurities of Ge and Gd are effective getters for technological impurities of oxygen and carbon in the melt of silicon;

-combined doping of REE and IVI allows to increase the thermal stability of dislocationless silicon monocrystals, suitable for semiconductor microelectronics.

R E F E R E N C E S

1.A.G.Dutov, Yu.A.Karpov, V.A.Komar, V.V.Petrov,V.S.Prosolovich, B.M.Turovsky, S.A.Chesnokov,Yu.N.Yankovsky Izv. AN SSSR. Neorgan.mat.25,1589,(1989).

408

2.Yu.M.Babitsky, N.I.Gorbacheva, P.M.Grinshtein, M.A.Ilyin, M.G.Milvidsky, B.M.Turovsky. Fiz. Tekhn. Poluprovodn..18,1309, (1984).

3.D.I.Brinkevich, N.I.Gorbacheva, V.V.Petrov, V.S. Prosolo-vich,B.M.Turovsky Isv.AN SSSR.Neorgan.mat. 25, 1376,(1989).

4.E.P.Bochkarev,V.P.Grishin,Yu.A.Karpov ,M.I.Marunina- Pro-perties of doped semiconductors.(Nauka,Moskow,1977)p.88.

5.W.Kaiser, H.L.Frisch,H.Reiss.Phys.Rev.112,1546, 1958.

6.A.Kanamori,M.Kanamori.J.Appl.Phys.50,8095,(1979).

7.A.R.Bean,R.C.Newman.J.Phys.Chem.Sol.33,255,(1972).

8.V.P.Markevich,L.I.Murin.Phys.Stat.Sol.(a),111,K149,(1989).

EPITAXIAL LAYERS Si:(Sn,Yb) PRODUCED BY THE CRYSTALLIZATION FROM THE MELT-SOLUTION ON THE BASIS OF Sn.

D.I.Brinkevich,N.M.Kazuchits,V.V.Petrov.
Department of Physics.Belorussian State University. Minsk, Republic Belarus.

ABSTRACT

Epitaxial layers (EL) Si:Sn doped with Yb in the process of liquid phase epitaxy were studied by optical microscopy and photoluminescence (PL) methods.At low concentration of lanthanoid ($0,01 < N_{Yb} < 0,1$ weight %) the good planarity of the interface and high quality of the surface are detected. At $N_{Yb} > 0,1$ weight % microirregularities are presented.

In EL Si:(Sn, Yb) irradiated by 4,5 MeV-electrons the suppression of the generation of radiaton defects,responsible for G- and C-lines of PL, has been found. This effect has been explainedwithin the score of the model takeng into the consideration gettering propering ofYb in reference to the impurities of O and C as deformation fields,attributed to the presence of Sn atoms.

INTRODUCTION

Doping by rare-earth elements (REE) can be used for purposeful control of the properties of semiconductor materials, susceptible to the presence of structural defects [1]. In particular, introducing of REE into monocrystal silicon allows to increase the stability of its main electrophysical parameters to radiation effect [2].

EXPERIMENTAL ASPECTS

The purpose of the given work is to study the processes of defect formation in epitaxial layers (EL) of silicon of n- and p-type conductivity,doped by Yb in the process of crystallisation from the melt-solution (M-S) on the Sn basis. EL were grown in the graphite cassets with the help of shift technology in the temperature range of 900-1150 $^{\circ}$C. Forced cooling of M-S was made with the rate of 0,1..1,0 K .$\min^{-1}$. Concentration of tin (N_{Sn}), according to the neutron-activation analyses made up (3-6).10^{19} cm^{-3}. The content of ytterbium (N_{Yb}) in M-S varied from 0 to 6.0 weight %. Resistivity of EL,found by four probes method,changed from 0,04 to 10 Ohm.cm. The thickness of EL varied in the range from 2 to 20 μ. PL spectra were measured at temperature of 4,2-77 K optical excitation was made by the arc xenon lamp.The impurity content of EL was studied with the help of the local X-ray probe microanalyses with 70 $\mathring{A}$ resolution. Irradiation by electrons with the energy of 4,2 MeV by fluencies (F_e) 2.10^{15}-3.10^{17} cm^2 was made at 300 K.

RESULTS AND DISCUSSIONS

Introducing of REE into the melt-solution allows to increase the wetting of the substrate and consequently, to decrease the temperature of the film growth. At low concentrations of Yb in M-S (0,01-0,1%) the grown layers had good planarity of the interface and high quality surface morphology. At the increase of $N_{Yb} > 0,1\%$ the microroughness was observed on the surface of EL, that is connected with the nonuniformity of the distribution of lanthanoid and with the increased solubility of the substrate on those areas where the content of Yb was high.

The analyses of the data, obtained by the method of the local probe microanalyses, allows to come to the following conclusions:

-Sn in EL was distributed nonuniformly. The local regions ($\sim$ 50 μ) where N_{Sn} reached the values (8-10).10^{19} cm^{-3} occurred:

-concentration of technological impurities of oxygen N_0 and carbon N_c in EL is 3 times as low than in the substrate. The above impurities were distributed non-uniformly;

-addition of the impurity Yb of the concentration to 0,1 weight% into M-S increases the homogeneity of the distribution of Sn and decreases N_c and N_0 in EL;

-at $N_{Yb} \geq 0.2..0.4$ weight % the mentioned impurity forms in EL the inclusions of the second phase with the dimensions to 10 μ. At low N_{Yb}(0.01..0.1 weight %) the inclusions of the second phase were not observed.Precipitation of the second phase at high concentrations REE was attributed to the low values of the effective coefficients of the distribution of lanthanoids in silicon (for Yb - 6.10^{-8}):

-the conditions of the growth (the speed of the cooling (V_c), the temperature of growth (T_g) and so on) substantially influenced the impurity content of EL. The decrease of V_c increased the homogeneity of the impurity distribution. The increase of T_g and the decrease of V_c resulted in the growth of N_c and N_o in EL.

It was stated by the method of optical microscopy combined with the selective etching, that the non-informly distributed growth defects of package and dislocation grouped into the form of the pulled clusters with the width ~ 50 μ are present in the epitaxial layers.The density of dislocations in the was clusters - 8.10^5 cm^{-2}, their average density on the surface ~ 1.10^3 cm^{-2}. Diislocations in Si:Sn,Yb were distributed uniformly, their density (~$2.5.120^4$ cm^{-2}) was higher, than in the layers being not doped by Yb. The density of the package defects did not exhibit dependency on N_{Yb}. The density of the grooves of etched pits in Si:Sn,Yb monotone increased with the removal from the surface of EL and was as high as ~ 1.10^7 cm^{-2} in the depth of the substrate.

The typical spectra of PL have the line of the free exiton (FE) and the set of lines,connected with the dislocations (D1-D4) [3]. In the substrate spectra, grinded on the nonoperative side before the epitaxy, FE line is absent, and the dislocation radiation (in D1 region) is considerably more intensive (~ three times) than in EL. Pre-epitaxial polishing of the "nonoperative" side of the substrates results in the decrease of the intensity of the dislocation lines. Lines of PL connected with the Yb impurity were not detected. The ratio of the intensity of the lines D1 to FE in Yb doped layers was critically higher than in Si:Sn. This is the evidence of the deterioration of the structural perfectness of EL with addition of

Yb with the concentration > 0,1 weight % into M-S.

Dislocations are not located at the EL-substrate interface, but "penetrate" the whole substrate.This conclusion is confirmed by the fact that the intensity of dislocation lines of PL practically did not change with moving away from the surface into the depth of the substrate.

As in the dislocation set the radiation in the region D1 dominates,one can come to the conclusion that dislocations were introduced directly in the process of epitaxy at high temperatures (900...1050°C), when the centers D3-D4 lines were annealed [4]. This conclusion is confirmed by the fact that in PL spectra of the original plates (before the epitaxy) only the line of the free exiton FE was present, and the dislocation lines did not occur.

After the electron irradiation C- and G- lines of PL exhibeted [5]. The values of τ in the EL agreed with the analogous values in the substrate. The intensity of the lines FE and D1-D4 decreased with the growth of Φ_e in the substrates as well as in EL.The given effect can be attributed to the decrease of τ and the "loss" of the dislocations of the recombination activity due to their interaction with the radiation defects (RD). At irradiation $\Phi_e \geq 10^{16}$ cm^{-2} in EL spectra the wide structureless band of radiation in the region of 0,75-0,95 eV (Fig.1 e)exhibited, its intensity increased with the increase of Φ_e and Sn concentration. It did not occur in the spectra of substrated (Fig.1 b). The introduction of the Yb impurity into M-S had no substantial effect on its intensity. The obtained results allow to conclude that the given band is connected with the centre caused by the Sn impurity.

Note the main peculiarities of PL spectra of the irradiated structures:

1) Considerable (more than two order) difference in the intensities C- and G- lines (I_c, I_g) for spectra of EL and substrates takes place.

2) The efficiency of introducing of RD into EL depends on the conditions of their obtaining. At the decrease of V_c from 0,25 to 0,1 K.min^{-1} or at the increase of T_g from 1173 to 1323 K I_c and I_g increased by the factor of 2.

3) Introducing of Yb impurity into M-S with the concentration 0,1 weight % decreased I_c, I_g as well as the value of their ratio comparing to EL Si:Sn. At the higher values N_{Yb} differences disappeared.

It has been stated before, that G- and C- lines were caused by the inner transitions. The former is connected with the centre, containing carbon, $(C_I - C_S)$; the latter one is connected with the complex, consisting of divacancy and the atoms of carbon and oxygen (C-O-W) [5].The decrease of I_c and I_g in EL as compared to the corresponding values in substrate is cause by the decrease of content of oxygen (N_O) and carbon (N_c) impurities. According to the data, obtained by the method of the local probe microanalysis by the scanning electron microscope "Nanolab-7" the average values run to: in EL–$N_o \approx 1.10^{17}$ cm^{-3}, $N_c \leq 2.10^{16}$ cm^{-3}; in the substrate $N_o \approx 1,2.10^{18}$ cm^{-3}, $N_c = 8.10^{16}$ cm^{-3}.

Besides, when interpreting the results one should take into account effects, connected with the deformation fields, attributed to the interface as well as Sn impurity. The presence of the con-

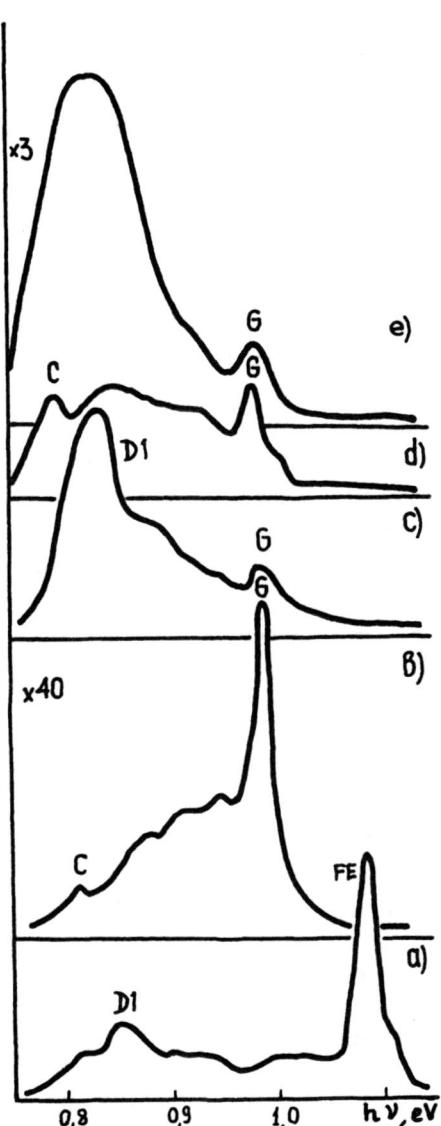

Fig.1. PL spectra of substrate (b) and EL Si:Sn(a,c-e) F_e, cm^{-2}: 0(a); 2,5.10^{15}(b,c) 8.10^{15}(d); 3.10^{16}(e).

traction deformations in EL results in the decrease of concentrations of defects of interstitial ty pe and to the increase of the concentration of vacancy complexes [6]. As a result of this, I_g must decrease, I_c -increase, and the I_c/I_g ration for EL should exceed the corresponding value for the substrate.

The decrease of V_c and the increase of T_g stimulates the processes of diffusion of technological impurities in EL from the container and the substrate, that must lead to the increase of N_c and N_0 in EL, and finally, to the increase of concentration of RD with their participation. On the other hand as in the process of epitaxy N_c and N_0 do not vary in the substrate, varying of T_g and V_c should not have an impact on the intensity of G- and C-lines in the substrate spectrum.

It should be noted that the Yb impurity has the double influence on the properties of EL. It interact with oxygen and carbon in M-S [2] and prevents their input into EL, that leads to the decrease of I_c and I_g. Besides, Yb atoms, forming additional deformations of compression in Si that enable the increase of I_c/I_g ratio at $N_{Yb} \geq 0,2$ weight %.

REFERENCES

1. O.M.Alimov, V.V.Petrov, T.D.Kharchenko and V.Yu.Yavid, Fiz. Tekh. Poluprovodn.**26**,1914 (1992).

2. Yu.A.Karpov,V.V.Petrov,V.S.Prosolovich and V.D.Tkachev, Fiz. Tekh. Poluprovodn.**17**,1530 (1983)

3. N.A.Drozdov,A.A.Patrin, and V.D.Tkachev, Pisma ZhETP,**23**, 651 (1976)

4. Yu.A.Osipjan,A.M.Rtishchev and E.A.Steinman, Fiz.tverd.tela, **26**,1772 (1984)

5. L.N.Safronov,Radiation Effects in Semiconductors,(Nauka,Moskow,1979-in Russian),p.101

6. E.V.Solpovjova,G.V.Lazareva, B.M.Leiferov et all, Fiz. Tekh. Poluprovodn.**18**,1573 (1984)

IC COMPATIBLE PROCESSING OF Si:Er FOR OPTOELECTRONICS

F. Y. G. REN*, J. MICHEL*, Q. SUN-PADUANO*, B. ZHENG*, H. KITAGAWA*[1],
D.C. JACOBSON**, J.M. POATE**, and L. C. KIMERLING*

* MIT, Dept. of Materials Science and Engineering, Cambridge, MA 02139
** AT&T Bell Laboratories, Murray Hill, NJ 07974

ABSTRACT

We have fabricated the first Si:Er LED, operating at 300 K, based on an understanding of the Si-Er-O materials system. Er-doped Si (Si:Er) provides an exciting opportunity for the monolithic integration of Si based opto-electronics. In this paper, Er-Si reactivity, and Er diffusivity and solubility have been studied to establish Si:Er process compatibility with a silicon IC fabline. Er_3Si_5 is the most stable silicid formed; and it can be oxidized into Er_2O_3 at high temperature under any oxidizing conditions. Among Er compounds, Er_2O_3 luminesces and Er_3Si_5 and ErN do not. The diffusivity of Er in Si is low and SIMS analysis yields a diffusivity $D(Er) \sim 10^{-12} cm^2/s$ at 1300 C and $\sim 10^{-15} cm^2/s$ at 900 C, and a migration enthalpy of $\Delta H_m(Er) \sim 4.6$ eV. The equilibrium solubility of Er in Si is in the range of 10^{16} cm^{-3} at 1300 C. The Si:Er LED performance is compared with GaAs LEDs to demonstrate its feasibility.

INTRODUCTION

The limiting factors to increased functionality of integrated circuits (ICs) are interconnection density, interconnect driver-related power dissipation and systems bandwidth. The integration of optoelectronic devices for interconnection provides an immediate solution of the above in the form of multiplexed outputs; absence of interconnect line capacitance and resistance; and an unlimited ($\sim 10^{15}$ Hz) bandwidth capacity. In addition, new capabilities in parallel architectures, immunity from electromagnetic interference and package integration present the opportunity for breakthrough applications. Hence, optical interconnection is a component on every silicon technology roadmap.

The purpose of this research is to create an IC compatible process technology for optical interconnection. We report here a summary of our work on the light emitting diode (LED), optical driver. Rare earth centers emit with a sharp linewidth [1] required for high data rates. For Er:Si, $\Delta\lambda = 0.1$ Å at 4.2 K and 100 Å at 300 K [2]. This paper will review our results on Er:Si reactivity, diffusivity and solubility which establish silicon fabline compatibility; and will then describe the first reported room temperature sharpline electroluminescence at 1.54 μm from a silicon LED.

[1] permanent address: Fukuoka Institute of Technology, Dept. Electr. Mat. Engineering, Fukuoka 811-02, Japan

SILICON:ERBIUM REACTIVITY

The phase stability of Si:Er in typical process ambients was determined by the evaporation of pure erbium on etched silicon substrates, followed by heat treatment in controlled ambients.

Figure 1 shows the thin film X-ray diffraction data for vacuum, air and nitrogen ambients. Polycrystalline Er_2O_3 and ErN form easily in air (450 C, 1 h) and N_2 (650 C, 1 h), respectively. In vacuum, a multiphase combination of silicides is often observed with some fraction of Er_2O_3, unless extreme care is taken. The best condition for single phase silicide formation, Er_3Si_5, was found to be (450C, 3 h) under a vacuum of 10^{-5} Torr.

The silicide, Er_3Si_5, is the most stable compound in the series ErSi, $ErSi_2$, Er_3Si_5. The reacted film exhibits a rough morphology on Si(100), but a planar highly textured Er_3Si_5 (001) on Si(111). The lattice mismatch between $Er_3Si_5(001)$ and Si(111) is small (1.2%) [3], providing a basis for this interface to be low energy. This conclusion was tested by reacting metallic erbium with a substrate of vicinal Si (111), miscut 4° toward <110> at elevated temperatures. Figure 2 shows a scanning electron micrograph of a cross section of a reacted sample. The lower part of the picture is a schematic of the SEM observations revealing consistency with Si(111) being the terrace interface. We conclude that the formation of an (001) Er_3Si_5 texture and of Si(111) terrace upon heat treatment in vacuum at 1160 C define this interface to be lowest in energy for the Si:Er system.

Heat treatment of Er on SiO_2 in vacuum and Er_3Si_5 in an oxygen containing ambient in the range 450 < T < 1100 C yields Er_2O_3. Thus, erbium reacts similar to titanium with respect to silicon [4]. Figure 3 shows the

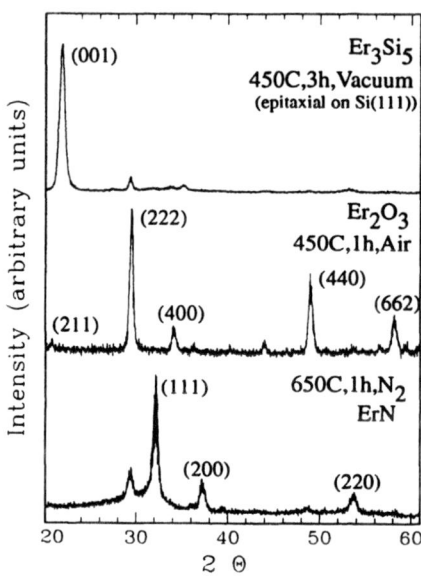

Figure 1. Comparison of X-ray diffraction spectra of the different erbium phases formed during the annealing of Er/Si under different annealing conditions.

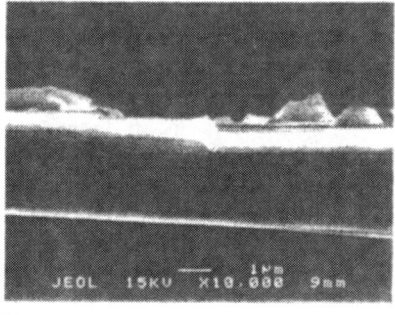

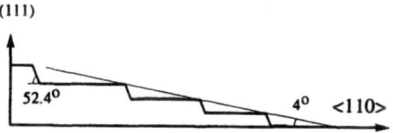

Figure 2. Crossectional SEM micrograph of a vicinal Si (111) surface.

relevant ternary phase diagram for the Er–Si–O system. Er_2O_3, the dominant phase, will capture the erbium activity under any oxidizing condition.

Of the reaction products Er_2O_3, Er_3Si_5 and ErN, only Er_2O_3 is optically active with emission at $\lambda=1.54$ μm. We assume that both Er_3Si_5 and ErN are conductive and absorbing. This conclusion is strengthened by the fact that in silicon both O and N provide optically activating ligand fields for erbium.

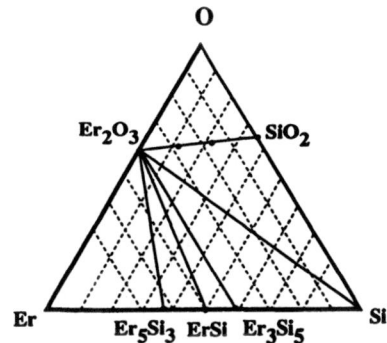

Figure 3. Proposed ternary phase diagram of the Er-Si-O system between 450 C and 1100 C based on X-ray phase identification.

Figure 4 shows a comparison of the optical spectra of Si:Er (900 C, 30 min) compound and poly-crystalline Er_2O_3 (450 C). There are two reproducible differences: 1) the primary emission shifts to shorter wavelengths for Si:Er; and 2) the multiple crystal field split emission bands for Er:Si are absent for Er_2O_3. This difference is critical in determining whether erbium emission originates in a surface oxide or bulk silicon.

ERBIUM DIFFUSIVITY AND SOLUBILITY IN SILICON

The internal quantum efficiency of Si:Er for light emission is dependent on the dominant role of erbium in the generation-recombination processes in the silicon host. Thus, it is necessary to establish that Er diffusivity is small enough during processing to prohibit cross-contamination of adjacent devices, wafers and process equipment. In addition, a high solubility of Er in Si is desirable because the optical power out is proportional to [Er] [5]. We describe below the experiments which define these parameters and discuss their consequences.

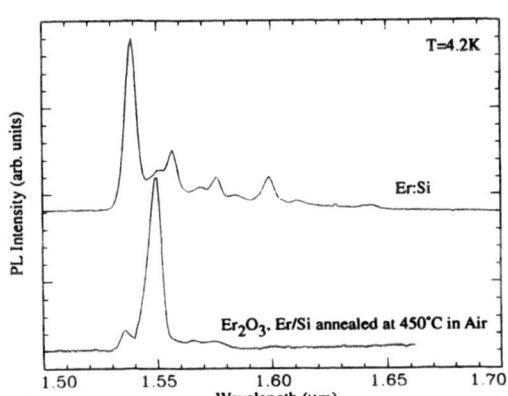

Figure 4. Comparison of photoluminescence spectra of Er_2O_3 and Er:Si.

Figure 5 shows a typical Secondary Ion Mass Spectroscopy (SIMS) profile of Er implanted silicon. The erbium was implanted to a peak concentration of 3×10^{18} cm^{-3} at an energy of 5.25 MeV. The data, after heat treatment at 850 C, 1/2 hour, represent no change from the as-implanted profile. A subsequent heat treatment of 1300 C, 1 hour, results in contraction of the profile rather than the broadening commonly observed for Group III and V dopants in silicon

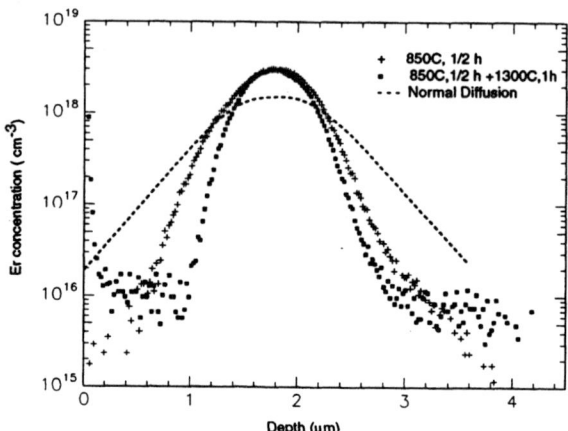

Figure 5. SIMS depth profiles of Er:Si before and after diffusion annealing at 1300 C for 1 h. The dotted line represents an expected normal diffusion profile

(dashed lines). A second aspect of the profile is a constant [Er] ~ 10^{16} cm^{-3} away from the implanted distribution.

The narrowing of the distribution unambiguously connotes precipitation at the peak of the distribution. Thus, the precipitates, as well as the surfaces, act as sinks for the [Er] to approach its equilibrium solubility. By assuming that complete point defect equilibrium is reached by 1300C, 1 hour, we define the flat portion of the distribution as the solid solubility ([Er]$_{sol}$) of Er in Si. Since [Er]$_{sol}$ is retrograde at high temperatures, [Er]$_{max}$ is somewhat greater than 10^{16} cm^{-3} (to be published). However, [Er]$_{max}$ is significantly less than the value of 2×10^{18} cm^{-3} at 900 C observed as an onset of precipitation by Eaglesham et al. [6].

It is important to note that Eaglesham et al. do not claim to measure solid solubility, but rather the threshold concentration for the onset of precipitation. This threshold may be controlled by either kinetic or equilibrium factors.

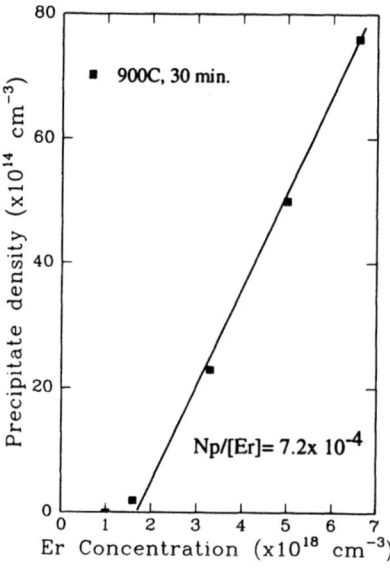

Figure 6. Precipitate density of Si:Er for a 900 C anneal versus [Er]. The line shows a fit using homogeneous nucleation theory

Figure 6 shows our fit (line) of the Eaglesham et al. [6] data (filled squares) to a diffusion limited precipitation model. The data exhibit two key features: an 'incubation' stage of $[Er] \approx 2 \times 10^{18}$ cm^{-3} at 900 C, 30 min; and a linear increase in precipitation density with [Er] beyond that concentration. The linear increase with [Er] is consistent with a nucleation limited precipitate density. The incubation stage represents the concentration-dependent, minimum time-at-temperature for Er atoms to associate.

The observed value of 10^{18} cm^{-3} sets a lower limit of the diffusivity $D(Er) > 6 \times 10^{-16}$ cm^2/s at 900 C. Analysis of the SIMS profiles over a wide range of temperatures has yielded a preliminary migration enthalpy, $\Delta H_m(Er) \approx 4.6$ eV, and a value of $D(Er) = 10^{-15}$ cm^2/s at 900 C.

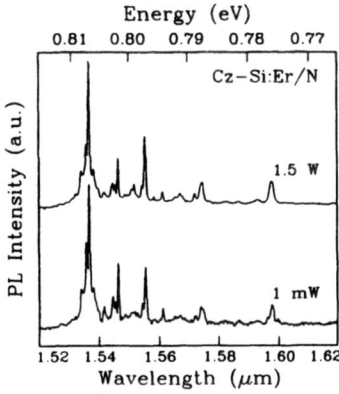

Figure 7. High resolution Si:Er spectra for different excitation powers.

Fitting the linearly increasing portion of the data with standard nucleation theory [7] provides a self correcting check for D(Er) and an estimate of the barrier to nucleation F*. The density of precipitates is given by equations 1(a) and 1(b).

$$N_p \sim \int I(T,t)\, dt \tag{1a}$$

$$= a_c \left(\frac{2D(Er)}{a_0^2} \right) [Er]\, t \exp\left[-\frac{F^*}{kT} \right] \tag{1b}$$

where I is the nucleation rate, a_c is a constant taken here as unity, and a_0 is the silicon interlattice distance (2.78 Å).

The slope of the data $N_p/[Er] \approx 7.2 \times 10^{-4}$ is very sensitive to D(Er) and F* in the analysis. Using $D(Er) = 10^{-15}$ cm^2/s, we estimate that approximately 100 hours are required to reach equilibrium at 900 C. A two parameter fit yields $D(Er) = 10^{-15}$ cm^2/s and $F^* = 1.57$ eV. The nucleation barrier is similar to values discussed for oxygen precipitation in silicon [8].

The values for D(Er) are similar to those observed for Ge in silicon [9] and $[Er]_{sol}$ is in the range observed for S in silicon [10]. The process relevant conclusions are as follows: 1) Er is a slow diffuser with moderate solubility and should not contaminate during silicon IC processing; 2) implanted erbium can be maintained in a metastable solid solution at concentrations two orders of magnitude higher than its solid solubility.

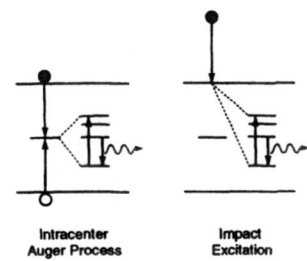

Figure 8. Possible excitation processes for Si:Er

Si:Er LIGHT EMITTING DIODES

Light emission at $\lambda = 1.54$ μm in the Si:Er materials system comes from relaxation of the $Er^{3+}(4f^{11})$ first excited state ($^4I_{13/2}$) to the first ground state ($^4I_{15/2}$). The 4f states are "core" states which are decoupled from "valence" states which comprise the valence band, conduction band and impurity gap states. Evidence of this decoupling is given in Figure 7 where the characteristic Si:Er emission spectrum is shown to be independent of pump power. No "hot lines" arise after the standard spectrum is saturated. In the absence of direct coupling to a gap state, excitation of the $4f^{11}$ manifold can only occur through an interactive Auger process or through impact ionization [11] (Figure 8). Impact ionization is relatively inefficient because only very hot electrons with energies greater than 0.8 eV (1.54 μm) can contribute. Previous photo-luminescence studies based on "back-side" excitation of "front-side" implanted silicon eliminate impact ionization as a necessary mechanism [2].

Recent total energy calculations [12] together with the observed donor activity of Si:Er suggest that Er^{3+} occupies a tetrahedral interstitial site. In addition, high internal quantum efficiencies are only observed when the Er is surrounded by a strongly electronegative ligand field. This requirement arises because the $^4I_{13/2} \rightarrow$ $^4I_{15/2}$ transition is dipole forbidden in isolated Er^{3+}. The ligand field is required to break the inversion symmetry of the site and to admix states of opposite parity. Crystal field splittings of the emission spectrum are observed whose number depends on the site symmetry and whose magnitude depends on the magnitude of the crystal field. Figure 9 shows the spectral dependence on the ligand field for the roles of oxygen and nitrogen

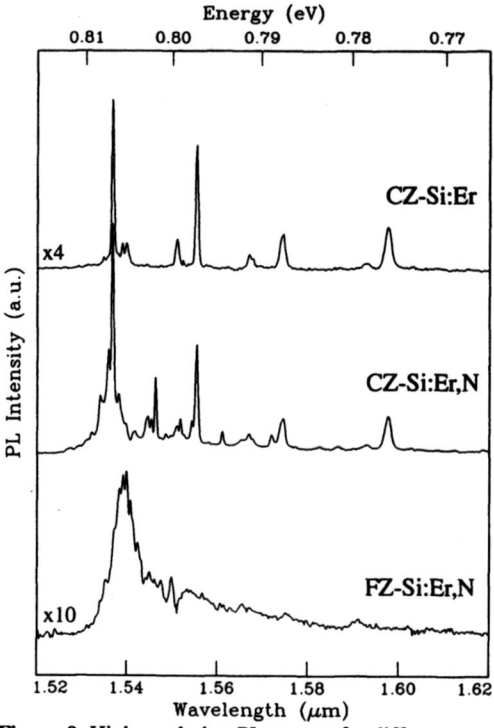

Figure 9. High resolution PL spectra for different co-implanted Si:Er samples.

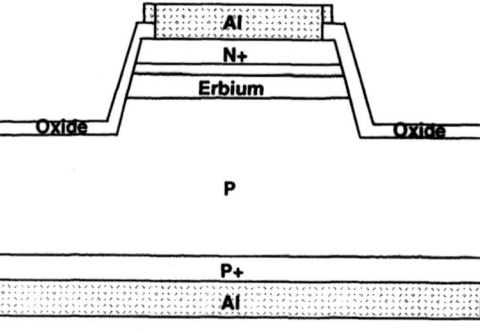

Figure 10. Schematic cross section of the Er:Si LED

(coimplanted with Er).

The process for our LEDs was designed based on the following understanding of the electronic excitation mechanism:

1) the Er ligand field determines optical activity,
2) oxygen is an effective ligand,
3) energy transfer to the f-manifold by Auger mechanism is most likely.

Figure 10 shows a schematic cross sectional cut of the device structure. We have constructed oxide passivated mesa diodes in both surface and edge emitting geometries. The erbium was implanted at an energy of 4.5 MeV together with a range of energies of oxygen co-implants designed to overlap the Er distribution.

Figure 11 shows the 300 K light intensity vs forward bias injection current for a typical diode. The light output saturates, as expected, because of the long excited state lifetime (~ 1 ms) [13] of the Er center. This saturation shows the relative ease of population inversion under electronic injection. The inset confirms that sharp line emission ($\Delta\lambda = 100$ Å) is maintained at 300 K. A comparison with the 100 K spectrum is also shown. At 300 K the intensity is reduced significantly (x50) and a background feature at $\lambda = 1.53\ \mu m$ arises (due to dislocations introduced by processing). In the next generation design, carrier confinement schemes and improved process paths should further enhance the impressive results.

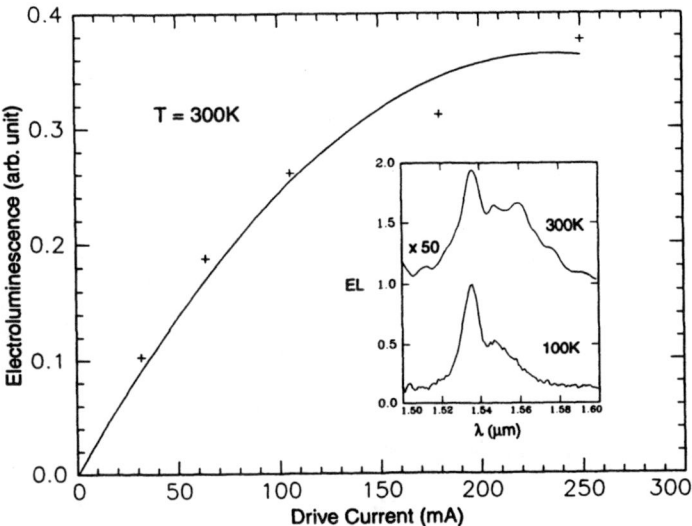

Figure 11. Si:Er LED intensity as function of drive current. The inset shows the EL spectra at 100 K and 300K.

COMPARISON OF Si:Er AND GaAs HOMOJUNCTION LEDS

In conclusion, Table I shows a comparison of the observed silicon LED performance with a GaAs homojunction LED at 77 K. The silicon LED offers significant advantages in the lack of self-absorption and sharper linewidth. With the additional compatibility of off-chip optical amplification at $\lambda = 1.54\,\mu m$, we see no fundamental limits to the introduction of Si:Er LEDs for optoelectronic interconnection.

Table I. LED Performance comparison between GaAs and Er:Si at 77 K

	GaAs	Er:Si
Wavelength (μm)	0.88	1.54
Absorption Coefficient (cm^{-1})	10^4	2
Internal Quantum Efficiency	0.5	0.5
External Quantum Efficiency	1.0%	1.4%
Linewidth (Å)	1000	100

ACKNOWLEDGEMENTS

The MIT group acknowledges partial support by Rome Laboratories, Hanscom, DARPA and AFOSR under contract F19628-92K-0012, and SRC under contract 92-SP-309.

REFERENCES

[1]   H.Ennen, J. Schneider, G. Pomrenke, and A. Axmann, Appl. Phys. Lett. **43**, 943 (1983).

[2]   J. Michel, J.L. Benton, R.F. Ferrante, D.C. Jacobson, D.J. Eaglesham, E.A. Fitzgerald, Y.-H. Xie, J.M. Poate and L.C. Kimerling, J. Appl. Phys. **70**, 2672 (1991).

[3]   J.A. Knapp and S.T. Picraux, Appl. Phys. Lett. **48**, 466 (1986).

[4]   S.P. Murarka, Silicides for VLSI Applications, (Academic Press, Orlando, 1983), p.142.

[5]   Y.-H. Xie, E.A. Fitzgerald and Y.-J. Mii, J. Appl. Phys. **70**, 3223 (1991).

[6]   D.J. Eaglesham, J. Michel, E.A. Fitzgerald, D.C. Jacobson, J.M. Poate, J.L.Benton, A. Polman, Y.-H. Xie and L.C. Kimerling, Appl. Phys. Lett. **58**, 2797 (1991).

[7]   J.P. Hirth and G.M. Pound, Progress in Materials Science 11 (Macmillan, New York, 1963) p. 15.

[8] P.E. Freeland, K,A. Jackson, C.W.Lowe, and J.R. Patel, Appl. Phys. Lett. **30**, 31 (1977).

[9] S.M. Sze, Physics of Semiconductor Devices, 2nd ed. (John Wiley & Sons, New York, 1981) p. 69.

[10] F.A. Trumbore, Bell Syst. Tech. J. **39**, 205 (1960).

[11] I.N. Yassievich and L.C. Kimerling, Semicond. Sci. Technol. **7**, 1 (1993).

[12] M. Lannoo and C. Delerue, (MRS Proceedings) this volume.

[13] P.B.Klein and G.S. Pomrenke, Electron. Letters **24**, 1503 (1988).

OXYGEN-ENHANCED 1.54μm PHOTOLUMINESCENCE OF Er^{+3} IN SILICON

F. ARNAUD D'AVITAYA[a*]), Y. CAMPIDELLI[a], J.A. CHROBOCZEK[a],
P.N. FAVENNEC[b], H. L'HARIDON[b], D. MOUTONNET[b], AND A. WASIELA[c]

[a] Centre National d'Etudes des Télécommunications, 38243 Meylan,France,

[b] Centre National d'Etudes des Télécommunications, 22301 Lannion, France,

[c] Laboratoire de Spectrométrie Physique, Université Joseph Fourier, 38402 St. Martin d'Hères, France.

ABSTRACT

Favennec et al. (Jap. J. Appl. Phys. 29, L524, 1990) reported that the 1.54μm photoluminescence of Si implanted with Er^{+3} is activated by oxygen impurities. We observe a significant enhancement in the luminescence in Er-doped silicon epitaxial layers MBE-grown with intentional oxygen contamination. The PL is shown to be a bulk property of the material as it persisted after a partial layer removal by wet etching.

INTRODUCTION

It has been reported by Favennec et al. [1] and Michel et al. [2] that the photoluminescence (PL) in the 1.54μm wave-length region is detectable in Er-implanted silicon crystals solely when the oxygen impurities and the Er^{+3} ions coexist in the Si matrix. This finding has been corroborated in [1] by two sets of experimental data. First a correlation has been established between the intensity of the 1.54μm PL in Si:Er and the residual oxygen concentration in Si substrates used for Er implantation. Thus Czochralski-grown Si substrates (known to be oxygen-rich) gave a much stronger PL intensity than the zone-refined crystals or CVD-grown Si layers (almost oxygen-free), after the Er implantation and a subsequent defect annealing. Secondly, a more direct confirmation of the oxygen-induced activation of the 1.54μm PL in Si:Er has been furnished by a series of experiments in which Er and O were both implanted into CVD Si epilayers. The layers showed appreciable PL and both elements were found to re-distribute during annealing, producing characteristic density peaks in their respective SIMS profiles, occurring at the same depths for Er and O and having comparable amplitudes. This fact has been interpreted [1] as evidence that a part of implanted O ions form optically active Er-O complexes in Si. The hypothesis of the Er-O complex formation has not been directly confirmed so far, although it is consistent with the large crystal-field splitting of the 4f multiplets of the Er^{+3} in the PL spectra [3]. This suggests ionic bonding of the optically-active Er ions embedded in the crystalline lattice. The idea of the oxygen-rare earth complex formation is not new; e.g. Petrov et al. [4] used this concept to account for their transport data in Si:Er. Furthermore correlations in SIMS profiles of oxygen and other elements have been reported (cf. e.g. [5]).

In this paper we present further evidence that oxygen impurities activate the PL of Er^{+3} ions in Si. We have studied the PL properties of MBE-grown Si samples, Er-doped during growth. Some of them were prepared with a controlled oxygen contamination. We found that the 1.54μm PL is detected only in the samples which contained oxygen. The growth experiments were followed by the PL measurements on samples subjected to etching. They demonstrate that the luminescence comes from the interior of the layers and rule out the possibility of it originating in the surface oxide, or a near-surface region.

SAMPLE PREPARATION

Epitaxial layers of Si were MBE grown maintaining a constant flux of Si and Er, calibrated so as to give the atomic Er concentration of the order of 10^{18}cm^{-3}. The layers were several kÅ thick and had good overall crystallographic properties. They were smooth, mirror-like, and showed a distinct 7x7 low energy electron diffraction pattern on the (111) surface. The ratio of intensities for Rutherford back-scattering, measured at the channeling and the randomizing orientations was typically below 10%, indicating a small defect concentration in Si and that the impurities were mainly substitutional. However the transport properties of the layers demonstrated that Er^{+3} cannot be treated as a simple acceptor, providing one hole per atom, although electrical transport in the layers was found to be hole-dominated. We interpreted the Hall data using a single-carrier transport model. The room temperature carrier concentration thus determined in the layers was typically of the order of $10^{15} cm^{-3}$, the hole mobility about $25 Vcm^2 V^{-1} s^{-1}$, and the resistivity $75\Omega cm$. These values are characteristic of trap-dominated conduction due to residual impurities in a compensated semiconductor. The samples doped with Er alone showed no appreciable PL in the $1.54\mu m$ wavelength region, although the near-band gap luminescence was visible at higher laser excitations, with the characteristic electron-hole droplet lines dominating the spectrum at sufficiently low temperatures.

The activation of the PL of Er^{+3} in Si by oxygen implantation described in ref. [1] suggested that the incorporating of oxygen into the MBE layers could lead to a similar effect. In order to introduce oxygen into the layers during the MBE epitaxy, we modified the growth chamber by introducing a capillary enabling the layers to be grown in a jet of gas, and preserving the overall vacuum level in the system at below 10^{-8} Torr. By maintaining the Si and Er flux at the same level as in the preceding set of experiments, conducted without the presence of oxygen, we obtained samples with an atomic Er concentration of about $10^{18} cm^{-3}$. Evaluation of the O content was not straightforward. However, from the appearance of a weak signal observable in the vicinity of 500eV in the Auger spectrum taken on the layers *in situ*, it could be evaluated to be above $10^{19} cm^{-3}$. The atomic composition of the MBE Si:Er:O samples thus fabricated was, therefore, similar to that of samples prepared by the Er and O co-implantation [1]

Another method we used for incorporating oxygen into Er-doped MBE layers involved amorphous Si:Er layers, grown on cold substrates. The samples were briefly exposed to an ambient atmosphere, in order to trap a small amount of oxygen in the amorphous layer. They were subsequently recrystallized either by rapid thermal annealing (RTA) or by furnace heating under an oxygen-free gas flow. This method of oxygen incorporation into the Er-rich layer does not offer much control of the O content, but was used for convenience and eliminated the necessity of contaminating the MBE chamber with oxygen. As we shall demonstrate below it gave comparable results to the other method.

RESULTS

The PL spectra taken at 1.8K on an Er-implanted Czochralski Si sample and on an Er-doped MBE sample, both containing about $10^{18} Er/cm^3$ are compared in Fig. 1. Note that the power of the primary Ar laser beam used for the MBE sample was about 4.5 times higher than that used for the implanted layer and produced only a weak, diffused PL signal in the $1.54\mu m$ spectral region. The implanted sample showed a reasonably strong luminescence with a resolved multiplet structure.

A comparison of the PL spectra measured at 77K on the Er-implanted sample (data at 1.8K shown in Fig. 1), and on two MBE-grown samples, containing both Er and O, is made in Fig. 2. The 0202 sample was MBE-grown in the oxygen jet, as described, and the 0602 sample was initially amorphous and recrystallized after exposure to air. The luminescence in MBE samples was easily measurable, but about an order of magnitude weaker than that observed in the Er-implanted sample. A comparison of the PL intensity near the 1.1eV indirect gap of Si with that in the spectral range of Er^{+3} ions (respectively near $1\mu m$ and $1.5\mu m$) was made. We found that the amplitude ratio of the most prominent peaks in the two spectral regions, measured at the same excitation level, at 1.8K, on the MBE0202 sample, was about 3:1. A separate study was devoted to optimizing the annealing conditions giving the strongest PL signal and the RTA-treatment at about 700°C was found to be optimal. The sample MBE0602 was prepared under these conditions.

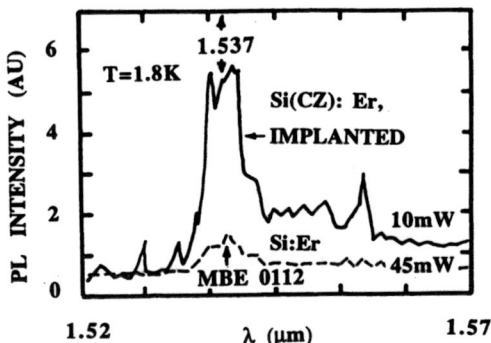

Figure 1
The 1.8K photoluminescence of Er-doped Si.
The Czochralski-grown Si are known to contain oxygen. The MBE0112 sample was grown without oxygen. Note different excitation power levels used for each sample.

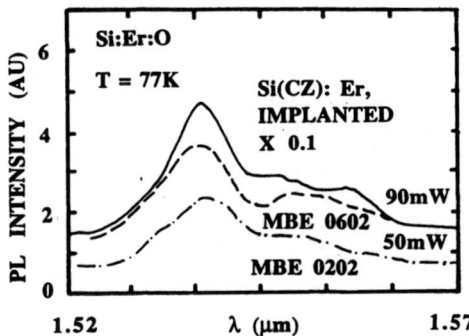

Figure 2
The 77K photoluminescence of Er-doped Si. Samples containing oxygen.
The Si(CZ) sample as in Fig. 1. The MBE0202 sample was grown in a jet of O_2 and the MBE0602 sample was initially amorphous and recrystallized after exposure to air.

Because Er^{+3} is known to luminesce when embedded in the SiO_2 matrix, we performed a series of experiments involving chemical treatment of the MBE-grown Si:Er:O samples in order to establish whether the PL we observed indeed originated in the doped Si.

First, we compared the luminescence of the samples before and after a dip in 30% HF. The HF etch is known to rapidly attack SiO_2 and to be neutral in relation to Si. An accepted criterion of oxide removal is the absence of wetting of the surface of a sample when dipped in the HF etch. We found that the samples subjected to such a treatment retained their luminescence properties.

Having established that the PL is not associated with the surface native oxide, we used a KOH etch, known to attack Si, in order to check whether the thinning of the doped layer affected the PL or not. We used a 33% KOH solution with a 3:1 isopropanol/methanol mixture as a wetting agent, at 75°C, stirring the solution during the reaction. The thickness of the removed layer was then directly measured. We found that the samples showed no change in the PL intensity after a 50% decrease in the thickness of the initial layer. It is worthwhile noting that the etch rate in the Si:Er:O samples was an order of magnitude higher than in the control samples [(111) zone-refined Si]; they were, respectively equal to about 0.2μm/min and 0.03μm/min. The KOH etch is known to be highly preferential and its etch rate on the (100) surfaces is about an order of magnitude higher than on the (111) planes, where it is primarily used for defect revelation. The increased etch rate in Si samples containing Er and O impurities suggests that the (100) planes were accessible on the nominally (111) surface of the specimens. This implies the presence of defects or steps at the surface.

As the formation of Er precipitates is known to limit the 1.54μm PL in III/V semiconductors, a few measurements of cathodoluminesce were carried out in order to check the homogeneity of Er^{+3} luminescence in the MBE Er-doped samples. The experiments revealed the presence of areas showing much stronger luminescence than the background. No systematic study has, however been devoted to this problem.

CONCLUSIONS

We found that the luminescence of Er^{+3} ions in Si can be enhanced if oxygen is introduced into Er-doped layers during the MBE growth, or during annealing of amorphous layers, previously exposed to atmospheric oxygen. At liquid He temperatures the PL of Er^{+3} in such samples is found to be 3 to 4 times weaker than the 1.1eV PL involving the indirect gap of Si. We have shown that the PL is not associated with the surface oxide layer nor with the near-surface oxygen-rich Si strata, but that it is a bulk property of Si:Er:O. This work confirms the previous reports[1,2,6,7] on oxygen activation of the 1.54μm PL of Er^{+3} ions in Si, detected in Er- and O-implanted samples. It is also consistent with the hypothesis put forward in ref. [1] that the PL might be associated with Er-O complexes in Si. However, we cannot exclude the possibility of the PL originating in regions having high concentrations of Er and O. Non-uniform cathodoluminescence suggests the formation of such regions.

ACKNOWLEDGEMENTS

We thank Annie-Claire Papadopoulo (Centre Nat. d'Etudes. Télécom.-Bagneux) for the cathodoluminescence analysis of our samples.

REFERENCES AND FOOTNOTES

*) present address: Université de Marseille-Luminy, 13288 Marseille, France.

1. P.N. Favennec, H. L'Haridon, D. Mutonnet, M. Salvi, and M. Gaunau, Jap. J. Appl. Phys. 29, L524 (1990).
2. J. Michel, J.L. Benton, R.F. Ferrante, D.C. Jacobson, D.J. Eagelesham, E.A. Fitzgerald, Y. H. Xie, I.M. Poate, and L.C. Kimmerling, J. Appl. Phys. 70, 2672 (1991).
3. F. Auzel, p. 59 in SPIE Vol. 1182, French-Israeli Workshop on Solid State Lasers, Jerusalem 1988.
4. V.V. Petrov, S.V. Prosolovitch, V.D. Tkachev, Yu. A. Karpov, and M.G. Milvidskii, Phys. Stat. Solidi A 88, K141 (1988).
5. A.E. Von Neida, S.J. Pearton, W.S. Hobson, and C.R. Abernathy, Appl. Phys. Letters, 54, 1540 (1989).
6. D.L. Adler, D.C. Jacobson, D.J. Eagelsham, M.A. Marcus, J.L. Beneton, J.M. Poate, and P.H. Citrin, Appl. Phys. Letters 61, 2181 (1992).
7. M. Efeoglou, J.H. Evans, T.E. Jackman, B. Hamilton, D.C. Houghton, J.M. Langer, A .R. Peaker, D. Perovic, I. Poole, N. Ravel, P.H. Hemment, and C.W. Chan, Semicond. Sci. Technol., 8, 236 (1993).

HIGH CONCENTRATIONS OF ERBIUM IN CRYSTAL SILICON BY THERMAL OR ION-BEAM-INDUCED EPITAXY OF ERBIUM-IMPLANTED AMORPHOUS SILICON

J. S. Custer, A. Polman, E. Snoeks, and G. N. van den Hoven
FOM Institute for Atomic and Molecular Physics
Kruislaan 407, 1098 SJ Amsterdam, the Netherlands

ABSTRACT

Solid phase epitaxy and ion-beam-induced epitaxial crystallization of Er-doped amorphous Si are used to incorporate high concentrations of Er in crystal Si. During solid phase epitaxy, substantial segregation and trapping of Er is observed, with maximum Er concentrations trapped in single crystal Si of up to 2×10^{20} /cm^3. Ion-beam-induced regrowth results in very little segregation, with Er concentrations of more than 5×10^{20} /cm^3 achievable. Photoluminescence from the incorporated Er is observed.

INTRODUCTION

There is substantial technological interest in achieving efficient light emission from crystal Si (c-Si). Unfortunately, because of its indirect band gap, Si exhibits very inefficient band-to-band luminescence. Ennen *et al.* have pointed out the potential of rare-earth ions as optical dopants in semiconductors, including Si [1,2]. Rare-earth ions in the correct charge state exhibit luminescent intra-4f transitions, which are shielded from the surroundings by filled outer electron shells. This produces a nearly host-independent luminescent transition. Erbium is of particular interest because Er^{3+} has a transition from the first excited state to the ground state at a wavelength of 1.5 μm, which is important in optical communication technology. There are three major steps to actually use Er-doped Si. First, high concentrations of Er have to be incorporated in Si. Second, the Er must be made optically active, that is have the 3+ charge state. Finally, the Er ions must be efficiently pumped by electrical excitation.

In this paper, we will concentrate on the first step, incorporating Er in c-Si. Because the solubility limit of Er in Si is not known, it is not clear *a priori* what concentrations can be easily achieved. By analogy to the transition metals, though, it is likely that the solubility of Er is relatively small ($\sim 10^{14} - 10^{16}$ Er/cm^3). This concentration is too low to make useful optical devices [3]. We demonstrate here that by using non-equilibrium crystal growth, at least 5×10^{20} Er/cm^3 (1 at.%) can be incorporated in c-Si. This is done by recrystallizing an Er-doped amorphous Si (a-Si) layer on c-Si, using either thermal solid phase epitaxy (SPE) [4] or ion-beam-induced epitaxial crystallization (IBIEC) [5]. Both methods are capable of trapping impurities in c-Si above their solid solubility limits [6-8]. Thermal SPE results in significant segregation and trapping of Er [9,10], with trapped concentrations of up to 2×10^{20} Er/cm^3 achievable. IBIEC at 320°C leads to little segregation and nearly complete trapping of Er at concentrations of at least 5×10^{20} Er/cm^3. Both types of samples exhibit photoluminescence (PL) characteristic of Er, and the PL intensity can be optimized by post-growth anneals.

EXPERIMENT

For the SPE sample, 9×10^{14} /cm^2 250 keV Er was implanted into c-Si. The IBIEC sample was made by a 4.0×10^{15} /cm^2 250 keV Er implant. Both implants were performed with the samples heat sunk to a copper block cooled by liquid nitrogen. The (100) c-Si substrates used in this work were Czochralski-grown and either P doped (1–1.5 Ωcm) (SPE samples) or B doped (5–10 Ωcm) (IBIEC samples).

The a-Si thicknesses and Er concentration profiles were measured with Rutherford backscattering spectrometry (RBS) using 2 MeV He and a backscattering angle of 110° was used to increase the depth resolution. Thermal anneals to induced SPE were done in a rapid thermal annealer (RTA) under flowing Ar. The indicated anneal times are measured once the actual anneal temperature was reached. Ion-beam-induced epitaxy was performed by irradiating with 1×10^{17} /cm^2

Mat. Res. Soc. Symp. Proc. Vol. 298. ©1993 Materials Research Society

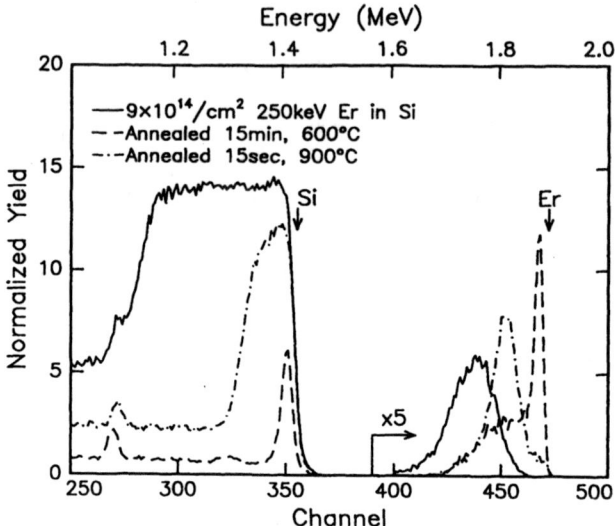

FIG 1. RBS/channeling spectra for samples as-implanted with 9×10^{14}/cm² 250 keV Er (solid line), which makes a surface amorphous layer, and after annealing either for 15 min at 600°C (dashed line) or for 15 sec at 900°C (dashed-dot line). The Er portions of the spectra are multiplied by a factor of 5.

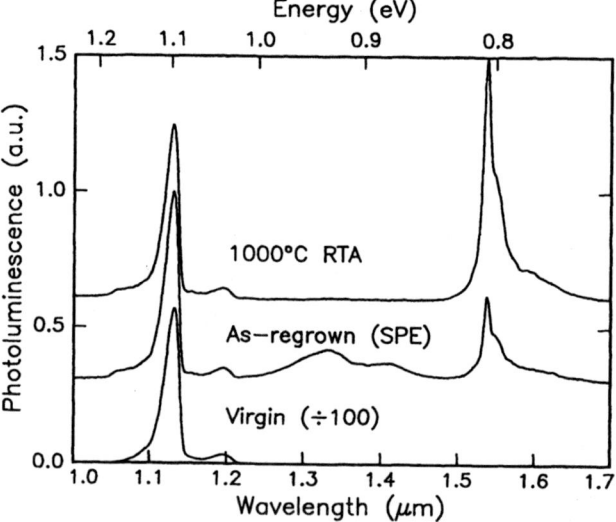

FIG 2. Photoluminescence spectra at 77 K for the 9×10^{14} Er/cm² sample as regrown at 600°C and after a post-regrowth 1000°C RTA anneal. A spectrum for a virgin c-Si wafer is shown for comparison, reduced by a factor of 100.

1 MeV Si with the substrates held at 320°C. The range of the 1 MeV Si, 1.2 μm, is well beyond the a-Si surface layer. After either SPE or IBIEC, samples were annealed in the RTA at up to 1000°C for times up to one minute.

Photoluminescence (PL) spectroscopy was performed using the 514 nm Ar ion laser line as an excitation source (50–250 mW focused into a $\sim$ 0.5 mm diameter spot). The samples were mounted in a vacuum dewar and cooled to liquid nitrogen temperature. The luminescence signal was spectrally analyzed with a 48 cm monochromator and detected with a liquid nitrogen cooled Ge detector. The pump beam was chopped at 10–20 Hz and the Ge detector output was recorded using a lock-in amplifier.

RESULTS

Figure 1 shows RBS/channeling spectra for the 9×10^{14} Er/cm^2 SPE sample as-implanted and after annealing for either 15 min at 600°C or 15 sec at 900°C. The as-implanted sample has an 160 nm thick a-Si surface layer with a Gaussian Er profile peaked (1.5×10^{20} Er/cm^3) at a depth of $\approx$ 70 nm. After the 600°C anneal, a thin disorder layer $\approx$ 10 nm thick remains at the surface, but the channeling minimum yield in the regrown layer is $\chi_{min} <$ 5 %, indicating that the crystal quality is good. The Er segregated with 65 % of the Er remaining trapped in the regrown crystal at concentrations up to 9×10^{19} Er/cm^3. During the 900°C anneal, the initial regrowth is epitaxial, but 60 nm from the surface the quality of the crystal begins to degrade through the introduction of twins [9]. Regrowth of a-Si layers implanted with different Er doses shows that at a given temperature there is a maximum trapped Er concentration above which twins form during growth. This maximum concentration is 2×10^{20} Er/cm^3 for SPE at 500°C, 1.2×10^{20} Er/cm^3 at 600°C, and only 6×10^{19} Er/cm^3 for the 900°C anneals [9,10].

Figure 2 shows PL spectra taken at liquid nitrogen temperature of the 9×10^{14} Er/cm^2 sample both after regrowth at 600°C and after a subsequent RTA anneal at 1000°C for 15 sec. In addition, the spectrum of a virgin (unimplanted) sample is also shown, with the signal divided by 100. All spectra are offset from each other for clarity. The features near 1.1–1.2 μm are the intrinsic phonon-assisted luminescence from c-Si [11]. In either Er-doped sample, the intrinsic c-Si luminescence is drastically reduced compared with the virgin sample, most likely as a result of reduced carrier lifetimes caused by carrier trapping at remaining structural defects. After SPE, the spectrum shows a feature near 1.55 μm which arises from the Er, as well as a broad band centered around 1.33 μm which corresponds to luminescence from various irradiation-induced defects [11]. Further annealing to 1000°C removes the defect band and increases the strength of the Er signal.

Figure 3 shows RBS/channeling spectra for the 4×10^{15} Er/cm^2 IBIEC sample as-implanted and after IBIEC with 1×10^{17}/cm^2 1 MeV Si at 320°C. The as-implanted sample has a 180 nm thick a-Si layer containing a Gaussian Er profile with a peak concentration of 5×10^{20} Er/cm^3. After regrowth, a thin disordered surface region (containing substantial C and O because of the long hot implant) remains on the surface. Behind this layer, the dechanneling yield is $\chi_{min} =$ 8 %, which is normal for IBIEC regrown material [5], but definitely not as good as for the SPE samples. This near-surface dechanneling comes from point defects in the regrown region, not from dislocations or other extended defects. However, starting at $\approx$ 700 nm from the surface (not seen in this scattering geometry) the dechanneling increases drastically because of a dense network of dislocations from the end-of-range damage of the 1 MeV Si irradiation [12]. The Er profile is barely affected by the IBIEC regrowth, although a little redistribution of Er towards the surface is evident.

Directly after IBIEC growth, the samples exhibit no measurable PL signal. Figure 4 shows PL spectra after post-IBIEC RTA anneals at either 600 or 1000°C for 1 min. The 600°C annealed sample exhibits a small Er-related feature near 1.54 μm. However, it lies on the tail of the large defect-related signal near 1.33 μm. Because the channeling data in Fig. 3 show that the crystal quality is not very good, it is not surprising that there is a substantial amount of defect-related signal in the spectrum.

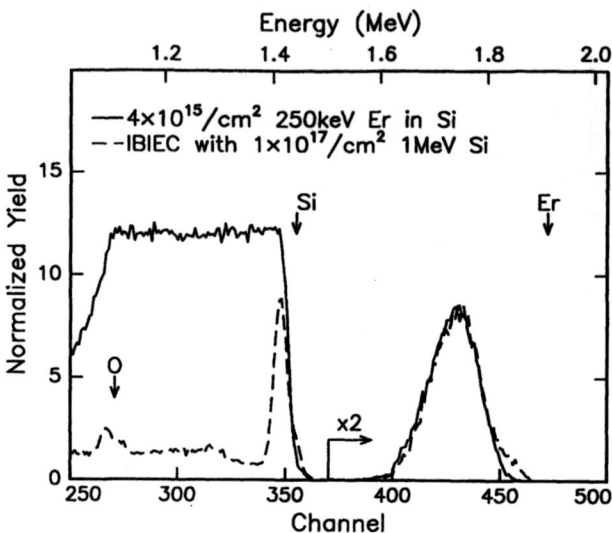

FIG 3. RBS/channeling spectra for the 4×10^{15}/cm² 250 keV Er sample as-implanted (solid line) and after IBIEC growth with 1×10^{17}/cm² 1 MeV Si at 320°C (dashed line). The Er portion of the spectrum has been multiplied by 2. Very little redistribution of Er is seen.

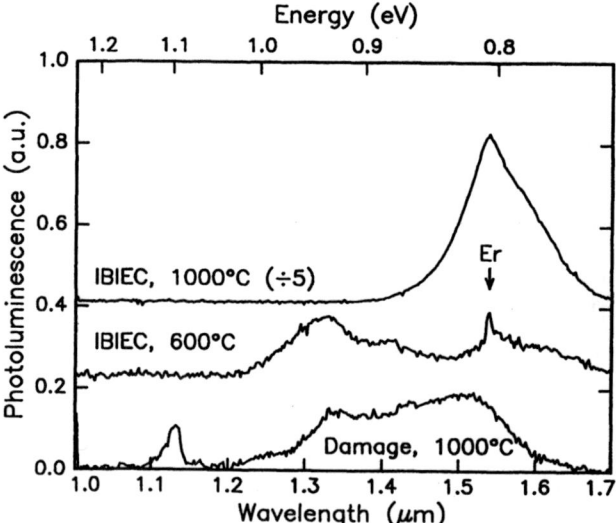

FIG 4. Photoluminescence spectra at 77 K for the IBIEC sample after RTA anneals at 600 and 1000°C. The spectrum of the 1000°C annealed sample has been divided by a factor of 5. The PL spectrum for low-dose Si-implanted Si after annealing at 1000°C also shown.

After annealing at 1000°C, the PL spectrum changes dramatically. The defect band near 1.3 μm disappears, as it does in the SPE sample. Instead, a large, broad feature centered near 1.5 μm appears, which appears very different from the Er signal in the 600°C annealed sample, or in the SPE samples. The question is if this signal is related to the Er at all. We believe that the majority of this signal is *not* from Er, but instead from implantation damage [11]. For comparison, a PL spectrum of a c-Si sample that has been lightly damaged (2×10^{14} /cm^2 1 MeV Si) and annealed at 1000°C is also shown in Fig. 4. This implant and annealing treatment will result in the formation of dislocations [13]. This sample exhibits the defect band near 1.3 μm (also observed for lower temperature anneals where dislocations are not formed), as well as a second band near 1.5 μm that is similar to what is observed for the IBIEC-regrown sample. Since this sample contains no Er, this signal is clearly not related to Er, and comes most likely from dislocation-related centers [11]. Since IBIEC results in a very high density of dislocations near the end-of-range of the 1 MeV Si, and the probe depth for the 514 nm pump light encompasses this region, the occurence of this signal is not surprising.

CONCLUSIONS

Thermal SPE of Er-doped a-Si results in significant segregation and trapping of Er at the moving interface. In this way, up to 2×10^{20} Er/cm^3 can be incorporated in c-Si. Post-regrowth RTA anneals can be used to increase the photoluminescence from the Er, and eliminate the defect-related PL. IBIEC of Er-doped a-Si results in trapping of at least 5×10^{20} Er/cm^3 in c-Si. Again, post-regrowth RTA anneals can be used to increase the PL signal. However, very little of the IBIEC-trapped Er is optically active compared with the SPE samples, and the PL response after high temperature anneals is dominated by the crystal damage inherent to the IBIEC process. Although we have solved the first problem, incorporating high concentrations of Er in c-Si, it appears that either the optical activation or excitation steps are limiting the photoluminescence response of the material.

This work is part of the research program of the Foundation for Fundamental Research on Matter (FOM) and was made possible by financial support from the Dutch Organization for the Advancement of Pure Research (NWO), the Organization for Technical Research (STW), and the Innovative Research Program (IOP Electro-Optics).

REFERENCES

1. H. Ennen, J. Schneider, G. Pomrenke, and A. Axmann, Appl. Phys. Lett. **43**, 943 (1983).
2. H. Ennen, G. Pomrenke, A. Axmann, K. Eisele, W. Haydl, and J. Schneider, Appl. Phys. Lett. **46**, 381 (1985).
3. Y.-H. Xie, E. A. Fitzgerald, and Y. J. Mii, J. Appl. Phys. **70**, 3223 (1991).
4. G. L. Olson and J. A. Roth, Mater. Sci. Rep. **3**, 1 (1988).
5. F. Priolo and E. Rimini, Mater. Sci. Rep. **5**, 319 (1990).
6. S. U. Campisano, J. M. Gibson, and J. M. Poate, Appl. Phys. Lett. **46**, 580 (1985).
7. F. Priolo, J. L. Batstone, J. M. Poate, J. Linnros, D. C. Jacobson, and Michael O. Thompson, Appl. Phys. Lett. **52**, 1043 (1988).
8. J. S. Custer, Michael O. Thompson, D. C. Jacobson, and J. M. Poate, Phys. Rev. B **44**, 8774 (1991).
9. A. Polman, J. S. Custer, E. Snoeks, and G. N. van den Hoven, Appl. Phys. Lett. **62**, 507 (1993).
10. A. Polman, J. S. Custer, E. Snoeks, and G. N. van den Hoven, Nucl. Instr. and Meth. B, in press.
11. G. Davies, Physics Reports **176**, 83 (1989).
12. R. G. Elliman, J. S. Williams, W. L. Brown, A. Leiberich, D. M. Maher, and R. V. Knoell, Nucl. Instrum. Methods **B19/20**, 435 (1987).
13. R. J. Schreutelkamp, J. S. Custer, J. R. Liefting, W. X. Lu, and F. W. Saris, Mater. Sci. Rep. **6**, 275 (1991).

OPTICAL DIRECT AND INDIRECT EXCITATION
OF Er³⁺ IONS IN SILICON

A. Majima*, S. Uekusa*, K. Ootake*, K. Abe*, and M. Kumagai**

*Meiji University, Kawasaki, Kanagawa, 214 Japan.
**Kanagawa High-Technology Foundation, Kawasaki, Kanagawa, 214 Japan.

ABSTRACT

Optical direct and indirect excitation of erbium (Er) ions in silicon substrates was performed in order to investigate the high efficiency of Er^{3+}-related $1.54\mu m$ emission ($^4I_{13/2} \rightarrow {}^4I_{15/2}$) for direct excitation that is not concerned with the indirect band gap and low quantum efficiency of a Si host. The samples were prepared by ion-implantation or thermal diffusion methods. In each sample, photoluminescence (PL) showed the peaks originating from $^4I_{13/2} \rightarrow {}^4I_{15/2}$ of Er^{3+} ions.

In Er thermally diffused samples, optical excitation for energy level $^4I_{11/2}$ of Er^{3+} ions was successfully effected by photoluminescence excitation spectroscopy (PLE). The PLE spectra consisted six peaks (963.1nm, 965.0nm, 976.1nm, 978.9nm and 980.9nm) which were caused by direct excitation ($^4I_{15/2} \rightarrow {}^4I_{11/2}$) of Er^{3+} ions. The emission directly excited is about 2 times more intense than the indirectly excited emission. The six peaks originating from the splitting of the $^4I_{11/2}$ levels meant that Er^{3+} ions were in the sites of noncubic symmetry. The samples prepared by Er ion-implantation did not show the effect.

INTRODUCTION

A emission in Si is weak because of the its indirect band gap, and the applications of Si have been limited to electrical devices and light absorbing devices. With development of optoelectronic integrated circuits (OEICs), it has been necessary to fabricate light emitting devices in/on Si substrate. In recent years an increasing amount of research has been focused on the realization of III-V/Si-structure OEICs. However, for systems of GaAs light-emitting devices on Si, there is a high lattice mismatch of about 4%. On the other hand, electroluminescence and photoluminescence from rare-earth ions incorporated into III-V compound semiconductors and silicon have been studied because of the sharp and temperature-independent emission due to intra-4f-shell transition of rare-earth ion [1-8]. Especially the $1.54\mu m$ luminescence peak from Er^{3+} ions have been attracted increasing attention because the wavelength is corresponding to minimum absorption of silica fibers. For the Si:Er system, Er^{3+}-related $1.54\mu m$ emission is weak, and it is hard to observe the emission at room temperature until now. However, It was reported that the Er doped amorphous silicon prepared from rf-sputtering technique showed Er^{3+}-related emission at room temperature [1]. Recent studies revealed that impurities (B, C, O, N and F) codoped in Si:Er enhance Er^{3+}-related emission [2]. This is due to the fact that these impurities increase the number of optically activated Er^{3+} ions. In this work, Er^{3+} ions were excited by direct and indirect excitation in order to investigate the Er^{3+}-related emission for direct excitation that is not concerned with the indirect band gap and low quantum efficiency of a Si host.

EXPERIMENTAL

In Er ion implantation, boron (B)-doped p-type CZ Si wafers were used. Er^+ ions were implanted at room temperature at energy of 2MeV to a dose of $1 \times 10^{13} cm^{-3}$. To avoid the channeling effect, samples were inclined $7°$ with respect to ion beam. After ion implantation, annealing was performed at a temperature of 900°C for 30 min. For Er-diffused samples, undoped n-type FZ Si substrates were used as starting materials. Er and Si were deposited on the substrates about 700Å thickness in vacuum (10^{-8} torr). Subsequently, the samples were kept at 600°C for 10 hours in vacuum (10^{-7}torr).

Photoluminescence (PL) spectra were measured by using an argon ion (Ar^+) laser (514.5nm or 488nm) and a tunable Ti:sapphire laser as the excitation light sources. PLE spectra were performed by using the Ti:sapphire laser over the wavelength range of λ=780 - 1000nm. The PL and PLE spectra are recorded by a 1-m double monochromator and a cooled Ge detector and analyzed using a conventional lock-in technique. PL and PLE spectra were measured as a function of excitation power or temperature.

RESULTS AND DISCUSSION

Typical low temperature PL spectrum at 10K of Er-implanted CZ Si is shown in Fig. 1. The emissions located at 1.5μm region were caused by the transition $^4I_{13/2} \rightarrow ^4I_{15/2}$ of Er^{3+} ion. The five PL peaks were assigned to the crystal fields splitting of the $^4I_{15/2}$ ground state into five levels. This shows that the Er^{3+} ion is surrounded by a crystal field of cubic symmetry [3]. The inset of Fig.1 shows the excitation power dependence of Er^{3+}-related emission intensity as well as band-edge emission. The sample was excited with an Ar^+ ion laser using the 514.5nm line which is near the excited state $^2H_{11/2}$ of Er^{3+} ion, and 488nm laser line. No difference in PL intensity was observed between the two laser lines. The full widths at half maximum (FWHMs) of lines of Er^{3+}- related PLE spectra are very narrow (as shown in Fig. 2), therefore the direct excitation to $^2H_{11/2}$ seems to be difficult. The excitation of the Er^{3+} ions is due to energy transfer from the recombination of electron-hole pairs in the Si host. The Er^{3+}-related emission intensity shows weak dependence against excitation intensity in comparison to GaAs:Er (Bantien *et al*) [4]. We assume that the efficiency of energy transfer to Er^{3+} ions is very low, or a small number of Er^{3+} ions exists in Si host. No Er^{3+}-related PLE spectra ($^4I_{15/2} \rightarrow ^4I_{11/2}$) was observed over the wavelength range of λ=960 - 990nm.

Figure 2 shows the typical low temperature PL spectrum at 10K of Er-diffused FZ Si. Five prominent peaks with other weak emissions were observed around 1.549μm. This spectrum is very similar to the GaAs:Er prepared by liquid phase epitaxy (LPE) [4] and thermal diffusion [5], but different from those in ion-implanted Si:Er [6] and molecular beam epitaxy (MBE)-grown Si:Er [7]. This fact suggests that the semiconductors such as Er doped GaAs and Si under equilibrium growth show same Er^{3+}-related spectra. The inset of Fig. 2 shows the PLE spectrum of Er^{3+}-related PL for the Er thermally diffused sample. The PL intensity of Er^{3+} was detected at the $1.549\,\mu m$ line. The spectrum consisted of six sharp peaks (963.1nm, 965.0nm, 976.1nm, 978.9nm and 980.9nm) that were caused by the direct excitation $(^4I_{15/2} \rightarrow {}^4I_{11/2})$ of Er^{3+} ion. This fact indicates that Er^{3+}-related emission is enhanced by direct excitation. The six peaks originating from the splitting of the $^4I_{11/2}$ level meant that Er^{3+} ions were in the site of noncubic symmetry. When Er^{3+} ion is in a site of cubic symmetry, $^4I_{11/2}$ state splits into four levels. Single-crystal X-ray diffraction was carried out to investigate crystal structure and revealed that surface of Si did not have the diamond structure. Therefore the Er diffused samples might have amorphous-like structure with a lower absorption coefficient than crystal Si.

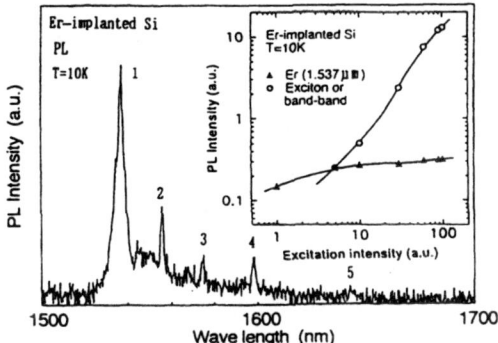

Fig. 1. PL spectrum of Er-implanted CZ Si. Inset shows excitation power dependence of Er^{3+}-related emission (1.537μm line) intensity.

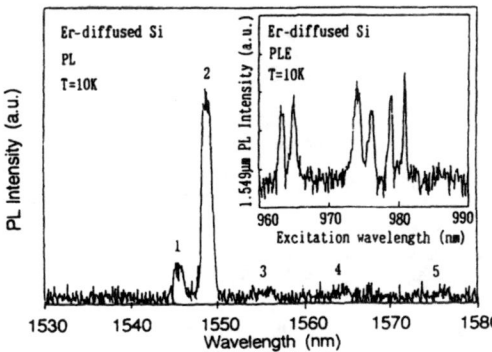

Fig. 2. PL spectrum of Er thermally diffused FZ Si. Inset shows PLE spectrum of Er^{3+}-related emission (1.549μm line) intensity.

Figure 3 shows the PLE spectra of the Er center in thermal diffused sample (20mol% Er in the Si powder), where the PLE spectra was obtained by monitoring the main PL peak ($\lambda_{PL}=1.549\mu m$) intensity. Figure 3 (a) and (b) show the spectrum under different excitation power. Under low excitation power as shown in Fig.3 (a), six PLE lines in the $0.98\mu m$ region ($^4I_{15/2} \rightarrow {}^4I_{11/2}$) and three weak lines in the $0.80\mu m$ region ($^4I_{15/2} \rightarrow {}^4I_{9/2}$) were observed. As shown in Fig. 2, these thermal diffused samples, the Er center is surrounded by a crystal field of noncubic symmetry. If the Er center is surrounded by a crystal field of noncubic symmetry, five PLE lines in $0.80\mu m$ region are expected. We assume that other two lines in this region are hidden behind base line, i.e., indirect excitation-related emission since the absorption coefficient of Si in $0.80\mu m$ region is an order higher than in $0.98\mu m$ region. For that reason, under high excitation power [Fig. 3 (b)], the base line intensity of PLE spectrum in $0.80\mu m$ region was increased, on the other hand, the PLE spectrum in $0.98\mu m$ region is very similar to that under low excitation. In $0.98\mu m$ region, the lines on long-wavelength side tend to intense, but it is not clear that it is caused by absorption coefficient of Si host.

Figure 4 shows the excitation power dependence of Er^{3+}-related PL peak ($\lambda_{PL}=1.549\mu m$). The optical direct excitation ($\lambda_{ex}=980.9nm$) and indirect excitation ($\lambda_{ex}=970.0nm$) were performed, respectively. The emission directly excited is always about 2 times more intense than the indirectly excited emission in each excitation power, but does not show

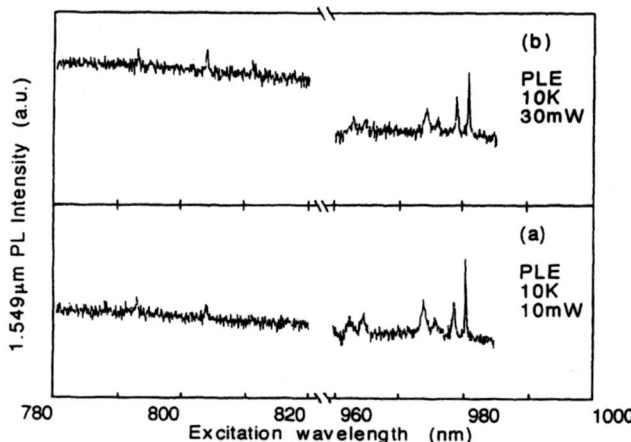

Fig. 3. PLE spectra of Er-thermally diffused FZ Si (Er: 20 mol% powder) taken at two different excitation power: (a) 10mW excitation; (b) 30mW excitation. The Er^{3+}-related PL was detected at the 1.549μm line.

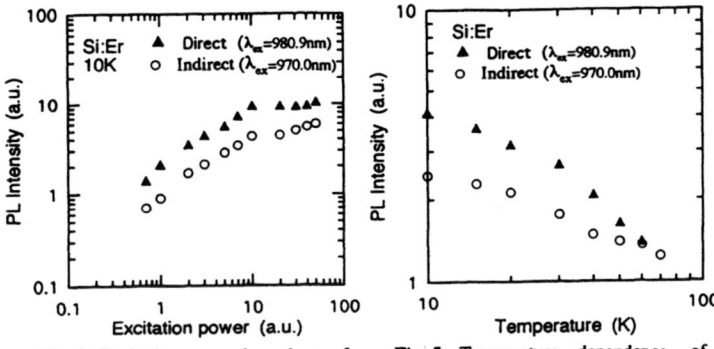

Fig. 4. Excitation power dependence of Er^{3+}-related PL peak (λ_{PL}=1.549μm).

Fig. 5. Temperature dependence of Er^{3+}-related of PL peak (λ_{PL}=1.549μm).

strong dependence on the excitation power. This indicates that the Er^{3+} ion is well excited by optical direct excitation, however, when the excitation energy is above the energy gap (Eg) of Si host, this efficient excitation is disturbed by absorption coefficient of host material even if the absorption coefficient was considerable low.

Figure 5 shows temperature dependence of the PL peak intensity (λ_{PL}=1.549μm) of direct excitation (λ_{ex}=980.9nm) and indirect excitation (λ_{ex}=970.0nm). These two intensities decrease with increasing temperature, and at a temerature of around 60K the direct excitation peaks were hidden behind the indirect excitation-related emission. We expected that the directly excited PL does not depend more on temperature than indirectly excited, however, the Er-related intensity was hardly observed about at 100K. It was reported that Er^{3+} ions excited directly by impact excitation (without energy transfer from the recombination of electron-hole pairs), the emission intensity decreases very slowly with increasing temperature [8]. This reason is assumed that the excitation does not depend on nonradiative process (e.g., Auger recombination) which increases with increasing temperature. In our case, the emission intensity decreases rapidly because the absorption coefficient of Si host in 0.98μm region increases with increasing temperature. In the region, the absorption coefficient of Si varies critically with excitation wavelength, and temperature.

CONCLUSION

In summary, the optical direct and indirect excitation of Er^{3+} ions in Si was performed to investigate the high efficiency of Er^{3+}-related emission. The excitation wavelength was chosen to be equal to the excited states of Er^{3+} ion, which is above bandgap-energy of Si host. For the samples prepared by ion-implantation, no difference in Er^{3+}-related PL intensity was observed between two laser lines, 488nm and 514.5nm. For the samples prepared by thermal diffusion, optical excitation for energy level

$^4I_{11/2}$ of Er^{3+} ions was successfully effected by photoluminescence excitation spectroscopy. This effect was caused by low absorption coefficient of Si host.

ACKNOWLEDGMENTS

This work is supported in part by a Grant-in-Aid for Scientific Research from the Ministry of Education, Science and Culture, Japan.

REFERENCES

[1]   T. Oestereich, C. Swiatkowski, and I. Broser, Appl. Phys. Lett. **56**, 446 (1990).

[2]   J. Michel, J. L. Benton, R. F. Ferrante, D. C. Jacobson, D. J. Eaglesham, E. A. Fitzgerald, Y.-H. Xie, J. M. Poate, and L. C. Kimerling, J. Appl. Phys. **70**, 2672 (1991).

[3]   Y. S. Tang, K. C. Heasman, W. P. Gillin, and B. J. Sealy, Appl. Phys. Lett. **55**, 432 (1989).

[4]   F. Bantien, E. Bauser, and J. Weber, J. Appl. Phys. **61**, 2803 (1987).

[5]   Xinwei Zhao, Kazuhiko Hirakawa, and Toshiaki Ikoma, Appl. Phys. Lett. **54**, 712 (1989).

[6]   H. Ennen, J. Schneider, G. Pomrenke, and A. Axmann, Appl. Phys. Lett. **43**, 943 (1983).

[7]   H. Ennen, G. Pomrenke, A. Axmann, K. Eisele, W. Haydl, and J. Schneider, Appl. Phys. Lett. **46**, 381 (1985).

[8]   Hideo Isshiki, Riichiro Saito, Tadamasa Kimura, and Toshiaki Ikoma, J. Appl. Phys. **70**, 6993 (1991).

EFFECT OF THE Er^{3+} CONCENTRATION ON THE LUMINESCENCE OF $Ca_{1-x}Er_xF_{2+x}$ THIN FILMS EPITAXIALLY GROWN ON Si(100)

A.S. BARRIERE, S. RAOUX, P.N. FAVENNEC*, H. L'HARIDON** and
D. MOUTONNET**
LEMME, Univ. Bordeaux I, 351 c. de la Libération, 33405 Talence, France,
*CNET/DIR/SAS, **CNET/LAB/OCM, France Télécom, 2 route de Tregastel, BP 40, 22301
Lannion, France.

ABSTRACT

$Ca_{1-x}Er_xF_{2+x}$ thin films, with a substitution rate, x, varying from 1 to 20%, were deposited on Si(100) substrates by sublimation of high purity solid solution powders under ultra-high-vacuum. Rutherford backscattering studies have shown that the films have the composition of the initial solid solution powders, are quite homogeneous and are epitaxially grown on the substrates.

The optical properties of these films were studied by means of cathodoluminescence and photoluminescence. At room temperature, the emissions due to the de-excitations from the $^4S_{3/2}$, $^4F_{9/2}$, $^4I_{11/2}$ and $^4I_{13/2}$ excited levels to the $^4I_{15/2}$ ground state of Er^{3+} ($4f^{11}$) ions are easily detected (λ = 0.548, 0.66, 0.98 and 1.53 µm)

The strong 1.53 µm infrared luminescence, which presents evident potential applications for optical communications, is maximum for an erbium substitution rate included between 15 and 17%. These Er concentrations are three or four orders of magnitude greater than the optimum ones in the case of Er-doped semiconductors, which are close to 10^{18} cm^{-3}. In the visible range, the luminescences are also important.They allow us to detect high energy ion or electron beams. However their maximum efficiencies were observed for a relatively low erbium concentration, close to 1%. These different behaviours are explained by the cross relaxation phenomena, which depopulate the higher levels to the benefit of the $^4I_{13/2} \rightarrow {}^4I_{15/2}$ transition.

The energy distribution of the Stark sublevels of the $^4I_{15/2}$ state, which results from crystal field splitting, was deduced from a photoluminescence study at 2K. The obtained results show that the environment of the luminescent centres does not change with the erbium concentration.

At last, it must be noted that the refractive index of the layers increases with the erbium concentration, leading to the realization of optical guides. Consequently opto-electronic components could be developed from such erbium doped heterostructures.

INTRODUCTION

Rare earth (RE) elements have partially filled 4f shells, which are well screened by outer closed orbitals $5s^2$ and $5p^6$. Consequently, the intrashell transitions of 4f electrons give rise to identical sharpe emission spectra in various host materials [1]. Among RE, erbium, which has a luminescence peak centered at 1.53 µm (0.811 eV), which lies within the spectral range for minimum absorption in silica-based optical fibers, seems to be particularly attractive [2.3].

In other respects, it is well known that calcium fluoride (CaF_2) thin films, can be epitaxially grown on silicon substrates. Indeed the crystalline structure of CaF_2 (fcc fluorine) is very similar to that of Si (fcc diamond) and its lattice parameter (5.462Å) is close to that of silicon (5.431Å) [4]. Moreover it must be noticed that the solubility limit of ErF_3 in CaF_2 is very large (40%) [5].

These considerations led us to study the growth of $Ca_{1-x}Er_xF_{2+x}$ thin films under ultrahigh vacuum (UHV) on Si(100) substrates, x varying from 1% to 20% and the optical properties of the obtained structures. In this case the host material seems to be very attractive to have a strong luminescence intensity. It presents a wide gap ($Eg = 11.9$ eV) [6], the low phonon energy in this material can enhance the probabilities of photonic de-excitations, compared with another matrix, and the solid solution thin films can be grown with a high degree of crystal quality and purity.

The texture, structure and composition of the obtained layers were respectively studied by means of scanning electron microscopy (SEM), X-ray diffraction (DX) and nuclear analysis. Their optical properties were deduced from cathodoluminescence (CL) and photoluminescence (PL) measurements, performed from room temperature (RT) to 2K.

GROWTH AND CHARACTERIZATION

Thin films of $Ca_{1-x}Er_xF_{2+x}$ were grown on Si(100) substrates, classicaly cleaned (degreasing, etching (1min in hot HNO_3), de-oxidization (HF-ethanol, 10%)) under a dry nitrogen atmosphere, before outgassed at 800°C under vacuum. The initial solid solution powders, prepared by direct synthesis of binary fluorides (CaF_2 and ErF_3) at 1150°C and controlled from X-ray diffraction [7], were evaporated under UHV at 1250°C from platinum crucibles. The growth rate was 0.1 nms^{-1} and the temperature of the substrates was 550°C. The thicknesses of the films did not exceed 1.5 μm.

A SEM study of the structures showed that for high erbium substitution rates the thin film surfaces are rugged, while, for x lower than 17%, the surfaces are smooth. For such Er concentrations, the X-ray diffraction patterns showed that the layers are crystallized. Only one diffraction peak was revealed indicating that the (100) planes of the fluoride are parallel to the substrate [7].

The composition of the layers was mainly deduced from Rutherford backscattering (RBS) of 2MeV $^4He^+$ particles. This technique and equipment are described in Ref. 8. The scattering angle in the laboratory coordinates was $\theta_L = 160°$. The surface barrier detector has a resolution of 13.5 keV, checked with a ^{241}Am source. The experimental equipment allows us to record the RBS spectra of single crystal targets under either random or channeling conditions [9,10]. For classical RBS investigations the spot of the incident beam was typically 1 mm^2. It can be reduced to 1.5x1.5 μm^2, using a nuclear microprobe, to study the spatial distribution of the basic components at the surface of the samples. During the measurements, the pressure in the analysis chamber was lower than 10^{-7} Torr and the samples remained at room temperature.

For example, RBS spectra, random (α) and aligned (β), of a 130 nm thick $Ca_{0.9}Er_{0.1}F_{2.1}$ thin film, deposited on a Si(100) substrate at 550°C are presented in fig. 1-a. The peaks occuring at 1.873, 1.355 and 0.873 MeV correspond, respectively, to

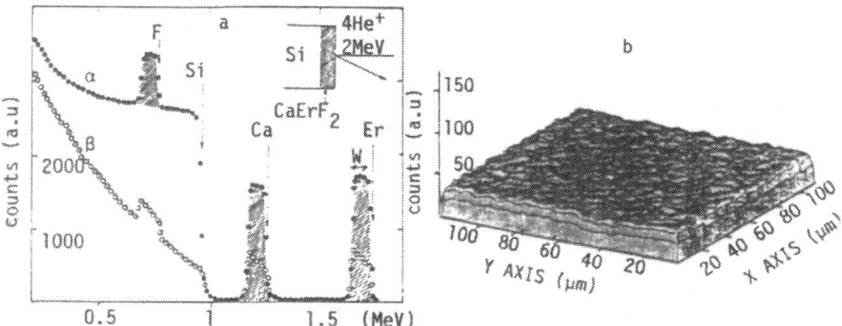

Fig. 1. (a) RBS spectra (α) random and (β) aligned of a 130nm thick $Ca_{0.9}Er_{0.1}F_{2.1}/Si(100)$ structure ; (b) surface distribution of erbium on a 120x120 μm^2 area of the fluoride layers.

erbium, calcium and fluorine of the layer. The gobal atomic concentration ratio of the different basic components, calculated from a comparison of the areas of RBS peaks, correspond well (whatever the theoretical substitution rate, x_{th}, varying from 1 to 20% to $N(F)/N(Ca) = (2+x_{th})/(1-x_{th})$ and $N(Er)/N(Ca) = x_{th}/(1-x_{th})$. No significant impurity, in particular oxygen, was revealed in the bulk of the layers with this analysis technique, even if the films are deposited on light substrates as carbon. Moreover, no heavy impurities were detected from particule induced X-ray emission (PIXE) studies.

To establish the in depth concentration profiles $N(F)/N(Ca)$ and $N(Er)/N(Ca)$, thin films of increasing thicknesses were grown on silicon substrates and the areas of RBS peaks of the different components were compared. In all cases, the atomic concentration ratios were found to be close to the expected theoretical values. The surface distribution of erbium atoms was studied by using the nuclear microprobe previously mentioned. Spectrum b in Fig. 1 was obtained on the basis of an energy window, W, corresponding to the erbium response, on the RBS pattern (fig. 1-a). It illustrates the surface distribution of this element on a 120x120 µm² sample area and shows that its content does not vary significantly in the different points of the studied structure.

The measure of the ratio H_A/H_R of the heights of two spectra taken in the near-surface region for aligned and random orientations, referred to as the minimum yield χ_{min}, allows us to estimate the quality of the epitaxy of a deposited thin film at the surface of a single-crystal substrate [11,12]. The random (α) and aligned (β) spectra of the structure previously studied are reported in fig. 1-a. They show that a channeling of the incident particles with the (100) axis of the substrate is observed in the fluoride layer. Therefore, an epitaxy of the solid solution thin films on Si(100) seems to be obtained. However, near the external surface of the structure χ_{min} has not been found lower than 25%, althought the relative mismatch between the solid solution and the silicon substrate does not exceed 1.3% at room temperature. Such an observation could be partially due to the presence of supplementary F⁻ions in interstitial sites of these materials.

OPTICAL PROPERTIES

The energy distribution of the $4f^{11}$ levels of Er^{3+} ions in CaF_2 environment was deduced from optical absorption measurements, performed at RT on the solid solution powders, by using a Varian-Cary 2415 spectrometer. Among the most intensive absorptions we have noted those corresponding to the transitions between the $^4I_{15/2}$ ground state and the $^4F_{7/2}$ and $^2H_{11/2}$ excited levels (λ = 484 and 517 nm). Consequently the 488 and 514.5 nm lines of an Ar⁺ ion laser were chosen as excitation source for PL studies. The Cl spectra were taken on a JEOL 840 scanning electron (SEM). The beam parameters were : 10 keV, 10 nA, focussed on 2.5x2.5 nm². For photon detection a cooled GaAs photo-multiplier and a cooled germanium detector were used in the visible and the near-infrared ranges, respectively.

For example, the PL and Cl spectra of a 240 nm-thick $Ca_{0.87}Er_{0.13}F_{2.13}$ epitaxial layer on a Si(100) substrate, performed at RT, are presented in Fig. 2 (full and dashed lines respectively). They show that in the visible range, the $^4S_{3/2}$ and $^4F_{9/2} \rightarrow {}^4I_{15/2}$ transitions (λ = 533 and 650 nm) and in the near-infrared domain the $^4I_{11/2}$ and $^4I_{13/2} \rightarrow {}^4I_{15/2}$ transitions (λ = 980 and 1530 nm) give strong light emissions. It is interesting to note that, in a good agreement with the nuclear micro-probe analysis, previously mentionned, the SEM-CL spectra show that the luminescence is homogeneous in all points of the surface of the layers. Moreover, in this case, we have remarked that the luminescence intensity at RT is only 30% reduced compared to that observed at 10K.

In the following we will only present some PL observations. First, we have reported in fig. 3, the evolution of the $^4S_{3/2} \rightarrow {}^4I_{15/2}$, $^4F_{9/2} \rightarrow {}^4I_{15/2}$ and $^4I_{13/2} \rightarrow {}^4I_{15/2}$ PL intensities as a function of the erbium substitution rate, x, in $Ca_{1-x}Er_xF_{2+x}$ thin films grown on Si(100) substrates. Each experimental point corresponds to the integrated surface

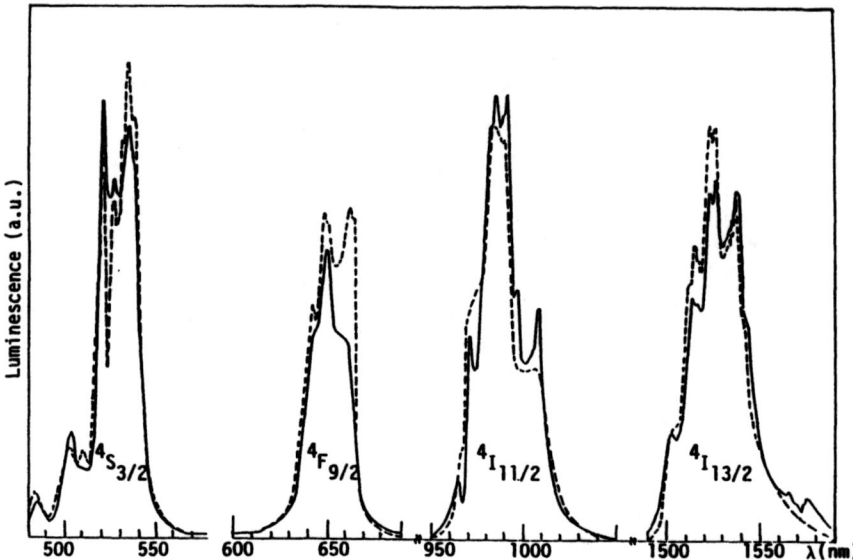

Fig. 2 - PL (full lines) and Cl (dashed lines) of a $Ca_{0.87}Er_{0.13}F_{2.13}/Si(100)$ structure at RT.

of the observed peaks. In the visible range, the PL intensity continuously decreases from x = 1% to x = 20%, while for the 1.53 μm transitions, the higher PL intensities have been found for x varying from 15% to 17%. These different behaviours are controled by the cross relaxation phenomena which depopulate the higher states while the $^4I_{13/2}$ levels is enriched.

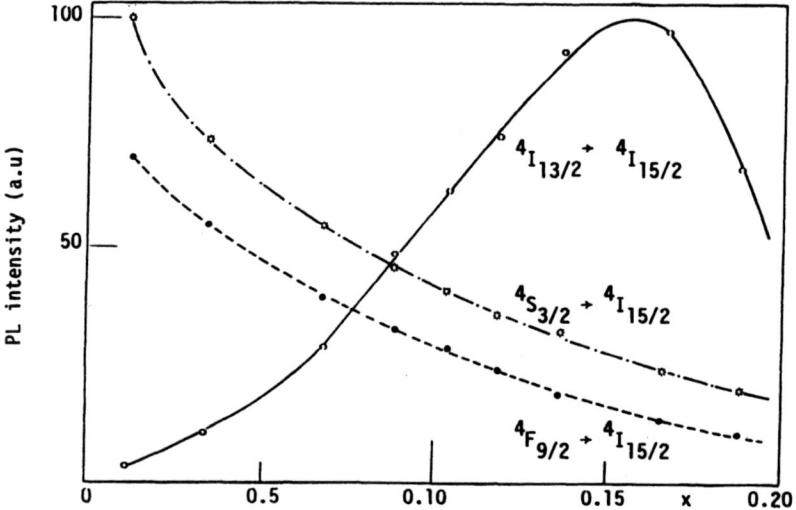

Fig. 3 - 1.53 μm and visible PL intensities, at RT, as a function of the erbium concentration, in $Ca_{1-x}Er_xF_{2+x}$ thin films grown on Si(100).

Such results seem very attractive since, in these solid solution thin films, the maximum of the 1.53 μm PL intensity is obtained for an erbium concentration of three order of magnitude greater than in the case of erbium-doped classical semiconductors (Si, GaAs, InP, ...).

The crystal field splitting (Stark effect) of the $^4I_{15/2}$ ground state was studied by means of PL measurements, performed at 2K, as a function of the erbium substitution rate, x. In fig. 4 are represented the PL spectra of three $Ca_{1-x}Er_xF_{2+x}/InP(100)$ structures. The mixed, full and dashed lines, correspond respectively to x = 4, 6 and 15%.

For low erbium concentrations, five principal lines occur on the spectra, labelled 1 , 2, 3, 4 and 5. This could be in agreement with a crystal field of cubic symetry, as present at a Ca substitutional site in the CaF_2 lattice, for which the free ion spin-orbit $^4I_{15/2}$ states of Er^{3+} ($4f^{11}$) split into five Kramers doublet states [13].

In other respects, we can remark that for x ≤ 4%, the most intensive luminescence peak, labelled 5 in fig. 4, corresponds to transitions from the back sublevel of the $^4I_{13/2}$ excited state to the lower sublevel of the $^4I_{15/2}$ fundamental state (λ = 1.518 μm, 818 meV). For x = 6% the same PL peaks, pointed out at the same energies, are observed, but their relative intensities are changed. The most intensive one corresponds now to that which is labelled 3, in fig. 4 (λ = 1.543 μm, 804 meV). At last, for high erbium concentrations (x = 15%), the peak 5, in fig. 4, has completely disappeared, but the energies of the other lines remain constante. The most

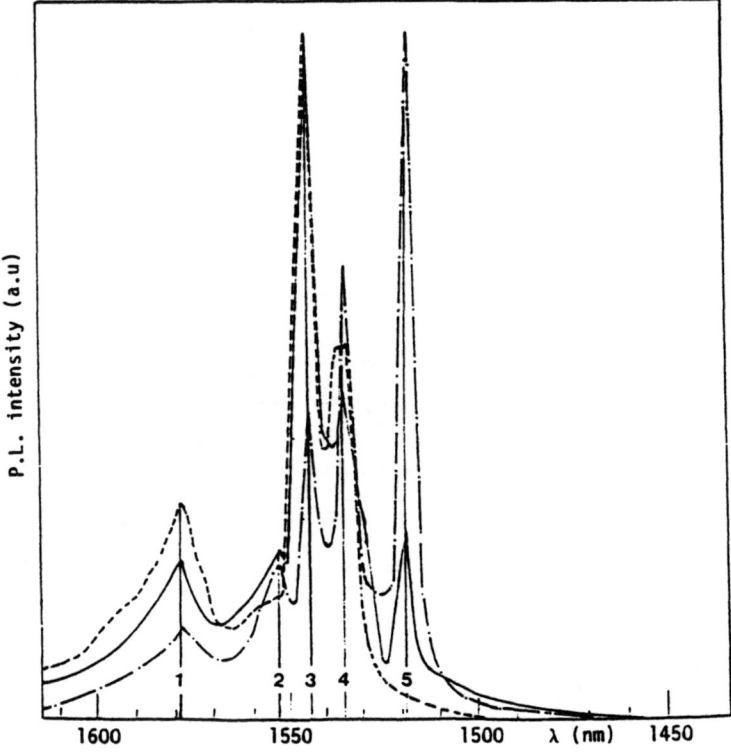

Fig. 4 - 1.53 μm PL spectra of $Ca_{1-x}Er_xF_{2+x}/Si(100)$ structures, berformed at 2K. Mixed, full and dashed lines correspond to x = 4, 6 and 15%.

intensive luminescence corresponds again to a de-excitation on the third fundamental sublevel. These results shown that the environment of the luminescent sites remains unchanged when x increases. On the other hand the evolution of the different line intensities as a function of the erbium substitution rate can be interpreted in terms of a reabsorption of the emitted light corresponding to the de-excitation on the ground sublevel of the $^4I_{15/2}$ state, due to the strong Er-Er coupling at high erbium concentration.

At last we must report that an ellipsometric study, performed at RT, on $Ca_{1-x}Er_xF_{2+x}$ thin films has shown that the refractive index, n, of the layers increases with the erbium substitution rate. For 1.3 and 0.6328 μm incident radiations, n increases from 1.425 to 1.466 and from 1.470 to 1.497, respectively, when x varies from 4% to 19%.

CONCLUSION

We have shown that homogeneous $Ca_{1-x}Er_xF_{2+x}$ thin films can be grown on Si(100) substrates with a high degree of crystal quality and purity. CL and PL studies showed that these films present strong luminescences, at room temperature, in the visible and the near-infrared spectral domains, corresponding to intra-4f-shell transitions of Er^{3+} ($4f^{11}$) ions. The $^4I_{13/2} \rightarrow$ $^4I_{15/2}$ (1.53 μm) transition, which lies within the spectral range for minimum absorption in silica-based optical fibers, presents evident potential applications for optical communications. Its maximum of luminescence intensity has been found for an erbium substitutions rate as high as 16%. The emissions in the visible range can be used, in particular, to visualize the spot of high energy electron or ion beams. At last, we report that the evolution of the refractive index of the layers as a function of the erbium concentration can lead to the realization of plane optical guides [14].

REFERENCES

[1]   P.N. Favennec, H. L'Haridon, M. Salvi, D. Moutonnet and Y. Le Guillou, Electron. Lett. 25 (1989) 719.
[2]   W.F. Krupke and I.B. Gruber, J. Chem. Phys. 39 (1963) 1024.
[3]   W.F. Krupke and J.B. Gruber, J. Chem. Phys. 41 (1964) 1225.
[4]   T. Asano and H. Ishiwara, Thin Solid Films, 93 (1982) 143.
[5]   B.P. Sobolev and P.P. Fedorov, J. Less Com. Metals, 60 (1978) 33.
[6]   A.S. Barrière, G. Couturier, G. Gevers, J. Grannec, H. Ricard and C. Sribi, Surf. Science, 168 (1986) 688.
[7]   A.S. Barrière, B. Mombelli, B. Porté, S. Raoux, H. Guégan, M. Reau, H. L'Haridon and D. Moutonnet, J. Appl. Phys. 73 (3) (1993) 1.
[8]   G. Gevers, A.S. Barrière, J. Grannec, L. Lozano and B. Blanchard, Phys. Stat. Solidi, A81 (1984) 105.
[9]   W.K. Chu, J.W. Mayer and M.A. Nicolet, Backscattering Spectrometry, (Acad., N.Y. 1976) p 223.
[10]   A.S. Barrière, A. Elfajri, H. Guégan, B. Mombelli and S. Raoux, J. Appl. Phys. 71 (1992) 709.
[11]   J.H. Barret, Phys. Rev. B3 (1971) 1527.
[12]   D.S. Gemmel, Rev. Mod. Phys. 46 (1974) 129.
[13]   H. Ennen and J. Schneider, "Luminescence of rare earth ions in III-V semiconductors" 13th I.C.D.S., Colorado, 1984.
[14]   P.N. Favennec, H. L'Haridon, D. Moutonnet, A.S. Barrière, S. Raoux and B. Mombelli - Patent Nr 92403099.2, nov. 18 (1992).

CORRELATION OF ELECTRICAL, STRUCTURAL, AND OPTICAL PROPERTIES OF ERBIUM IN SILICON

J. L. Benton, D. J. Eaglesham, M. Almonte, P. H. Citrin, M. A. Marcus, D. L. Adler, D. C. Jacobson and J. M. Poate
AT&T Bell Laboratories, Murray Hill, New Jersey 07974

ABSTRACT

An understanding of the electrical, structural, and optical properites of Er in Si is necessary to evaluate this system as an opto-electronic material. Extended x-ray absorption fine structure,EXAFS, measurements of Er-implanted Si show that the optically active impurity complex is Er surrounded by an O cage of 6 atoms. The Er photoluminescence intensity is a square root function of excitation power, while the free exciton intensity increases linearly. The square root dependence of the $1.54\mu m$-intensity is independent of measurement temperature and independent of co-implanted species. Ion-implantation of Er in Si introduces donor activity, but spreading resistance carrier concentration profiles indicate that these donors do not effect the optical activity of the Er.

INTRODUCTION

The interest in erbium doping of silicon is an outgrowth of the successful materials science which resulted in the new generation of lightwave communications based on Er-doped fibers and fiber amplifiers. Lightwave networks employing Er doped components offer a new technology with the potential for powerful communications services. It is a natural scientific step, therefore, to introduce Er into silicon in an attempt to invent a compatible silicon optoelectronic device. Light from Si, especially at $1.54\mu m$, is an important scientific advance, but further work in this area should be measured against technological benchmarks.

A careful evaluation of materials parameters has suggested that the erbium-silicon system is not well suited for light emitting diodes, amplifiers or modulators, and that the best chance for its commercial success will reside in lasers. [1] Achievement of a laser based on electroluminescence of Er in Si requires increasing the impurity incorporation beyond the measured solid solubility of $1\times10^{18}\,cm^{-3}$ and demands, as well, an understanding of the excitation mechanism of the Er defect. With advances on these two fronts, it might be possible to create and maintain the inverted population necessary for stimulated emission.

The results presented in this study add some understanding to the issue of Er excitation. The extended x-ray absorption fine-structure (EXAFS) experiments indicate that the optically active Er defect in Si is a complex of Er and six oxygen atoms. The Er luminescence exhibits a square-root function of excitation laser power. This result eliminates exciton bonding as the rate limiting step and suggests that either a back-transfer mechanism or Auger process limits the efficiency of the Er emission. Previous work [2] established that high energy ion implantation of Er into Si introduced donor activity. Although the donor concentration correlates with the Er photoluminescence intensity, this paper shows that the donor defects are not involved in the excitation process.

EXPERIMENTAL PROCEDURE

Er^+ ions were implanted into room-temperature n- and p-type silicon at energies of 0.2 to 5.25 MeV. The concentration of oxygen in the samples was varied by using float zoned (FZ) Si, with $[O]\approx=1\times10^{16}\,cm^{-3}$, Czochralski-grown (CZ) Si, with $[O]\approx=1\times10^{18}\,cm^{-3}$, and Si implanted with additional oxygen to doses as high as $1\times10^{19}\,cm^{-3}$. Post-implantation heat treatments were performed in vacuum at temperatures ranging between 900-1350°C.

Mat. Res. Soc. Symp. Proc. Vol. 298. ©1993 Materials Research Society

The donor activity associated with the implanted Er was monitored by four standard measurements, room-temperature Hall effect, spreading resistance probe (SRP), capacitance-voltage (C-V) profiling, and deep level transient spectroscopy (DLTS) [2] The 514 nm line of an argon ion laser at a power of 400 mW excited the Er photoluminescence (PL) which was detected with a LN_2 cooled germanium detector. The PL spectra were generated with a 0.75-m Spex monochromator.

Multiple Er ion implants were used to prepare samples for EXAFS measurements, which produced an Er concentration of approximately $5 \times 10^{17} cm^{-3}$ uniformly distributed from the silicon sample surface to a depth of ~2 μm. Er L_3 -edge EXAFS data were obtained with fluorescence-yield detection at the National Synchrotron Light Source. All x-ray absorption measurements were obtained from samples at 50K. Bulk standard compounds of $ErSi_2$ and Er_2O_3 were used as references in the determination of coordination numbers and bond lengths.

RESULTS

Local Structure of Optically Active Er

The microscopic structure of the optically active Er species in Si was resolved by EXAFS experiments. Er incorporation in Si is limited by precipitation, with the solid solubility of approximately $1 \times 10^{18} cm^{-3}$ at 900°C. The Er precipitates, which have a $ErSi_2$ structure, are found in both FZ and CZ silicon and are not the source of the 1.54 μm luminescence.[3] The presence of oxygen impurities in the CZ silicon increases the Er luminescence intensity by more than two orders of magnitude compared with FZ material.[4] Therefore, both CZ and FZ silicon Er-implanted substrates were measured by EXAFS along with bulk compounds of $ErSi_2$ and Er_2O_3 . [5]

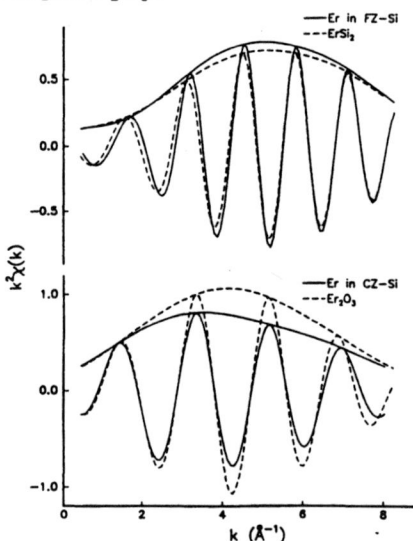

Figure 1. Filtered, back-transformed EXAFS data from first neighbor shells of Er implanted into FZ Si (top data) and Er in CZ Si (bottom data). Comparision with reference samples of erbia and Er silicide show that the optically active Er is surrounded by 6 oxygen atoms.

The filtered EXAFS data are shown in Figure 1. Analysis of the amplitude functions in each pair of spectra identify both the type and number of nearest neighbor atoms surrounding the absosrbing Er atoms. the local environment of Er in FZ Si closely resembles the 12 Si atoms in $ErSi_2$, and the first coordination shell of atoms surrounding Er in CZ Si resembles the 6 oxygen atoms in Er_2O_3. The frequency information in Figure 1 allows the average distance between absorbing Er and its nearest neighbors to be determined, giving for Er-O, 2.25Å and for Er-Si, 3.0Å. Since the EXAFS experiments indicate that greater than 80% of the incorporated Er is measured in the CZ and FZ samples, it is clear that the local environment around the Er determines its optical activity.

Excitation Mechanism of Er Luminescence

The excitation dependence of a PL line measured as a function of laser power was examined to give an indication of the excitation mechanism.[6] Figure 2 shows the excitation laser-power dependence of both the Er signal at 1.54 μm and the free exciton signal at 1.23 μm in the same boron-doped Si sample. The laser power was held constant at 400 mW and the excitation level was reduced by insertion of neutral density filters. The free exciton intensity is linear in power for all measured temperatures, as expected. The Er signal intensity, however, exhibits a square-root dependence on excitation power, which eliminates exciton bonding as the rate limiting step for the Er photoluminescence. The data at 10K, 40K, 60K, and 80K, presented in Figure 2, illustrate that the square- root function is independent of temperature. Additional measurements indicate that this function is also independent of co-implanted species.

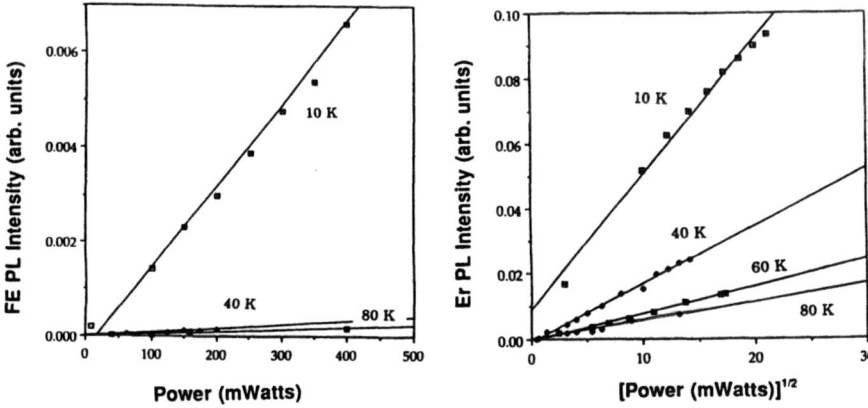

Figure 2.
Excitation power dependence of the free exciton (FE) luminescence is linear in laser power at measurement temperatures of 10K, 40K, and 80K. The 1.54 μm Er photoluminescence (PL) intensity varies as the square root of excitation laser power.

This square-root dependence of Er PL intensity could result from a three-level excitation mechanism or it could indicate the presence of an alternate nonradiative deexcitation mechanism. Back transfer mechanisms have been proposed for Er luminescence in InP and evidence has been presented for localized Auger quenching.[7] [8] In MBE Si:Er, the PL intensity was also reported to be sublinear with excitation power.[9]

Electrical Activity of Er in Si

Determination of the optimal annealing conditions for Er implantation into Si were made by monitoring the PL intensity at 1.54 μm as a function of heat treatment temperature and time. The process was optimized at 900°C for 30 minutes.[4] Deep level transient spectroscopy (DLTS) of ion-implanted Er in n-type Si at low doses $(10^{14} - 10^{16} \text{cm}^{-3})$ detected 9 levels related to Er, and carrier concentration profiling by C-V measurements and spreading resistance probe (SRP) correlated the presence of donors with the implanted Er concentration.[2] This excess donor concentration increased with increasing Er concentration and reached a maximum at approximately $4 \times 10^{16} \, cm^{-3}$, as shown in Figure 3. C-V profiling determined the donor concentrations related to low doses of Er, and SRP and Hall effect measurements recorded the mobile carrier concentrations at higher ion-implantation doses. Since the Er luminescence intensity showed a similar relationship to the implanted Er dose (open circles in Figure 3), it was suggested that these donor states might provide the pathway for excitation of the Er.[2] Further investigation of the electrical activity of the Er:Si system indicates that this is not the case.

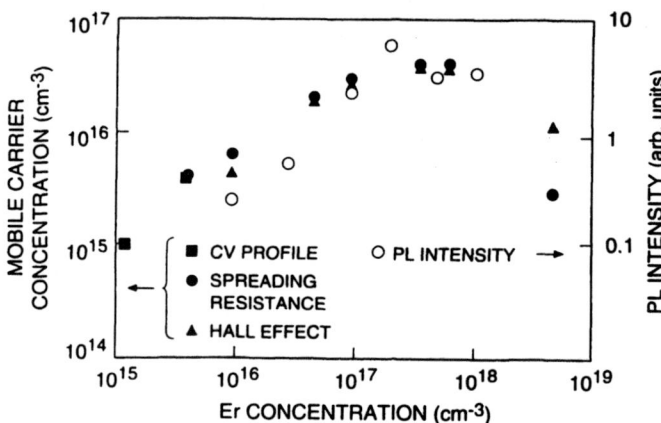

Figure 3. Excess donor concentration, as measured by C-V profiling, SRP, and Hall effect, plotted as a function of implanted Er ion concentration in CZ, n-type Si (left axis). Corresponding 1.54 μm PL intensity is included for comparison.

The samples represented in Figure 4 were given a series of 500 keV implants at $7 \times 10^{17} \, cm^{-3}$, each followed by a heat treatment at 900 °C for 30 min, totaling a dose of $2.1 \times 10^{18} \, cm^{-3}$. Both FZ and CZ, boron doped silicon was monitored. The donors introduced by the Er implantation compensated the material and the dip in the SRP profiles at 0.66 μm marks the p-n junction. The maximum donor concentration is approximately three times greater in the oxygen-rich CZ material, consistent with previous results.[2] However, the Er PL signal is a factor of 30 greater in the CZ sample. Both samples were processed and measured concurrently. The presence of high concentrations of Er-related donors in silicon is not sufficient for efficient luminescence. The data in Figure 4 prove that the Er-related donors alone do not produce luminescence at 1.54 μm, and the data in Figure 5 prove that these donors are not even necessary for the PL production. Spreading resistance carrier concentration profiles were measured on boron doped FZ silicon after three different sets of implant and annealing conditions. The profiles represented by the dashed and dashed-dotted lines are similar and show no evidence of donor activity. The first sample (dashed line) received a 5.25 MeV Er dose of $7 \times 10^{17} \, cm^{-3}$ followed by a vacuum anneal at 1100 °C for 30 min; the second sample (dotted-dashed line) received an additional oxygen implant, $2 \times 10^{18} \, cm^{-3}$ and subsequent thermal treatment of 1100°C for 30 min. The additional co-implantation of oxygen increased the Er luminescence intensity by two orders of magnitude. The third sample (solid line) received the identical Er and O implants as the second sample, but only the final anneal was given. The carrier concentration profile for this case shows donor compensation of the boron dopant, a p-n junction at 2 μm, and a maximum donor concentration of 4×10^{15} cm^{-3}. The Er PL intensity was the same for the two samples containing both Er and O ion implants. The Er related donors are not providing the pathway for the energy transfer necessary to excite the Er luminescence.

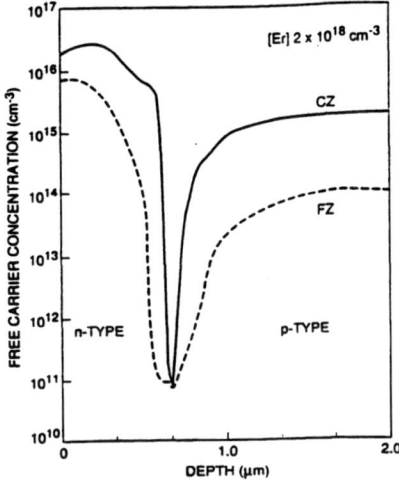

Figure 4. Spreading resistance profiles converted to free carrier concentrations of 2×10^{18} cm^{-3} Er ions implanted into both CZ and FZ p-type Si at 500 keV. Donors are created in both samples but only the Si with high oxygen concentration has an intense Er PL signal.

The initial impetus for these experiments, represented in Figures 4 and 5, was that proper choice of processing might reduce the implantation damage and thus allow the formation of greater concentrations of the Er-O complexes, which would increase the luminescence intensity. However, up to 10 sequential implants and anneals at 900 °C at doses of both 3×10^{17} cm^{-3} and 7×10^{17} cm^{-3} yielded the same Er PL intensities to within a factor of three.

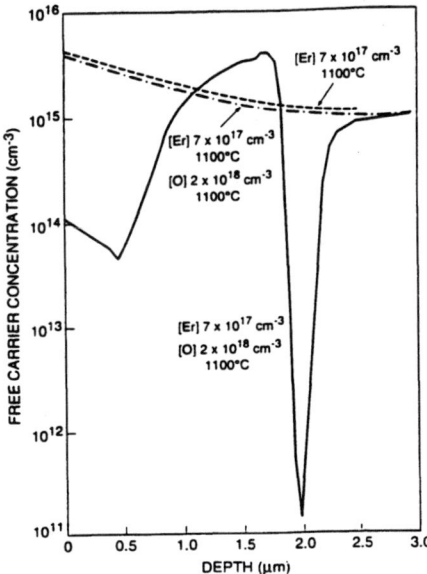

Figure 5. Free carrier concentration profiles measured by SRP in p-type, FZ silicon. Er ion-implantation alone gives the dashed data with no evidence of donor formation and low 1.54 μm PL intensity. After co-implantation of O (dashed-dotted data), the PL intensity increases by two orders of magnitude but the electrical activity remains constant. The solid line shows the formation of a p-n junction at 2 μm.

CONCLUSIONS

Erbium is an effective "microscopic getter" of oxygen in silicon, forming optically active point defects of Er surrounded by six oxygen atoms. A complete understanding of the excitation mechanism of the Er 1.54 μm luminescence in Si requires an investigation of the electrical activity of these Er-O defects. Ion-implanted Er introduces a related donor in both n-type and p-type Si with the electrical activity saturating at $4\times10^{16}\,\mathrm{cm}^{-3}$. These donor-related defects recover, however, under proper annealing conditions without effecting the Er photoluminescence intensity. The PL intensity of the Er signal is proportional to the square root of the excitation laser power. This power dependence suggests that a three-level system or a back- transfer mechanism controls the Er luminescence efficiency.

REFERENCES

1. Y. H. Xie, E. A. Fitzgerald and Y. J. Mii *J. J. Appl. Phys.* **70**(6), 3223 (1991).

2. J. L. Benton, J. Michel, L. C. Kimerling, D. C. Jacobson, Y. H. Xie, D. J. Eaglesham, E. A. Fitzgerald and J. M. Poate, *J. Appl. Phys.* **70** (5), 2667 (1991).

3. D. J. Eaglesham, J. Michel, E.A. Fitzgerald, D.C. Jacobson, J. M. Poate, J. L. Benton, A. Polman, Y. H. Xie and L. C. Kimerling, *Appl. Phys. Lett.* **58**, 2797 (1991).

4. J. Michel, J. L. Benton, R. F. Ferrante, D. C. Jacobson, D. J. Eaglesham, E. A. Fitzgerald, Y. H. Xie, J. M. Poate and L. C. Kimerling, *J. Appl. Phys.* **70** 2672 (1991).

5. D. L. Adler, D. C. Jacobson, D. J. Eaglesham, M. A. Marcus, J. L. Benton, J. M. Poate, and P. H. Citrin, *Appl. Phys. Lett.* **61** (18) 2181 (1991).

6. J. Weber, W. Schmid and R. Sauer *Phys. Rev. B*, **21**, (6) 2401 (1980).

7. A. Taguchi, M. Taniguchi and K. Takahei, *Appl. Phys. Lett.*, **60**, 5604 (1991).

8. B.J. Heijmink Liesert, M. Godlewski, T. Gregorkiewicz, and C. A. J. Ammerlaan, *Appl. Phys. Lett.*, **59** (25) 3279 (1991).

9. H. Efeoglu, J. H. Evans, J. M Langer, A. R. Peaker, N. L. Rowell, J-P Noel, D. D. Perovic, T.E. Jackman and D. C. Houghton *Mat. Res. Soc. Symp. Proc.* **220**, 367 (1991).

1.54 μm PHOTOLUMINESCENCE OF ERBIUM IMPLANTED HYDROGENATED AMORPHOUS SILICON

M. Kechouane, N. Beldi, T. Mohammed-Brahim
Lab "Les Couches Minces", USTBH, Institut de Physique, BP 32, el alia, ALGER, ALGERIE
H. L'Haridon, M. Salvi, M. Gauneau
FRANCE TELECOM/CNET, LAB OCM, BP 40, 22301 LANNION, FRANCE
P.N. Favennec FRANCE TELECOM/CNET, SAS, BP 40, 22301 LANNION, FRANCE

ABSTRACT

The luminescence of erbium implanted in hydrogenated amorphous silicon (a-Si : H) is presented. For the first time an intense and relatively sharp luminescence at 1.54 mm is observed in a-Si : H, using implanted erbium. The Er^+ emission intensity strongly depends on the hydrogen film content and on the measurement temperature. A direct correlation between the optical gap energy of the semiconductor and the hydrogen film content is observed.

INTRODUCTION

In the last few years, erbium doping of III-V compounds semiconductors and silicon was extensively studied [1, 3]. Great effort on various semiconductor materials was made to detect the luminescence of erbium, introduced in the host matrix by doping or ion implantation [3-5]. The erbium luminescence at 1.54 μm, associated with $^4I_{13/2} -->$ $^4I_{15/2}$ transitions in the 4f shells of Er^{3+} ions, was found to be independent of the host material, nor on both its temperature and the band gap energy of the semiconductor [3]. A wide band-gap is necessary to obtain an intense room temperature emission. Thus, hydrogenated amorphous silicon appears as an alternative material insofar as it offers the semiconductor properties necessary for industrial and electro-optical applications. Research effort on this material is greatly motivated by the interest of improving solar cells efficiency; A recent study (6) has shown for the first time an erbium luminescence in hydrogenated amorphous silicon, using erbium as a dopant. In this study the erbium ions are introduced by the ion implantation technique and this paper deals with the erbium luminescence in implanted hydrogenated amorphous silicon. In particular, the intensity variation of the erbium emission with the hydrogen film content is studied.

EXPERIMENTAL

Samples preparation

Undoped a-Si : H films were prepared, using a high rate dc sputtering system. The target and the substrate holder were in parallel plate configuration, at a distance set to 33 mm. The sputtering target, bonded to a wafer cooled copper backing plate, was a 76 mm diameter disk of high purity undoped silicon (99.99995 %). An electrical resistance is used to heat the substrate holder, the temperature being measured and controlled with a type k thermocouple fixed on the substrate holder. Prior to sputtering, the chamber was pumped out and then heated to about 100°C to degas it to a pressure less than 10^{-4} Pa. A gas mixture of hydrogen and argon of 5N purity was introduced directly in the plasma region. The conditions of the deposition were chosen so as to obtain samples with good electrical characteristics ; typical conditions were :

deposition temperature :	T_d	~	280°C
power	: P	~	180 W
Argon pressure	: P_{Ar}	~	0.93 Pa

The hydrogen content C_H was varied by changing the hydrogen flow in the discharge chamber. Films with various thicknesses were deposited on two kinds of substrates : fused quartz plates and crystalline silicon. The optical parameters (thickness and gap) were determined from the transmission measurements in the 0.4 µm - 2.5 µm range. Implantation of erbium ions was carried out at room temperature, using a beam 7° off the normal of the sample surface and at an energy of 330 keV. Doses were varied from 10^{13} to $10^{15} Er^+ cm^{-2}$. After implantation the samples were annealed at 600°C for 15 s, under high purity oxygen flowing. An argon laser source (0.4880 µm line) was used to excite the luminescence and a liquid nitrogen-cooled germanium detector to detect the signal. For measurements at 77 K, the samples were directly immersed in liquid nitrogen.

RESULTS AND DISCUSSIONS

After erbium implantation in hydrogenated amorphous silicon and annealing, an intense and sharply luminescence spectrum was measured around 1.54 µm. Fig. 1 shows typical spectra in the 1.5 µm - 1.6 µm range, recorded at 77 K and at room temperature (R.T). The decrease of the peak intensity, from 77 K to R.T, is around a factor of 40. As far as we know, this is the first experimental evidence of erbium luminescence in implanted a-Si : H.

Fig. 2 shows a comparison at 77K of the luminescence spectrum of erbium implanted into crystalline silicon at 300 keV with a dose of 10^{13} cm^{-2} and the luminescence spectrum in a-Si : H with the same dose and energy. It may be seen that the luminescence peak in a-Si : H is higher and broader than that in c : Si. The broadness is due to the amorphous nature of our material. In addition, the PL spectra have the same luminescence line at 1.536 µm, ; however a second luminescence line at 1.545 µm is observed in the a-Si : H spectrum.

Table 1 summarizes the results of SIMS (Secondary Ion Mass Spectroscopy) and transmission measurements made on various samples in order to determine the hydrogen content C_H and the optical gap energy of the semiconductor, respectively Eg, the optical gap, was deduced from the linear relation between $(\alpha h\upsilon)^{1/2}$ and $h\upsilon$, where α and υ represent respectively the absorption coefficient and the frequency of light.

Sample	C_H %	Eg (eV)
K_1	7	1.68
K_2	18	1.79
K_3	7	1.64
K_4	2	1.53

Table 1 : Hydrogen content C_H and optical gap energy in various a-Si : H   substrates

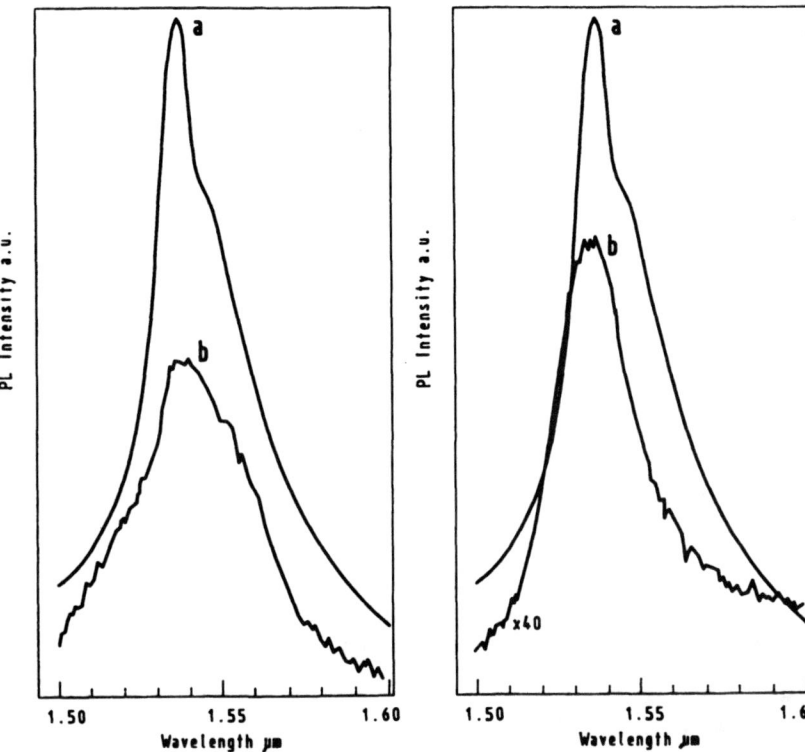

Fig. 1 : PL spectra for erbium implanted
a-Si:H (sample K1) (energy : 300 keV
dose : 10^{14} Er cm^{-2}, annealing : 850°C 15s)
a : spectrum at 77K
b : spectrum at room temperature

Fig.2 : a) PL spectrum for erbium
implanted a-Si:H compared to b) PL
spectrum for erbium implanted
crystalline silicon (dose : 10^{13}
Er+ cm^{-2}, energy : 300 keV,
annealing 900°C 30 min)

A recent study showed that, with the same samples studied by E.R.D (Elastic Recoil
Detection and transmission measurement), the optical gap Eg increases linearly when C_H
varies from 5 to 15 % (Fig. 3). This linear behaviour may be described by : Eg(eV) =
$(0.014 * C_H + 1.55)$.

Favennec et al [3] showed that there is a direct correlation between the efficiency of
optical centres induced by the implanted erbium element and the band gap energy. Thus in
order to increase the luminescence intensity versus temperature, the band gap energy may
be adjusted by varying the hydrogen content in the a-Si : H films.

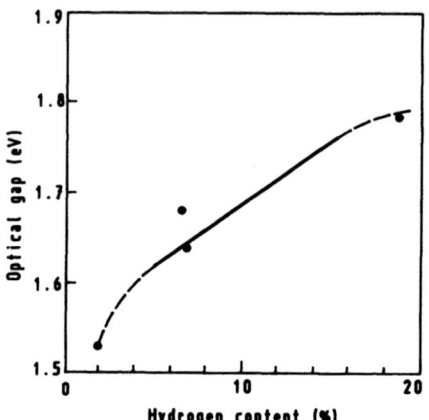

Fig. 3 : Optical gap variation with hydrogen film content ; full line : obtained from transmission and ERD measurements in Ref. (7)

In conclusion ion implantation seems a very attractive technology to introduce erbium impurities into a-Si : H. An intense luminescence is observed at 1.54 µm, even at room temperature. By increasing the hydrogen content from 5 to 15 % in the a-Si : H films, it may be possible to adjust the optical gap energy value from 1.62 to 1.76 eV, and so to increase the luminescence intensity at room temperature. These results offer new possibilities for applications to optical devices, using wide gap a-Si : H films.

References

[1] H. Ennen, J. Schneider, G. Pomrenke and A. Axman
     Appl. Phys. Lett. 43, 943 (1983)
[2] P.S. Whitney, K. Uwai, H. Nagagome and K. Takakei,
     Electron. Lett. 24, 740 (1988)
[3] P.N. Favennec, H. L'Haridon, M. Salvi, D. Moutonnet and Y. Le Guillou
     Electron. Lett. 25, 718 (1989)
[4] C. Rochaix, A. Rolland, P.N. Favennec, B. Lambert, A. Le Corre, H. L'Haridon
     and M. Salvi
     Jap. J. of Appl. Phys., 27 L 2348 (1988)
[5] G. Pomrenke, H. Ennen and W. Haydl
     J. Appl. Phys. 59, 610 (1986)
[6] T. Oestereich, G. Swiatkowski and I. Broser
     Appl. Phys. Lett., 56, 446 (1990)
[7] N. Beldi, A. Rahal, D. Hamouli, M. Aoucher, T. Mohammed-Brahim,
     D. Mencaraglia, Z. Djebour,. Chahed and Y. Bouizem
     Proceedings of the 11th European Photovoltaïc. Conference, Montreux
     (Switzerland), October 1992

Author Index

Subject Index

CPSIA information can be obtained at www.ICGtesting.com
Printed in the USA
LVOW12s0842230514

386805LV00012BA/524/P